THE ANATOMY OF PALMS

THE ANATOMY OF PALMS

Arecaceae — Palmae

AUTHORS

P. BARRY TOMLINSON

Harvard University, Harvard Forest, Petersham, Massachusetts 01366 USA, and the Kampong of the National Tropical Botanical Garden, Miami, Florida 33133 USA

JAMES W. HORN

Fairchild Tropical Botanic Garden, Coral Gables, Florida 33156 USA

JACK B. FISHER

Fairchild Tropical Botanic Garden, Coral Gables, Florida 33156 USA

OXFORD

UNIVERSITY PRESS

OXFORD

UNIVERSITY PRESS

Great Clarendon Street, Oxford OX2 6DP

Oxford University Press is a department of the University of Oxford.
It furthers the University's objective of excellence in research, scholarship,
and education by publishing worldwide in

Oxford New York

Auckland Cape Town Dar es Salaam Hong Kong Karachi
Kuala Lumpur Madrid Melbourne Mexico City Nairobi
New Delhi Shanghai Taipei Toronto

With offices in

Argentina Austria Brazil Chile Czech Republic France Greece
Guatemala Hungary Italy Japan Poland Portugal Singapore
South Korea Switzerland Thailand Turkey Ukraine Vietnam

Oxford is a registered trade mark of Oxford University Press
in the UK and in certain other countries

Published in the United States
by Oxford University Press Inc., New York

British Library Cataloguing in Publication Data

Data available

Library of Congress Cataloging in Publication Data

Data available

Typeset by SPI Publisher Services, Pondicherry, India
Printed in China
on acid-free paper through
Asia Pacific Offset

ISBN 978–0–19–955892–6

10 9 8 7 6 5 4 3 2 1

PREFACE

This book

There is no complete description of the total anatomy of a palm and yet we claim to present information about all 183 genera which are included in the palm family (Arecaceae). This paradox arises because palms are large organisms of considerable structural complexity that changes with age so that a full description of anatomical variation within a single individual has never been attempted. Nevertheless, our seemingly reductionist approach, which deals with small samples from relatively standard regions, must serve to generate the overview we present in this book. So immediately we make apologies for the evident deficiencies of our work. Furthermore, we deal only with vegetative anatomy: inflorescence, flower, fruit, and seed represent different topics.

The palms are such a distinctive group of plants that it is appropriate to describe their unique anatomical features in detail. They are unusual monocotyledons because they have developed the tree habit in the absence of secondary growth, a feature shared by a few other groups of monocotyledons (e.g. Pandanaceae, bamboos). Because palms cannot increase stem thickness by secondary growth they are developmentally constrained in terms of habit (aerial stems normally unbranched and with a terminal cluster of large leaves), giving them their distinctive physiognomy. They are, with few exceptions, entirely tropical in their distribution and they have become tropical icons, most familiar in the economically important coconut palm.

The present book provides an overview of the vegetative anatomy of palms in a systematic context, using a recent classification based on molecular evidence (Dransfield et al. 2005, 2008b). This allows some development of ideas about the distribution of structural features in the relation to the putative phylogeny of palms. The book is an amplification of the first extensive account in English of the anatomy of palms as Volume II (Palmae) in the series 'Anatomy of Monocotyledons' originating under the editorship of C.R. Metcalfe (Tomlinson 1961).

Justification for this newer treatment is determined by several factors. First, is the availability of much living palm material in cultivation in the tropical and subtropical United States, notably at Fairchild Tropical Botanic Garden and the Montgomery Botanical Center in Coral Gables, Florida together with the National Tropical Botanical Garden, Kalaheo, Kauai, Hawaii. Second, the much greater understanding of the structure and development of palms achieved during the five decades since 1961. Much of this work has been summarized in *The Structural Biology of Palms* (Tomlinson 1990). Third, there has been the availability of a detailed overview of palms in *Genera Palmarum* (Uhl and Dransfield 1987), now replaced by its second edition *Genera Palmarum—The Evolution and Classification of Palms* (Dransfield et al. 2008b); these books are a prime source of information on the morphology, geography, phylogeny, and classification of palms.

It should be emphasized that information is presented in a concise format, but its documentation is supported by extensive photographic illustration, in an attempt to minimize textural description. The intent is to present a broad overview that illustrates the distinctive structural features of palms in the context not only of systematics but also ecology and possible evolution.

Information will have use in several other disciplines and applied fields. Comparison with other monocotyledons emphasizes the many unique features of palms and the peculiar difficulties of carrying out structural and developmental research on a group of plants that have been described as 'unruly monsters'. The special methods we have used and which are outlined here could be applied to many categories of plants.

Palms are important in the understanding of plant biomechanics because they illustrate how tall trees can be made structurally sound in the absence of secondary thickening. Palms as fossils, notably petrifactions, can only be interpreted on the basis of information derived from extant plants but the peculiar difficulties of doing this with palms needs constant emphasis. Structural features are the basis for functional processes so that anatomical information is essential for knowledge of how the palm 'works'. Abnormal or pathological growth processes can only be recognized when a normal condition has been described. A natural extension for this approach is in the field of agronomy and horticulture because palms constitute a major economic resource. A somewhat more esoteric application is in forensic science because small fragments of palms may need to be identified and can provide evidence in criminal cases. Knowledge about the use of palms in ancient cultures can be much improved if small fragments can be identified. Palms have multiple uses in the local economy of those modern countries in which they grow; this is the field of ethnobotany, and this enormously diverse use of palm materials must have an important anatomical base from which to assess its value depending on its source.

Palm anatomy in context

Three major factors contribute to the anatomical diversity of palms. Distribution of histological features can be correlated with systematic (i.e. phylogenetic) position, reflecting the ancestral origin of characters. This is a somewhat speculative field, but one most easily observed when histological features coincide in their distribution with the morphological and molecular evidence used in classification. However, the situation in which the palm grows can lead to adaptive structural responses, as in the contrast between palms of wet or dry habitats, of shaded and sunlight environments, or from low versus high altitudes. Finally, the size of palm structures, related to their biomechanical properties, influences structural variation. Such differences can be seen where a genus includes species of different overall size or habit, as in the distinction between caulescent (trunk-forming) versus acaulescent palms without a visible above-ground trunk, as in a few rhizomatous palms. This variation is best appreciated in palms with the climbing habit (notably the economically important rattans), in which distinctive features unique to them can be seen and relate directly to their growth form. Comparative study is possible here because the climbing habit has evolved independently in several groups of palms that are not closely related. This invokes the concept of parallelism (homoplasy) in the evolution of structural features.

However, we cannot claim to provide answers to all the intriguing problems of palm structure and function. The objective is to create a new foundation for the study of this fascinating and distinctive family.

The sources of information and their limitations

We have examined some material of the lamina of all 183 currently accepted genera of palms in Dransfield et al. (2008b) together with a fairly comprehensive study of leaf axis, stem, and root sufficient to provide an overview of structural features that can be useful taxonomically. In addition to much material assembled since we started the project in 2006 we have had access to a large collection of microscope slides retained at Harvard Forest, the residuium of material processed during extended periods of palm study in the 1960s to 2000. Much of this included an extensive collection of slides of leaf axis, stem, and root material. In turn, much of this was based on fluid-preserved samples made by Professor H.E. Moore as adjuncts to herbarium material now housed at the Bailey Hortorium of Cornell University. This puts the time frame of our recent activity into an extended prior period and emphasizes the extensive early collaborative nature of the effort that has resulted in this book. Without this existing resource we could not have completed the project in the four years during which we had direct support from the National Science Foundation or met the terms of the grant by means of which we were financed. It is important to make clear the history leading up to our project completion because it illustrates the lengthy time it takes scientific effort to come to fruition.

Documentation

All recently collected material both original fluid-preserved specimens and the temporary mounted sections derived from it are preserved at Fairchild Tropical Botanic Garden (FTBG) and should become the source for continuing research on palm anatomy. The Harvard Forest collections are also still useful although many sections have deteriorated with age and would require remounting.

An apologia

Having said this we must offer an explanation as to why our results are obviously very incomplete and can easily be criticized. But we must constantly emphasize the fact that palms are bulky organisms with structures well beyond the range of practising anatomists. How can one represent the anatomy of a palm of the gigantic proportions of a *Corypha*, with stems easily 1 m in basal diameter and leaves so large that at least two people are necessary to support one when detached? To reduce this monster (Fig. 45C) to six small images (Fig. 60A–E) derived from sections of a size that can be examined microscopically is a travesty of reductionism and can be repeated throughout our presentations. This is a prime demonstration of the difficulty of transposing the traditional approaches of plant anatomy to the exuberance of tropical plant form. The little bits we examine would seem to bear little reference to the whole, were it not for the results we present, which speak for themselves because they do allow us to make perceptive statements about anatomical variation throughout the palm family in relation to its phylogeny and hence classification. The results should be of significance to evolutionary biologists seeking to account for changes in a large tropical plant family that must have occurred in evolutionary time.

The immediate constraint has been the impossibility of presenting all the information we have about palm anatomy in a single monographic study. We have examined some material of about 300 species and taken of the order of 2500 photographs of which we present about 800 but do not attempt a description of every genus. Our presentation has thus been quite arbitrary and its incompleteness should be recognized. From the start we have used the term 'big picture' to account for our approach. Consequently our work must be seen as a beginning. It should provide a base-line for the very necessary and more comprehensive studies, taxa by taxa, of palm anatomy that awaits the enthusiastic (and brave!) botanists of the future.

P.B.T.
J.W.H.
J.B.F.
1 February 2010

ACKNOWLEDGEMENTS

Information, both new and old, is derived from a great diversity of sources. Older information, the early literature, is from Tomlinson (1961) although we have not cited this in detail. This may appear to be a disservice to earlier workers but we have space restrictions to consider and prefer to present a broad picture that allows one to appreciate the trends that can be seen in the evolution of anatomical characters, especially in the lamina. Our own studies have relied extensively on the living collections already assembled in Botanical Gardens in the United States. These have been in Coral Gables, Florida; Fairchild Tropical Botanic Garden (FTBG), 10901 Old Cutler Road and The Montgomery Botanical Center (MBC), 11901 Old Cutler Road; in Hawaii, at the National Tropical Botanical Garden (NTBG), 3530 Papalina Road, Kalaheo, Kauai; in California, at the San Francisco Botanical Garden and the Botanical Garden of the University of California, Berkeley. We thank the Directors of these institutions and their staff who have assisted in many different ways.

Additional material from the Herbarium of the Royal Botanic Gardens, Kew, Richmond, England has been supplied by John Dransfield and Bill Baker and from the Bailey Hortorium of Cornell University, Ithaca, New York; its processing for anatomical study is described under 'Methods'.

An important source has been the slide collection assembled by P.B. Tomlinson at Harvard Forest, Petersham, Massachusetts, over the course of many years of detailed study on the structure and development of palms. For continued use of laboratory facilities there we thank David Foster, its Director. Hands-on activity in the preparation of material for microscopic examination, often of difficult specimens, was in earlier days largely the responsibility of Leslie Niyogi and Monika Mattmüller, but the bulk of the sections and other preparations we used recently was at the hands of Karen Laubengayer, graduate student at Florida International University (FIU) assisted by volunteer Brenda Whitney, under the supervision of Jay Horn. Most slide making and image capture was centred at FTBG during the last four years.

For advice on matters systematic, nomenclatural, and phylogenetic we have relied extensively on the expertise of Carl Lewis (FTBG), Larry Noblick (MBC), and Scott Zona (FTBG and FIU). All images were made by the authors except for Fig. 4 (original plate supplied by J. Nowak) and Fig. 66F (photo by L. Raz). We thank them for their cooperation.

Financial support for the recent research was derived in 2006–2009 from National Science Foundation (Washington, DC) under grant 0515683 with additional support from the Crum Professorship in Tropical Botany of NTBG. Early research at Harvard Forest was supported by the Maria Moors Cabot Foundation of Harvard University. In a project of this magnitude carried out in a relatively short period of time and in several institutions we freely recognize that there will have been mistakes and oversights; all these are our responsibilities.

For advice on editorial matters and pictorial layout we thank Ian Sherman and Helen Eaton at Oxford University Press.

JH thanks his partner, David Peterson, for love and support, Christine Bacon for sharing her dissertation work, and Kenneth Wurdack for unstinting encouragement and work schedule flexibility. JF thanks his wife Karen for her continued love and support.

BARRY TOMLINSON, JAY HORN, AND JACK FISHER
February 2010

TABLE OF CONTENTS

LIST OF FIGURES

PART 2
Systematic descriptions
Subfamily Calamoideae

Subfamily Nypoideae

Subfamily Coryphoideae

Subfamily Ceroxyloideae

Subfamily Arecoideae

CLASSIFICATION

Based on Dransfield et al. (2008b) and Bacon et al. (in subm.).

I CALAMOIDEAE
  Eugeissoneae

                              Eugeissona

  Lepidocaryeae
    Ancistrophyllinae

                              Oncocalamus
                              Eremospatha
                              Laccosperma

    Raphiinae

                              Raphia

    Mauritiinae

                              Lepidocaryum
                              Mauritia
                              Mauritiella

  Calameae
    Korthalsiinae

                              Korthalsia

    Salaccinae

                              Eleiodoxa
                              Salacca

    Metroxylinae

                              Metroxylon

    Pigafettinae

                              Pigafetta

    Plectocomiinae

                              Plectocomia
                              Myrialepis
                              Plectocomiopsis

    Calaminae

                              Calamus
                              Retispatha
                              Daemonorops
                              Ceratolobus
                              Pogonotium

II NYPOIDEAE

                              Nypa

III CORYPHOIDEAE
  Sabaleae

                              Sabal

  Cryosophileae

                              Schippia

Trithrinax
Zombia
Coccothrinax
Hemithrinax
Leucothrinax
Thrinax
Chelyocarpus
Cryosophila
Itaya

Phoeniceae

Phoenix

Trachycarpeae
  Rhapidinae

Chamaerops
Guihaia
Trachycarpus
Rhapidophyllum
Maxburretia
Rhapis

  Livistoninae

Livistona
Licuala
Johannesteijsmannia
Pholidocarpus
Saribus (including *Pritchardiopsis* & *Livistona* spp.)

Unplaced members of Trachycarpeae

Acoelorraphe
Serenoa
Brahea
Colpothrinax
Copernicia
Pritchardia
Washingtonia

Chuniophoeniceae

Chuniophoenix
Kerriodoxa
Nannorrhops
Tahina

Caryoteae

Arenga
Caryota
Wallichia

Corypheae

Corypha

Borasseae
  Hyphaeninae

Bismarckia
Satranala
Hyphaene
Medemia

  Lataniinae

Latania
Lodoicea
Borassodendron
Borassus

IV CEROXYLOIDEAE
Cyclospatheae

Pseudophoenix

Ceroxyleae

Ceroxylon
Juania
Oraniopsis
Ravenea

Phytelepheae

Ammandra
Aphandra
Phytelephas

V ARECOIDEAE
Iriarteeae

Iriartella
Dictyocaryum
Iriartea
Socratea
Wettinia

Chamaedoreeae

Hyophorbe
Wendlandiella
Synechanthus
Chamaedorea
Gaussia

Podococceae

Podococcus

Oranieae

Orania

Sclerospermeae

Sclerosperma

Roystoneeae

Roystonea

Reinhardtieae

Reinhardtia

Cocoseae
Attaleinae

Beccariophoenix
Jubaeopsis
Voanioala
Allagoptera
Attalea
Butia
Cocos
Jubaea
Lytocaryum
Syagrus
Parajubaea

Bactridinae

Acrocomia
Astrocaryum
Aiphanes
Bactris
Desmoncus

Elaeidinae

Barcella
Elaeis

Manicarieae

        Manicaria

Euterpeae

        Hyospathe
        Euterpe
        Prestoea
        Neonicholsonia
        Oenocarpus

Geonomateae

        Welfia
        Pholidostachys
        Calyptrogyne
        Calyptronoma
        Asterogyne
        Geonoma

Leopoldinieae

        Leopoldinia

Pelagodoxeae

        Pelagodoxa
        Sommieria

Areceae
  Archontophoenicinae

        Actinorhytis
        Archontophoenix
        Actinokentia
        Chambeyronia
        Kentiopsis

  Arecinae

        Areca
        Nenga
        Pinanga

  Basseliniinae

        Basselinia
        Burretiokentia
        Cyphophoenix
        Cyphosperma
        Lepidorrhachis
        Physokentia

  Carpoxylinae

        Carpoxylon
        Satakentia
        Neoveitchia

  Clinospermatinae

        Cyphokentia
        Clinosperma

  Dypsidinae

        Dypsis
        Lemurophoenix
        Marojejya
        Masoala

  Linospadicinae

        Calyptrocalyx
        Linospadix
        Howea
        Laccospadix

  Oncospermatinae

        Oncosperma
        Deckenia

ABBREVIATIONS

±	more or less
ETOH	ethanol
fb(s)	(non-vascular) fibre bundle(s)
g.c.	guard cell(s)
HF	hydrofluoric acid
I.S.	inner sheath
l.s.c.	lateral subsidiary cell(s)
LS	longitudinal section
MRP	matrix representation with parsimony
mxy	metaxylem
O.S.	outer sheath
phl	phloem
pxy	protoxylem
s.st.ch.	substomatal chamber
sl.	slightly
spp.	species
SV	surface view
TS	transverse section
t.s.c.	terminal subsidiary cell(s)
v.	very
vb(s)	vascular bundle(s)
xyl	xylem

POSTLUDE

We apologize for these introductory remarks because they may appear more biographical than necessary, but the reader needs some explanation for this book's origin and the history of subsequent events that lead up to the present—especially as one of us (P.B.T.) has appeared, Rip Van Winkle-like, almost 50 years after the publication of a book on the systematic anatomy of palms (Tomlinson 1961), which the present volume replaces. This fortunate longevity can only be explained on the principle enunciated by the centenarian when asked to account for his long life span: 'Because I was born a long time ago and was young for a long time.'

The story begins when P.B.T. toward the end of the period of study on the systematic anatomy of gingers at the Royal Botanic Gardens, Kew, in preparation for a Ph.D. thesis under the supervision of Dr C.R. Metcalfe and Professor Irene Manton, asked if he might continue with work on the palms. The request was based on the availability of palm material being disposed of in preparation for the renovation of the venerable and iconic Kew Palm House built in 1855. Dr. Metcalfe agreed, with the further suggestion that this post-doctoral research should lead to the second volume (on palms) to follow the first volume, on grasses (Metcalfe 1960), in the developing series 'Anatomy of Monocotyledons'. This series itself was intended to result in a companion series to the existing two volumes 'Anatomy of Dicotyledons' (Metcalfe and Chalk 1950). For this brashness, much was to follow.

With fellowship support arranged by Professor Manton, a period of study at the Singapore Botanic Gardens succeeded by a Lectureship at the University College of the Gold Coast (now University of Ghana), much investigation of palms (and many other plants) in tropical surroundings became possible and the project was completed in a relatively short time (Tomlinson 1961) in a volume that described for the first time in English the vegetative anatomy of palms in a systematic context. But that was not an end but a beginning.

An offer to establish a research programme at Fairchild Tropical Garden, Coral Gables, Florida, was accepted, with the sole requirement to work on that Garden's rich living collection, especially of palms. The scope of the palm research thus was greatly broadened, especially now under the guidance of Professor H.E. Moore, Jr., of Cornell University, the doyen of palm systematists; but not neglecting many other groups. A stimulus was the developing interest of the distinguished plant physiologist Martin Zimmermann at Harvard Forest, in the structure and development of the palm stem. Many new observations on palms could be made as a result of his technical wizardry and acute powers of observation and experimentation. Under his leadership the vascular pattern in the stems of monocotyledons was finally resolved. This led to a move by P.B.T. to Harvard in 1971.

Fairchild Garden still remained a focus for research, but now in association with Jack Fisher, who assumed the responsibilities of its Research Scientist, broadening the scope of studies especially in the direction of large-scale experiments on whole plants. Important observations on palm anatomy and morphology were made, but now increasingly studied during expeditions to the deep tropics.

By now the scientific study of palms was broadened beyond the coterie of these early pioneers and many young investigators entered the discipline, often most appropriately based on extensive field work (correctly so, as palms can only be represented in a fragmentary way when prepared for the herbarium). This extensive research was summarized in a newer volume *The Structural Biology of Palms* (Tomlinson 1990) dealing now extensively with habit, vegetative morphology, and anatomy, but also including a limited overview of what is known of the anatomy of flowers, fruits, and seeds. However, the information was not presented in any detailed systematic context.

With the subsequent advent of molecular and more explicit approaches to palm phylogeny and classification a new vista of the interrelationships among palm groups was revealed and a more refined systematic framework was progressively established, without shaking too much of the earlier structural foundations based on morphology alone. Indeed, the well established 'naturalness' of most palm groups was reinterpreted as evidence for their monophyletic origins, but not without some surprises that could not be anticipated from morphology alone.

This development had repercussions on the monographic generic circumscription of all palms in *Genera Palmarum* (Uhl and Dransfield 1987), based on H.E. Moore's initial extensive work. As research on palms continued, much broadened to include many new workers from tropical countries, the need for a new *Genera Palmarum* was recognized and in a surprisingly short period has been achieved (Dransfield et al. 2008b).

For vegetative anatomy the challenge was now to use this new phylogenetic framework as the basis for a more extensive survey, but based on the rich resources of palms cultivated in Hawaii and Florida in the tropical and subtropical United States. Three years' support from the National Science Foundation has seen this seemingly mammoth task completed in the present volume, the result of dedication, extensive collaboration, and strong institutional support. Into it is distilled the knowledge gained in two lifetimes, but the major outcome is surely not the knowledge presented, but the continual revelation of new pathways still to be trod. Palms will continue to attract the attention of future generations of botanists because, as tropical icons, they surely can reveal biological processes likely to be unexplained from only a temperate-based perspective.

REFERENCES

Dransfield, J., Uhl, N. W., Asmussen, C. B., Baker, W. J., Harley, M. M. and Lewis, C. C. (2005). A new phylogenetic classification of the palm family, Arecaceae. *Kew Bulletin*, 60, 559–569.

Dransfield, J., Uhl, N. W., Asmussen, C. B., Baker, W. J., Harley, M. M. and Lewis, C. E. (2008b). *Genera Palmarum. The evolution and classification of palms*. Kew Publishing, Royal Botanic Gardens, Kew, UK.

Metcalfe, C. R. (1960). *Anatomy of the Monocotyledon.s* Vol. I. *Gramineae*. Clarendon Press, Oxford.

Metcalfe, C. R. and Chalk, L. (1950). *Anatomy of the Dicotyledons*. 2 vols. Clarendon Press, Oxford.

Tomlinson, P. B. (1961). *Anatomy of the Monocotyledons*. Vol. II: *Palmae* (ed. C. R. Metcalfe). Clarendon Press, Oxford.

Uhl, N. W. and J. Dransfield. (1987). *Genera Palmarum*. L. H. Bailey Hortorium and International Palm Society, Lawrence, Kansas, USA.

INTRODUCTION TO PALM STRUCTURE (Part 1)

CHAPTER 1 ANATOMICAL METHODS

INTRODUCTION

As we have emphasized, by virtue of their texture and frequent large proportions, palms are difficult objects to study using conventional anatomical techniques and have largely been avoided by plant anatomists. Their tissues are very fibrous, often highly lignified and they include numerous silica cells (stegmata), all of which rapidly dull microtome knives. Their organs are large and complex and do not lend themselves to the reductionist approach of the orthodox plant anatomist. Special methods have been devised to deal with particularly difficult objects (Tomlinson 1990).

Despite these obstacles, palms were studied very early by plant anatomists of distinction and were suggested by the doyen of them all, Hugo von Mohl (1849), to be organisms that should be studied in detail because they could provide insights into the structure of monocotyledons generally. Von Mohl's observations (1849) were based on material supplied by Martius, obtained during his travels in South America, supplemented by 'stove house' (i.e. glasshouse) specimens cultivated in Europe. The results were published in *Historia Naturalis Palmarum* (Martius 1823–50) and are astonishing in their detail and accuracy, although there is no indication of the methods used. We have included a few samples to illustrate his work (Figs. 1 and 2). These images are as accurate and reliable as anything we have produced with our modern photographic methods; compared with them, our facile work pales into insignificance.

In the present survey, in which many taxa have been examined, simple methods that allowed rapid observation were essential. That is not to say that more elaborate methods using standard procedures to observe microscopic features were inappropriate, they certainly have been and can be exemplified by the pioneering ultrastructural research on palm phloem by Parthasarathy (1974), the many detailed studies of palm flowers and their development by Uhl (e.g. 1988), and the study of palm chromosomes by Read (e.g. 1966).

Nevertheless the beginner needs to give some thought to procedure if standard methods are used. Paraffin and plastic embedding produces poor results unless material is first desilicified in hydrofluoric acid, otherwise one gets results suggesting that it is filled with fine sand, which is indeed the case with the ubiquitous silica bodies. The best results, if embedding

is considered, are provided by the celloidin method, unfortunately now little used because of the tedium involved (see later section).

Appropriate for the study of such intractable objects, a great deal of innovation has gone into the preparation of microscope slides and especially into their examination. Methods include the cinematographic analysis of sequential sections of palm stems that has resulted in a detailed understanding of stem vascular construction in palms (Zimmermann and Mattmüller 1982). This allows presentation of anatomical features as if in motion and greatly facilitates the understanding of otherwise complex three-dimensional organization. Azeotropic distillation has been used to dehydrate bulky palm specimens prior to celloidin embedding, greatly speeding up the process (Tomlinson 1959a). However, in the present work, which has still relied extensively on earlier observations made by these methods, the basic approach has been to cut sections from unembedded material using a sliding microtome and staining the sections with standard histochemical reagents, but without making them permanent (Fig. 3). Using these important tools, rapid results can be achieved in a laboratory provided only with the simplest equipment. We have more recently added the use of cleared and stained whole segments of the lamina so that details of the transverse venation are revealed (Fig. 11). We add an illustration from the literature that shows how the simple methods we advocate can be useful in significant study of palm leaf development (Fig. 4, after Nowak et al. 2007).

The message has been more important than the medium and is not transcended by elaborate and expensive modern technology. We feel that Hugo von Mohl would approve.

MATERIALS EXAMINED

Leaves and other vegetative organs were examined in samples that were variously collected fresh, preserved by fixation, or dried as herbarium specimens and later revived for sectioning. The sources are varied, and these are noted in the Acknowledgements. The materials we examined are listed online as: http://www.fairchildgarden.org/palm-anatomy. Information on accession numbers, vouchers, and full species nomenclature (including some synonymies) is present there.

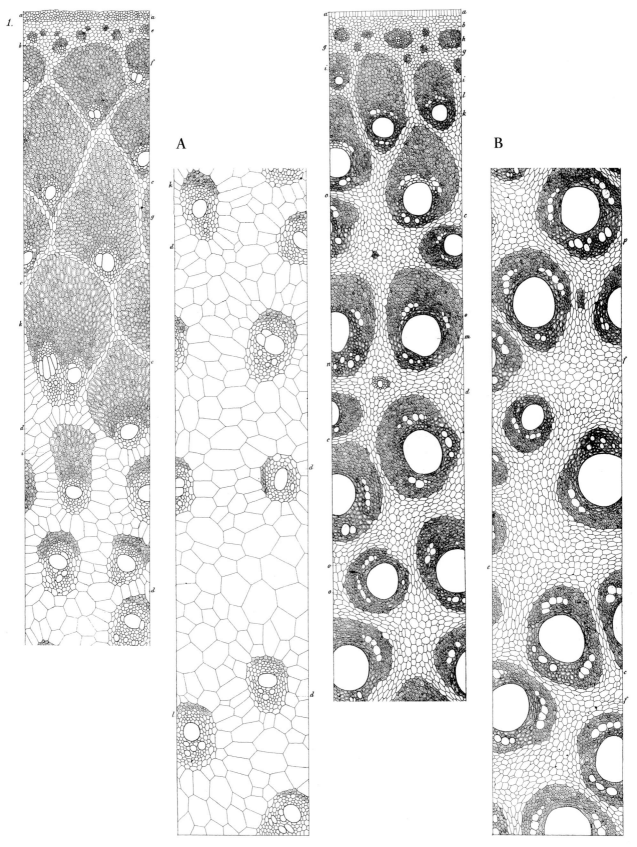

Fig. 1 Stem sections of palms published by the first palm anatomist, Hugo von Mohl. **A.** *Chamaedorea linearis* (originally as *Kunthia montana*), stem from periphery to centre, TS. **B.** *Daemonorops draco* (originally as *Calamus draco*), stem from periphery to centre, TS. From H. von Mohl (1849) *De Palmarum Structura* ex C.F.P. Martius (1823–50) *Historia Naturalis Palmarum*. 3 vols. Munich. (Courtesy of Library of the Fairchild Tropical Botanic Garden, Coral Gables, FL.)

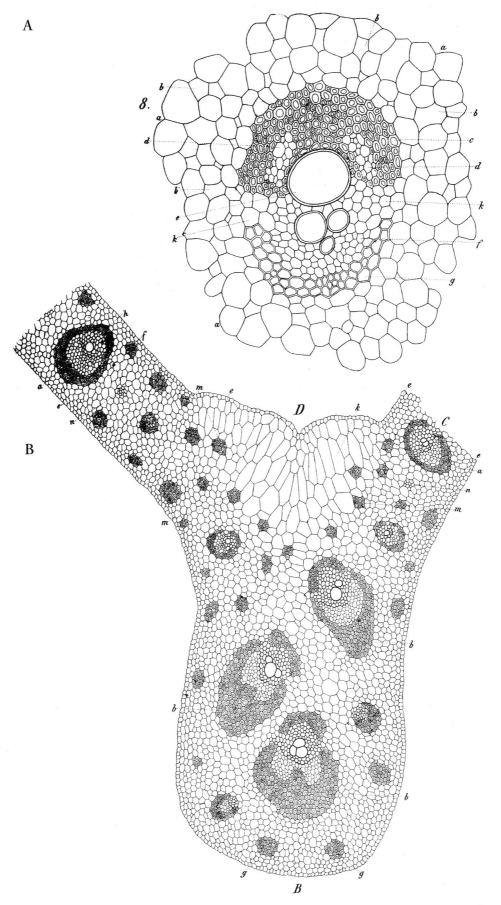

Fig. 2 Leaf sections of palms published by the first palm anatomist, Hugo von Mohl. **A.** *Daemonorops draco*, petiole vb, TS (originally as *Calamus draco*). **B.** *Chamaerops humilis*, lamina abaxial rib, TS. From Tabula K in H. von Mohl (1849) *De Palmarum Structura* ex C.F.P. Martius (1823–50) *Historia Naturalis Palmarum*. 3 vols. Munich. (Courtesy of Library of the Fairchild Tropical Botanic Garden, Coral Gables, FL.)

Fig. 3 Bleaching and staining of sections. *Bismarckia nobilis*, lamina with different treatments and stains, TS. **A.** Fresh and unstained (bar = 250 μm). **B.** Bleached and unstained (bar = 250 μm). **C.** Bleached and stained with toluidine blue for general histology (bar = 250 μm). **D.** Bleached and stained with Sudan IV for lipids (bar = 250 μm). **E.** Bleached and stained with phloroglucinol + HCl for lignin (bar = 250 μm). **F.** Stomata stained with Sudan IV for lipids (cutin and suberin) (bar = 25 μm). **G.** Stomata stained with I₂KI for starch in guard cells (bar = 25 μm). **H.** Stomata stained with toluidine blue (bar = 25 μm).

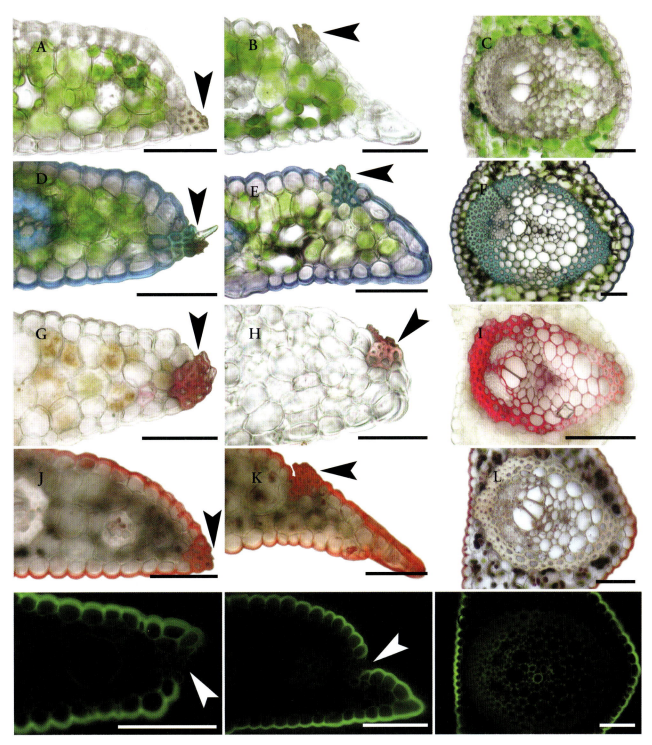

Fig. 4 Fresh hand sections of mature leaflets of *Chamaedorea elegans* at three equivalent regions. Figs. A, B, D, E, G, H, J, K, M, N submarginal cell aggregate (arrow head) indicative of previous tissue continuity variously seen discoloured, suberized, cutinized, and lignified. Figs. C, F, I, L, O vascular tissue of a large vein similarly treated. **A–C.** Unstained leaflets. **D–F.** Stained with toluidine blue O. **G–I.** Stained with phloroglucinol + HCl. **J–L.** Stained with Sudan III. **M–O.** Stained with fluorol yellow, viewed with UV light. (all Bars = 50 μm). (From Nowak et al. 2007. © 2007 University of Chicago Press.)

PRINCIPAL METHODS USED

Selection and fixation of material

Fresh material

The bulk of the material examined has come either from cultivated specimens or material supplied by field collectors (including ourselves). Documentation of sources is thus often provided by herbarium records, accession numbers, and/or planting records. Documentation is most precise since it is usually associated with a herbarium specimen, which if annotated provides a most recent name. In this respect, a great deal of material has come from the collections assembled by H.E. Moore, whose specimens are in the Bailey Hortorium of Cornell University (BH). Professor Moore made every effort to provide additional material of the palms he collected specifically for structural and developmental study by a number of colleagues. Material was mostly fixed in the field. Lamina samples were taken at a 'standard level' as defined in Chapter 3.

Revival of dried material

Material has come either from herbarium and museum specimens or stem samples dried in the field. Dr J.G. Wessels-Boer provided stem material of Surinam palms in this way. Such material is easily transported once thoroughly dried, but it is often necessary to re-hydrate it prior to sectioning, and is a requirement for lamina material. Here material is either boiled for several minutes in water to which a household detergent or a little domestic bleach (sodium hypochlorite) is added, or soaked in a surfactant such as 5% Contrad 70® (Decon Labs). It can then be stored in 70% ethanol (ETOH). A 1:1 mixture of 95% ETOH and glycerine can help soften stored material. Leaf material revived in this way usually gives results almost indistinguishable from that of fresh material except that tannin deposits may increase in the initial drying. Archaeological and ethnobotanical fragments are revived in this way.

Fixation

The standard fixative has been formalin–acetic–alcohol (FAA) as follows: 5% commercial (40%) formaldehyde; 10% glacial acetic; 85% of 70% ETOH.

These proportions may vary in the hands of different investigators. Quality of fixation is not important in relation to the ways in which material is subsequently processed so that any fluid that prevents specimen decay (other alcohols, formalin alone) is acceptable. The field worker may have to be very inventive in locating preservative chemicals, whose nature should always be recorded, provided they are legal!

Desilicification

Where microtoming is used, the removal of silica is a necessary standard procedure, otherwise small silica particles are dragged through a thin section as they catch against the edge of a microtome knife. Standard methods, as used for example in Martens and Uhl (1980), are not always the best. Some thick sections of unprocessed material will still be needed so that the shape of silica bodies *in situ* can be determined. Preparations without prior removal of silica, such as surface layers prepared by scraping, cleared specimens, and macerated material are all useful in the study of silica distribution, especially if mounted permanently in standard synthetic resins because of contrasted refractive indices of resin and silica. The original position of silica in desilicified material is easily established because only silica is removed, but not the silica cells (stegmata) themselves (Fig. 23F). Hydrofluoric acid (HF) is used to dissolve silica but should only be handled in a fume hood; because it dissolves glass, plastic bottles have to be used. One batch of HF will process many samples.

Material is soaked overnight in HF/tap water (1:1), or longer, as when large stem blocks are processed. Subsequently the material must be well washed in running water before it is stored in 70% ETOH ready for sectioning. Specimens are kept separate by wrapping them individually in cheesecloth.

Maceration

The following rapid method has been used routinely. Material (preferably fixed, with the fixative removed by washing well in running water) is cut into small slivers, boiled for 4–5 minutes in 10% potassium hydroxide (KOH) (most easily in a boiling tube over a Bunsen burner or alcohol lamp), washed well in repeated changes of water, and placed in 20% chromic acid (chromium trioxide). The length of time in the acid depends on the size and texture of the tissue—20–25 minutes for lamina material, but sometimes much longer (e.g. overnight for stem tissue with the further help of a 65°C oven). Nevertheless, the hardest palm tissues seem totally resistant to this treatment, illustrating the propensity for palm stems to be preserved as fossils. Maceration is complete when the material can be broken up with a mounted needle. Over-maceration can destroy most cell types, but is often a useful source for the resistant epidermal cuticle and reveals cell shapes in surface view. After washing well in water, small samples of the material are teased apart on a microscope slide with a little added glycerine/water solution (1:1). Unused macerations can be stored in 70% ETOH for later examinations. For the study of tracheary elements, it is possible to pick out large vessel elements from a teased-apart sample under the dissecting microscope so that they can be stained and measured independent of other cell types. Examples are illustrated in Fig. 21D–L.

Permanent preparations of macerated material can be made by dehydrating and staining in the usual way but using a hand centrifuge (yes, a hand centrifuge!) to spin down the cells before each solution change. It is helpful to first neutralize any retained acid with a little ammonia. Of the diversity of stains used, safranin or Delafield's haematoxylin have been the most reliable, but the choice is a matter of individual preference to be established empirically. It is worthy of comment that despite the vigour of this process cytological details of previously living cells, notably nuclei, may still be seen in the finished product.

Epidermal preparations

The lamina epidermis of palms does not 'peel' as a single layer so special approaches are needed to study it in surface view. This is done by scraping away unwanted tissue from a small piece of lamina laid flat on a ceramic tile or Petri dish lid.

The sample may be fresh or fixed. The most useful tool is a scalpel with a rounded blade which need not be very sharp. The specimen is scraped down to the surface layers, rarely to the epidermis alone, so that subjacent hypodermal layers, which are often very fibrous, are also included. The preparation may then include several layers that can be studied in surface view. Adaxial and abaxial surfaces are prepared separately in the same way. The preparation can be examined directly in water, but is usually bleached in sodium hypochlorite for 15–20 minutes and after thorough washing in tap water it can be processed for staining, as if it were a section. Sudan IV is an appropriate lipid stain for the cuticle. Safranin is a convenient stain for permanent preparations, but experiments can be made with other stains. Curled preparations can be a problem but can be alleviated by flattening them with the thumb and then tying them to a microscope slide with cotton thread so that slide and section can be processed as a unit. Epidermal preparations of this type are usually quite thick, but when complemented by thin sections can give a picture of stomatal structure and other surface features.

As suggested earlier, macerated material can produce fragments of the surface layers, which can be isolated and mounted separately. Over-maceration will digest all of the leaf tissue except the cuticle, which can be studied for epidermal cell outlines but without cellular detail.

Pinna clearing

Details of venation pattern and silica distribution in fibres of all veins are easily seen in pinna parts cleared of protoplasm. Examples are shown in Fig. 11C–K and throughout the later text. Portions of pinna (including midrib and margin) are submerged in 5% NaOH in a covered Petri dish and warmed to 55°C in an oven for 1–3 days or until relatively clear. The parts are washed, bleached in commercial sodium hypochlorite solution, washed, dehydrated in 50% ETOH, stained with Johansen's safranin, destained, and dehydrated in an ETOH series ending with xylene or Citrosolve®. The parts are mounted on large glass sides with Permount® and cover glasses but require to be kept flat for days or weeks before they dry to a storable state.

Sectioning of unembedded material

Often described as 'freehand' sectioning because the process corresponds to material cut with a razor or razor blades, rapid results can be obtained with a sliding microtome working on fixed, desilicified material. Sections as thin as 10 μm can be produced, but anything up to 30 μm can be considered ideal. There is no 'technical correctness' and one accepts the best results one can achieve. In producing long sequences of sections for analysis of narrow stems (e.g. rattans), sections as thick as 200 μm are

acceptable because a complete stem section is necessary (Tomlinson et al. 2001).

Large samples (about a 2.5-cm cube) of stem, leaf axis, and larger roots can be sectioned without support, other than the specimen holder of the microtome.

The lamina, with its larger ribs, needs to be supported in the specimen holder by material such as elderberry pith, carrot tissue, or very soft balsa wood. Several small squares of lamina folded together to make a tight wad can be clamped so that one cut produces several sections. Sections should be cut in longitudinal (axial) as well as transverse planes. Once a series of sections has been prepared they can be stored in 70% ETOH in small vials for a subsequent diversity of procedures, as outlined in the following sections.

Staining and mounting sections

Temporary preparations

It is useful to mount some sections, especially of the lamina, without staining as this preserves cell contents. Fresh sections mounted in water show the natural distribution of chlorophyll and the fewest artefacts (Figs. 3A and 4A–C). Mounting in glycerine/water (1:1) under a cover glass makes a wet preparation that can be stored flat for a long period (years in our experience). An alternative is to mount in commercial corn syrup ('Karo' syrup), diluted slightly with water if it is too viscous so that sections essentially remain hydrated. A little phenol (2 g per litre) should be added to sterilize the syrup and prevent microbial growth. The medium dries and becomes solid so that the slide can be stored indefinitely in a dry atmosphere. We have worked with 40-year-old slides in this way, i.e. the preparation is hardly 'temporary!' The advantage is that material can be reprocessed for further use by soaking the slide in warm water to redissolve the syrup. One then has a hydrated lamina section to work with again.

Permanent preparations

Sections can be stained, dehydrated, and mounted in Canada Balsam (now little used), 'Permount', or some equivalent synthetic resin to produce slides for long-term storage. Some resins may become clouded with age so that experience is necessary in their choice. The following stain combinations have been used: safranin and fast green; safranin and alcian green or alcian blue; safranin and Delafield's haematoxylin; safranin and gentian violet. For sections produced from specimens embedded in celloidin, because the embedding matrix is not removed, it can give an intense background colour as an eye-pleasing artefact (Fig. 16F).

Specifically for the identification of callose in phloem, tannic acid and resorcin blue is a good combination that works well as a general double stain (Fig. 22). Bismarck brown is a convenient and rapidly applied single stain. We caution, however, that in a survey such as ours, where the objective is a large amount of comparative data, that experimenting with a diversity of stains, enjoyable as it is, can be very time-consuming.

Temporary stains

One must distinguish general stains (which render tissues more visible and contrasted) from histochemical tests (which identify particular cell components and cell walls by means of a colour reaction). The standard general stain is aqueous toluidine blue, a polychromatic or metachromatic stain, because it produces contrasted blues, purples, and even greens (Figs. 3C, H and 4D–F). The stain is instantaneous, its intensity being determined by the concentration used (0.01%, 0.05%, 0.1%); over-staining should be avoided. Specimens are rinsed well in tap water after staining and mounted in the standard glycerine/water (1:1) mixture. Such preparations dry out slowly and may be stored flat for several years. They are easily reconstituted in water, to be either restained or made permanent. However, toluidine blue is not a permanent stain in any of the methods used here and it loses its brilliance in a few weeks.

Histochemical tests

These must be carried out on a separate series of sections and are not normally retained for any length of time. It is useful for cell wall tests to bleach a section for 5–10 minutes in sodium hypochlorite (domestic bleach, diluted 1:1 with water) to remove cell contents (Fig. 3B). Bleached sections of this kind also produce the best results with the stain combinations mentioned earlier; colours are greatly enhanced and contrasted. Standard tests are as follows (Figs. 3 and 4):

- **Starch:** aqueous iodine/potassium iodide (I_2KI) (Fig. 3G).
- **Lipids:** (notably suberin and cutin): saturated Sudan IV in 95% ETOH (Figs. 3D, F and 4J–L).
- **Lignin:** phloroglucinol (saturated solution in 95% ETOH solution) with concentrated hydrochloric acid (HCl). The reagents may be used in sequence or mixed 1:1.
- **Cellulose:** I_2KI followed by 70% sulphuric acid (H_2SO_4), gives a blue colour. Here the section (bleached) is mounted in I_2KI and a drop of the acid drawn under the cover glass until it meets the section. At the right concentration, cellulose walls turn blue. The prolonged action of the acid digests all walls, but cuticle and suberized deposits persist as yellow-brown tissue replicas. This is a technique used by paleobotanists to isolate cuticles from fossil material.
- **Tannin:** can be identified because it turns blue with iron salts (as ink!), but in unstained fixed material is easily identified as black, brown, or yellow cell deposits, variously amorphous or globular. Tannin cells generally differentiate early but undergo no later changes, e.g. in size. We have not emphasized them in our descriptions.

Celloidin embedding

Palm tissues are usually very heterogenous and fibrous so that sections are often very friable. Celloidin embedding provides a solution although its application is time-consuming. Material is embedded (or at least encased) in Parlodion® (cellulose acetate). Because the solvent is ether this requires care as the combination can be explosive. However, results can be spectacular.

Rather than pass bulky material slowly through a graded series of ethanol solutions of decreasing water content, material can be dehydrated rapidly by boiling under reflux in a mixture of water, ethanol, and benzene in a form of azeotropic distillation (Tomlinson 1959a). Water and benzene are progressively removed as a constant boiling mixture. Material is transferred to absolute ethanol and finally to ether. Infiltration is performed by passing the material through a graded series of celloidin dissolved in ether at concentrations of 2%, 4%, 8%, and 12% celloidin. This must be done in completely sealed bottles (the lid held on by a metal clamp) and promoted in a warm (65°C) oven.

The celloidin matrix is hardened by plunging the specimen into chloroform and transferring for storage in 70% ETOH after 10 minutes. Blocks should be trimmed square for storage. The celloidin matrix picks up tannin and other coloured substances from the tissue so that it too may stain in permanent preparations, as is seen frequently in our illustrations. This is not totally disadvantageous because the stained matrix identifies air spaces and empty cell lumens, e.g. of tracheary elements (Fig. 21A). The matrix can, of course, be removed by ether, but the section then tends to fragment.

Handling loose sections
Larger sections

Loose sections (both embedded and unembedded) of larger palm organs, together with epidermal scrapes, can most conveniently be handled by tying them to a microscope slide by means of cotton thread, with a little glass wool ('angel's hair') added to separate thread from specimen and prevent unstained stripes. Such slides can then be stained and made permanent in the manner of slide series produced by wax embedding, transferring slide from solution to solution rather than as individual sections. The thread is cut free before placing the coverglass.

Lamina sections

Because these are delicate and not easily tied to a slide a method devised by Moore (1957) has proved useful in making permanent preparations, as follows:

1) Bleach sections in commercial sodium hypochlorite for 3–4 minutes. This removes most cell contents and clarifies tissue distribution, but it may be omitted if cell contents need to be retained (cf. Fig. 3A–C).
2) Wash well in several changes of water.
3) Float sections in water onto a labelled microscope slide smeared with a drop of egg albumen adhesive (as used in the paraffin methods). Arrange sections neatly in parallel series, and stand on end to dry.
4) When sections are almost dry and begin to turn white, complete the drying in warm air well above a Bunsen burner or alcohol lamp.
5) Pass the slide three times *rapidly* through the flame, section side up and plunge immediately into a staining jar of 95% ETOH.

6) The sections should remain attached to the slide and can be made permanent by handling them as one would sections prepared by the paraffin method. Preferred stains are safranin and fast green, or safranin and Delafield's haemotoxylin. The method eliminates the tedium of handling delicate leaf sections, which tend to curl. The sections should be thin enough that they do not collapse when a cover glass is added.

7) Alternatively, loose sections can be transferred through a staining series in a plastic or porcelain container with small holes on the bottom.

A detriment to these rather vigorous methods applied to bleached sections is that phloem tissue is disorganized and can only be recognized by its location. It often remains unstained in sections stained with toluidine blue.

Artefacts

All plant histology involves artefacts, beginning with the fixation process. The subsequent diversity of methods introduces many more. They should be understood to avoid misinterpretation.

- Torn sections. Thin sections of heterogeneous materials tear easily when sliced with the microtome knife. Normally the artefact is clear but it may be difficult to distinguish them from naturally occurring air lacunae. Cutting thicker sections or sections in different planes can resolve the difficulty.

- Displaced cell contents. The action of the microtome knife can displace cells or cell contents into anomalous positions. This is particularly apparent with the functionally empty cell lumen of tracheary elements in which one may see starch grains and cell debris.

- Incomplete structures. Trichome morphology is not well understood in palms because distal thin-walled cells are either lost as the leaf matures or are displaced as material is processed, e.g. in making epidermal preparations. Only the persistent and often sclerotic proximal hair base is then retained. The thin-walled part of the indumentum may form a complete covering to the lamina visible as a continuous sheet of cells in which any original pattern is lost. Ideally one should study the development of trichome structure in palms, beginning at least with the unexpanded spear leaf, but this is time-consuming and has been insufficiently explored in our survey.

The root epidermis of palms is normally lost as the root matures, often followed by other surface layers. This loss must always be recognized in comparative study.

- Stained matrix. As mentioned earlier, in the celloidin method the matrix is not normally removed in stained preparations, producing a coloured background.

- Loss of silica. Complete removal of silica is achieved with HF treatment. However, during pinna clearing in 5% NaOH at temperatures above 70°C, silica bodies were sometimes completely lost.

- Wound response. Common responses are the development of a callus or periderm formed by cell division, the formation of mucilage and tannin, and the formation of tyloses in vessels (Fig. 58I). One may find the remains of protein sheaths from insect proboscises in internal tissues (Tsai and Fisher 1993).

Image processing

Digital images were mostly taken with a Nikon Coolpix 4500 digital camera through compound and dissecting microscopes (Olympus BHZ and BH2, Wild M5 and M20, and Leitz Orthlux II) or using the camera directly with macro close-up focus for macrophotography of pinna clearings and plant organs. The images were cleaned of background spots, adjusted for colour levels, white balance, and uniform contrast using Adobe Photoshop. Manipulations were applied to the entire image, never to selected areas. A stage micrometer image was included with each original image to follow changes in magnification during processing. A scale bar was included in the final image when multiple image plates were assembled using Adobe Illustrator. Other than cleaning the background, pixels within the object of concern were not deleted.

CHAPTER 2 PALM CONSTRUCTION AND CLASSIFICATION

THE UNIQUENESS OF PALMS

Although every group of organisms is unique in a genetic sense, some combine structural and developmental features in such a way that they become easily identified objects. Palms are tall plants but with primary growth only. Stems are self-supporting because they become progressively lignified with age but their vascular and mechanical tissues cannot be augmented by secondary addition.

This absence of secondary growth poses a number of metabolic and mechanical problems that have been solved in ingenious ways. Conventional modern trees, represented by conifers and dicotyledonous hardwoods, continually make new vascular tissues via a secondary vascular cambium so that each vegetative axis increases in thickness with age, adding secondary xylem (wood) internally, serving the functions of water transport and mechanical strength, and secondary phloem externally, serving the functions of assimilate transport. However, these transport functions are temporary; conducting elements of the wood are short-lived and soon lose their ability to transport water as they become air-filled, although they still retain their mechanical function via their thick, lignified cell walls. Phloem tissues, external to the cambium, are progressively incorporated into the bark of the tree and sloughed off. The tree sustains its vitality by continually generating new conducting and mechanical elements. We are still familiar with the idea that trees can be very long-lived as individuals, easily upwards of thousands of years. The paradox is that their constituent metabolically active tissues are short-lived; living cells in wood rarely have a lifespan of more than a few decades. How does this compare with a palm that, lacking a vascular cambium, makes a primary vascular system in a stem that must serve the tree throughout its total life, in some instances measured in centuries (Tomlinson et al. 2009)?

The palm solves the functional problems in two principal ways. First, by developing abundant mechanical fibres throughout its tissues, seen most distinctively in the structure of leaf and stem fibres. These are associated either with vascular bundles (as 'vascular fibres') or independent of vascular bundles ('non-vascular fibres'). Second, palms distribute mechanical and conducting tissues in the context of 'fibro-vascular bundles' throughout the stem and leaf in such a way that the amount of this vascular tissue occupies the maximum volume in minimum space. In a sense the conventional tree (conifer or hardwood) is 'two-dimensional', its conducting vascular tissue occupies the **surface** of the hollow cylinder represented by the bifacial vascular cambium, but a palm is 'three-dimensional', its vascular tissue occupies the whole **volume** of a solid cylinder. This difference is, of course, the basic distinction between dicotyledons with a ring of vascular bundles in the stem, and monocotyledons in which the vascular bundles are said to be 'scattered' throughout the stem, an unfortunate concept because the vascular system of monocotyledons is highly organized, as we describe later.

But these two principles are only partial solutions because there still remains the problem of longevity of tissues, because in palms, once established, the primary tissues cannot be replaced.

Research on palm vascular tissues has shown that the cytological and cell wall features of palm xylem and phloem do not differ in any significant way from similar tissues in conventional trees and can be described using the standard terminology found in any plant anatomy text. **Tracheary elements** (tracheids and vessels) are structured according to their time of origin within a tissue (protoxylem or metaxylem) and lose their cell contents at functional maturity so that they are easily modelled as narrow capillaries. **Sieve-tube cells** and associated companion cells develop the same cytological peculiarities found in all vascular plants, sieve cells in particular becoming enucleate at functional maturity. The answer seems to be that they are long-lived because they are selectively long-lived compared with equivalent tissues of the leaf. Individual sieve tubes that function for centuries may be hard concept for plant physiologists to grasp. This longevity of primary tissues also exists in most tissues of the root.

However, these vascular tissues do not exist in isolation, they are embedded in ground tissue which itself heightens the paradox because these differentiated parenchyma cells themselves are as long-lived as the organ in which they are situated; again for palm stems this can be centuries.

Furthermore, these long-lived **ground tissue cells** demonstrate their vitality in many ways: they retain a normal nucleated and vacuolated cytoplasm; they can divide and so multiply and in many examples produce a measurable increase in stem thickness because they also expand their volume; their cell walls can thicken with the formation of new deposits of cellulose and lignin, possibly also making new pit connections with adjacent cells; they can continue to accumulate starch grains as carbohydrate reserve and then reverse the process as starch is later hydrolysed and mobilized, most obviously in palms that have once-flowering (hapaxanthic) stems. Perhaps the most significant long-term cell function is found in the **vascular fibres** of the palm stem, as they continue to thicken and lignify their walls for many years—so increasing stem stiffness as the tree grows taller without significantly increasing in diameter, a property unmatched by conventional trees. It is possible to find the cell nucleus within the much narrowed cell lumen of mature fibres in palm stems, although this requires diligent searching.

With this appreciation in mind, our concept of the uniqueness of palms takes on greater biological significance because they are not merely tropical icons (the coconut palm!) but icons that show the distinctive ability of plant cells to sustain vital processes, seemingly indefinitely.

These conceits need to be mitigated by consideration of the inherent difficulties encountered by the plant anatomist who looks into palm structure. Simply stated, how does one represent on a single microscope slide the anatomy of a leaf which, as in *Raphia*, can be 25 m long, or a stem that is 1 m in diameter and 18 m tall, as in *Corypha*? The problem is just as acute in smaller organs as with leaflets or leaf segments, which show quantitative variation throughout their length, often easily more than a metre long. This in part accounts for our surprising lack of measurements in our description because the 'standard length' we use is itself an arbitrary position. Where changes along a stem have been studied they are considerable. There are equally fundamental structural changes along the complete length of a leaf axis, most obvious in the change from sheath to petiole, but also in the further transition from petiole (with fibre bundles) to the rachis (without fibre bundles). As we show, this aspect of palm anatomy has not been fully explored.

The inevitable conclusion is that a small sample of any vegetative palm organ cannot represent its anatomical variation in the same way that a small sample of wood, when sectioned, gives a good overview of the overall construction of the secondary xylem of a conventional tree. Our descriptions are thus limited because we have dealt little with differences between different species other than general summary statements or where there is some striking variation. For all large genera our results should serve as a guide to future studies based on complete sampling.

A SYNOPSIS OF THE PALM FAMILY

Plants mostly with woody self-supporting aerially unbranched trunks without secondary vascular tissues, or rarely branched dichotomously. Stems of all palms retaining living tissues throughout their lifespan (often centuries) by a unique retention of metabolic activity of all cell types and growth of ground tissue (*sustained primary growth*). Stems never branching sympodially in relation to flowering. Rhizomatous, stoloniferous or acaulescent forms uncommon, rhizomes rarely with dimorphic (horizontal versus erect) axes. Stems rarely branching by dichotomy, but characteristic of the underground axis of *Nypa*; diffusely branched in the horizontal trunk of *Serenoa*. Scandent forms

common (15% of all palm species) but independently derived in perhaps seven lineages and always grapnel climbers, the supporting organ, either an extended leaf rachis (cirrus) armed with hooks or reflexed leaflet spines (acanthophylls), or a modified inflorescence (flagellum) armed with hooks. Germination hypogeal (cryptocotylar), the seedling axis adjacent to the seed or remote by extension of the cotyledonary middle piece. Seedling axis unbranched in single-stemmed palms, branching eventually from basal axillary suckers in multiple-stemmed palms. Radicle short-lived. Establishment growth producing an obconical axis and a broad surface for adventitious root development, the axis sometimes inverted but re-erected (*saxophone-shape*) in later development. Leaves large, forming a terminal crown, individual leaves long-lived; the individual leaves complex with a distinct sheathing base, always forming initially a closed tube, a longer or shorter petiole, and a distal rachis supporting the multiplicate blade. Leaf blade (lamina) initially undivided, but in most palms becoming secondarily divided to form leaf segments (palmate leaves) or leaflets (pinnate leaves) depending on the amount of extension of the distal part of the leaf axis, with costapalmate leaves somewhat intermediate. Lamina segments separating to form V-shaped (induplicate) or Λ-shaped (reduplicate) units. Leaf base splitting or eroding in various ways to accommodate internal expansion of younger organs while preserving mechanical requirements as the leaf ages. Root system adventitious, individual roots without secondary growth but considerably extended. Aerial roots (stilt-roots) in some palms supporting an extended obconical axis. Lateral roots often developed as aerating pneumathodes.

Vegetative axis at flowering producing successive flowering units (branched or rarely unbranched inflorescences), either one (rarely more) per leaf axil as a continuous series (pleonanthic) without loss of vegetative development or axillary units aggregated distally to form a terminal (often gigantic) panicle resulting in the eventual death of the axis (hapaxanthic). Axillary inflorescence units in the caryotoid palms unusual because of their basipetal order of final expansion. Inflorescence unit in all palms each with a basal prophyll and usually one or more basal sterile protective bracts and distal fertile bracts subtending up to four further branch orders; the primary axis rarely unbranched and spicate. Flowers borne on ultimate axes of all orders (rachillae). Flowers on the rachillae either solitary but most commonly aggregated (as dyads, triads, or polyads) in the axil of an ultimate bract representing a condensed cymose complex (cincinnus), each flower with an attendant bracteole. Flowers actinomorphic sometimes bisexual (hermaphrodite) but usually unisexual with varying degrees of dimorphism and monoecy or dioecy. Pollination either by wind or probably most commonly by insects (beetles, flies, bees), possibly never by large animals; the flowers then often nectariferous, odorous, or pollen rich. Flower parts rarely dimerous, most commonly trimerous and sometimes polymerous; stamens 3, 6, or more numerous (multistaminate), with a biseriate perianth, the petals never conspicuous except in aggregate, the sepals entirely protective. Gynoecium superior with one, most commonly three but occasionally more carpels; the carpels free in putative basal groups but fused in varying degrees as a derived condition, most commonly with three carpels only one of which is functional. Fruits indehiscent, moderately to very large, water- or animal-dispersed; essentially as a fibrous drupe, the endocarp sometimes thick and woody, the mesocarp sometimes fleshy. Seed number determined by number of functional carpels. Endosperm abundant, at an early stage liquid (persisting in *Cocos*); embryo small.

Anatomically palms are distinguished by their extensive development of fibres (vascular and non-vascular) which provide mechanical stability in the absence of secondary growth. Stem with basic arrangement of vascular bundles ('*Rhapis* principle'—as found in most monocotyledons) with an axial system interconnecting with lateral appendages (leaves and inflorescence units) in a constant and predictable way. Stem vascular bundles simple, never compound (i.e. never with more than one aggregate of each kind of vascular tissue (French and Tomlinson 1986)), usually compact peripherally to form an extensive peripheral mechanical layer surrounding a less dense centre. Silica cells (stegmata) abundant in aerial parts (absent from many roots) in association with vascular and non-vascular fibres, each silica cell with a single either spherical or hat-shaped silica body. Raphide sacs common in ground tissues, fewest in or absent from roots, but there these sacs commonly form linear aggregates of elongated cells that become mucilage canals. Trichomes (hairs) on the lamina various, but mostly a basal series of thick-walled persistent cells developing a distal, often shield-like series of thin-walled ephemeral cells. Tannin abundant in the ground tissue of most organs. Stomatal developmental pattern constant throughout the family.

THE HABIT OF PALMS

From the unique properties of the stem we recognize that palms are limited in their construction by the lack of secondary development of vascular tissue although, as mentioned, a distinctive kind of diameter growth can occur, described later as sustained primary growth. The structural constraint of palm stem development nevertheless permits a wide variation in stem diameter, from less than 1 cm, as in smaller species of *Chamaedorea*, up to 1 m, as in *Corypha* and *Jubaea*. Consequently there is an allometric relation between the size of axis and appendages (Tomlinson 2006), but there are limits to branching because an axis of fixed mechanical and transport capacity cannot support or supply an ever-expanding crown. Branching, if it does occur, is therefore almost always restricted to the base of the parent axis with each branch producing its own new root system. The characteristic mature axis diameter for each species is achieved by the greater or lesser gradual increase in size of the embryonic axis, itself never more than a few millimetres in diameter, associated with increase in size and elaboration of each successive leaf and its associated internode. This process is termed *establishment growth* and results in an obconical shape to the beginning stem.

Establishment growth

The seedling axis produced by obconical growth is normally subterranean and not visible, but sometimes it is elaborately modified in order to maintain the crown underground, temporally or permanently. For example, the seedling axis may turn vertically downward for some distance before reversing its direction of growth to return to the surface, the descending proximal part of the axis applied to the ascending and expanding distal part to produce a 'saxophone-like' shape. This buries the axis deep in the substrate, forming an extensive subterranean structure with a large mass of adventitious roots. This saxophone stage is found in a number of unrelated palms, e.g. *Rhopalostylis* (Tomlinson and Esler 1973), *Sabal*, and *Syagrus*. The large genus *Syagrus* shows several variants of this form in relation to differing ecological conditions (L. Noblick, pers. comm.).

In the stilt-rooted palms (Arecoideae: tribe Iriarteeae) the seedling stem grows erect in the establishment phase and its internodes elongate so that development of the axis proceeds above ground, resulting in a stem that may be described as tapering downward, but developmentally, of course, it expands upwards. The obconical axis is then very obvious, and becomes supported by successive adventitious roots of successively increasing diameter so as to form eventually a cone of obliquely descending roots adequate for mechanical support and vascular supply (Fig. 19F). This very precise correlation between increasing stem and root diameter is found in many monocotyledons during the post-seedling phase of development. Notable and obvious examples can be found in climbing aroids. Once the diameter of the expanding stem has reached its normal adult size, the stem continues its uniform or 'maintenance' growth, internodes becoming longer and a visible trunk is developed. The final developmental change of the palm is the sexual phase with production of inflorescences.

Seedling roots

The seedling radicle of palms, incapable of secondary growth, is insignificant in sustaining the hydraulic needs of the increasingly widening stem of the establishment phase and is replaced by adventitious roots, normally over the whole expanding surface of the obconical stem. Roots increase not only in numbers, but also in diameter as is most visible in many stilt-palms (Fig. 19F). Once the mature stem diameter is reached, root production appears to cease and the root requirements of the adult palm appears to have been met. But the ability of the stem to produce new roots has not been lost, as can be seen in two ways.

First, roots can appear distally, as for example in damaged trunks, or rarely as a design feature, as in some species of *Chamaedorea*, even though the roots may never contact the soil. In *Thrinax* and its relatives, a tight cluster of superficial roots develops at the stem base ('root boss'). In *Cryosophila*, aerial roots can develop over the whole stem surface but become modified as protective spines of determinate growth and have no absorbing function; see also in *Raphia* (Cardon 1978).

Second, in the transplanting of large nursery-grown palms (*Bismarckia*, *Phoenix*, *Roystonea*, *Washingtonia*) the transplant is first root-pruned in preparation for the new setting. In order for the palm to survive, a new subterranean root system must develop, but how this happens has never been investigated anatomically (Broschat and Donselman 1984).

Branching

An axis of fixed diameter and consequent vascular capacity, as in a palm, can only branch vegetatively in a limited manner because any multiplication of the crown cannot be accommodated by a proportionate increase in basal vascular supply (Holttum 1955). True dichotomy, an equal division of the shoot apex to produce two equal crowns, occurs in a number of unrelated palms, most familiarly in *Hyphaene* (Schoute 1909), but see also *Nannorrhops, Nypa*, and several arecoid palms (Fisher and Maidman 1999). In *Nypa* (Fig. 42) the branching system is expressed continuously because the massive axis, embedded in estuarine mud, is horizontal and produces adventitious roots along its length. The basal part of old leaves persist and may function as pneumatophores, as suggested in Tomlinson (1986).

We must in principle distinguish single- from multiple-stemmed palms. Single-stemmed palms, like *Cocos, Roystonea, Washingtonia,* because they do not branch, retain the single apical meristem of the seedling throughout their lifespan. The apex persists because it is deeply enclosed within the leaf bases of the crown. Otherwise, and commonly, palms branch basally from lateral suckers, originally from the seedling axis but successively by further lateral buds of these daughter axes. At ground level each new shoot produces its own independent adventitious root system. This results in the clustering or multiple-stemmed habit that may be a generic or specific character and may even occur among some individuals of an otherwise single-stemmed species. Multiple-stemmed palms are usually shorter and with narrower trunks than single-stemmed palms (cf. *Phoenix reclinata* and *Phoenix canariensis*).

Basal suckers normally grow erect immediately, hence are clustered; rarely do they grow horizontally from the parent axis for some distance before they grow erect to repeat the basal branching. This produces the rhizomatous habit, most familiar in the commonly cultivated *Rhapis excelsa*. In this diminutive palm there is pronounced axis dimorphism, the rhizomatous portion producing only scale leaves, the erect axis producing foliage leaves and inflorescences. Repeated developmental expression produces a sympodial system, rare in palms. The stoloniferous habit, where the branches are above the soil surface, as in *Chamaedorea stolonifera* is equally rare. We may contrast this limited sympodial development with the sympodial rhizomatous habit of many monocotyledons, notably in the Zingiberales (Holttum 1955). The difference seems correlated with the presence of woody or lignified axes (palms) versus fleshy and little lignified axes (gingers). As described earlier, the continued lignification of the palm trunk makes the tree-habit possible because erect stems are progressively mechanically stable as their height increases. This histological feature may be an ancestral feature for the palm family.

Rhizomes

The relation between branching and habit described earlier is but one method for the production of essentially horizontal or underground stems that may be generally designated as 'rhizomatous' (Tomlinson and Zimmermann 1966a). The following are variants.

1) Sub-erect, short-stemmed palms, the crown remaining at ground level. The stem may branch as in the clustering habit, e.g. *Rhapidophyllum*. Examples with quite massive trunks occur in the phytelephantoid palms. Bernal (1998) describes the reclining trunks of *Phytelephas* as potentially immortal, although old tissues progressively decompose.

2) Persistent juvenile (paedomorphic) forms, which remain at ground level and usually unbranched, as in low growing species of *Sabal* (e.g. *S. minor*). The juvenile 'saxophone' shape contributes to this form.

3) Monopodially branched stems that remain horizontal at ground level, producing an irregular sequence of both vegetative and reproductive branches, as in *Serenoa repens*. The habit is very much determined by location. In fire-prone habitats stems are horizontal, as in pine rocklands, but in cultivated specimens axes become erect as a more bush-like form.

4) The *Nypa*-habit, as described previously; where horizontal growth is established in the seedling and persists as a dichotomizing system with lateral inflorescences. The crown is distinctive; despite the spiral phyllotaxis, all leaves grow vertical by appropriate differential growth.

5) The rhizomatous habit of *Rhapis*, as described earlier with dimorphic axes produced sympodially.

This diversity of intrinsically rhizomatous forms is frequently associated with distinctive anatomical features, as described in the systematic section. However, the subject still remains insufficiently explored anatomically.

Flowering

We do not deal with the anatomy of reproductive axes in this book, although this is an important component of the palm habit. Palms flower in two main contrasted ways with reference to the position of inflorescences on individual axes:

1) **Pleonanthy**—in which inflorescences are produced in the axils of foliage leaves, so that the monopodial habit is sustained. The inflorescence may mature within the leafy crown, i.e. *interfoliar*, as in *Cocos*, or below the leafy crown, i.e. *infrafoliar*, after the subtending leaf is shed, as in many arecoid palms, e.g. *Roystonea*. The biological significance of this process is that the palm indefinitely continues its vegetative development and simultaneous production of flowering axes.

2) **Hapaxanthy**—here the stem continues its vegetative growth without producing inflorescences for an extended period of time before making a dramatic switch to a flowering condition, producing an often massive terminal panicle, whereupon the crown collapses and the axis dies (Fig. 45C). Popular descriptions refer to these as 'suicide palms'. This is the so-called 'suprafoliar inflorescence', although it is obvious that one branch of the terminal panicle is the homologue of a single lateral inflorescence in a pleonanthic

palm, hence the more precise term 'inflorescence unit' to refer to the equivalent parts.

Conspicuous hapaxanthy is not common in palms, e.g. in *Corypha, Metroxylon* (Fig. 29E), *Tahina*, and a few climbing palms. It is less conspicuous in *Raphia* (Fig. 29F) because the inflorescence units are borne within the massive leafy crown. The caryotoid palms (Coryphoideae: Caryoteae), with few exceptions, are particularly distinctive because the reproductive phase involves maturation of axillary inflorescence units in a basipetal sequence. The inflorescence buds are produced in vegetative development in a normal acropetal sequence but remain dormant until flowering commences, but first with the expansion of the youngest, i.e. *distal* inflorescences and continues basipetally towards the stem base; consequently fruiting axes will appear above younger, flowering axes.

The physiological switch from the vegetative to flowering state in hapaxanthic palms represent a massive redeployment of stored reserves, visible in the abundant starch of the stem, which has few biochemical parallels in the plant kingdom. A relevant and familiar result is the extraction of reserves for human food as in the sago palms (*Metroxylon*). The pre- and post-flowering study of the anatomy of hapaxanthic palms has not been investigated. One might ask, for example, how does the vascular supply to inflorescences in the stems of caryotoid palms differ from that described for pleonanthic palms?

CONCLUSIONS

From this brief overview it is clear that the constraints on habit of exclusively primary growth have been overcome in many ingenious ways by palms. In comparison with other trees, palms may still be interpreted as developmentally restricted. For example, they include examples of only four of the architectural tree models of the typological series described in Hallé et al. (1978). These are Corner's, Holttum's, Schoute's, and Tomlinson's model (Dransfield 1978; Fisher and Maidman 1999). However, these limited descriptions by no means encompass the total habit diversity of palms and, being independent of size, give no indication of range of stem diameters over two orders of magnitude. Most fundamentally, palm axes are, with few exceptions, monopodial in organization in contrast to most other woody and tree-like monocotyledons, which can be continuously sympodial in development as in Pandanaceae, *Dracaena*, and *Cordyline*, and with or without secondary growth. The monopodial palm habit is presumably ancestral and combined with lignified stems produces physiognomically distinctive trees. Most remarkable of all is that they have been claimed to be the longest-lived of all woody plants because their fully differentiated stem cells of many types remain metabolically active longer than in any other seed plant (Tomlinson 2006).

The organography of palms is covered in the next chapters.

CHAPTER 3 LEAF LAMINA

INTRODUCTION

The palm leaf provides a complex expression of diversity within a single organ, related to the primary function of photosynthesis. Complexity arises as a consequence of the frequent large size and longevity of palm leaves; for example, coconut leaves on average have a lifespan of about 10 years, of which 5 years may be spent enclosed during early development (cf. Fig. 5A–C). The palm leaf also has an unusual method of blade development because an initially simple, undissected structure becomes compound, first becoming folded by differential growth and then, in most examples, becoming segmented by the separation of an unfolded marginal strip obvious in Fig. 5D, F, followed by separation into leaflets or leaf segments to become either a pinnate or a palmate blade—the difference determined by the extent to which the blade-bearing portion of the leaf axis is or is not extended. Incomplete splits can be represented by holes in the blade (Fig. 5G) as in 'window palms' and also easily seen in juvenile coconut leaves.

That the method of leaf development in palms differed from that in all other plants was early recognized by von Mohl (1849) but not fully elucidated until the work of Kaplan et al. (1982a, b). More recently the research has been extended to a study of leaflet separation in the small palm *Chamaedorea* that involves apoptosis or programmed cell death in precise regions (Nowak et al. 2007, 2008, 2009; cf. Fig. 4). Although this research has settled much of the early controversy about the development of the palm leaf (Dengler et al. 1982; Kaplan et al. 1982a, b; Dengler and Dengler 1984) much remains to be understood because the palm leaf is extremely diversified in its morphology.

BASIC CONSTRUCTION

The leaf axis in a palm is divided into three components: 1) a basal **sheath**, which is always initially a closed tube (Fig. 5B, C); 2) a longer or shorter **petiole**; and 3) the continuation of the axis as the **rachis** into the segmented **blade**. These parts have been presented in Tomlinson (1990) in terms of their functions and mechanical properties. The sheath can be seen as a barrel which supports the distal part of the leaf by clasping the stem, accommodating the considerable expansion of enclosed leaves (Fig. 5A) and transmitting forces applied to the blade through the petiole and into the stem. A long petiole is a flexible tapered cantilever that extends the blade away from the axis in order to reduce drag and mutual shading. The blade is a corrugated surface that is initially compressed within the crown but ultimately expands into a dissected surface. The mechanical efficiency of corrugated construction is most obvious in the palmate leaf.

The primary typification distinguishes **palmate** (fan: Fig. 45A) from those with **pinnate** (feather: Fig. 45G) leaves, which differ in principle in the extent to which the rachis is either condensed (palmate) or extended (pinnate), but with an intermediate **costapalmate** type in which the rachis is extended somewhat into the unsegmented portion of the palmate blade (Fig. 45H).

The blade itself, as an initially continuous structure, is supported by prominent ribs, which may project adaxially or abaxially, and usually with a rib at each fold of the plicate surface. Splitting of the blade into leaflets or segments may occur in one of two contrasted ways. When the split occurs on an abaxial fold, the leaflet or segment is Λ-shaped in section or **reduplicate** (roof-shaped). When the segment or leaflet is separated along an adaxial fold it is V-shaped (gutter-shaped) in section. This basic distinction is generally consistent within a subfamily so that we may distinguish **reduplicate**-leaved from **induplicate**-leaved palms.

STANDARD LEVEL

Anatomical description of the palm **leaf blade** requires some introductory commentary because it is obviously limited by the size and complexity of the palm leaf. We emphasize considerably the anatomy of the lamina, but choose a 'standard level' for comparison, i.e. the middle region of a leaflet from the mid-section of a pinnate leaf or the middle section of a lateral fold or free segment of a palmate leaf, depending on the extent of the unsplit portion of the blade. The standard level has obvious limitations in description because of the decreasing thickness of the lamina from its insertion on the rachis to the tip of the leaflet or leaf segment. Number of hypodermal and mesophyll layers together with the size of vascular bundles and amount of fibrous tissue are thus hardly constant. We have therefore been very conservative in our discussion of species differences by avoiding overemphasis on quantitative characters.

Fig. 5 Shoot development and structure. **A.** *Rhapis excelsa*, crown with shoot apex, LS (bar = 3 mm). **B.** *Rhapis excelsa*, shoot with leaf bases at level of shoot apex, TS (bar = 5 mm). **C.** *Rhapis excelsa*, shoot apex and youngest leaf primordia, TS (bar = 500 μm). **D.** *Archontophoenix alexandrae*, expanded leaf retaining marginal strip. **E.** *Zombia antillarum*, leaf sheath with two contrarotating series of fibro-vascular bundles; region of overlap at side of the sheath opposite the petiole, shown at left. **F.** *Archontophoenix alexandrae*, close up view of leaf marginal strip. **G.** *Reinhardtia gracilis*, mature leaf lamina with incomplete segmentation, hence 'window palm'. **H.** *Cocos nucifera*, cut surface of mature trunk revealing numerous vascular bundles uniformly scattered within the ground tissue of the central cylinder (bar = 5 cm).

The **petiole** also has been described in terms of a 'standard level', i.e. half way along its length, although it is clear that several features do change throughout the length of the leaf axis, notably the presence of fibre bundles in the ground tissue, which disappear distally. Limited information about the anatomy of the **leaf sheath** is presented because we have sectioned very little material and there is little discussion of it by earlier botanists. The fact that the sheath has a long developmental history and is a complex structure is suggested by Fig. 5B, C, E. In relation to the biomechanics of the whole organ the structure of the leaf axis has been insufficiently studied. We do not know how the mechanics of this structure influences at its maturity the stability of the leaf, both in development and during wind storms. The major changes the leaf sheath has undergone during palm diversification, outlined later, remains a significant evolutionary topic.

SIZE AND SHAPE

Despite basic constructional principles, palm leaves show enormous diversity, even within a single taxon, but nevertheless including numerous diagnostic features such as leaflet shape and disposition, degree of segmentation (in palmate leaves), as well as ancillary modifications such as spines and grapnels. A primary source of variation is the long series of progressively larger juvenile leaves developed during establishment growth. Adult size is a major component of leaf diversity because this varies from less than 1 m long, as in treelet palms like *Chamadedorea*, *Reinhardtia*, and *Rhapis*, up to the giant leaf of *Raphia regalis*, measured at 25.9 m long (Hallé 1977). The vast scope of biomechanical requirements can only be imagined. The range of leaf size starts with the relatively few palms that retain essentially simple or unsegmented leaves; these are normally small as in species of *Asterogyne*, *Chamaedorea*, and *Geonoma*, but can be up to 5 m long in *Manicaria*, *Pelagodoxa*, *Salacca*, and *Verschaffeltia*, although here the blade as it ages eventually may become segmented by wind. Such small as well as large leaves can occur within a single genus, e.g. *Bactris simplicifrons* (small) and *B. militaris* (large). The diamond-shaped leaf of *Johannesteijsmannia* is a striking example of an undivided but large palmate leaf, up to 4 m long.

The bipinnate leaf of *Caryota* (Fig. 45D) is the most extreme elaboration of palm leaf blade morphology and can reach a length of several metres, as in *Caryota no*. Other examples of morphological variation include the depth of segmentation in palmate leaves, which determines the size of the undivided portion of the blade (*palman*), the extent to which the leaflets of pinnate leaves are clustered irregularly along the rachis, a feature that still requires developmental explanation, and the shape of the leaflet itself, ranging from narrowly linear to broadly diamond-shaped (*Korthalsia*). The leaflet margin may be irregularly toothed or incised and with a ragged tip ('praemorse') when 'fishtail' describes its shape (*Caryota*, *Socratea*). The praemorse condition can be diagnostic of certain groups, e.g. Ptychospermatinae. Finally, there can be a whip-like extension of the apex of the rachis (**cirrus**) in many climbing palms with an obvious grapnel function, the grapnel hooks being either recurved distal leaflets or aggregates of recurved spines resembling cat's claws.

PETIOLE

Petiole length varies from almost nothing where the blade is sessile on the leaf base to the long petioles in many fan palms, up to 4 m in *Corypha*. Such long petioles impart additional flexibility to the whole leaf, but also occur in small leaves in relation to overall blade size, as in *Rhapis*. The petiole is most commonly grooved adaxially and rounded abaxially, in part for developmental reasons (Fig. 5C), but it can be rhombic in section as in small fan palms or even terete, as in *Nypa* where it may help to maintain the erect posture of the blade (cf. Fig. 12A–E). Petiole anatomy is relatively uniform in palms, a statement likely to be modified when the subject is addressed more completely.

LEAF SHEATH

The base to the palm leaf is an organ that has been typified in order to describe its considerable morphological and anatomical diversity. The typology uses examples in different genera based on their external appearance but with some indication of the early changes that result in visible features of mature leaves. All leaf bases in palms are initially hollow, complete cylinders (Fig. 5B) but to this must be added distinctive differences in anatomy, mostly seen in the general course of vascular and fibre bundles (cf. Figs. 5E and 13I, J). Although the leaf sheath is a closed tube, at least initially in development (Fig. 5B), it is normally thickened on the dorsal side (corresponding to the side below the petiole) and thinner on the ventral side (the side opposite the petiole). This distinction is most obvious in early leaf development (Fig. 5C). A more detailed account, using a typological approach, is included later. The designations used are not entirely arbitrary because several major groups show a common type. The *Veitchia*-type is dominant in the Areceae (Arecoideae). The *Hyphaene*-type is largely restricted to fan palms of the Coryphoideae (e.g. *Sabal* and Borasseae). The *Zombia*-type is restricted to that genus, although spines develop from the leaf sheath fibres in *Rhapidophyllum* and incipiently in some other Rhapidinae, but similarly in Caryoteae. The *Calamus*-type is associated with cane-like stems and is most obvious in the Calamineae (Calamoideae) but elsewhere is represented, for example, by *Chamaedorea*.

Further research requires much more detailed anatomical study based on destructive sampling because developmental information is necessary and can only be obtained by sacrificing the whole palm crown. A detailed study of *Cocos* (Tomlinson 1964) shows the main features of the contrarotating vascular system, which seems to occur in several palm groups (e.g. Dassanayake and Sivakadachchan 1972). At the moment the distribution of types is not capable of a phylogenetic interpretation, indicative of the complex construction and developmental plasticity of the palm leaf base.

LAMINA ANATOMY (Figs. 3, 6, 7, 10, and 11)

Our observations on the anatomy of the palm leaf blade (lamina) are restricted to a standard level described earlier. We provide here only enough information to make understandable our later detailed observations. Lamina anatomy is best described in terms of its several major components. These include the limiting surface layers of **epidermis** and **hypodermis** and a central **mesophyll**. The mesophyll is largely composed of chlorenchyma, the photosynthetic tissue. The mesophyll and hypodermal layers usually include discrete **fibre bundles** (non-vascular fibres) whose distribution is diagnostic for different taxa. The vascular system includes parallel **longitudinal veins** or vascular bundles, interconnected by **transverse veins** (transverse commissures) of varying density.

Lamina symmetry

A primary division is between palms with **dorsiventral** lamina symmetry (Fig. 6A in TS) and **isolateral** symmetry (Fig. 6C in TS), the latter rather uncommon. We largely illustrate from transverse sections (TS) rather than longitudinal sections (LS) because these are more easily interpreted (cf. Fig. 6A with B) and Fig. 6C with D). In dorsiventral symmetry, upper (adaxial) and lower surfaces (abaxial) are unequal (Fig. 6E, F) with stomata largely restricted to the abaxial surface and with upper and lower mesophyll layers dissimilar. In isolateral symmetry, stomata are equally distributed on upper and lower surfaces and upper and lower mesophyll layers are also similar. In a strict sense, lamina symmetry is never precisely isolateral because vascular tissues always have the same orientation, with xylem uppermost and phloem lowermost, i.e. there are no inverted vascular bundles. In our illustrations we always present lamina images with the adaxial surface uppermost on the page.

Epidermis

The stomatal complex and its guard cells are treated in detail in a later separate section. Epidermal cells are normally distributed in longitudinal files, the result of intercalary growth by regular transverse divisions in the developing lamina. Development of oblique end walls may obscure this pattern. The epidermis is almost always shallow compared with the hypodermis and flat because the outer wall is rarely papillose; papillose cells otherwise may be associated only with stomata. Cell shape is otherwise always described as seen in surface view (SV). Wall thickness varies, but usually the outer wall is thickest, the exception being in some species of the Livistoninae. In dorsiventral leaves, cell shape in SV is most regular and diagnostically useful. A major division among all palms is between groups in which the adaxial epidermal cells are rectangular and longitudinally extended (Fig. 7B–E) versus those in which, because of their oblique end walls, cells are rhombohedral to spindle-shaped (Fig. 7F–H), the angle of obliquity often changing to produce domains of differently oriented cells.

Abaxial epidermal cells are less regular and usually with a distinction between **costal** regions of longitudinal cell files below longitudinal veins or hypodermal fibre bundles and **intercostal** regions to which stomata are restricted. Costal cells usually resemble cells of the upper epidermis; intercostal cell shape is irregular because of the presence of stomata. In isolateral leaves this irregularity is found on both surfaces. In some Livistoninae there is a differentiation along a cell file into **long** and **short** cells, a condition recalling that in panicoid grasses, but clearly a derived character in palms. Files of epidermal cells above the anticlinal walls of hypodermal cells may be either wider than elsewhere or deeper and with thinner walls (Fig. 60K, L), but again an uncommon and derived character.

Anticlinal walls that are **sinuous** in surface view are distinctive of many Calamoideae (Fig. 7J, K) and some Coryphoideae (Fig. 7D, E), but rarely occur elsewhere. The wall may be sinuous but exceptionally only the cutinized anticlinal wall is markedly sinuous (e.g. *Plectocomiopsis* (Fig. 39D), *Salacca*). A presumed benefit is that cells are interlocked in the manner of pieces in a jigsaw puzzle. Where sinuous walls are less pronounced we describe walls simply as undulate, with a portion of the wall curved to some extent (Fig. 38H).

Cuticle and wax

The epidermis of the palm leaf is protected by outer deposits of polymerized lipids (fatty) substances, which have hydrophobic properties. Two distinct layers are present. **Epicuticular waxes** are secreted through the outer wall of the epidermis and form a distinct outer loose layer, i.e. one which is easily detached. **Cutin** is secreted into the outer epidermis where it accumulates to form the **cuticle** but cannot be detached because it impregnates the cellulose wall layers. The cuticle can be isolated by strong oxidizing reagents (e.g. concentrated sulphuric acid), which digests the underlying cellulose. Cuticular material in plants is thus resistant to decay and easily forms fossils, when it may be identified if it retains the shape of the original cells.

Wax and cuticle have different physical properties; wax is crystalline and birefringent but does not stain with Sudan IV (Fig. 23N) so that it can easily be recognized (e.g. adaxial surface in Fig. 58A). On the other hand the cuticle is amorphous, not birefringent and stains with Sudan IV (Fig. 3F). Because wax is easily detached it may not persist through the sectioning process and is best studied by scanning electron microscopy (SEM) techniques using fresh material (Barthlott and Thiessen 1998; Barthlott et al. 1998). Carnauba wax from *Copernicia prunifera* is an important commercial product used in high quality polishes and is obtained by beating cut leaves to remove the thick epicuticular wax deposits.

Palm wax has been little studied, but falls into the 'Strelitzia-type', i.e. rod-like category of Barthlott et al. (1998), usually seen with the light microscope as a continuous ('crust') layer. It has been examined in the Coryphoid palms, as then understood, by Taylor (1999). Localized wax production may occlude the outer stomatal chamber, especially in sunken stomata. It is often well developed in abaxial grooves (Fig. 23M, N) in those few palms that have them.

The cuticle, as seen in sections stained with Sudan IV is much more variable in thickness. It may form a thin limiting layer, but

Fig. 6 Lamina anatomy, adaxial surface oriented toward top of page. **A**. *Tahina spectabilis*, lamina, TS (bar = 100 μm). **B**. *Tahina spectabilis*, lamina, LS (bar = 100 μm). **C**. *Hyphaene thebaica*, lamina, TS (bar = 250 μm). **D**. *Hyphaene dichotoma*, lamina, LS (bar = 250 μm). **E**. *Rhopalostylis sapida*, adaxial epidermis, SV (bar = 100 μm). **F**. *Rhopalostylis sapida*, abaxial epidermis, SV (bar = 100 μm). **G**. *Wallichia disticha*, lamina, TS (bar = 50 μm). **H**. *Euterpe* sp., lamina, TS (bar = 200 μm). **I**. *Hyphaene* sp., trichome, LS (bar = 50 μm). **J**. *Hyphaene dichotoma*, trichome, SV (bar = 500 μm). **K**. *Wettinia aequalis*, trichomes, TS (bar = 100 μm). **L**. *Arenga undulatifolia*, trichome, SV (bar = 25 μm). **M**. *Elaeis oleifera*, trichome, LS (bar = 100 μm). **N**. *Bactris gasipaes*, lamina, abaxial trichome, SV (bar = 50 μm).

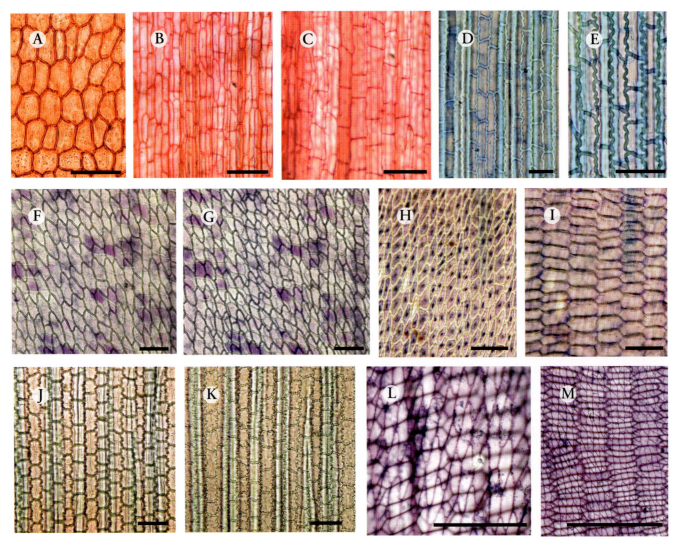

Fig. 7 Adaxial epidermis and hypodermis of lamina surface, all SV (from Horn et al. 2009); longitudinal files vertical except in A. **A**. *Juania australis,* non-elongated, hexagonal epidermal cells (bar = 50 μm). (**B, C**). *Trachycarpus fortunei,* elongated, rectangular epidermal cells with straight walls; rectangular hypodermal cells. **B**. Epidermis (bar = 50 μm). **C**. Same position with different focal plane to show hypodermis (bar = 50 μm). (**D, E**). *Rhapis excelsa,* elongate rectangular epidermal cells with sinuous walls; same preparation with different focal planes to show (**D**) hypodermis and (**E**) epidermis (bar = 50 μm). (**F, G**). *Dypsis lutescens,* oblique hexangular epidermal and hypodermal cells. **F**. Epidermis (bar = 50 μm). **G**. Same position with different focal plane to show hypodermis (bar = 50 μm). (**H, I**). *Arenga microcarpa,* oblique and elongated hexagonal cells of epidermis and hypodermis; **H**. Epidermis; fresh specimen shows nuclei in epidermal cells (bar = 50 μm). **I**. Same position with different focal plane to show hypodermis. (**J, K**). *Plectocomia muelleri,* rectangular epidermal cells with sinuous walls and hypodermis with fibres (light vertical bands) and rectangular cells. **J**. Epidermis (bar = 50 μm). **K**. Same position with different focal plane to show hypodermis (bar = 50 μm). **L**. *Reinhardtia gracilis,* oblique and hexagonal (spindle-shaped) epidermal cells with transverse hexagonal hypodermal cells (bar = 100 μm). **M**. *Socratea exorrhiza,* longitudinal and elongated epidermal cells with transverse hexagonal hypodermal cells (bar = 200 μm).

will then usually penetrate the anticlinal walls of the epidermal cells, so as to outline their shape in surface view. At the other extreme, the epidermis may be wholly cutinized and even extend into the anticlinal walls of hypodermal cells. The two contrasted states may, to a large extent, typify given groups of palms, but more equally have relevance to drought resistance. Where the anticlinal walls of epidermal cells are sinuous in surface view, the cuticle may exaggerate the undulations in the outer cell wall, as in many calamoid palms.

The guard cells of the stomata in most palms each include two horizontal cutinized ledges seen in TS as horn-like projections into the stomatal pore (Fig. 3F). Commonly also, the cuticle can extend into the walls of cells surrounding the substomatal chambers (Figs. 51F and 69O). Details of these features and their structural diversity are given elsewhere where stomatal structure is described.

Trichomes (hairs) (Figs. 6I–N and 8)

The term **trichome** is preferred here, as in botany generally, because the diversity of structures referred to in plants greatly exceeds the general concept invoked by 'hair'. Although there are scattered references to trichomes in palms (Tomlinson 1961), no comprehensive survey of their diversity and development has

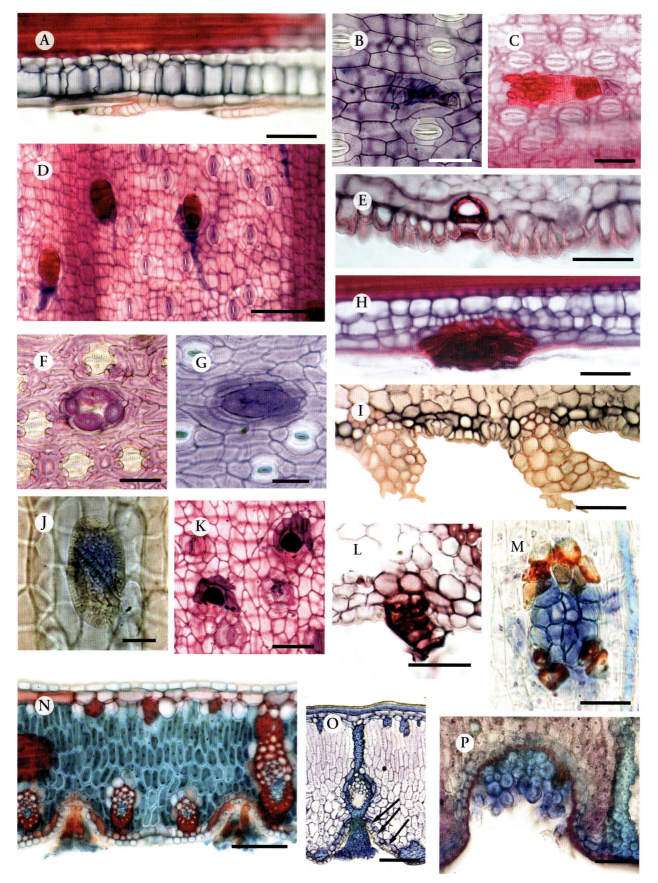

Fig. 8 Trichomes on lamina. **A**. *Bactris campestris*, LS (bar = 50 μm). **B**. *Desmoncus oxyacanthos*, SV (bar = 50 μm). **C**. *Daemonorops* sp., SV (bar = 50 μm). **D**. *Ptychosperma macarthurii*, SV (bar = 100 μm). **E**. *Iriartea deltoidea*, TS (bar = 50 μm). **F**. *Arenga undulatifolia*, SV (bar = 50 μm). **G**. *Rhopalostylis sapida*, SV (bar = 50 μm). **H**. *Juania australis*, LS (bar = 50 μm). **I**. *Chamaerops humilis*, TS (bar = 50 μm). **J**. *Wettinia fascicularis*, SV (bar = 50 μm). **K**. *Reinhardtia gracilis*, TS (bar = 100 μm). **L**. *Rhopalostylis sapida*, TS (bar = 50 μm). **M**. *Roystonea oleracea*, SV (bar = 100 μm). **N**. *Cocos nucifera*, TS (bar = 100 μm). **O**. *Ceroxylon quindiuense*, TS (bar = 100 μm). **P**. *Parajubaea cocoides*, TS (bar = 50 μm).

23

been undertaken. Nevertheless, amidst this limited knowledge there is something of a common structural pattern in the form of a basal group of cells within the epidermis that gives rise to a group of thin-walled cells; these may be visible with a ×10 hand lens. In a few palms these distal cells may persist at maturity (Fig. 62E), but they are usually eroded early and completely so that a later systematically useful description can only refer to the persistent trichome. We can still appreciate that trichome structure can be diagnostic for many groups, but must accept a great deal of uncertainty.

A common feature of all trichomes is the group of cutinized, often lignified, basal cells, which can also include tannins, so accounting for their persistence. (The illustrations in Figs. 6I–N and 8 provide a sampler.) The distal cells are usually thin-walled and lose their cell contents at maturity. Where they persist as a continuous indumentum their cell patterning can be quite spectacular.

Distribution of trichomes is different on the two surfaces of dorsiventral leaves, few or absent adaxially and so mostly restricted to the abaxial surface. Here they can be uniformly distributed but are often restricted to costal bands. In ribbed leaves hairs may occur only within the furrows as in *Ceroxylon* (Fig. 7O).

Outline of trichome types

1) In spiny palms there may be a transition from prominent spines to short, pointed microscopic hair-like trichomes, which are nevertheless a painful and persistent nuisance, as in Bactridinae (Figs. 6N and 8A, B).
2) Linear hairs are characteristic of the Geonomateae (Fig. 83I).
3) Similar trichomes but smaller and fewer-celled occur in many Calamoideae, but the distal cells may be more filamentous than shield-like and with a more massive base (Fig. 8C).
4) In the Arecoideae basal cells are few, but large and arranged in discrete tiers (Fig. 8D, G, K–M).
5) The tribe Caryoteae can be identified by its trichomes. Each has a superficial cylinder of four to six narrow sclerotic cells surrounding one to three (commonly two) central thin-walled cells (Figs. 6L and 8F).
6) Distal cells in the Iriarteeae are commonly filamentous, but the base is distinctive in the few large cells with conspicuously pitted walls (Figs. 6K and 8E).
7) A more generalized trichome is found in the Coryphoideae in which the base is massive, ellipsoidal, multicellular and commonly sunken within the epidermis; the distal expanse of thin-walled cells is shield-like (Figs. 6I, J and 8I). Similar trichomes are found in Ceroxyloideae (Fig. 8H) and can be deeply sunken, the base top-shaped within the abaxial groove (Fig. 8O).
8) In Attaleinae and Elaeidinae there is commonly a single or few inflated basal cells extended into a distal shield; in *Cocos* (Fig. 8N) and *Elaeis* (Fig. 6M) the base is deeply sunken.
9) Where the lamina is grooved there can be basal hairs producing a filling of thick-walled cells (Fig. 8P).
10) Other genera with distinctive hairs include *Nypa* (Fig. 43E, G more hydathode-like than trichomes), *Orania*, and Pelagodoxeae (producing a continuous indumentum).

This overview is deficient in many respects but shows that palm trichomes are difficult to categorize. It gives evidence for extensive parallel development within different groups and most significantly the existence of precise diagnostic characters in many large taxa. Clearly a more extensive developmental study of palm trichomes needs to be made and should include those on leaf axis and even stem. Ghose (1979) and Ghose and Davis (1973) provide the only recent development study. They make the suggestion that trichomes and stomata may be similar in very early development. Henderson and de Nevers (1988) illustrate hairs on the rachilla of *Prestoea*, which suggests diagnostic difference. In this and other later studies SEM techniques have proved useful.

Functions of trichomes

One can merely speculate about a function for the trichomes in palm leaves and a solution even then is difficult because of their diversity. A simple suggestion is that the loose ephemeral distal cells serve as a lubricant that either facilitates the separation of the initially tightly packed leaflets or leaf segments as the blade expands or minimizes friction when organs growing at different rates slide past each other. The amount of material produced can be considerable, as on the outer surface of the leaf axis in *Nannorrhops* where it accumulates in such dried quantity that it can be used as tinder. In principle, however, there is no acceptable explanation in the absence of precise investigation and experimentation.

The stomatal complex

The stomatal complex in palms is characterized by a very consistent developmental pattern involving 10 cell divisions, which produces a total of 12 cells (Tomlinson 1974). The result is a group of cells surrounding the pair of guard cells (g.c.). They are identifiable as: (i) two lateral subsidiary cells (l.s.c.); (ii) two terminal subsidiary cells (t.s.c.); and (iii) two neighbour cells usually trapezoidal in surface view. The first cell type is always differentiated as a pair of relatively thin-walled cells that enclose the guard cells; the terminal subsidiary cells exhibit a wide range of cell forms; they may not be identifiable, or may be differentiated in ways that can partly obscure the outer aperture of the complex. The neighbour cells are usually recognizable only by their shape, but can be also differentiated in cell wall features.

Irregularities in the stomatal complex can be caused in various ways: when stomata are very crowded, cell divisions in one complex may serve for an adjacent complex; early cell divisions may be missing; but most frequently the divisions that make distinct terminal subsidiary cells can be missing.

Development (Fig. 9 stages 1–6)

The stomatal complex is initiated as a small densely cytoplasmic cell (meristemoid, or guard cell mother cell), possibly by an

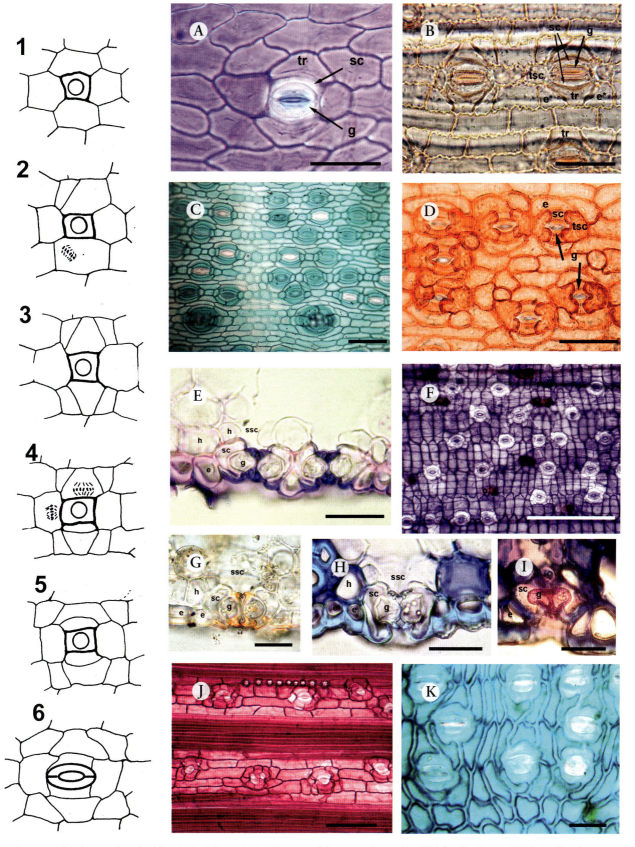

Fig. 9 Stomata. All in lamina abaxial epidermis. **1–6**. Stages in development of the stomatal complex, SV (after Tomlinson, 1990). **A.** *Rhopalostylis sapida*, SV (bar = 50 μm). **B.** *Mauritiella aculeata*, SV (bar = 50 μm). **C.** *Borassus aethiopum*, SV (bar = 100 μm). **D.** *Juania australis*, SV (bar = 50 μm). **E.** *Arenga microcarpa*, TS (bar = 25 μm). **F.** *Wettinia* sp., SV (bar = 200 μm). **G.** *Caryota mitis*, TS (bar = 25 μm). **H.** *Oraniopsis appendiculata*, TS (bar = 25 μm). **I.** *Sabal causiarum*, TS (bar = 20 μm). **J.** *Phoenix dactylifera*, hypodermis at level of substomatal chamber, SV (bar = 50 μm). **K.** *Borassus aethiopum*, at level of substomatal chamber, SV (bar = 50 μm). e, epidermal cell; g, guard cell; h, hypodermal cell; sc, subsidiary cell; ssc, substomatal chamber; tsc, terminal subsidiary cell; tr, trapezoidal neighbour cell.

unequal transverse division in one cell of a file of protodermal cells (stage 1). It remains undivided until the last division of the complex, which is a longitudinal division to form the pair of guard cells (stage 6). Within each cell adjacent to the meristemoid, but in the adjacent protodermal cell file, there are two successive oblique divisions, the resulting walls of which do not intersect (non-intersecting oblique cells) (Fig. 9 stages 2 and 3). These cut out a trapezoidal cell, the neighbour cell, within each of which a longitudinal division cuts out the lateral subsidiary cell (stage 4). This may then be followed by a transverse division in each of the cells adjacent to the meristemoid and in the same file to form short terminal subsidiary cells. The final division is that within the meristemoid, which forms the two guard cells (stage 6). The resulting precise geometric pattern may be obscured as the protodermal cells expand, especially in those taxa in which expansion produces oblique cells. Rarely there may be a second division in the neighbour cell to form a second lateral subsidiary cell (e.g. *Medemia*).

This pattern of stomatal development is certainly basic to all palms and we have seen no conspicuous exceptions. It is not, however, unique to palms and occurs, for example, in Pandanaceae but its general distribution within other families of monocotyledons remains incompletely explored (Tomlinson 1974).

Guard cells

The size of **guard cells** varies appreciably in palms but is largely independent of the size of epidermal cells. In most palms each guard cell has two prominent cutinized ledges which are best seen in TS (Figs. 3F and 9G) although their reaction to stains varies (cf. Fig. 9E, H). These are normally equal but in some palms the outer ledge is a larger than the inner (Fig. 8I). In a few genera there is only one outer ledge (e.g. *Synechanthus*). A special condition should be noted in *Nypa*; here there is a series of small ledges between the two larger outer ledges (Fig. 43J). In the tribe Caryoteae the outer surface of the two ledges is finely corrugated in a highly diagnostic manner so that they look like cockle shells (Fig. 61F).

In most palms the guard cells are at the same level as the normal epidermal cells, but because the lateral subsidiary cells are deeper than the normal epidermal cells they and sometimes the guard cells extend below the epidermis. Truly sunken stomata in which the guard cells are positioned well below the epidermis occur in a number of coryphoid palms (e.g. Fig. 63D). This produces an outer stomatal cavity which can be occluded by wax. More complete occlusion of the outer cavity is produced in some palms by enlargement of the outer wall of the terminal subsidiary forming irregular lobes which overarch the outer stomatal cavity (Fig. 9D. *Juania*). If the neighbouring cells are similarly lobed, the enclosure is more complete (e.g. *Zombia*, some *Arenga* spp.). An unusual condition is found in the subtribe Pelagodoxeae in which the terminal and neighbour cells are thick-walled and lignified (Fig. 85B).

Stomatal distribution

In most palms stomata are restricted to the abaxial surface of the lamina, although in these circumstances stomata may still occur sporadically on the adaxial surface, especially in relation to the larger veins or ribs. In a few palms stomata are equally distributed on both surfaces of the lamina, a condition we have already distinguished as isolateral. The stomata do not necessarily occur in continuous regular longitudinal files in the manner common in some grasses. Where distinct costal bands of epidermal cells are differentiated below major veins, stomata are restricted to intercostal regions. In the extreme condition the bands are so narrow that they can accommodate only one or two irregular files of stomata.

Hypodermis

In most palms the one or more layers of cells immediately below the epidermis are differentiated as larger cells, usually colourless, i.e. without chloroplasts. They are also usually thin-walled and unlignified. The few taxa with leaves that lack a differentiated hypodermis typically have a thin lamina and occur in shaded forest environments as in species of *Chamaedorea* and *Geonoma*. In contrast, a multiseriate hypodermis at one or both surfaces is found in plants of exposed, dry situations. The inner layers in such a hypodermis are irregular or less well differentiated.

In many palms, shape of hypodermal cells in surface view (SV), from which perspective they are usually described, varies from square to rectangular and then longitudinally extended (Fig. 7C). However, in many Arecoideae the hypodermal cells are transversely extended and narrowly hexagonal because the cells of adjacent files interlock in a way which may be mechanically efficient (Fig. 7I). This condition is almost always associated with obliquely extended epidermal cells (Fig. 7L, M) so that an additional mechanical benefit may come from the two-ply construction that resists stress in different directions (Horn et al. 2009). This suggestion is supported by the observation that such leaflets are rather deficient in mechanical fibres.

The hypodermis below stomata is interrupted by enlarged intercellular spaces (**substomatal chambers** = s.st.ch.), which have a fairly consistent configuration. A single chamber may be subdivided by lobed cells. The stomatal initial is normally located in a file of epidermal cells that lies beneath the anticlinal walls of the immediate hypodermal layer (Fig. 9F). As the stomatal apparatus develops, the hypodermal cells separate and most commonly divide transversely to produce four L-shaped cells enclosing a single chamber in a characteristic standard resulting pattern (cf. Fig. 55C, D). However, considerable diagnostic variation can occur. This includes two C-shaped cells whose inner lobes meet so as to partly occlude the chamber. There may be more than four cells, resulting from increased numbers of cell divisions as in Bactridinae. Several lobed cells produced in this way may further subdivide the chamber as in *Phoenix* (Fig. 9J). Where the hypodermis is multiseriate, the substomatal cells in different layers can be different and so produce chambers of some complexity, as in the Borasseae (Fig. 63G, H).

The substomatal chamber is continuous with the intercellular space system of the mesophyll and so provides a passage for gas exchange through cells which, near the surface, are looser than elsewhere.

Expansion cells

Blade expansion of the developing lamina requires that its initially tightly-packed folds flatten into a broad expanse. This involves general cell expansion, but more precisely and anatomically more obvious in the hypodermal cells of the ribs associated with upper and lower folds. Hypodermal cells become greatly enlarged as files of **expansion cells** that are visible as translucent bands, or in costapalmate leaves as translucent regions. Their position is determined by leaf type. In pinnate induplicate leaves the expansion cells usually form a single band within the fold, either adaxial to an abaxial rib (Figs. 52A, C and 60J) or abaxial to an adaxial rib (Fig. 58D), whereas in reduplicate leaves there are usually two parallel bands of expansion cells on the abaxial side (Figs. 52B and 76H). Expansion cell bands may include fibre bundles, as in many Coryphoideae. *Phoenix* leaflets lack a median rib to the leaflet but still include a series of wide expansion cells (Fig. 55B). This 'phoenicoid' fold can also occur in the leaf segments of fan palms in other genera of Coryphoideae. Expansion cells are otherwise found in the lamina independent of the ribs and usually associated with irregularities in the folded leaf.

Mesophyll

Although the mesophyll is primarily the photosynthetic (green = chlorenchymatous) tissue of the palm lamina (Figs. 3A; 4A–C; and 79B) it is also the ground tissue that supports veins and fibre bundles and includes often numerous tannin cells and raphide sacs, but only occasionally elongated sclereids. The organization of the central cell layers of the lamina is best seen in longitudinal section because this shows the files of cells developed as the lamina elongates by transverse cell division followed by cell expansion (Fig. 12D). Dorsiventral symmetry is often reflected in the difference between one or more adaxial **palisade** layers of compact anticlinally extended cells and abaxial layers of isodiametric cells (Fig. 37A). In the absence of a distinct palisade layer, adaxial mesophyll cells may be smaller and more compact than those elsewhere. Rarely is the mesophyll composed of transversely extended cells, as in *Sclerosperma* (Fig. 75A). However, abaxial cells are not usually differentiated as 'spongy' mesophyll cells as in dicotyledonous leaves because they are not lobed except near the substomatal chambers. In a number of Coryphoideae the palisade cells can have regular papillose protuberances, a condition referred to as 'peg parenchyma' in other monocot families (e.g. Cutler 1969).

In isolateral leaves the mesophyll cells towards each surface are equally differentiated with small, often palisade-like, cells and intercellular spaces in contrast to the central larger and more compact cells of the mid-mesophyll (Fig. 3B–E). Where this is pronounced, the central cells may contain few chloroplasts (Fig. 3A). In this construction the opposed mesophyll surface layers are mirror images of each other (Figs. 62C and 78B).

Fibre bundles

An important mechanical component of the lamina is the **non-vascular fibre** that occurs either singly or more often in aggregates as axially-extended **fibre bundles** or strands. They complement the mechanical tissues of the veins, which consist of **vascular fibres**, other sclerotic tissue, and the tracheary elements. In palm leaves non-vascular fibres are variously lignified or unlignified, whereas vascular components are always lignified (Fig. 3E). An additional but infrequent component is lignified or unlignified **fibre-sclereids** as isolated fibres that ramify among the mesophyll cells, sometimes randomly but otherwise with a preferred transverse or vertical orientation (Fig. 33A). In some examples these cells can be traced to the sheathing fibres of the transverse veins and are their terminal extensions. Fibre sclereids are thought to be homologous with the fibrous sheath cells of the transverse veins because a clear transition series can be observed between the two types within a leaf (e.g., *Daemonorops*), and, additionally, they may be topographically continuous (e.g. *Licuala*; Tomlinson 1959b).

The distribution of both non-vascular and vascular fibres in the palm leaf varies considerably and can be highly diagnostic. Figure 14 provides a sample range. Least commonly, non-vascular fibres are absent and then usually from the leaves of understorey palms (e.g. some but not all species of *Chamaedorea* and *Geonoma*) and some rattans. Otherwise two major patterns can be recognized:

1) Lamina with the non-vascular fibres mostly restricted to or most common either within or in contact with the **hypodermal layers**. Additional, but fewer, fibres may occur in the mesophyll. This condition is characteristic of the Calamoideae, Coryphoideae, and Ceroxyloideae suggestive of an ancestral state for palms generally.
2) Lamina with the non-vascular fibres restricted to or most common in the **mesophyll** and rarely in contact with the hypodermis. This condition is characteristic of Arecoideae, with some conspicuous exceptions (e.g. Attaleinae, and is suggestive of a derived condition within this subfamily.

The overall abundance of fibres in the leaf lamina is very variable in palms, even in closely related taxa; Fig. 59 provides a striking example. In several taxa, fibres may make up as much as 50% of total lamina cell volume so that photosynthetic tissue is minimal (Fig. 52F). For palms as a whole, the overall trend is one in which there is a progressive decrease in fibre volume in successively divergent lineages. This minimization may be compensated in a mechanical sense by the cross-ply texture of the surface layers, as suggested in Horn et al. (2009).

Another variable is the size of fibre strands themselves, which can be very large where they connect veins to surface layers (Figs. 10I and 79A). Independent fibre bundles can also be large and may exceed the diameter of the smaller longitudinal veins (Figs. 6G and 61A).

Fibres have been treated so far as uniform entities but they vary appreciably in wall thickness. Where the walls are thin, and the lumen correspondingly wide (Fig. 60L), fibres are commonly septate by cell division, with thin transverse partition walls at regular intervals. In the Coryphoideae septate fibres may be diagnostically useful (Fig. 47C). In the Bactridinae non-vascular fibres have very narrow lumens.

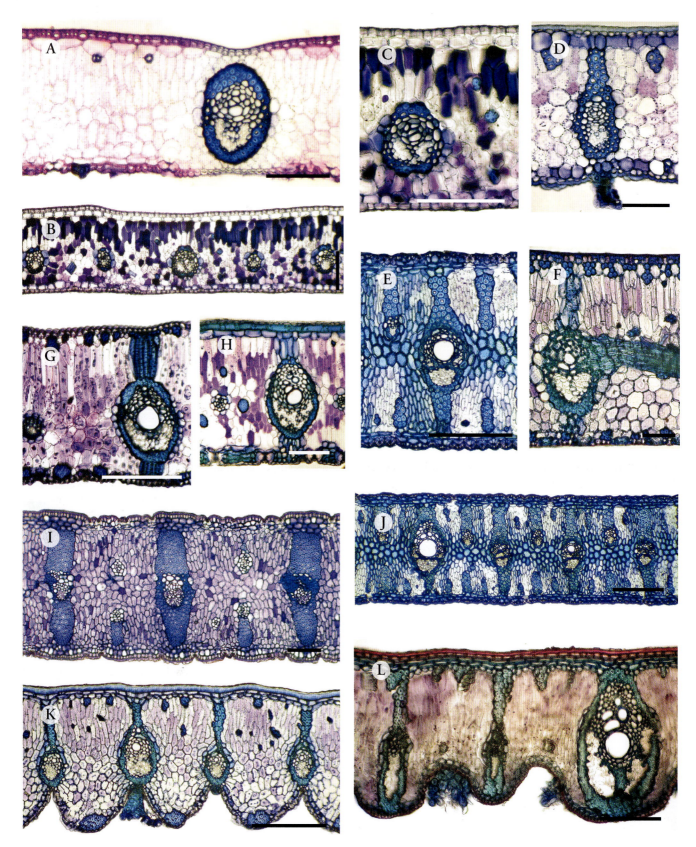

Fig. 10 Lamina in TS, adaxial surface oriented toward top of page. From Horn et al. 2009. (**A–C**). Vascular bundles free of surface layers (bar = 100 μm). **A**. *Arenga microcarpa*. **B**, **C**. *Dypsis lutescens*. (**D**, **E**). Vascular bundles attached to surface(s) by vertically extended fibre bundles. **D**. *Oraniopsis appendiculata* (bar = 100 μm). **E**. *Bismarckia nobilis* (bar = 100 μm). (**F–H**). Vascular bundles attached to adaxial surface by palisade of ad/abaxially elongate sclereids. **F**. *Metroxylon vitiense* (bar = 50 μm). **G**. *Mauritiella armata* (bar = 100 μm). **H**. *Nypa fruticans* (bar = 100 μm). (**I**, **J**). Examples of convergent evolution of isobilateral lamina anatomy. **I**. *Butia capitata* in Arecoideae (Cocoseae: Attaleinae), (bar = 100 μm). **J**. *Bismarckia nobilis* in Coryphoideae (Borasseae), (bar = 200 μm). K, L. Examples of convergent evolution of fibre bridges and stomatal furrows in leaves of Andean palms (bar = 200 μm). **K**. *Ceroxylon quindiuense* in Ceroxyloideae (Ceroxyleae), (bar = 200 μm). **L**. *Parajubaea cocoides* in Arecoideae (Cocoseae: Attaleinae), (bar = 200 μm). (From Horn et al. 2009.)

Vascular bundles (veins) and ribs

The distinction between vein and rib is not always clear, and they form something of an anatomical continuum or transition. Both can be prominent above or below the lamina surface, but rib should be restricted to the prominent vascularized structure that mark the segmental folds of the multiplicate blade with veins referring to the vascular bundles within the intervening portions of the lamina. Marginal ribs often also occur in many pinnate leaf segments in monocotyledons, as 'parallel' although a better term might be 'divergent' in fan palms or in feather palms with the 'fishtail' type of leaflet (e.g. *Caryota, Iriartea, Korthalsia*) or where the leaflet has a narrow insertion. Strictly also, venation should refer only to vascular bundles and not include parallel fibre bundles which may be visible superficially. We have emphasized vein structure in our analysis because of the systematic features our clearing methods reveal, especially in the transverse vein system (Fig. 11).

Longitudinal veins

A size hierarchy for veins is determined as they successively differentiate as procambial strands in the developing lamina, although the sequence has not been quantified in any detail. The result is a few major veins, more numerous intermediate veins, and most numerous minor veins, although the hierarchy varies considerably in different taxa; compare *Elaeis* (Fig. 81A) with *Leopoldinia* (Fig. 84A). This size hierarchy is matched by a decreasing complexity of the vein tissues. Each vascular bundle has the same components, an **outer parenchymatous sheath** (O.S.), and an **inner sclerotic sheath** (I.S.) surrounding the **vascular tissues** themselves.

The outer sheath consists of a single series of thin-walled colourless cells, or at most with a few chloroplasts, the cell shape varying from cubical to tabular, when the long axis of the cell may be parallel or perpendicular to the vein. The outer sheath can be continuous around the minor veins (Fig. 37A), or incomplete, as is common in larger veins where it may be absent above (Fig. 6A) or below (Fig. 39A), depending on the location of the vein in the mesophyll. In larger veins the outer sheath is normally present only laterally when it may be continuous with cells of either hypodermis. When a vein is continuous with both surfaces via sclerotic or fibre 'bridges', the outer sheath is discontinuous because it is restricted to the vasculated portion of the vein (Fig. 10F), and may be insignificant (Fig. 10I).

Cell walls of the outer sheath are not conspicuously pitted and are never lignified, although in some safranin-stained preparations a 'pseudo-Casparian' strip may appear because the common wall, as seen in surface view in TS, takes up a general stain selectively. This condition seems common in Livistoninae and some Arecoideae. We have referred to it in our notes as 'endodermoid', but there is no Casparian strip.

The inner sheath, in contrast, is always thick-walled and lignified and normally forms a complete investment to the vein, even though it may be as narrow as a single cell layer in minor veins, but massive in major veins. It consists of fibres either above and below, or below only or above only, but is completed at mid-level by short, pitted, elongated sclerenchyma cells. In this region the transverse veins are connected (Figs. 10F and 51A).

Vascular tissue itself includes adaxial xylem and abaxial phloem. Xylem in major veins includes usually one (Fig. 6G) but sometimes more (Fig. 6H) wide metaxylem vessels plus associated narrow protoxylem tracheids indicative of early vascular differentiation in the developing leaf, with progressive reduction to few elements in minor veins. Metaphloem of larger veins is well developed (Fig. 6G) and often divided by sclerotic phloem parenchyma into two (Fig. 6H) or more (Fig. 34C) separate strands. Multistranded phloem of this type is common in the Geonomateae. Where two phloem strands occur, the separating sclerenchyma usually represents the position of the original protophloem, now obliterated. Narrow elements at the margin of the phloem represent the result of insertion of phloem from transverse veins (Fig. 51A).

The distribution of longitudinal veins within the mesophyll and their association with surface layers varies in highly diagnostic ways, as is illustrated in the sampling of Fig. 10. The overall impression is that veins tend to be associated with surface layers more commonly in earlier-diverging groups like Calamoideae and especially Coryphoideae than in strongly nested groups like Arecoideae, although size and ecological characters are other significant associations. Connections can be made directly to surface layers or via bundle sheath extensions, described as 'bridges' or 'girders', made up either of lignified fibres like those of the inner sheath, or unlignified fibres which are comparable to the non-vascular fibres. Larger veins can connect to both surfaces, but smaller veins may extend only to one surface. Minor veins are usually independent of surface layers.

A major diagnostic difference in the girders involves the presence of anticlinally extended **sclerotic cells** with relatively wide cell lumina (Fig. 10G, H) contrasted with **fibres** (Fig. 10I–L). The distribution of presence or absence and type of bridging cell (Fig. 28B) is emphasized in Horn et al. (2009) and shows a clear transition from early to later diverging groups.

Transverse veins

These lack an outer parenchymatous sheath but retain a sheath of sclerotic cells that varies from little to quite massive, depending on the type of sclerenchyma. Where massive, the cells are thick-walled fibres (Fig. 6B); where limited (Fig. 6D), the cells are elongated with only moderately thickened walls and wide lumina, often one cell thick. The vascular tissue is reduced and irregular, often including several separate strands (Fig. 11A). *Borassus* is unique because the sclerotic sheath extends to both surfaces to form conspicuous transverse girders, which compartmentalizes the mesophyll (Fig. 64B, D, F). In contrast, the transverse vein system in most Arecoideae is diffuse and with limited sheathing tissue, again suggesting the trend towards limited sclerification of the lamina in more recently divergent lineages.

A major feature of transverse veins is their position in relation to the surface layers, which may be adaxial and either running above many of the longitudinal veins, or in the mid-mesophyll at the level of the longitudinal veins, or abaxial, below the longitudinal veins. Our clearing techniques (Fig. 11) demonstrate how this,

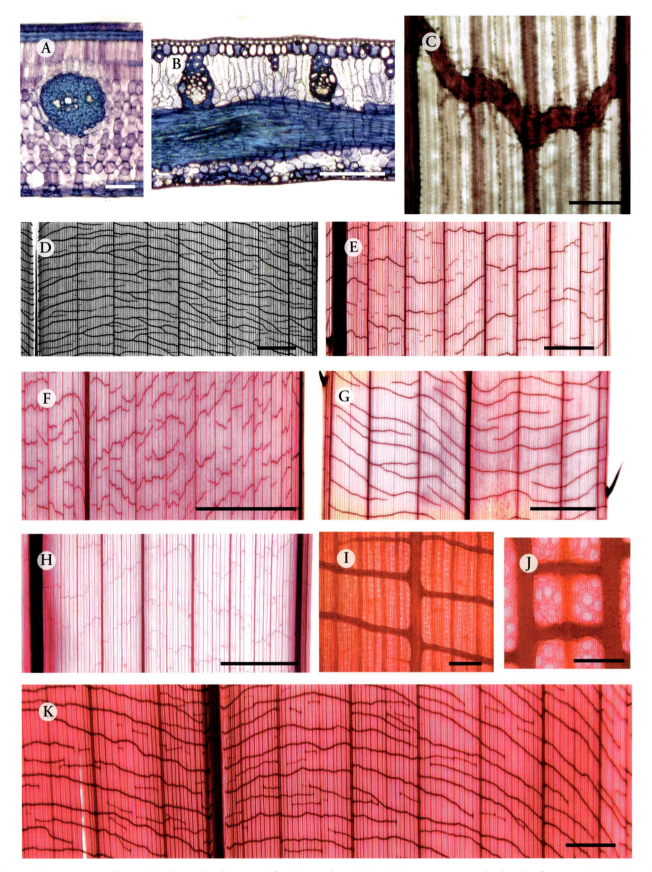

Fig. 11 Transverse veins of lamina: histology and architecture; (after Horn et al. 2009). (**A, B**). Lamina sections with adaxial surface oriented toward top of page. **A**. *Metroxylon vitiense*, transverse vein in TS, lamina cut in LS (bar = 50 μm). **B**. *Borassodendron machadonis*, transverse vein in LS, longitudinal veins in TS (bar = 100 μm). **C–K**. Cleared lamina, all adaxial SV. **C**. *Acanthophoenix rubra*, transverse vein sheathed by sclerotic parenchyma (bar = 100 μm). **D**. *Borassodendron machadonis* (bar = 5 mm). **E**. *Metroxylon vitiense*, midrib on left (bar = 1 mm). **F**. *Zombia antillarum*, margin on right (bar = 5 mm). **G**. *Laccosperma secundiflorum*, margins right and left, midrib centre (bar = 5 mm). **H**. *Dypsis lutescens*, margin right (bar = 2 mm). **I**. *Borassus heineanus* (bar = 250 μm). **J**. *Borassus flabellifer* (bar = 200 μm). **K**. *Attalea allenii*, midrib left center (bar = 2 mm).

together with the size of the transverse veins, produces characteristic vein patterns. Veins may be either almost exclusively long (Fig. 11D, G, K), or very short (Fig. 11E), and either more or less orthogonal (Fig. 11I) or very irregular (Fig. 11F). This irregularity we have referred to in these times of recession as the 'stock market pattern'. Where the transverse veins have a sheath with little sclerenchyma, they are inconspicuous (Fig. 11H). *Borassus* remains as the most distinctive (Figs. 11J and 64D).

Ribs

In principle, as a corrugated structure, there is potentially a projecting supporting enlarged portion of the lamina, a **rib**, either at the top (adaxial) and bottom (abaxial) of each successive fold. This is most likely to be seen in the unsegmented portion of the leaf blade (**palman**) of palmate leaves. In fact, the common condition is for there to be a single **abaxial midrib** in each segment of induplicate leaves and an **adaxial midrib** in each segment of reduplicate leaves. This is often modified, as when the individual segments have multiple folds or when there is a marginal as well as a midrib in each segment (e.g. Figs. 43H and 87D). Despite this variability two main configurations distinguish the structure of midribs (and to a certain extent marginal ribs) in palms.

The basic structural component is one large abaxial vascular bundle whose abaxial orientation is determined by the position of its xylem and phloem, together with smaller adaxial bundles of varying orientation (Fig. 69L). Rib structure corresponds to two contrasted types of considerable diagnostic consistency:

1) Vascular bundles of the rib independent of each other and not enclosed by a common sclerotic cylinder. Found exclusively in induplicate-leaved palms (Figs. 51C and 62B).
2) Vascular bundles of the rib enclosed by a more or less continuous sclerotic sheath. Found in all reduplicate-leaved palms (Figs. 43B; 76H; 79H; and 84C) and a few induplicate-leaved palms (Fig. 52A, B).

Consequentially, this contrast is not wholly consistent with the type of leaf segmentation. In Coryphoideae, which are almost all induplicate leaved (except *Guihaia*), there is considerable diversity as shown in Fig. 47A. For this subfamily, types that can be described as 'Coryphoid' as well as 'Arecoid' both exist.

Rib anatomy can in some respect be seen as a reduced version of petiolar anatomy, as shown in a progressive reduction distally along the narrowing rachis of pinnate leaves or the intruded costa of costapalmate leaves. This is understandable in view of the common mechanical function of all these structures. Structurally and possibly also developmentally there is thus a continuum from the vascular system of the leaf axis as a whole to the longitudinal veins of the lamina.

CHAPTER 4 LEAF AXIS

The determinate nature of the palm leaf is contrasted with the indeterminate stem, but both have similar functions of mechanical support, conduction, and sometimes carbohydrate storage. However, the leaf axis is a complex organ divided into three discrete and disparate regions: leaf sheath, petiole, and rachis, each with a different function—basal support, blade extension, and blade support respectively. Although this leads to different anatomical features there is a structural continuity from one to the other because they are developmentally continuous as a unit and vascular bundles extend from one region to another (Fig. 5A–C). The leaf sheath alone is anatomically diverse in relation to changes along its length. It is also difficult to compare the extended rachis of pinnate leaves with its condensed equivalent in a palmate or costapalmate leaf. The petiole–rachis transition is abrupt in palmate leaves, but extended in costapalmate and especially pinnate leaves, and has not been studied.

We use the '**standard level**' approach comparing a topographically identical level along the axis among different taxa. For the petiole this is the distal part toward the blade, for the rachis of pinnate leaves it is at a mid-level along its length, whereas more arbitrary levels apply to the leaf sheath. Regretfully, we do not enter into a detailed discussion of the many changes that occur along the leaf axis because the leaf axis in palms has been little studied both by ourselves and by earlier anatomists. An introduction is provided in Tomlinson (1964). Here we begin with the petiole as a specific part of the leaf axis for comparative study, a quite arbitrary decision, but based on the petiole's inherent simplicity and accessibility.

PETIOLE (Fig. 12)

Both stem and petiole are supporting organs and have mechanical similarities. The petiole is simpler because it has no lateral supply to appendages, but being oriented away from the vertical its symmetry is usually dorsiventral (Fig. 12A, C, D, G). Vascular bundles apparently run continuously from base to apex and are only lost distally as they diverge into the lamina as the rachis tapers. However, this three-dimensional aspect has been investigated neither structurally nor developmentally. We present only a few diagnostic anatomical features.

Surface layers

The epidermis is uniseriate, cutinized, and somewhat thick-walled, bearing most similarity to the adaxial surface of the lamina. Where stomata and hairs are present they are of the same basic construction as on the abaxial surface of the lamina, but we have not investigated them in detail. In contrast, a detailed attempt to distinguish different taxa within *Rhopalostylis* in New Zealand on the basis of petiolar scales seems rather inconclusive (Salter and Delmiglio 2005). The hypodermis may be differentiated as colourless or sclerotic layers but some layers close to the surface are always chlorenchymatous. These surface layers, always narrow relative to the central ground tissue, have the status of a stem cortex, but without vascular or much fibrous tissue.

Peripheral sclerenchyma

The major mechanical tissue of the petiole is a continuous or discontinuous layer of fibres abruptly delimited from the surface layers (Fig. 12F), but in the absence of radial symmetry these are often most extensive abaxially, the most conspicuous exception being the petiole of *Nypa* where the leaf stands more nearly vertical (Fig. 12E). Sclerenchyma is formed by a peripheral series of vascular bundles, each with a well-developed fibre sheath and often with a transition to bundles composed exclusively of fibres (Fig. 44A). Bundles are more massive and confluent on the abaxial side, often with the appearance of additional small strands of vascular tissue embedded within them. There is commonly a median abaxial 'keel' bundle of some complexity (Figs.12G and 87H), which probably represents the first vascular bundle differentiated in the leaf primordium (Fig. 5C). The sclerotic layer is less well developed in adaxial bundles, but continues a similar configuration of larger and smaller vascular bundles that form a hierarchy of alternating sizes (Fig. 12B). Although there is no radial symmetry, it is still partly reflected in the inverted construction of the adaxial bundles, i.e. with the phloem towards the periphery.

V-arrangement

A distinctive feature of the majority of many palm petioles is a regular array of one or more series of vascular bundles that form a deep (Fig. 12A) or shallow (Fig. 12C, D) 'V' in the central

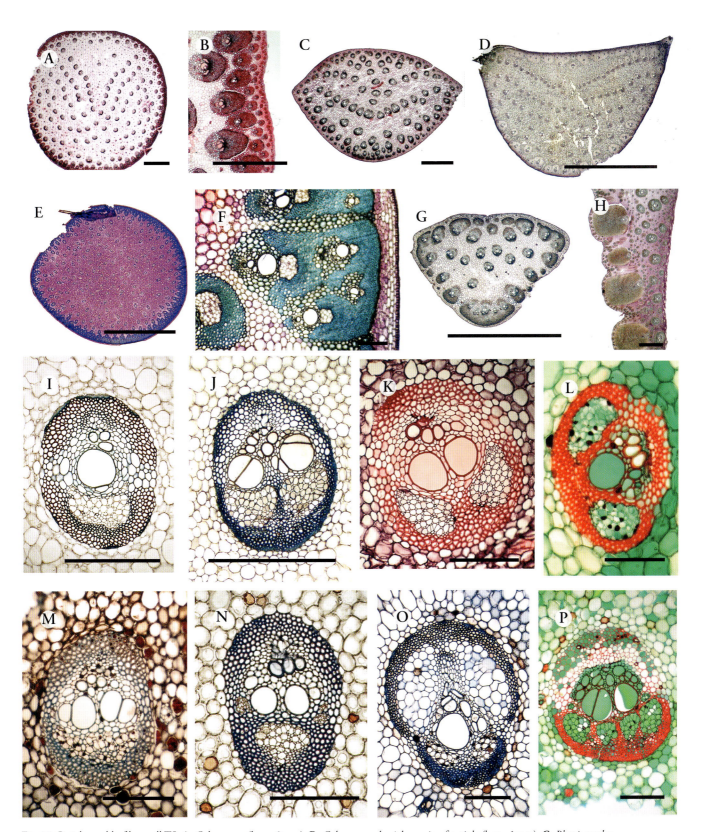

Fig. 12 Petioles and leaf base, all TS. **A**. *Calamus* sp. (bar = 2 mm). **B**. *Calamus* sp, abaxial margin of petiole (bar = 1 mm). **C**. *Rhapis excelsa* (bar = 1 mm). **D**. *Geonoma interrupta* (bar = 2 mm). **E**. *Caryota mitis* (bar = 10 mm). **F**. *Caryota mitis*, edge of petiole, showing peripheral sclerotic cylinder (bar = 200 μm). **G**. *Reinhadtia gracilis* (bar = 2 mm). **H**. *Cocos nucifera*, immature sheath (bar = 2 mm). **I–P** individual vbs. **I**. *Lodoicea maldivica* (bar = 500 μm). **J**. *Phytelephas aequatorialis* (bar = 100 μm). **K**. *Pseudophoenix sargentii* (bar = 100 μm). **L**. *Licuala grandis* (bar = 100 μm). **M**. *Phoenix* sp. (bar = 200 μm). **N**. *Sabal etonia* (bar = 200 μm). **O**. *Plectocomia muelleri* (bar = 200 μm). **P**. *Manicaria saccifera* (bar = 200 μm).

tissue, as seen in TS. It can be seen most directly in a transverse cut of the petiole examined with a hand lens; staining for lignin enhances this. The 'V' can be seen in its entirety in a small petiole, but small central blocks of a larger petiole should also reveal its presence. Figure 12E illustrates this visual approach in a petiole without an obvious 'V'. Adaxial inverted vascular bundles commonly occupy the sinus of the 'V' and thus correspond in orientation to the dominant vascular bundles immediately below the adaxial surface. The distribution of this 'V' throughout palms is shown in Fig. 27B, suggesting that a uniformly scattered arrangement of bundles is the ancestral state for the family, with notable independent origins of the V-arrangement of bundles in the clade of Ceroxyloideae + Arecoideae and within Calamoideae. However, its recognition may be somewhat subjective.

Vascular bundles (Fig. 12I–P)

The individual central vascular bundles of the petiole offer diagnostically useful information in terms of the amount and distribution of vascular tissues. Although there can be many vascular bundles (hundreds in a *Corypha* petiole) their construction is very uniform because it largely remains unchanged throughout most of the leaf axis, uninterrupted in the petiole by any appendicular attachments. This is a consequence of developmental extension of vascular bundles as a result of intercalary growth, and largely seen as a relatively constant number of protoxylem elements in each bundle. There is still an obvious correlation with size, seen as fewer and narrower bundles in small petioles. Fibrous sheathing tissue partially encloses the phloem but is more limited around the xylem. There is always a lateral median sinus interrupting the bundle sheath on each side and through which connections from transverse commissures penetrate.

Four major types of petiolar vascular bundle can be recognized in the central tissue according to the number of wide metaxylem vessels and phloem strands. A distinction has to be made between two widely separated vessels and a vessel–vessel overlap where one begins and the other ends. In this latter situation one vessel simply replaces another to from a linear series, or 'pipe'.

1) 1 wide metaxylem (mxy) vessel: 1 phloem (phl) strand, e.g. *Nypa* (Fig. 44E), *Lodoicea* (Fig. 12I).
2) 1 wide mxy vessel: 2 phl strands, e.g. *Calamus, Cocos, Licuala* (Fig. 12L), *Plectocomia* (Fig. 12O).
3) 2 wide mxy vessels: 1 phl strand, e.g. *Sabal* (Fig. 12N).
4) 2 wide mxy vessels: 2 phl strands, e.g. *Phytelephas* (Fig. 12J), *Pseudophoenix* (Fig. 12K).

Multiple numbers of either metaxylem vessels (e.g. *Phoenix* Fig. 12M), or phloem strands can occur (e.g. *Manicaria*, Fig. 12P) as a fairly constant feature. Clearly, however, within the many petiolar vascular bundles, some variation is to be expected. The distribution of numbers of phloem strands in the vascular bundles of palm petioles (Fig. 28A) suggests that the double-stranded condition is a derived character state outside of Calamoideae.

Other minor conditions include vascular bundles with more than two wide metaxylem vessels or some range in numbers that may make it difficult to categorize the type of vascular bundle in a given sample. The phloem strands can be discrete and widely separated, but often the intervening sclerotic sinus is narrow or even interrupted so that two strands are not totally separated. An element of subjectivity is introduced. Paradoxically, developmental study suggests that the initial condition of the phloem in most vascular bundles of palms is the production of a single strand, initially of protophloem, subsequently of metaphloem, as described under the section on phloem. The protophloem elements are stretched and collapsed whereupon they are replaced by sclerotic elements, usually fibres. This will produce a narrow sinus in the phloem strand, but if it continues a partition of varying width results, separating the phloem into two strands. The mature condition then bears little obvious relation to the initial condition.

A distinct 'Gestalt' appearance can be perceived in many taxa and is illustrated in our few examples, but the information is difficult to describe or quantify. An exception appears in the striking appearance of vascular bundles in the petioles of many Calamoideae (Fig. 32). Here the tissue surrounding the protoxylem consists of inflated cells within a very narrow sheath. These cells are apparently empty of cell contents at maturity and become air-filled. Their function is unknown.

Ground tissue

The ground tissue within which the vascular system is embedded is uniformly parenchymatous although it may contain air spaces, fibre and small vascular strands, and the almost ubiquitous raphide sacs or raphide canals. Stegmata are universally present in association with fibres.

Commonly aerenchyma of an irregular kind forms by collapse of cells in narrow central regions creating significant large air spaces. Only in a few palms of wet places, notably *Nypa*, does the ground tissue become regularly lacunose throughout. In *Nypa* the petiole base is submerged at high tide, or embedded in mud and probably functions in gas exchange (especially oxygen) like a snorkel, even in leaves where the blade has been lost. Narrow fibre bundles are frequently encountered in palm petioles, uniformly distributed within the ground tissue. Although a distinctive character we do not emphasize it as they are not necessarily consistently present throughout the petiole. Rather it seems they are more common in the leaf base (and continuous into the stem cortex) but extending incompletely in a distal direction. Similarly narrow vascular bundles with limited vascular fibres can also appear basally although their longitudinal distribution is unclear.

Transverse veins (commissures)

The axial vascular systems of the palm petiole do not anastomose, which might suggest that transport is exclusively longitudinal. However, interconnection is made between the axial bundles by narrow transverse commissures, each with a single strand of narrow xylem elements and a few phloem cells. The sheath is narrow and not usually sclerotic. Their course can be very irregular so that they run in and out of the plane of a transverse section. They are presumably important in allowing conduction among adjacent axial bundles, especially when the petiole is damaged. They bear comparison to the unusual transverse commissures of rattan stems where it is known that such bundle interconnections differentiate late from mature ground parenchyma cells (Tomlinson et al. 2001).

Summary of petiolar vascular bundle types

Vascular bundle types in palm petioles in a systematic context are as follows:

In general, for all subfamilies 1 mxy + 2 phl is the most common condition (synapomorphy), other conditions are variable and presumed derived.

1) **Calamoideae**: 1 mxy: 2 phl. A highly consistent petiolar anatomy.
2) **Nypoideae**: 1 mxy: 1 phl. An unusual condition.
3) **Coryphoideae**: all have 1 phl except Livistonae; the number of mxy vessels is sometimes rather variable as in Cryosophileae.

   Sabaleae: 2 mxy: 1 phl.
   Cryosophileae: 2 or more mxy: 1 phl (this combination is rather unusual in palms).
   Phoeniceae: 1–2–many mxy: 1 phl.
   Livistoneae:
   　Raphidinae: 1–2–many mxy: 2 phl.
   　Livistoninae: 1 mxy: 2 phl.
   　Unplaced: 1 mxy: 2 phl.
   Chuniophoeniceae: many mxy: 1 phl.
   Caryoteae: 1 mxy: 1 phl (an unusual combination for the subfamily).
   Corypheae: 1–2 mxy: 1 phl
   Borasseae:
   　Hyphaeninae: 2 mxy: 1 phl.
   　Lataniinae: 1 mxy: 1 phl (as in Caryotae).

4) **Ceroxyloideae**
   Cyclospatheae: 2 mxy: 2 phl (this condition is unusual in palms).
   Phytelepheae: 2 mxy: 2 phl (this condition is unusual in palms).
   Ceroxyleae: 2+ mxy: 1 phl.

5) **Arecoideae**
   Basal members of the Arecoideae (Iriarteeae and Chamaedoreeae) have: 1 or 2 mxy: 1 phl. Most later divergent groups have 1 mxy: 2 phl, but with occasional exceptions, e.g. Leopoldinieae: 2 mxy: 1–2 phl.

   Iriarteeae: 2 mxy: 1 phl.
   Chamaedoreeae: 2+ mxy: 1 phl.
   Podococceae: no data.
   Sclerospermeae: no data.
   Roystoneeae: 1 mxy: 2 phl.
   Reinhardtieae: 1 mxy: 2 phl.
   Cocoseae:
   　Attaleinae (except *Jubaea*): 1 mxy: 2 phl. *Jubaea*: many mxy: 2 phl.

   Bactridinae: 1 mxy: 2 phl.
   Elaeidinae: 1 mxy: 2 phl.
   Manicarieae: many mxy: 2 phl.
   Euterpeae: 1 mxy: 2+ phl.
   Geonomateae: 1–2 mxy: 2 phl.
   Pelagodoxeae: no data.
   Areceae:
   　Archontophoeniceae: 1–2 mxy: 2 phl.
   　Arecineae: 1+ mxy: 1–2 phl.
   　Basseliniinae: 1+ mxy: 1–2 phl.
   　Carpoxylinae: 1 mxy: 2 phl.
   　Clinospermatinae: no data.
   　Dypsidinae: 1 (–3) mxy: 2 (+) phl.
   　Linospadicinae: 1 (–2) mxy: 1 (–2) phl.
   　Oncospermatinae: 1+ mxy: 2 phl.
   　Ptychospermatinae: 1 mxy: 2 phl.
   　Rhopalostylidinae: 1 mxy: 2 phl.
   　Vershaffeltiinae: 1 mxy: 2+ phl.
   　Unplaced Areceae: 1 mxy: 1+ phl (unusual in palms).

LEAF BASE

The general condition in monocotyledons

In this section we examine unique attributes of leaf morphology in palms as a consequence of their frequently massive construction and great diversity, but it is worthwhile to comment on monocotyledons generally, because this is rarely done. Leaf morphology in monocotyledons differs from that of dicotyledons, for the most part, in the possession of a basal sheathing attachment that usually encircles the whole stem circumference. This organization, together with the vascular attachment, is part of the distinctive method of shoot construction in monocotyledons that should be thought of as the result of a unified developmental process. Since monocotyledonous leaves develop by intercalary growth, with the leaf base normally maturing last, the leaf base plays a key role in the mechanical and conductive efficiency of the shoot because older leaves that have completed their maturation support younger leaves which have not (e.g. Fig. 5B). Nevertheless, monocotyledonous leaves vary widely in the extent to which a discrete basal organ (leaf sheath) is developed. In some groups (e.g. many Liliflorae) there is no leaf sheath and this is often correlated with the geophytic habit. Leaves on underground parts are often reduced to scales and even the leaves of above-ground, ephemeral parts are simple, unspecialized and often with a narrow insertion. Leaves of bulbous monocotyledons may, for example, show whole organ specialization as protective scales or storage organs (Arber 1925). More commonly, however, the leaf base is specialized as a tubular structure and the leaf as a whole is differentiated into sheath, petiole, and blade. The leaf sheath then plays an important (mainly mechanical) role and may even be regarded as a separate organ category in view of its structural and developmental distinctiveness. In many Zingiberales, exemplified by the banana, leaf sheaths collectively form a 'pseudostem' that supports leaf blades and ultimately the inflorescence.

Open and closed leaf sheaths

The leaf sheath in monocotyledons may be described as either 'open' or 'closed'. Open leaf sheaths are discontinuous circumferentially with a longitudinal ventral slit, the opposite margins often overlapping appreciably; in transverse section they are crescentic. Closed leaf sheaths are continuous circumferentially, without a ventral slit; in transverse section they are annular (Fig. 5B) a condition which arises early in leaf development (Fig. 5C). The difference between open and closed sheaths may only be slight developmentally and relates to the time of onset of intercalary (basal) growth of the whole leaf. If this begins before the annulus of leaf insertion is complete, an open leaf sheath results, if the annulus is complete, a closed sheath results. Despite this seemingly slight developmental switch, many taxa of monocotyledons can be characterized by one or the other type of leaf sheath, often at the family level, but the character is rarely used by systematists. Palms always have closed leaf sheaths. Dahlgren and Clifford (1982) comment on the taxonomic distribution of a discrete petiole, but not the contrast between open and closed sheaths, probably because the former feature is more readily observed in herbarium specimens. However, presence or absence of a discrete petiole is a less constant taxonomic feature than the type of leaf base. In *Cordyline*, for example, presence or absence of a petiole varies, even in different cultivars of one species (*C. terminalis*); a petiole is present in only a few species of *Pandanus*; both genera are otherwise characterized by open leaf sheaths.

The functional difference between open and closed leaf sheaths may not be very obvious and may even be counterintuitive. Collectively, either type can form a closed tube, with outer leaf bases supporting and enclosing inner. However, pseudostems in Zingiberales are in all families except Costaceae, always made of open leaf sheaths; those in banana incompletely encircle the stem. Open leaf sheaths are found in Bromeliaceae, even though, as tank or cistern epiphytes, they collectively retain water. This is the result of tight adherence between overlapping leaf margins. An open leaf sheath by itself could not form a water-containing structure. Mechanical constraints of closed leaf sheaths, especially when they are thick and woody at maturity, constrain the expansion of younger structures. The palms offer ingenious solutions to this problem by structural devices that function after the leaf base is mature. A closed tubular leaf sheath may also restrict the development of an axillary structure; again the palms can solve the problem in unique ways.

The palm leaf sheath

The palm leaf base is always a closed tube, although this may be evident only by dissection of whole crowns because late development changes occur that rupture previously contiguous tissues. A clear distinction exists between this type ('not obviously tubular') and types in which the closed leaf sheath is evident because leaves either persist as a whole or are regularly abscised as a unit ('obviously tubular') and require special mechanisms by means of which they become detached (see *Veitchia*-type later in this section; Fig. 13). There is an approximate, but not wholly consistent correlation between the two major categories and the

major systematic groups of palms, since obviously tubular leaf sheaths are virtually restricted to (but not universally present in) the subfamily Arecoideae.

A categorization of leaf sheath types was proposed by Tomlinson (1962b), types being named after representative genera. As a typology this categorization has value in raising questions as to evolutionary origins of the categories selected. The following key and subsequent details provide a concise overview of a topic that still needs extensive anatomical investigation.

Key to leaf base types in palms (after Tomlinson 1962b, 1990)

1. Leaf base obviously a closed tube at maturity, vbs with an axial orientation (Fig. 13J), the leaf not shredding partly into constituent fibres 2
1A. Leaf base not obviously a closed tube at maturity, vascular bundles with a helical orientation (Fig. 5E), the ventral tissues variously eroding as the leaf ages 3
2. Leaf deciduous as a single unit (Fig. 13I, J) by means of abscission layers (Fig. 13A–D) that first open the tube and then separate it at the node . . . . *Veitchia*-type
2A. Leaf persistent, usually after shedding the blade, the leaf base eventually lost without precise abscission . . . . *Calamus*-type
3. Leaf base eroding ventrally, splitting widely on the dorsal side (Fig. 13K) via a weak zone (Fig. 13E–H) to produce a gradually widening dorsal cleft which accommodates internal expansion . . . . *Hyphaene*-type
3A. Leaf base eroding ventrally in various ways, but not splitting dorsally 4
4. Contrarotating fibre and vascular system persistent and developing as distal recurved spines (Fig. 24A) . . . . . *Zombia*-type
4A. Fibre system not persistent as spines 5
5. Ventral tissue persisting as a sackcloth-like material and separating as a single piece from the dorsal margin . . . *Cocos*-type
5A. Ventral tissue fragmenting and not persisting as a single piece 6
6. Ventral tissue splitting mechanically into fibres along a single dorsal line . . . . *Caryota*-type
6A. Ventral tissues eroding completely into fibres 7
7. Residual fibres of ventral tissue long persistent as a fibrous mass clothing the trunk . . . . *Trachycarpus*-type
7A. Residual fibres of ventral tissues eroding rapidly, the trunk without persistent fibres . . . . *Phoenix*-type

Obviously tubular sheaths

1) *Calamus*-type. This is found in climbing palms and many palms with narrow cane-like stems. The tight adherence of the sheath to the stem and extensive overlapping of successive leaves is particularly important in climbing palms as it transmits the tension of the supporting leaf rachis to the axis. Commonly the leaf sheath is extended distally beyond the insertion of the petiole as a tubular ligule. Again in climbing palms this may be mechanically important as it directs the

Fig. 13 Development of the leaf base split. **A**. *Hyophorbe verschaffeltii*, leaf insertion with evident nodal separation layer; arrows = leaf abscission zone, radial LS (bar = 1 mm). **B**. *Ptychosperma macarthurii*, leaf insertion with evident nodal separation layer; arrows = leaf abscission zone, radial LS (bar = 2 mm). **C**. *Ptychosperma macarthurii*, Ventral side of leaf sheath in region of abscission sinus, TS (bar = 500 µm). **D**. *Hyophorbe verschaffeltii*, ventral side of leaf sheath with abscission sinus, TS (bar = 100 µm). **E**. *Corypha utan*, immature leaf sheath in region of dorsal split indicated by shallow sinus and obscure future cleft, TS (bar = 3 mm). **F**. *Corypha utan*, cleft initiation by cell necrosis along abaxial (dorsal) side of leaf sheath, TS (bar = 1 mm). **G**. *Corypha utan*, leaf sheath in region of future cleft with separated necrotic cells, TS (bar = 1 mm). **H**. *Corypha utan*, details of cleft margin in F with incipient *etagen* periderm, TS (bar = 500 µm). **I**. *Hyophorbe lagenicaulis,* ventral side of leaf sheath with prominent vertical separation sinus. **J**. *Ptychosperma* sp., ventral side of leaf sheath with advanced separation of leaf base. **K**. *Corypha utan*, surface of mature palm with dorsal cleft in each leaf base widened by internal stem expansion.

extending spear leaf in a strictly axial direction so maintaining the leaf in as vertical a position as possible.

2) *Veitchia*-type (Fig. 13I, J). The very numerous members of this type all have pinnate leaves and are characterized by a closed tubular sheath that is eventually abscised precisely, both circumferentially at the node and longitudinally by means of a ventral slit. Since the oldest attached tubular sheath encloses all younger organs it is conspicuous as a green tube below the crown of pinnate leaves known as the 'crownshaft' (Fig. 72F). The trunk is clean, with precise annular leaf scars and the inflorescences do not expand until the subtending leaf has fallen. All these features make these palms attractive so that they are prime horticultural subjects, the Royal Palms (*Roystonea* spp.) being striking examples (Fig. 72A).

Not-obviously tubular sheath

Palms of this type all have closed sheaths, but as the leaf matures there is progressive loss of ventral tissue so that at maturity the sheath appears to be open on the ventral side. In one type, changes also occur on the dorsal side. The fibrous sheath of fibro-vascular bundles persists and produces tissue of distinctive textures. The 'sacking' of coconut (*Cocos*) is perhaps the most familiar example.

3) *Hyphaene*-type (Fig. 13K). The leaf sheath is short, ventral tissues are lost, and the widening of the leaf base is accommodated in final stages of leaf maturation by the development of a dorsal cleft in a precisely defined region of the leaf base, as described by Schoute (1915). It characterizes several members of the Coryphoideae, notably *Sabal*, *Corypha* (Fig. 13K), and Borasseae. Leaves of this type, although they may shed the lamina through a break in the petiole will often persist on the trunk as a woody lattice-work of leaf bases (Fig. 45H).

4) *Phoenix*-type. The leaf sheath expands in late stages simply by erosion of ventral tissue that does not persist. It characterizes the genus *Phoenix*, but can be seen in varying degrees in palms with a well-developed dorsal portion. Leaves do not abscise cleanly, but the axis breaks to leave a short woody stump on the stem surface, the stumps collectively revealing the original phyllotactic pattern.

5) *Cocos*-type. This type is found in a number of cocosoid palms, most obviously in *Cocos* itself and it is the only example of a leaf base type that has been studied in anatomical detail

(Tomlinson 1964). The ventral ground tissue breaks down and dries up, but the fibro-vascular system persists. Internal expansion is accommodated by separation of the ventral tissue from the woody dorsal portion. Ultimately the 'pseudo-petiole' so formed abscises cleanly from the stem to leave a rhombohedral scar connected circumferentially by the scar of the ventral tissue. The sacking often falls independently of the leaf and may be studied for its distinctive vasculature, as described later.

6) *Trachycarpus*-type. Breakdown of ventral tissue in this type does not result in its loss; rather it persists as a fibrous mass that permanently clothes the stem (Fig. 45F). The leaf blade is lost from old leaves again by a mechanical breakage of the petiole, its basal portion persisting within the fibrous material. This type occurs most extensively in the coryphoid group; in species of *Coccothrinax*, for example, the persistent fibrous mass is very conspicuous, notably so in *C. crinita*.

7) *Caryota*-type. In this genus the leaf base is a woody cylinder of quite uniform thickness. Internal stresses are accommodated by a single ventral split with fibrous remains persisting along the margins of the slit.

8) *Zombia*-type. This is highly specialized and produces a distinctive type of armament. Ventral tissue remains woody, but separates between groups of vascular bundles to form a network of contrarotating strands (Fig. 5E). A secondary function is the spiny distal projection of fibro-vascular bundles from the ligule, which become recurved as the leaf sheath widens. Spines from persistent leaf bases collectively form an effective barrier (Fig. 24A). This type is unique to *Z. antillarum*, but erect spiny projections of similar origin occur in *Rhapidophyllum*, while in *Arenga pinnata* and a few related coryphoid palms the larger vascular strands persist as needle-like, but not sharp, projections.

These eight types by no means describe the range of variation to be found, but provide points of reference against which other palms can be measured. The essential feature of all these different types is that leaf base morphology is modified in late stages to accommodate the enlargement of enclosed tissues and organs in a passive manner. Future research requires precisely worked out examples, concentrating on anatomical detail, measurement of quantitative changes, and contrasted developmental methods. The subject is indicative of the way in which tropical examples have no counterpart in temperate regions and remain largely unexplored by botanists.

CHAPTER 5 STEM ANATOMY

The overall construction of the palm stem is exemplary of all monocotyledons although it is represented in a group of trees with a habit that is rare among other groups of woody monocotyledons like bamboos, pandans, and arborescent members of Asparagales (e.g. *Cordyline, Dracaena, Xanthorrhoea, Yucca*). This leads us to Hugo von Mohl's dictum that palms exhibit most clearly the characters of monocotyledons and 'therefore afford the most favourable means of acquiring satisfactory ideas about them'.

TOPOGRAPHY

The palm stem is composed of three discrete regions (Figs. 14A, B and 15A, B). First is the **epidermis**, always a single layer of cells and therefore structurally insignificant; it is not dealt with in detail here. Second is the **cortex**, a relatively narrow peripheral region that can be as few as 10 cells wide in narrow stems, but up to several centimetres in wider trunks. The cortex includes numerous fibre bundles and is traversed by leaf traces. It is usually abruptly delimited internally from the wide **central cylinder** by a marked discontinuity in the density of vascular bundles, but not by any specialized limiting layer. The central cylinder includes peripheral crowded vascular bundles, each provided with a well-developed fibrous sheath, and a transition to the more diffusely distributed central vascular bundles, each with a less-well developed fibrous sheath (cf. A and B in Figs. 14 and 15). This topographic distinction is based on developmental processes that are well understood (Tomlinson 1990) but are not our concern here.

HISTOLOGY

The structural differentiation of cell types within the palm stem reflects functional differences and consists of a matrix of **vascular bundles** (more completely fibro-vascular bundles) embedded in a matrix of parenchyma cells or **ground tissue**. Stem texture is neither radially nor axially uniform because outer tissues are the most sclerotic at any level and basal stem regions become increasingly sclerotic with age.

Each cell type has a discrete function. The ground parenchyma matrix is primarily for carbohydrate storage and so includes starch in varying amounts. In hapaxanthic axes, which flower terminally, starch accumulation is considerable (e.g. *Metroxylon*) and is metabolized in the reproductive phase. Water in cell vacuoles may also represent an important reserve. Cell walls of ground parenchyma cells may become progressively thickened and lignified with age and so add to stem stiffness.

Vascular bundles are histologically complex and serve the dual function of transport (via xylem and phloem) and support (via sheathing fibres and sclerotic parenchyma). The fibre sheath is most developed external to the phloem and is seen in section as a fibre 'cap.' Phloem includes sieve-tubes, companion cells, and phloem parenchyma, and is described elsewhere. Xylem includes tracheary elements (tracheids and vessels, also described in detail elsewhere), with the important distinction made between **protoxylem** (the first-formed tracheary elements, which mature during tissue extension) and **metaxylem** (the later-formed elements, initiated during tissue extension but not completing maturation until tissue extension has ceased). Protoxylem is possibly always composed of tracheids alone, metaxylem of vessels. Fisher et al. (2002) found paint-filled protoxylem elements (not lacunae) indicating that some protoxylem elements were longer than 20 mm in *Calamus* and *Daemonorops*. Protoxylem and metaxylem vary along a single vascular bundle as determined by positional and developmental differences, as described later. Differences between tracheids and vessels in terms of wall sculpturing and pitting reflect these developmental features. Tracheary elements in plants assume their function as water-conducting cells after they lose their cell contents so that water moves from cell lumen to cell lumen, through pitted walls as part of the **apoplast**, whose activity is non-metabolic. In contrast to the apoplast, the living cells of the stem (**symplast**) retain their metabolic functions throughout the total lifespan of the palm, which may amount to centuries. The uniqueness of palms is thus seen to be substantially metabolic.

Additional cell types in stems include raphide sacs, tannin cells, and silica cells, all described elsewhere.

THE VASCULAR SYSTEM

The construction of the stem vascular system of palms, and as a consequence of all monocotyledons, has been written about in such detail (e.g. Tomlinson 1990, 1995) that only a brief review is presented here. This knowledge also explains the limits imposed

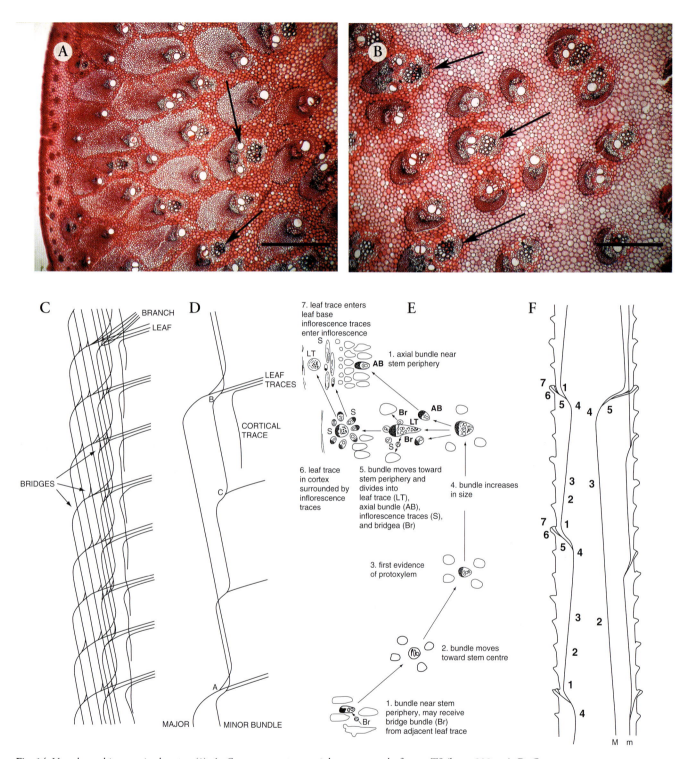

Fig. 14 Vascular architecture in the stem (1). **A**. *Geonoma* sp., stem periphery, arrow = leaf trace, TS (bar = 200 μm). **B**. *Geonoma* sp., stem centre, arrow = leaf trace, TS (bar = 200 μm). **C**. Diagram of vbs and leaf traces of stem, LS. **D**. Diagram of a new axial bundle connected to leaf trace three nodes above, LS. **E**. Same as D but diagrammed in TS. **F**. Same as E but diagrammed in LS with numbers as used in E. (C–F from Tomlinson 1990).

by the palm stem's distinctive features on the use of vascular anatomy as a source of systematic information. There is a common overall pattern found in all palms (which can be applied to many other monocotyledons) and how this pattern is developed has been elucidated. Not surprisingly, a clear understanding of structural principles has produced a better understanding of the hydraulic function of the vascular system and its mechanical properties. An exceptional condition is found in the climbing palms so that a separate description is given later; the basic vascular pattern can be over-ridden by habit considerations.

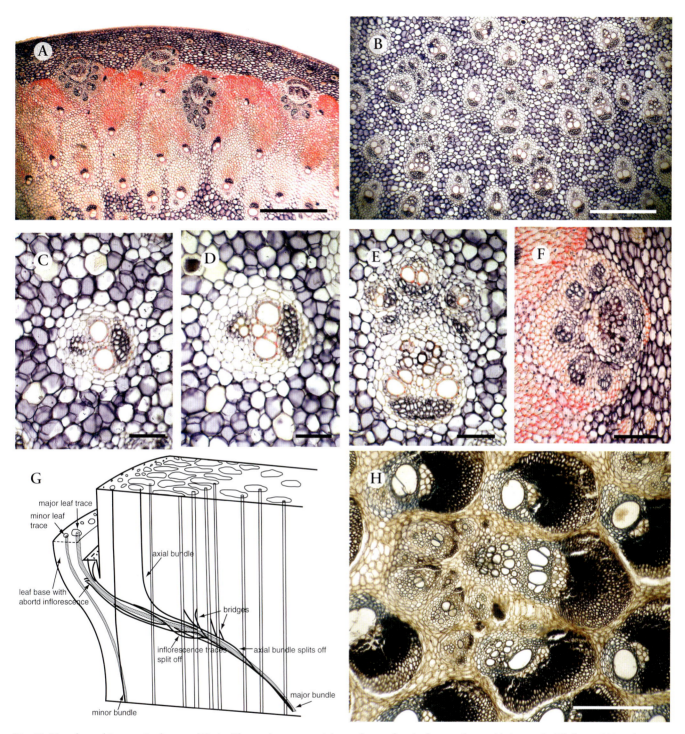

Fig. 15 Vascular architecture in the stem (2). **A**. *Chamaedorea* sp., periphery of stem, four leaf traces clustered below node, TS (bar = 500 μm). **B**. *Chamaedorea* sp., central region of stem, TS (bar = 500 μm). **C**. *Chamaedorea* sp., axial vascular bundle with little protoxylem, TS (bar = 100 μm). **D**. *Chamaedorea* sp., axial bundle with more developed protoxylem, TS (bar = 100 μm). **E**. *Chamaedorea* sp., leaf trace bundle with much protoxylem and adjacent inflorescence bundles, TS (bar = 100 μm). **F**. *Chamaedorea* sp., leaf trace without metaxylem and well developed inflorescence bundles (bar = 100 μm). **G**. Diagram of stem node with leaf traces and inflorescence bundles (from Zimmermann and Tomlinson 1965; 1967). **H**. *Prestoea acuminata*, leaf trace complex below node, with many inflorescence traces (satellite vbs), TS (bar = 500 μm).

THE VASCULAR PATTERN

Terms of motion are here used extensively although they describe fixed structures, but are quite appropriate when viewing analytical movies made by frame-by-frame photography of sequential sections (Zimmermann and Tomlinson 1966).

The diversity in the construction of bundles seen in a transverse section of a palm stem (Fig. 14C–F) is based on changes within each vascular bundle throughout its total length. Each

41

successive leaf (Fig. 14C) is supplied by a large number of **leaf traces** and at any one sectional level the supply to a series of many distal leaves is seen. If one follows a single bundle axially, the common plan is revealed (Fig. 14D). Considering a central vascular bundle the number of protoxylem elements reflects the distance the bundle still has to traverse axially in a distal direction before it exits the stem as a leaf trace (Fig. 14E, F), no or few elements indicates a long distance, many elements a short distance. Figure 15 shows the details. Peripheral vascular bundles of the central cylinder have a single wide metaxylem vessel and no protoxylem (Fig. 15A). Central regions include a mixture of vascular bundles with one or two metaxylem elements and varying amounts of protoxylem (Fig. 15B). Followed axially upwards the single metaxylem vessel may change little in appearance but sooner or later it separates into two vessels, and always at the initial divergence of an axial bundle at its level of abrupt departure into a leaf as a leaf trace. Close to its exit into the leaf metaxylem vessels become narrow, numerous, and form two strands. This is shown in the sequence of images C, D, E in Fig. 15. Protoxylem is extensively developed. This position along an axial bundle can represent the level at which a branch bundle separates.

Branch bundles can be **bridges** that connect to adjacent axial bundles, a new **continuing axial bundle**, or, in reproductive shoots, **inflorescence traces** ('satellite bundles') that supply axillary inflorescences (Fig. 15E, F). Normally there is only one continuing axial bundle that diverges from an outgoing leaf trace and provides axial vascular continuity in a distal direction. This one-to-one replacement thus maintains a constant number of vascular bundles at any one level. There can be one or more bridge bundles and often very numerous inflorescence traces of varying sizes (Fig. 15H). The complex configuration of just one departing leaf trace is shown diagrammatically in Fig. 15G.

In addition to changes in xylem elements there is a considerable change in the numbers of fibres in each bundle. This is seen in section as the contrast between peripheral bundles with many fibres, forming a massive phloem cap or sheath (Figs. 14A and 15A), and central bundles with few, although the size of the cap varies considerably and accounts for much of the variation in overall stem density (Figs. 14B and 15B). The concentration of fibrous tissue at the stem periphery maximizes mechanical efficiency; in many palms the stem, although solid, functions mechanically as if it were a hollow cylinder. These relationships were quantified in rattans by Bhat et al. (1990).

The overall course of vascular bundles is not strictly axial because each originates (as a branch from a leaf trace) at the stem periphery and diverges both in a radial direction, inward and then outward, i.e. the 'double curve' of Hugo von Mohl, its discoverer (Mohl 1849), and in a tangential direction as a shallow internal helix, discovered by Meneghini (1836). The gradual inward migration of each vascular bundle from its proximal level of origin at the stem periphery and its outward migration distally into the leaf base as a leaf trace accounts for the different concentration of vascular bundles in peripheral versus central regions (Fig. 14C). A further cause results from the continuous development over an extended period of many vascular bundles serving a single leaf. Major bundles originate early and show the largest radial displacement; minor bundles originate last and show the least radial displacement, with a continuous range of intermediate bundles in-between. The interval along any one bundle between departure of successive leaf traces is the '**leaf contact distance**'. Two examples illustrate this principle in the simplest possible way (Fig. 14D).

CONCLUSIONS

This summary description allows an appreciation of three-dimensional features, seen only in two dimensions in single sections, and allows a more complete functional interpretation. We now know that:

1) The peripheral crowding of vascular bundles in the palm stem is a consequence of the double curve combined with an extensive leaf–stem vascular continuum.

2) Axial hydraulic continuity is mediated through the metaxylem and its continuity between successive leaf contact distances. There are, nevertheless, hydraulic constrictions at the level of bundle branching because of the narrowing of metaxylem elements.

3) The xylem supply to each leaf is entirely made up of protoxylem tracheids. Continuity between protoxylem and metaxylem occurs only in the stem centre and represent the true 'vascular insertion' (Tomlinson and Vincent 1984).

4) This disparity between diameter of protoxylem and metaxylem tracheary elements means that axial movement of water, up the stem, is favoured over appendicular movement into the leaf.

This protects the axial supply, whereas leaves themselves are eventually shed and their vascular system is disrupted. This approach to stem hydraulics in palms is discussed extensively in Zimmermann (1973; 1983) and Zimmermann and Sperry (1983).

TEXTURAL DIFFERENCES

The internal texture of the mature palm stem varies considerably in different taxa, determined by diameter, age, and habit. One can contrast, for example, the relatively homogenous texture of a coconut trunk (Fig. 5H) with that of a royal palm, which has a very soft centre and a relatively narrow sclerotic peripheral region (Fig. 17E–L). Diminutive palms, as in those of the forest understorey, also tend to have homogenous stems. Rattans represent the ultimate in this respect and they are also rendered light because their vascular bundles are uniformly crowded, with wide vessels, which in dried canes are air-filled. Stem texture is thus determined by the distribution of vascular bundles, the extent to which ground tissues may become aerenchymatous with age, and their cells eventually become thick-walled and lignified. Stem stiffness is correlated with vascular bundle density, fibre area, and fibre wall thickness in *Cocos* (Kuo-Huang et al. 2004) and *Calamus* (Abasolo et al. 1999). The stems of rhizomatous, acaulescent (without a visible trunk), and low-growing palms are dealt with in individual accounts. Their special features reflect the palm's habit. The use of larger trunks for constructional timber and chipped-wood products should be noted.

ONE-VESSEL AND TWO-VESSEL PALMS

Some earlier literature has emphasized the difference between palms with either one or two wide metaxylem vessels in each central vascular bundle as seen in TS and this has some systematic validity, although it should be emphasized that in each type peripheral vascular bundles have one vessel consistently. The difference, however, is based on a quantitative difference between the proximal and distal portions of any one bundle within a single leaf contact distance. In 'one-vessel' palm stems the single vessel condition in a vascular bundle, which exists immediately where the axial bundle diverges from a leaf trace, is maintained over a much longer axial distance than in a 'two-vessel' palm. There is thus some subjectivity in establishing the difference. Furthermore, some palms have 'many-vessel' stems because the metaxylem includes several narrow vessels over much of its course (e.g. *Chamaerops*). The difference among 'one-, two-, and many-vessel' palms is evident only in the stem centre and it must be understood how this variation relates to the common structural pattern of all vascular bundles.

Phloem structural variation is discussed later in the context of the whole palm.

VARIATION ALONG A SINGLE AXIS

There is considerable structural variation within a single palm stem as a result of several continual developmental processes. This means that a single small sample cannot represent the total stem anatomy.

- Early development of the axis, i.e. in the initial 'establishment' phase of growth where the stem has an obconical shape means that the number of vascular bundles continually increases until a constant stem diameter is reached (Tomlinson and Zimmermann 1966b). This essentially juvenile phase can be said to persist in some acaulescent palms that can be interpreted as paedomorphic forms (e.g. some *Sabal* species).
- Radial differences at one level include those of the persistent cortex, outer layers of the central cylinder, and the transition to the stem centre, usually with more diffuse bundle distribution (as quantified by Kuo-Huang et al. 2004).
- There are changes in histological features in the axial direction as the palm grows taller. This has been explored in some detail (by measuring cell dimensions at successive intervals along a single axis), notably those of vascular fibres (living cells) and metaxylem vessel elements (dead cells). Changes reflect changes in the expression of differentiating primary

tissues as the stem grows taller. For example, distally fibres are shorter and metaxylem vessel elements are shorter and narrower (Swamy and Govindarajalu 1961; Tomlinson 1990).
- In many palms changes in the histology of stem tissues occur gradually at any one level as the tissue grows older. The fibres of the vascular bundles can continue to deposit cellulose cell wall material and accompanying lignin for many years. This process increases the stiffness of mechanical tissues as the palm grows taller (Rich 1987a, b; Bhat et al. 1990). This has the unfortunate result of increasing the difficulty of cutting thin stem sections. In addition there can be continuing changes in ground tissue cells that involve cell division, cell enlargement, cell wall deposition and metabolic changes such as lignin deposition and particularly starch storage, as we now discuss.

SUSTAINED PRIMARY GROWTH (Figs. 16 and 17)

Physiologically long-continued changes in the stems of palms are notable, if not unique, because they can occur in differentiated cells over periods measurable in centuries (Tomlinson 2006). The externally visible effect of this is that stem diameter may increase by a process that has been referred to as *diffuse secondary growth* (Tomlinson 1990), but **sustained primary growth** is a more informative and less ambiguous term emphasizing the observation that no new vascular tissue is added (Tomlinson et al. 2009). In contrast, secondary growth in conifers and dicotyledonous trees involves the continued addition of new vascular tissues from a vascular cambium. Nevertheless, as long suspected by earlier anatomists (e.g. Barsickow 1901), secondary increase in diameter of palm stems was measured extensively by Schoute (1912) in the Bogor (Buitenzorg) Botanic Gardens, Indonesia. He showed, in the relatively short period over which he made his observations, that there was a permanent increase in diameter of the stem near ground level in several cultivated palms. Most significantly he again confirmed that this expansion did not involve the formation of new vascular tissues at this level but was the result of division and cell enlargement of existing ground parenchyma cells and increase in wall thickness of vascular fibres. A more complete analysis based on natural populations of *Archontophoenix cunninghamiana* was provided by Waterhouse and Quinn (1978) who indicated stem volume increases of the order of 20 times in some individual specimens.

It should be emphasized that histological changes with age in palm stems can only be examined by comparing the top and bottom of a single stem because one cannot sample the same stem level at different ages. Cores cannot be extracted and even if they could, wound healing processes obscure any results. The changes observed can be spectacular and largely eliminate any possibility of simple character analysis for systematic purposes; Fig. 16A–E shows different aspects within the stem of one species.

In our own studies (Tomlinson et al. 2009) we have found evidence of several different kinds of stem enlargement involving

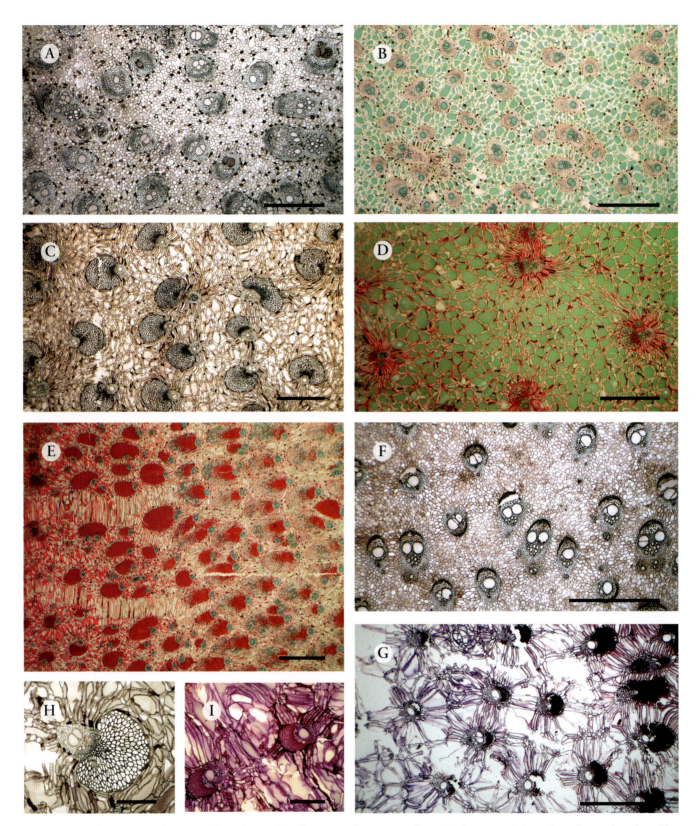

Fig. 16 Sustained primary growth (1). **A**. *Archontophoenix alexandrae*, centre of young part of stem, TS (bar = 2 mm). **B**. *Archontophoenix alexandrae*, older part of stem, TS (bar = 2 mm). **C**. *Archontophoenix alexandrae*, subperiphery of older part of stem, TS (bar = 1 mm). **D**. *Archontophoenix alexandrae*, centre of older part of stem, TS (bar = 2 mm). **E**. *Archontophoenix alexandrae*, periphery of older part of stem, TS (bar = 2 mm). **F**. *Socratea durissima*, centre of young part of stem, TS (bar = 2 mm). **G**. *Socratea exorrhiza*, centre–periphery transition of older part of stem, TS (bar = 2 mm). **H**. *Archontophoenix alexandrae*, detail of vb in stem centre, TS (bar = 500 μm). **I**. *Euterpe oleracea*, detail of vbs in stem centre, TS (bar = 500 μm).

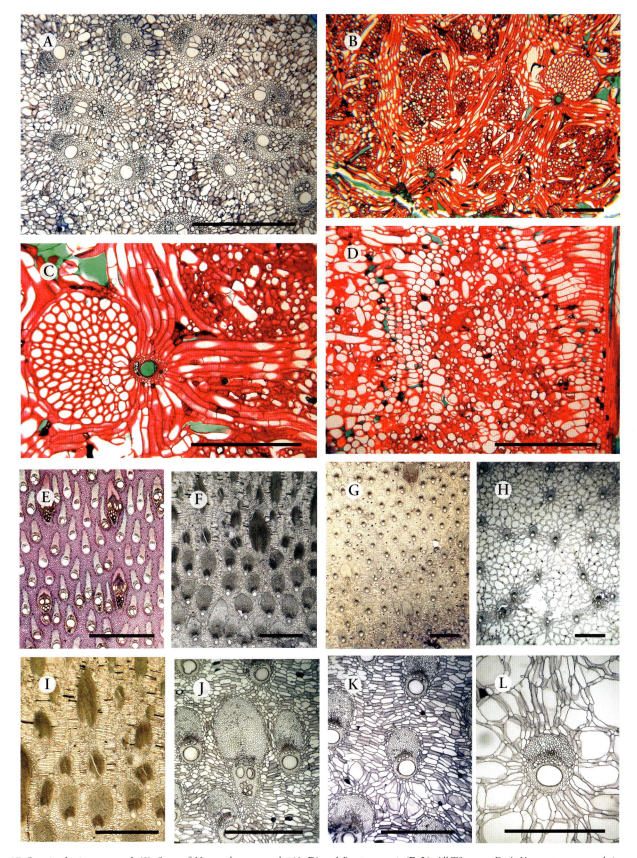

Fig. 17 Sustained primary growth (2). Stem of *Normanbya normanbyi* (**A–D**) and *Roystonea regia* (**E–L**). All TS except D. **A**. Young stem, ground tissue little expanded (bar = 2 mm). **B–D**. Mature stem. **B**. Ground tissue with proliferated islands of cells filling lacunae at first formed by transverse extension of ground tissue cells (bar = 2 mm). **C**. Single vb with enlarged bundle sheath fibres (bar = 2 mm). **D**. Ground tissue as in B, LS (bar = 2 mm). **E**. Stem periphery with contrasted leaf traces and axial vbs, fibres of bundle sheath immature (bar = 2 mm). **F**. Outermost central cylinder transitional to cortex (above), fibres of bundle sheath enlarged but still immature (bar = 2 mm). **G**. Periphery of central cylinder (bar = 2 mm). **H**. Stem centre with lacunose ground tissue (bar = 2 mm). **I**. Cortex and outer central cylinder, tangential expansion of ground tissue cells (bar = 2 mm). (**J–L**). Details of vbs of central cylinder. **J**. Leaf trace (bar = 1 mm). **K**. Central vb at transition to inner lacunose ground tissue (bar = 1 mm). **L**. Central ground tissue with wide lacunae surrounding a vb (bar = 1 mm).

cellular changes in **ground tissue**, some of which may be restricted to certain palm groups. These include:

- Horizontal enlargement of cells. In the stem periphery this expansion is entirely tangential and can account for some of the overall increase in stem diameter (Figs. 16E and 17I–K). In the stem centre, existing intercellular spaces become considerably widened and the ground tissue quite lacunose. Vascular bundles may become widely separated and appear in TS suspended by plates of elongated parenchyma cells; the changes are well illustrated by comparing Fig. 16F and G which contrasts early and late stages. Otherwise features are well exemplified in Figs. 16B–D, G–I, and 17H, L). Cells adjacent to vascular bundles remain little or not expanded.

- Cell division associated with cell enlargement, usually resulting in a septum within a horizontally extended cell (Fig. 17C, D).

- Uniform elongation of parenchyma, often subdivided, radiating from each central vascular bundle but localized in the trunk to form a marked swelling. This is striking in 'belly palms' (*Acrocomia*) and 'bottle palms' (*Hyophorbe*). In *Acrocomia* the ground tissue contains more water than the unswollen regions, and starch is concentrated in the unexpanded tissue next to vascular bundles (Fisher et al. 1996).

- Cell enlargement associated with collapse of intervening cells resulting again in lacunae, but not a very regularly configured ground tissue.

- Differential cell enlargement occurs so that relatively few cells increase in volume, often considerably and also at the expense of other cells which collapse. This is best seen in *Acoelorraphe* (Fig. 58H).

- In the extreme example of the stem of *Normanbya*, the original cavities produced by these several processes become completely filled by enlarging and dividing cells (Fig. 17A–D). This results in a mosaic of small cells filling the space between the plates of tissue that support the vascular bundles. In this example these processes do not necessarily result in large changes to the diameter of the stem but they do indicate considerable cellular activity of a kind that should not be described as secondary.

In addition to changes in ground parenchyma, continued maturation of **bundle sheath fibres** can involve cell expansion as well as cell wall deposition (Fig. 17E, F). We have no evidence that fibres also extend axially by possible tip growth even though the distal (younger) fibres in a palm stem have been shown to be shorter than basal (older) fibres. Concomitant with cell wall addition in fibres there is also increased lignification (Weiner et al. 1996). Both processes normally are initiated in central fibres close to the phloem and appear progressively in outer fibres (Figs. 16C and 17I) (as quantified in Rüggeberg et al. 2009). That such fibres retain cytoplasmic viability is indicated by the retention of nuclei, even though the fibre lumen may become very narrow. The end result can produce either fibres of differing diameters close to versus remote from the phloem (Fig. 16H), or fibres of uniform wall properties (Fig. 17C).

Systematic implications

The different kinds of late primary cellular activity may have different degrees of expression in different groups and are only expressed in the extreme in wider palm stems, i.e. cane-like palms show little or no such activity (Figs. 14A and 15A). The use of such features in the identification of fossil palms recalls the prescient remark of the Indian botanist K. N. Kaul (1935) who was of the opinion that ground tissue was of potential information in fossil palm identification, but no details were provided. Unfortunately, examination that would improve our understanding will require extensive destructive sampling because one would have to combine information obtained from sections along the full length of a trunk as well as at the same levels in individual of one species of different size (by implication, different ages). Until this is done the metabolic basis of cell longevity in palm stems will remain unexplained.

WOUND REACTIONS IN PALM STEMS

The intrinsic long-time viability of palm stem tissues has been demonstrated in the wounding experiments of Weiner and Liese (1995). They bored holes into the stem of a 35-year-old *Roystonea regia* that was 11 m tall in the greenhouse of the Old Hamburg Botanic Garden up to 21 days before it had to be cut down. Bore holes at different heights, i.e. different ages, were taken.

The tissue adjacent to the holes did not develop a protective 'barrier zone' or callus, at least over the post-wounding 21-day period but showed progressive responses similar to those found in dicotyledonous hardwoods: tylose development in metaxylem vessels and protoxylem tracheids plus 'slime' (pectiniophilic substances) in tracheary elements, sieve tubes, and intercellular spaces. Phenolic substances appear progressively in ground tissue cells and eventually fibres, although the prior existence of such substances, as would be likely before wounding, was not reported. Cells even in the basal (oldest) cells had retained nuclei. There was, in addition, tangential expansion of cells, which could become septate, indicative of cell division, although this is a feature found normally in unwounded tissues. Such is the availability of this common ornamental palm that a future study might take advantage of a much older specimen available for destruction.

Generally, surface injury results in necrosis, darkening (polyphenolics), and suberization of adjacent living cells. Deeper injuries cause tylosis formation and gums within interrupted metaxylem vessels (Fig. 41E). Tyloses also occur in vessels of roots (Fig. 58I).

STEM ANATOMY IN CLIMBING PALMS (Fig. 18)

The climbing habit has evolved in several unrelated groups of palms (Tomlinson and Fisher 2000) but is most extensively developed in Calamoideae, perhaps a reflection of this group's

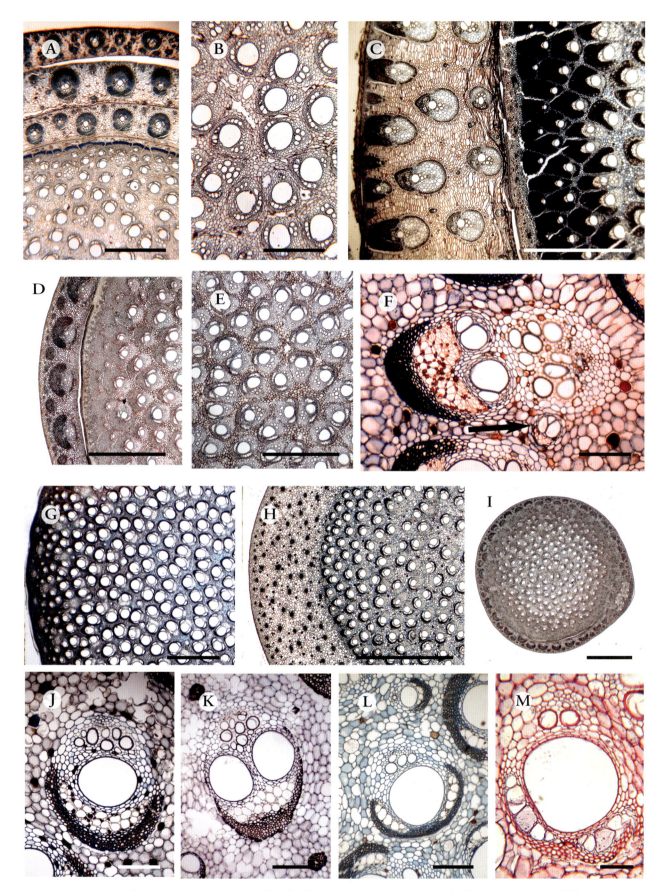

Fig. 18 Rattan stems. **A**. *Korthalsia laciniosa*, internode with three leaf bases, TS (bar = 1 mm). **B**. *Korthalsia laciniosa*, stem centre, TS (bar = 500 μm). **C**. *Laccosperma* sp., internode with leaf base, TS (bar = 2 mm). **D**. *Calamus muelleri*, internode with leaf sheath, TS (bar = 1 mm). **E**. *Calamus muelleri*, stem centre, TS (bar = 1 mm). **F**. *Laccosperma* sp., stem vb with protoxylem and possible bridge bundle (arrow), TS (bar = 200 μm). **G**. *Calamus* sp., aerial stem, periphery of internode, TS (bar = 2 mm). **H**. *Calamus deerratus*, rhizome internode with wide cortical region, TS (bar = 2 mm). **I**. *Calamus muelleri*, stem, node with partly attached leaf base and base of flagellum, TS (bar = 2 mm). **J**. *Plectocomia elongata*, stem vb, TS (bar = 200 μm). **K**. *Myrialepis paradoxa*, stem vb, TS (bar = 200 μm). **L**. *Calamus deerratus*, stem vb, TS (bar = 200 μm). **M**. *Korthalsia cheb*, stem vb, TS (bar = 100 μm).

early diversification. A common feature is grapnel development as the mechanism of climbing, either as distal reflexed leaflet spines (acanthophylls) on the terminal extension of the rachis (forming a cirrus), or as a modified inflorescence forming a whip-like flagellum armed with spines that are often recurved forming 'cat's claw' appendages. Aerial stems usually arise from a sympodial underground rhizome, rarely is the stem single and then it is hapaxanthic (*Plectocomia*). In *Korthalsia* the aerial stems are said to dichotomize and also flower terminally. Rhizomes have a wide cortex and narrow vascular bundles (Fig. 18G vs. H)

As is general in climbing plants, internodes are long, stems are narrow, and metaxylem vessels are disproportionately wide (Fig. 18J–M). Because these stems have exclusively primary growth the vascular transport must be not only efficient but must be long-lasting. However, there are paradoxical features, which do not seem very efficient, so that there are unexpected as well as expected features of construction.

Expected (efficient) features include:

- One (Fig. 18J) or two (Fig. 18K) metaxylem vessels in the central axial portion of each vascular bundle. The widest, as in *Korthalsia* and *Plectocomia*, may approach 400 μm in diameter (Fisher et al. 2002).
- Sieve tubes up to 50 μm wide (Fig. 18M).
- No secondary expansion of ground tissues, in agreement with the common condition of palm 'treelets', i.e. self-supporting palms with cane-like stems. The ground tissue is usually homogenous, as seen in TS, becoming thick-walled and lignified. Cell shape differences may be diagnostic, as emphasized by Weiner (1992) and Weiner and Liese (1990).
- Lack of continuity (= direct contact) between protoxylem and metaxylem (Fig. 18F, J–M). Most, if not all, climbing monocotyledons show this feature but its significance is not clear (Tomlinson and Fisher 2000; Tomlinson and Spangler 2002).

Calaminae

This subtribe (the 'true rattans') produces a striking anomaly in that there is no interconnection among vascular bundles that can provide a continuing axial pathway within the xylem and phloem as is found in the method of axial bundle branching from a leaf trace in other palms (Tomlinson et al. 2001, Tomlinson and Spangler 2002). New axial bundles appear de novo within the differentiating ground tissue and are continually generated over many internodes before they exit (without branching) as a leaf trace into a leaf base. The basal narrow extremities of axial bundles provide a distinctive marker for this kind of anomaly in the absence of verifying three-dimensional analysis. These are seen as narrow vascular bundles of varying diameter interspersed among the wide central vascular bundles, where they cannot be mistaken for bridges and other accessory bundles of the outgoing leaf traces.

Vascular interconnections do exist as narrow, late-forming transverse commissures that recall those found in the petiole of all palms. Radial transport is thus possible, but the connections are narrow and not extensive. There is a similar lack of continuity between the vascular system of the flagellum (and presumably inflorescence axes) and the axial vascular system. In stem sections the adnate flagellum vascular system is seen as an aggregate of narrow bundles on one side of the stem (Fig. 18I). This overall unusual vascular system is hardly maladaptive because *Calamus* itself is the largest genus of palms (approximately 350 species) and its stems can reach lengths approaching 200 m (Burkill 1935), indicating enormous functional success. Typologically the extreme modification of the *Calamus* stem can be seen as the end product of a reduction process in which there is progressive discontinuity and final disruption of the metaxylem pathway (Tomlinson 2006). Nevertheless, the system is demonstrably **efficient** because large amounts of water move along narrow stems, but also **safe** because the canes are long-lived. A possible benefit is that wide vessels can be of considerable length because they are not constrained within one leaf contact distance. This may reduce axial resistance to water flow.

The problem of hydraulic efficiency can only be resolved by continued detailed study.

Synoptic key to stem anatomy in climbing palms

1. Stem sieve tubes with compound oblique sieve plates; silica bodies either spherical or hat-shaped [**subfamily Arecoideae**] 2
1A. Stem sieve-tubes with simple, ± transverse sieve plates; silica bodies exclusively spherical [**subfamily Calamoideae**] 4
2. **Subfamily Arecoideae**
2. Silica bodies spherical; mxy vessel elements <80 μm diameter . . . *Dypsis scandens* [Arecoideae: Areceae: Dypsidinae]
2A. Silica bodies hat-shaped; mxy vessel elements >80 μm diameter 3
3. Peripheral ground tissue of central cylinder becoming sclerotic . . . . . *Desmoncus* [Arecoideae: Cocoseae: Bactridinae]
3A. Peripheral ground tissue of central cylinder not becoming distinctly sclerotic . . . . . . *Chamaedorea elatior* [Arecoideae: Chamaedoreeae]
4. **Subfamily Calamoideae**
4. Central vbs with 1 phl strand (Fig. 18J, K) epidermis silicified or not 5
4A. Central vbs with 2 phl strands (Fig. 18L, M); epidermis silicified (Fig. 41H). 11
5. Epidermis not silicified but sclerotic hypodermis including numerous stegmata (Ancistrophyllinae, Fig. 34G) 6
5A. Epidermis silicified, hypodermis not sclerotic and without stegmata; yellow cap often present [Plectocomiinae] 8
6. Central vbs with 1 wide mxy vessel; peripheral vbs crowded, with massive fibre caps (Fig. 18C); ground tissue with unaggregated long and short cells . . . . *Laccosperma*
6A. Central vbs with 2 wide mxy vessels; peripheral vbs not crowded, fibre caps not massive; ground tissue without long and short cells 7

7. Cortex including narrow raphide canals; ground tissue cells somewhat lobed in TS, creating wide intercellular spaces . . . . *Oncocalamus*

7A. Cortex without raphide canals; ground tissue cells rounded in TS, intercellular spaces narrow . . . . *Eremospatha*

8. Central vbs with 2 wide mxy vessels (Figs. 18K and 39H, I) . . . . *Myrialepis*

8A. Central vbs with 1 wide mxy vessel 9

9. Subperipheral vbs with 2 wide mxy vessels . . . . *Plectocomiopsis geminiflorus*

9A. Subperipheral vbs with 1 wide mxy vessel 10

10. Ground tissue in LS with anastomosing bands of short cells . . . . *Plectocomia*

10A. Ground tissue in LS with non-anastomosing bands of short cells . . . . *Plectocomiopsis* spp.

11. Outermost vbs of central cylinder with a 'yellow cap' of short sclerotic cells at the periphery of the phl fibre cap (Fig. 37M) . . . . [Korthalsiinae] . . . . *Korthalsia*

11A. Outermost vbs of central cylinder without a 'yellow cap' (Figs. 18 D and 41B, G) . . . . . [Calaminae] 12

12. Sieve tubes of phl strands consistently in 1 series as seen in TS 13

12A. Sieve tubes of phl strands usually in more than 1 series, as seen in TS 14

13. Ground tissue in TS of 'Type A' in Weiner (1992) . . . . *Calamus*

13A. Ground tissue in TS of 'Type B' in Weiner (1992) . . . . *Daemonorops*

14. Raphide canals absent, mxy vessels <150 μm diameter; stem slender (~1 cm) . . . . *Pogonotium*

14A. Raphide canals present, often wide; stem robust 15

15. Raphide canals up to 100 μm wide . . . . *Calospatha*

15A. Raphide canals relatively narrow, <70 μm wide 16

16. Outer vbs of central cylinder crowded, with massive fibre caps; ground tissue with wide intercellular spaces . . . . *Retispatha*

16A. Outer vbs of central cylinder diffuse, without massive fibre caps; ground tissue without wide intercellular spaces . . . . *Ceratolobus*

In view of the limited sampling of canes, this key may be less helpful in identification and more useful in listing diagnostic characters (see also Weiner and Liese 1990). The topic is important biologically because of the extensive parallelisms revealed, but obviously also because of the enormous economic importance of these unusual palms.

CHAPTER 6 ROOT

Palm roots are, of necessity, sampled close to the base of the trunk to guarantee correct identification. This can produce samples of different ages, representing either young, newly emerged roots a short distance from their attachment, or the proximal part of an old, well-established root, long extended distally away from their stem attachment. Sampling roots at different stages of development gives a more complete picture but suggests the need for caution in describing a single root, as structure can change with age. In particular, surface layers may erode and the diagnostic features they show may be absent.

MORPHOLOGY

The root system of palms, except for the seedling radicle, is entirely adventitious, i.e. originating within stem tissues and furthermore, lacks secondary growth so that, as with stems, individual roots are long-lived. We describe only anatomy of the first-order roots, i.e. those arising directly from stems. These are normally restricted to the stem base, but aerial roots of several kinds can be developed. Notable are the stilt-roots of a number of palms, especially in Iriarteeae and Verschaffeltiinae (Fig. 19F), associated with establishment growth, or the spine roots (McArthur and Steeves 1969) of *Cryosophila* (Fig. 24P). Root architecture in palms is otherwise conditioned largely by underground branching, which extends up to five orders of progressively diminishing size. In higher-order roots, anatomy changes progressively until the smallest feeder roots lack sclerenchyma and have reduced steles and lack a medulla as described in *Serenoa* (Fisher and Jayachandran 1999). Additional complexity involves the production of erect aerial roots (pneumatophores) often characteristic of palms of wet or seasonally flooded habitats, associated with specialized lateral roots as pneumathodes and pneumatorhiza, as summarized in Tomlinson (1990). Mycorrhizal associations are described by Fisher and Jayachandran (1999, 2005, 2008). 'Root spines' are said by Seubert (1997) to be common in Coryphoideae and the description apparently relates to short determinate second-order roots, a configuration more obvious on the aerial roots of Iriarteeae.

Contractile roots were described in *Corypha* in which the surface was transversely wrinkled and the root more parenchymatous than the non-contractile roots (Drabble 1904). In contrast to the more common non-contractile root, the contractile roots had collapse or 'mucilaginization' of cortex parenchyma, no cortex fibres or lignified cells, unlignified and square-shaped endodermal cells, and no lignified sheaths around each metaxylem vessel. Modern studies are needed to confirm these earlier descriptions.

DEVELOPMENT AND TERMINOLOGY

Root anatomy in palms has been extensively surveyed by Seubert (1996a, b, 1997, 1998a, b) who suggests that all genera (e.g. in Coryphoideae) can be distinguished from each other by microscopic root characteristics. Given that first-order roots of an individual palm will show structural plasticity one from another and correlated with size and age, our view is more conservative. Seubert's studies, the most intensive and complete of their kind, do show the way forward with respect to a number of critical problems that may be fairly said to have baffled previous approaches to the study of palm roots. Chief among these innovations is the more precise conceptualization of mature structure that is based on ontogenetic criteria. Histological zones are distinguished in relation to their pattern formation at the root apical meristem. This results in a terminology differing from that of earlier authors who used simple, but varied, topographic approaches. Seubert's research is illustrated entirely by line drawings, which are therefore difficult to compare with our sections, an addition to the problems outlined earlier.

She bases her method of description on the histogen concept of Hanstein (1870), as presented in Guttenberg (1968); a recent update is that of Heimsch and Seago (2008). As a consequence of this, she focuses on and elevates the relative patterns of histological differentiation among the most superficial layers of the root—epidermis (rhizodermis or velamen), exodermis, and outer cortex—to principal comparative importance. These layers are clearly defined in early development and, in mature structure, form a protective unit with a histological configuration that is related to both root size and age variation (when not obliterated by periderm). Hence, this region provides a rich source of variable characters that often can be repeatedly observed among first-order roots of a given species.

With regard to systematics, Seubert successfully applies her descriptive system across the entire family, providing data for nearly all genera. Thus uniformly codified, major patterns of

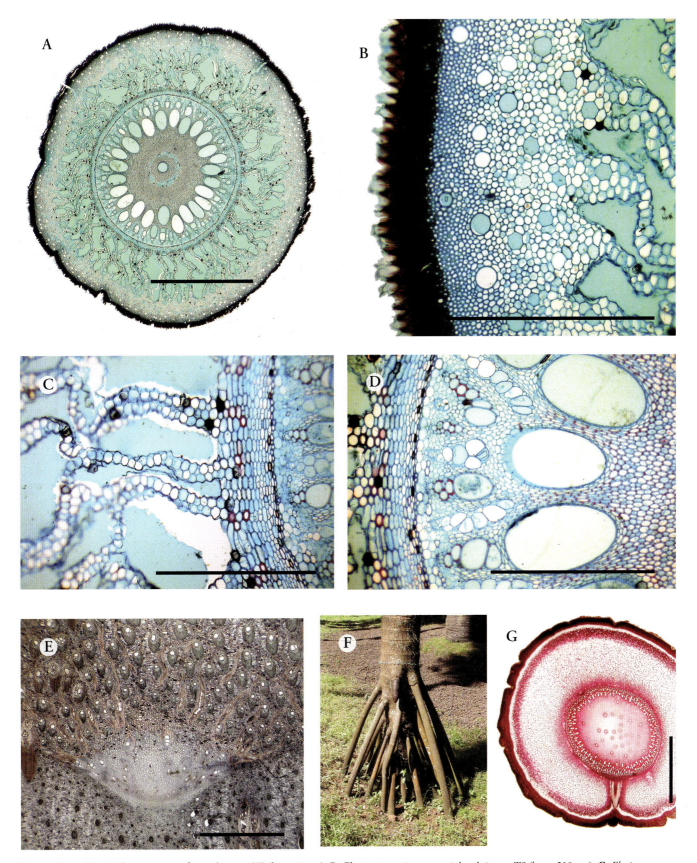

Fig. 19 Roots (1). **A**. *Elaeis guineensis*, first order root, TS (bar = 2 mm). **B**. *Elaeis guineensis*, root, peripheral tissues, TS (bar = 500 μm). **C**. *Elaeis guineensis*, root, inner cortex, and stele boundary, TS (bar = 500 μm). **D**. *Elaeis guineensis*, root, outer stele, TS (bar = 500 μm). **E**. *Elaeis guineensis*, TS of stem base with adventitious root primordium in LS (bar = 2 mm). **F**. *Verschaffeltia splendida*, stilt roots at base of stem. **G**. *Gaussia maya*, first-order root with second-order root initial, TS (bar = 5 mm).

variation emerge when her anatomical data are correlated with estimates of palm phylogeny. A short outline of Seubert's (1997) approach to the description of palm roots follows:

1) **Dermatogen**: gives rise both to the **rhizodermis** (= epidermis) which, if persistent and thick-walled, is the **velamen**; and also the **exodermis** (= hypodermis). She uses the term 'velamen' outside its standard anatomical use as a specialized multiseriate layer, as in the roots of aroids and orchids.

2) **Periblem**: gives rise to the root cortex, which is separable into **outer** and **inner cortex**. The **outer cortex** shows perhaps the greatest variation within palm roots; one to three histological zones may be differentiated within it. The **inner cortex** is characteristically differentiated into three zones: an often well-developed aerenchymatous middle zone partitions the outer and inner zones. The innermost layer of the inner cortex, as in all monocotyledonous roots, becomes the **endodermis**. The innovation here is to replace the traditional description of cortical regions based on their visible features.

3) **Plerome**: gives rise to the **vascular cylinder** (= stele), whose outer layer is the **pericycle**, which surrounds the normal polyarch arrangement of vascular tissues. The centre of the stele (**medulla**) includes additional vascular tissue and may be parenchymatous rather than sclerotic. This terminology conforms to standard usage.

OUR APPROACH

We choose to differ somewhat from the system Seubert proposed. The distinction between rhizodermis and a velamen, as defined by Seubert, is sometimes arbitrary in palm roots. Although significant functional differences may be attributed to these two developmentally homologous character states (especially at the extremes of their expression), their distribution is highly homoplasious in light of palm phylogeny, conveying very little grouping information. The term velamen is most commonly associated with the roots of epiphytic orchids and aroids (Kauff et al. 2000) where it has a specific function in water storage. Where sclerotic tissues of differing layers are juxtaposed, as at the root periphery, the complex functions as a unit and forms a defined protective entity.

DESCRIPTION

Although Seubert's work must be recognized as establishing with precision character state configurations which are indeed useful systematically, we retain in this book a simple nomenclature based on obvious topography: **surface layers**, **cortex**, and **stele**, as in Fig. 19A–D. This seems reasonable in the absence of much precise developmental information, their obvious differences, and the progressive changes that occur as the root ages.

SURFACE LAYERS

Epidermis

(This corresponds to the **rhizodermis** and **velamen** of Seubert.) While root hairs were thought to be a feature restricted to seedling roots, Seubert established that they are a prominent feature of mature root systems of many Coryphoideae (e.g. Fig. 20I) and Chamaedoreeae. Epidermal cells are usually larger than those of the adjacent exodermis (Fig. 20E). Most palm root epidermal cells have uniformly thin to moderately thickened walls; wall thickening may be uniform to unequal, with outer periclinal and peripheral portions of the anticlinal walls showing a relatively greater degree of thickness (Fig. 20H). The above variation is continuous and without a broad systematic pattern of distribution. Thin-walled epidermal cells do not persist at maturity (an exception is *Nypa*, Fig. 44K). Those with thickened walls are not always persistent. *Sabal* (Fig. 51K) and some Cryosophileae differ from other palms in having unequally thickened epidermal walls develop to such an extent that they almost occlude the cell lumen. A presumed multiseriate epidermis occurs in the clade inclusive of Corypheae and Borasseae and also in *Metroxylon* and *Salacca* (Fig. 38L).

Exodermis

This is the layer immediately within the epidermis. Cells of the exodermis are often clearly separated from those of adjacent layers. In Calamoideae (Fig. 38L) and Nypoideae (Fig. 44K), the exodermis is characteristically multiseriate and consists of lignified parenchyma with, at most, moderately thickened walls. The exodermis of Coryphoideae is uni- to multiseriate and most commonly includes thick-walled fibres. In some Cryosophileae these fibres are not separable from those of the outer cortex in TS, but in LS, the exodermal cells are shorter than those of the outer cortex, and have truncate, transverse end walls. Within Ceroxyloideae the exodermis is represented by one or two layers of thin to moderately thick-walled cells, except in Phytelepheae (Fig. 70O), in which the fibrous exodermal cells are like those mentioned earlier for Cryosophileae. As a whole, Arecoideae have the most varied exodermal structures, but a uniseriate exodermis is prevalent. Cells with strongly thickened outer periclinal walls are frequent and have a wide systematic distribution within the subfamily (but found outside Arecoideae in some Trachycapeae). Uncommonly, this tissue region may demonstrate a limited amount of tangential division, forming a 'suberoid exodermis' (Guttenberg 1968). Seubert (1996b) describes this phenomenon in *Oraniopsis*.

Periderm

In older roots, especially those which are eroded or damaged, cells at the periphery of the inner cortex (but never the epidermis) may divide tangentially and form a ligno-suberized layer of the 'etagen' type (i.e. with repeated tangential divisions in successive cells, as is characteristic of monocotyledonous protective layers). To this the term periderm seems appropriate and its dis-

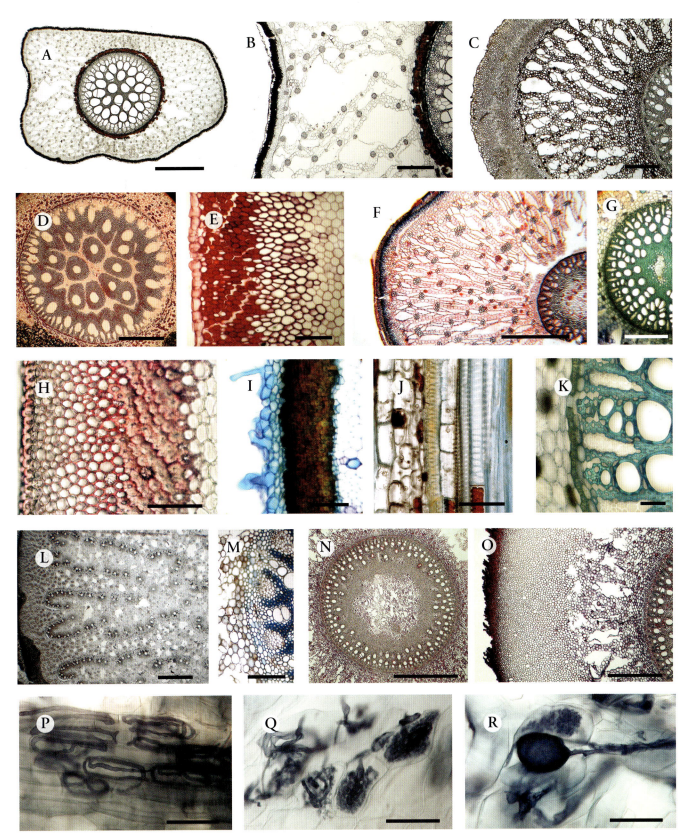

Fig. 20 Roots (2). **A**. *Arenga pinnata*, first-order root, TS (bar = 2 mm). **B**. *Arenga pinnata*, root, epidermis to stele, TS (bar = 50 μm). **C**. *Rhapidophyllum hystrix*, root, epidermis to stele, TS (bar = 500 μm). **D**. *Nannorrhops ritchiana*, root stele, TS (bar = 1 mm). **E**. *Nannorrhops ritchiana*, root periphery, TS (bar = 100 μm). **F**. *Raphia taedigera*, root, epidermis to stele, TS (bar = 1 mm). **G**. *Zombia antillarum*, root, stele, TS (bar = 500 μm). **H**. *Sabal palmetto*, root, periphery, TS (bar = 100 μm). **I**. *Corypha utan*, root, periphery with root hairs, TS (bar = 100 μm). **J**. *Brahea berlandieri*, root, outer stele, endodermis in centre, LS (bar = 100 μm). **K**. *Zombia antillarum*, root, outer stele, TS (bar = 50 μm). **L**. *Socratea exorrhiza*, outer half of fluted stele, TS (bar = 2 mm). **M**. *Socratea exorrhiza*, endodermis region of stele, TS (bar = 200 μm). **N**. *Cryosophila* sp., root, stele, TS (bar = 2 mm). **O**. *Cryosophila* sp., root surface to stele, TS (bar = 1 mm). **P**. *Coccothrinax argentata*, root cortex cells with coils of arbuscular mycorrhizal hyphae, LS (bar = 50 μm). **Q**. *Coccothrinax argentata*, root cortex cells with arbuscules of arbuscular mycorrhiza, LS (bar = 50 μm). **R**. *Coccothrinax argentata*, root cortex cells with vesicle of arbuscular mycorrhiza, LS (bar = 50 μm).

tribution is not very consistent taxonomically as summarized in Fig. 26A, but its absence characterizes Calamoideae, and independent losses of periderm production are synapomorphic for several major clades.

OUTER CORTEX

In many palms, the outer cortex provides the majority of the protective mechanical tissue of the root and in old roots, with the loss of outer surface layers, it can appear as a tannin-rich dark mass (Figs. 19B and 20I). Correspondingly, cells of this region are partly or, more often, uniformly sclerotic (Fig. 20E, H). Otherwise this is the most histologically complex and variable region of the root. Differentiation of up to three histological zones within the outer cortex is common. While always recognizable as an entity, the outer cortex is sometimes not sharply differentiated from internal tissues at maturity.

In Calamoideae (Figs. 33H; 34I; 35G; 37J; 38L; and 39J), the many-layered and well-defined outer cortex is either homogeneous and of thick-walled fibres, or divisible into two zones based on wall thickness of the sclerenchyma. *Nypa* is similar in this respect (Fig. 44K). Coryphoideae, like the previous two subfamilies, have a lignified outer cortical region, which nevertheless is more variable and complex than any other subfamily (Figs. 51K and 55H). Here up to three zones may be differentiated. A homogeneous, few-layered outer cortex characterizes most genera of the clade of Corypheae and Borasseae; two zones characterize Chuniophoeniceae. But as a whole, structural data from the outer cortex correlate imprecisely with coryphoid phylogeny. Furthermore, the uncertain distinction between the exodermis and outer cortex in several groups within the subfamily creates descriptive difficulties.

Within the clade including Ceroxyloideae and Arecoideae outer cortical structure is possibly more congruent with phylogeny. As in Coryphoideae, up to three layers may be differentiated. Lignified cells are always present, but these may occur in conjunction with one or two zones of unlignified tissue. Since the outer zone of the inner cortex may be lignified in many genera of this clade, the outer cortex sometimes appears continuous with the inner cortex.

INNER CORTEX

The inner cortex is characteristically separable into three zones, a peripheral compact layer, a central diffuse layer, and an internal compact layer. This histological configuration is plesiomorphic within palms. Instances of an almost homogeneous inner cortex, as in *Trithrinax, Podococcus,* and Geonomateae (Fig. 83L, M), are independent derivations of this state. This condition is a very distinctive state for roots of Chamedoreeae (Figs. 19G and 74K)

The compact layer may or may not be abruptly delimited from the outer sclerotic layers. It frequently contains numerous wide raphide canals, as described elsewhere, and these cells may

become thick-walled, in concert with cells of the surrounding tissue. Inner and outer zones of dense tissue are partitioned by a usually wide middle zone that is distinguished by its more or less aerenchymatous appearance (Fig. 19A); in extreme examples the cortex is almost completely aerenchymatous (Fig. 20C, F). Air spaces originate lysigenously, by the breakdown of cells accompanied by collapse of their walls (Drabble 1904; Seubert 1997). Aerenchyma is most extensive in the roots of palms of wet habitats, notably in *Nypa*, where it consists of radial air spaces, separated by narrow radial plates of collapsed or living cells (Fig. 44I). Their implied function is the development of an internal gas space that can facilitate respiratory activity. This function is presumably reduced in the denser cortex of ultimate absorbing roots of the branching system across which water and nutrients must be transported radially.

The innermost cortex is narrow and often with radial seriation of cells of progressively narrow diameter as the limiting layer of the endodermis is approached (Fig. 19C, D). Developmentally the endodermis is derived by the ultimate division of the innermost periblem layer, i.e. its cells are sister to the innermost cell layer of the cortex, as can be seen usually by their radial juxtaposition, but it is best described as part of the stele because of its regulatory function. This seemingly esoteric distinction illustrates the conflict between developmental and topographic considerations in root anatomical description.

Cortical fibres and sclerenchyma

The middle root cortex of palms can include ideoblastic elements of some complexity, notably sclerenchyma, but also mucilage canals, raphide sacs, tannin, and occasionally stegmata. Fibres can occur as either solitary or clustered cells, the latter forming fibre bundles of two main kinds (Drabble 1904). In Drabble's 'Raphia-type' the fibre bundles are circular in TS (Fig. 55I) and made up of thick-walled cells with the lumen almost occluded, the bundles often suspended within aerenchymatous tissue (Fig. 20A, B, F). Such strands can be associated with stegmata. This type is restricted to several tribes in Coryphoideae but only occasional in Calamoideae but extremely uncommon elsewhere (Fig. 27A). The indication here that 'Raphia-type' fibres do not actually occur in *Raphia* hinges on whether stegmata are always associated with the fibres in the subfamily, a point to be re-investigated. In the 'Kentia-type' fibre strands are made up of thin-walled cells with wide lumina, often restricted to the inner cortex but with a transition from solitary sclerenchyma cells nearer the periphery, usually with truncate end walls and themselves transitional to sclerenchyma of the outer cortex (Figs. 86M and 88L). Sometimes the innermost fibres are tannin filled (Fig. 88O). Fibre bundles of the 'Kentia-type' are widely distributed in Arecoideae and many Ceroxyloideae. The remaining subfamilies are typically without cortical fibre strands. Cortical sclerenchyma otherwise can include scattered short sclereids, as in the inner cortex of some Cryosophileae, or wide-lumened solitary fibres as a synapomorphy for roots of Chamaedoreeae (Fig. 74K). Such a concentration of fibrous tissue in the root cortex seems a distinctive character for many palms.

STELE

Endodermis

As described earlier, the endodermis is developmentally the innermost layer of the cortex. As it matures it shows progressive wall thickening to end with thick-walled pitted cells of two main types. Infrequently the endodermis remains thin-walled, most notably in *Nypa*. In most Calamoideae the walls are uniformly thickened ('O-shaped' in TS; Figs. 35H; 38O; and 36K); in other groups the walls are thickened unequally on the inner side ('U-shaped' in TS; Fig. 82J) so that they are eccentric, with the lumen often almost occluded. Pits in thick-walled endodermal cells are conspicuously developed, especially on the inner wall of U-cells where the pits may be 'branched', and are especially conspicuous in LS (Fig. 20J).

In the fluted stele of wide roots the endodermis is interrupted, with cortex and medulla becoming continuous, the histology of the cortex then appearing as if intruded into the stele (Figs. 20L and 86N). Obvious and persistent thin-walled passage cells generally do not occur in palms, but delayed thickening and lignification of endodermal cells opposite the phloem poles are found in younger roots (Fisher and Jayachandran 1999). The difference between O-cells and U-cells has taxonomic significance, a condition recalling that in the stem endodermis of *Potamogeton* (Tomlinson 1982). In palms, O-shaped cells characterize many Calamoideae and can be regarded as plesiomorphic for the family, whereas U-shaped cells are the derived condition in later divergent groups (Fig. 26B).

Pericycle

The peripheral stelar tissue is a less regular layer, one to three cells wide within the endodermis, and without any radial conformity because it is the outer layer of the stele proper. Cells are large, pitted, and often somewhat thick-walled, always separating vascular tissue from the endodermis (see LS view in Fig. 58K). It has an important role as the site of origin of lateral (branch) roots (Fig. 19G).

Vascular tissue

This is polyarch, in larger roots and especially aerial roots with over 100 alternating protoxylem and protophloem strands, continuous in regular or less regular ways with radial strands of early metaxylem and metaphloem. The vascular tissue is enclosed in a sclerotic conjunctive tissue, which may extend internally and gradually or abruptly to a parenchymatous medulla.

In many palm roots, especially those of larger diameter, there are additional internal wide metaxylem vessels, typically each enclosed by conjunctive tissue (Fig. 20D). Rarely the medulla is occupied fully by uniformly distributed wide vessels, as in some Caryoteae (Fig. 20A) and reported for *Corypha* (Tomlinson 1961). This could be cited as further evidence for the recently established proximity of these two taxa based on molecular analysis, even though they are otherwise dissimilar morphologically. This feature is known in other monocotyledons, as in Musaceae and especially Araceae (Huggett and Tomlinson 2010), but further work on its consistency is much needed, especially in palms. Phloem strands can also appear nearer the centre of the stele, consistently so in Chamaedoreeae (Fig. 74L).

Polystelic complexity

The stele itself can become essentially 'polystelic', notably in larger stilt-like aerial roots, correlated with increasing diameter through a series of transitional stages (Cormack 1896; Drabble 1904). In many palm roots, notably in those of the Arecoideae, the endodermis remains continuous but the stele becomes fluted, appearing lobed in TS. Progressively, in roots of increasing diameter, the lobes separate and become invaginated, with a discontinuous endodermis (Fig. 88N). Ultimately, central xylem strands appear in the medulla; the strands themselves often individually enclosed partly or wholly by an endodermis (Fig. 20L). Fluted steles of this type are uncommon in earlier diverging subfamilies of palms (e.g. only in *Licuala* of Coryphoideae) but present in larger roots of Ceroxyloideae and not uncommon in the Arecoideae. A partial explanation for this condition lies in the method of attachment of adventitious first-order roots to stem vascular bundles (Fig. 19E). 'Polystelic' roots of palms offer a challenge to the interpretation of development based on the existence of a plerome histogen because in them endodermal cell layers essentially overlap or even become almost concentric.

IDEOBLASTIC ELEMENTS

Mucilage canals appear as linear series of crystal-bearing raphide sacs, occur rather frequently in palms, and are particularly common in Calamoideae. They are easily recognized in TS as wide cells in the outer cortex, but their structure has to be confirmed in LS because cell contents are easily lost in sectioning. They may become thick-walled and lignified and simulate wide fibres.

Stegmata are rare in palm roots with the exception of their common association with fibres of the 'Raphia-type', the shape of the silica body corresponding to that of the aerial parts, i.e. spherical in Borasseae, hat-shaped in Caryoteae. Short files of silica cells occur in the stele of roots of several genera in the tribe Areceae usually at the boundary between thick-walled conjunctive cells and thin-walled medullary tissue.

Tannin occurs commonly in roots, usually in unspecialized parenchyma cells of both cortex and stele; they can be conspicuous among the modified thin-walled cells of the middle cortex where the cortex is otherwise little differentiated histologically (Fig. 84L). The outer sclerotic layers of old roots can become extensively tanniniferous.

VASCULAR TISSUES

Types of xylem (tracheary) elements and phloem (sieve) elements are described in detail elsewhere. In summary, the metaxylem elements of palm roots are more 'advanced' compared with those of the aerial organs, e.g. most commonly with simple perforation plates. Sieve plates, on the other hand, are 'primitive,' i.e. putatively ancestral because sieve-tube elements have long, oblique compound sieve plates in roots, whereas sieve plates in aerial parts are less oblique and vary from slightly oblique to transverse in their end wall position. Calamoideae show this in an extreme degree because root sieve plates are long and oblique, whereas stem sieve plates in aerial parts are, with few exceptions, transverse and simple.

ROOT REGENERATION

When first order roots are damaged, as during the transplanting of mature plants, new lateral roots can be initiated near the cut ends. Such 'branch' roots presumably arise from the pericyle as typical lateral roots (Fig. 19G), but the anatomy of regeneration has not been studied. Broschat and Donselman (1984) showed that species vary in ability to regenerate roots, either from cut roots or by initiation of new adventitious roots from the stem (Fig. 19E). The ease of transplanting adult palms depends upon their mode of root regeneration.

MYCORRHIZAL ASSOCIATION

An important observation by Fisher and Jayachandran (1999, 2005, 2008) was that palms possessed mycorrhizal associations with arbuscular mycorrhizal fungi (formerly called vesicular-arbuscular or VAM), a demonstration highly relevant to the palm horticultural industry. Figure 20P–R shows diagnostic fungal structures in *Coccothrinax argentata*. Except for the presence of intra- and intercellular hyphae in cells external to the endodermis, the histology of the thin, higher-order roots are unaffected by fungal colonization. The widest first-order roots emphasized in our survey usually lack mycorrhizae.

CHAPTER 7 VASCULAR TISSUES AND CELL INCLUSIONS

There are two basic types of vascular tissues that are involved with long distance transport in palms: **xylem** that carries mainly water, and **phloem** that carries mainly sugars and other organic compounds. The two tissues occur together in vascular bundles (Fig. 21A–C), which may contain only phloem when small (Fig. 33B). Ultrastructural features have been examined by Parthasarathy and Klotz (1976).

XYLEM AND TRACHEARY ELEMENTS (Fig. 21)

A basic recognition of the difference between **protoxylem** and **metaxylem** is necessary. Protoxylem is represented by tracheary elements originating within extending organs and is characterized by **tracheids** with imperforate tapered end walls and usually annular to spiral wall sculpturing (Fig. 21D). Metaxylem is, with few exceptions, characterized by the presence of **vessels,** usually much wider than tracheids, with individual **vessel elements** having perforated end walls and scalariform wall sculpturing. The individual vessel elements form a linear series of cells with cell lumen continuity from one element to another, the vessel itself. In TS of vascular bundles the difference in cell diameter and position between metaxylem and protoxylem is obvious (Fig. 21A–C). The earliest formed protoxylem elements become disrupted and are displaced by surrounding parenchyma cells (Fig. 21B, C). Narrow, late differentiating metaxylem vessels can also occur (Fig. 21C).

Most emphasis has been on the nature of the terminal end wall (**perforation plate**) of individual vessel elements because this shows most variation (Fig. 21E–J). Lateral walls are uniformly sculptured (Fig. 21K). Overlapping ends where two vessels are in contact show extended scalariform pitting, contrasted with the narrow series of pits that represent the outline of parenchyma cells (Fig. 21L). In principle, the most fundamental distinction is between perforation plates that are **simple** (without included wall thickening, e.g. Fig. 21I) and those that are **compound** (with included thickening bars, e.g. Fig. 21E–H, J). The latter are most often **scalariform,** although frequent **reticulate** perforation plates can occur in some genera (e.g. petiole of *Daemonorops*). The dimensions of the perforation plate are given in terms of its length, obliquity, and

number of thickening bars. The sequence shown in Fig. 21E–G shows some diversity of longer perforations, whereas the perforations in side view (Fig. 21H) and face view (Fig. 21J) show the absence of pit membranes.

Tracheary element distribution

The distribution of types of tracheary elements throughout the vegetative organs of palms was surveyed in a preliminary way in Tomlinson (1961), but subsequently and more extensively in the PhD thesis of Klotz (1977) with which this account begins. Summary information is presented in Klotz (1978a–c). He referred largely to the conspicuous wide elements of the metaxylem. His accounts supply a good deal of quantitative data and are discussed in terms of the taxonomic groupings of Moore (1973) together with inferences about correlations with habit and habitat. In principle, our observations do not conflict with Klotz's conclusions and we provide only a brief outline. In a more speculative mode he also discusses tracheary element morphology in an evolutionary context, with an implied trend from 'primitive' to 'derived' conditions, based on the principles established by Cheadle and co-workers (e.g. Cheadle 1944), i.e. that 'primitive' vessel elements most resemble tracheids. We can condense these many observations into the somewhat aphoristic statement that in palms metaxylem vessel elements have the following distribution in different organs:

- **Root**—most specialized ('advanced').
- **Stem**—intermediate specialization.
- **Leaf**—least specialized ('primitive').

Cheadle (1942, 1943a, b) interpreted results of this kind as an indication that vessels in monocotyledons originated first in roots, second in stems, and last in leaves. However, none of this early work took into consideration the variability within a single organ, especially the stem, or the developmental complexity of vascular bundles in palms, again especially in the stem. Functional attributes were presented simply as working hypotheses without experimental data, as were the ecological attributes of variation in xylem structure. The elementary suggestion was that vessel elements with simple perforation plates, as in the roots of many palms, were more efficient in conducting water than

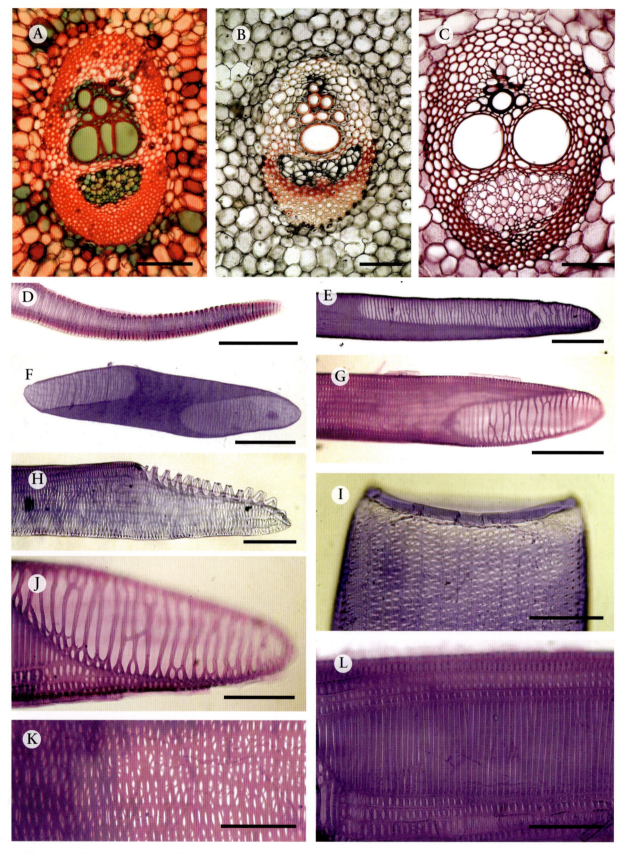

Fig. 21 Xylem and perforation plates. **A**. *Hyphaene* sp., stem vb, TS (bar = 200 μm). **B**. *Rhapis excelsa*, stem vb, TS (bar = 100 μm). **C**. *Cryosophila warscewiczii*, petiole vb, TS (bar = 100 μm). (**D–L**). Individual cells separated by maceration. **D**. *Thrinax radiata*, stem tracheid (bar = 100 μm). **E**. *Dypsis leptocheilos*, stem vessel element, scalariform perforation plate (bar = 250 μm). **F**. *Drymophloeus pachycladus*, stem vessel element, scalariform perforation plate (bar = 500 μm). **G**. *Raphia farinifera*, lamina midrib vessel element, scalariform perforation plate (bar = 200 μm). **H**. *Phoenix reclinata*, stem vessel element, scalariform perforation plate (bar = 100 μm). **I**. *Laccosperma acutiflorum*, stem vessel element, simple perforation plate (bar = 100 μm). **J**. *Raphia farinifera*, lamina midrib, scalariform perforation plate, (bar = 100 μm). **K**. *Drymophloeus pachycladus*, stem, vessel–parenchyma pitting (bar = 100 μm). **L**. *Laccosperma acutiflorum*, stem, vessel–vessel pitting (bar = 100 μm).

compound, scalariform perforation plates because the latter had an implied greater resistance to water flow. We now understand that safety, i.e. the ability of xylem elements to withstand cavitation induced by negative water pressure, is of equal importance in the overall efficiency of tracheary elements (Zimmermann 1983) and that a trade-off between safety and conductive efficiency must exist.

Pioneering work on palms in this area was initiated by Zimmermann and collaborators (Zimmermann et al. 1982; Sperry 1985, 1986). Furthermore, the putative association among 'primitive' characters in hypothetical 'primitive' groups of palms can now be examined in a more balanced context, especially with independent evidence about phylogeny based on molecular evidence. Variation in vessel dimensions throughout single palm stems is presented in detail in Tomlinson (1990), including the initial observations of Swamy and Govindarajalu (1961).

Vessel elements

Five broad categories can be recognized for wide metaxylem vessels in vascular bundles:

1) Metaxylem vessel elements exclusively with simple and usually transverse perforation plates. By Cheadle's and Klotz's criteria this would represent an advanced condition.
2) Metaxylem vessels including elements with a mixture of simple and scalariform perforation plates, on transverse to slightly oblique end walls.
3) Metaxylem vessels including elements only with scalariform perforation plates, usually on oblique end walls.

   a. End walls short, slightly oblique with fewer than 10 thickening bars;
   b. End walls long, oblique to very oblique with more than 10 thickening bars.

4) Metaxylem without vessels, including only tracheids. This would represent the primitive condition.

The distribution of these conditions throughout the taxonomic groups is not phylogenetically informative and is presented in summary form in the subsequent detailed systematic descriptions. However, sieve-tube distribution does have phylogenetic significance as seen in the following section on phloem.

PHLOEM AND SIEVE ELEMENTS (Fig. 22)

The conducting elements of the phloem, i.e. sieve tubes (Behnke 1981), are distinctive in all monocotyledons because, in the absence of secondary growth, they are all primary and function throughout the lifespan of the organ in which they develop. For stems, whose age can sometimes be measured in centuries with reasonable precision, this demonstrates the remarkable metabolic persistence of sieve tubes in angiosperms, a feature little appreciated by plant physiologists. The only exception to this generalization is found in those few monocotyledons with secondary growth by means of a distinctive tiered (etagen) vascular cambium whose derivatives differentiate as secondary ground parenchyma and secondary vascular bundles, as, for example, in *Cordyline* and *Dracaena*. *Dracaena* is the only monocotyledon with secondary growth in roots. Roots in palms are also long-lived but have no secondary growth. Their maximum age has not been determined.

The phenomenon of long-lived sieve tubes has received some attention by plant anatomists, but most work has been on dicotyledonous phloem in which phloem is short-lived, being situated outside the vascular cambium in most examples. The curious cytology of sieve tubes involves loss of nuclei at functional maturity and a highly modified cytoskeleton. So the question remains, are the centuries-old functioning sieve tubes of palms distinctive from the ephemeral sieve tubes of dicotyledons?

An extensive survey of anatomy and ultrastructure in the metaphloem of palms in the 1960s and 1970s demonstrated clearly that sieve tubes in palms do not differ in any fundamental cytological way from those in other angiosperms (Parthasarathy 1968, 1974, 1980). To this was added a detailed survey of the dimensions of metaphloem elements throughout single stems (Parthasarathy and Tomlinson 1967). Within stems, loss of function of sieve tubes does occur in leaf traces (as a consequence of leaf senescence and abscission). This occurs beyond the departure of a leaf trace as a branch from an axial bundle and involves necrosis and collapse of sieve tubes and their replacement by sclerotic elements. Axial continuity, as with metaxylem, is maintained entirely within metaphloem of the vascular system of the central cylinder. Sieve tubes, of course, may become necrotic because of disease or damage. However, they cannot be regenerated.

The most important, but limited, variables of palm sieve tubes that might be systematically useful are the structure of sieve plates and their end wall inclination, which range from simple to compound and from very oblique to transverse (Parthasarathy, 1968). Subsequent comparative work has verified and extended the earlier generalizations summarized in Table 1.

These results are also in agreement with those presenting an overview of all monocotyledons but in a phylogenetic context, based on the putative ancestral condition suggested by the structure of gymnospermous sieve cells (Cheadle and Whitford 1941; Cheadle 1948; Cheadle and Uhl 1948). In this approach, sieve tubes with long oblique compound sieve plates are regarded as representing the ancestral condition whereas simple sieve plates on transverse end walls represent a derived condition. Intermediate conditions are determined by the degree of inclination of end walls and the distribution and numbers of sieve areas. This is comparable to the situation with regard to tracheary elements although the features seen in sieve tubes are more variable. A corollary of this is that different organs show different degrees of advancement; roots being least specialized, stem and leaves increasingly advanced, as is shown most strikingly in most Calamoideae. However, with the two earliest diverging subfamilies possessing simple sieve plates in their stems and petioles, there is indeed the possibility that this 'advanced' character state may be plesiomorphic within the family. Further character analysis is needed.

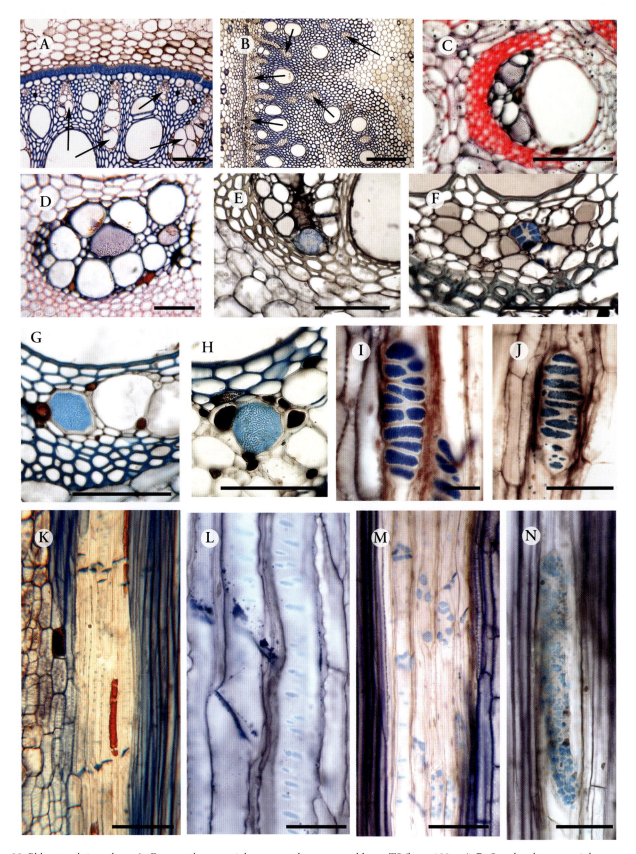

Fig. 22 Phloem and sieve plates. **A**. *Eremospatha* sp., periphery root stele, arrows = phloem, TS (bar = 100 μm). **B**. *Synechanthus* sp., periphery root stele, arrows = phloem, TS (bar = 200 μm). **C**. *Calamus deerratus*, rhizome vb, simple sieve plates, TS (bar = 100 μm). **D**. *Raphia farinifera*, petiole, simple sieve plates TS (bar = 50 μm). **E**. *Bactris gasipaes*, stem, compound sieve plate, TS (bar = 100 μm). **F**. *Wettinia* sp., stem, compound 'arecoid' sieve plate, TS (bar = 100 μm). **G**. *Daemonorops angustifolia*, stem, simple sieve plate TS (bar = 100 μm). **H**. *Plectocomia elongata*, stem, simple sieve plate TS (bar = 100 μm). **I**. *Nannorrhops ritchiana*, stem, compound sieve plate, LS (bar = 25 μm). **J**. *Pseudophoenix sargentii*, stem, compound sieve plate, LS (bar = 50 μm). **K**. *Salacca zalacca*, stem, transverse sieve plates, LS (bar = 50 μm). **L**. *Wettinia* sp., stem, oblique sieve plates and lateral sieve areas, LS (bar = 50 μm). **M**. *Orania paraguanensis*, petiole, oblique compound sieve plates and lateral sieve areas (in face view), LS (bar = 100 μm). **N**. *Plectocomiopsis geminiflora*, root, oblique compound sieve plate LS (bar = 50 μm).

Table 1 Sieve plate types in palms

Subfamily	Organ		
	Root	Stem	Leaf
Calamoideae	Compound v. oblique	Simple ± transverse	Simple ± transverse
Nypoideae	?	Simple ± transverse	Compound ± transverse
Coryphoideae	Compound oblique	Compound less oblique	Compound ± transverse
Ceroxyloideae	Compound oblique	Compound less oblique	Compound ± transverse
Arecoideae	Compound oblique	Compound less oblique	Compound ± transverse

Histology

Phloem tissue always occurs within vascular bundles. The **protophloem** is obliterated during organ enlargement, but may remain as a centre of crushed cells. Maturing after organ enlargement, **metaphloem** only is permanent. Sieve tubes are accompanied by companion cells as in all angiosperms. Phloem parenchyma cells may become sclerotic. This may result in a network of sclerotic elements distributed throughout the phloem. More significant, for systematic purposes, is that in petioles the phloem may be represented by either one or two strands within a single vascular bundle, as described earlier (Fig. 12I–P). In vascular differentiation within a bundle, protophloem elements are lost by cell extension and disruption during tissue elongation, the original site becoming sclerotic and represented by a median sinus of sclerenchyma. Sclerenchyma may be so extensive as to separate the phloem tissue into two discrete strands. As a consequence there are intermediate conditions and incompletely separated strands may make diagnosis somewhat arbitrary. The distribution of one- versus two-stranded petiolar vascular bundles is summarized in Chapter 4.

The condition of the phloem of stem vascular bundles is somewhat different because the histology of a bundle changes along its length according to the consistently established vascular pattern (*Rhapis*-principle). Nevertheless, the same developmental relation between protophloem and metaphloem exists as in petioles and again a median sclerotic sinus may be evident, but insufficient to produce two strands. A conspicuous exception is found in the stem vascular bundles of rattans (climbing palms in the Calamoideae, e.g. Calaminae and Korthalsiinae) in which all genera have two widely separated phloem strands, which come to lie lateral to the single wide metaxylem vessel (Figs. 18L, M and 41D). The initial presence of a single protophloem strand in this type of vascular bundle is not obvious without developmental study. The arrangement of sieve tubes in a single or double series in each phloem strand can be diagnostically useful.

Where vascular bundles branch, as in stems, or where transverse commissures are attached to longitudinal veins, as in the leaf axis and lamina, the evident phloem connections are seen as small lateral strands within the limiting sclerenchyma of the bundle sheath. An unusual example, where these associated phloem elements are narrow, is seen in Ancistrophyllinae.

Some cellular details

A few examples serve to illustrate phloem diversity in palms (Fig. 22); all images are from permanent sections stained with a tannic acid/resorcin blue combination, which is also useful as a double stain, but specifically identifies callose deposits on sieve-tube walls. Figure 22G, H illustrates the periphery of the root stele, with phloem strands (arrows) of varying size alternating with the xylem poles. Figure 22 shows in face view, from transverse sections, simple sieve plates in stem (C, G, H) and petiole (D) of vascular bundles of four calamoid palms, the most 'advanced' condition in palms. These may be contrasted with the long oblique stem sieve plates in Fig. 22 I, J, N, in face view, from a longitudinal section.

An intermediate and very common type of sieve plate—compound, but transverse and with few sieve areas—is shown in Fig. 22E; that in Fig. 22F is distinctive ('pie slice') and is common in arecoid palms ('Arecoid-type'), but not restricted to them.

Figure 22K illustrates coincident, almost transverse end walls of a population of sieve tubes from a stem vascular bundle, a condition that arises as procambial cells are derived by longitudinal division from one or a few initials early in vascular development. Figure 22L illustrates lateral sieve areas of sieve tube walls, and oblique sieve plates, whereas Fig. 22M shows how the size of the individual sieve areas on an oblique sieve plate may be little bigger than lateral sieve areas. The long oblique sieve plate in a calamoid root (Fig. 22N), representing the least specialized condition, is thus sharply contrasted with the advanced condition shown in Fig. 22G, H.

CELL INCLUSIONS (Figs. 23 and 25)

This section outlines features of cell types not themselves organized into discrete tissues, but distributed within other, mainly ground, tissues. They may still have a specific location, as with silica cells which are always associated with fibres.

Silica cells as stegmata

Silica, in the form of polymerized monosilicic acid $(Si(OH)_4)$ is common throughout stem and leaf in palms, but infrequent in roots. Silicon itself is not an element involved in plant metabolism and its presence in certain groups but not in others,

61

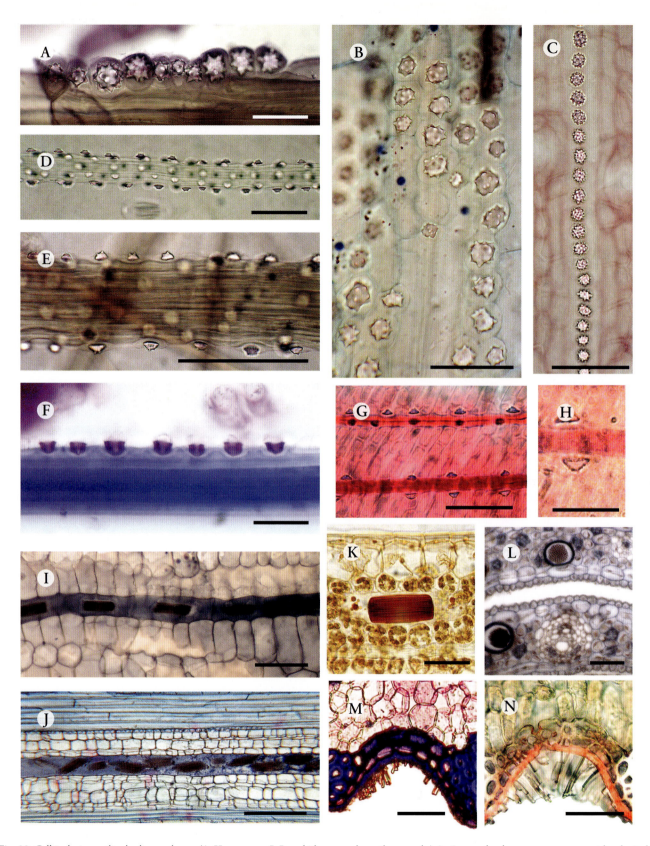

Fig. 23 Cell inclusions, silica bodies, and wax. (**A–H**, stegmata; **I–J**, raphide sacs and mucilage canals.) **A**. *Ammandra decasperma*, stegmata with spherical silica bodies, stem maceration (bar = 50 μm). **B**. *Podococcus barteri*, stegmata with spherical silica bodies, stem maceration (bar = 50 μm). **C**. *Juania australis*, stegmata with spherical silica bodies, lamina maceration (bar = 50 μm). **D**. *Bactris hondurensis*, stegmata with hat-shaped silica bodies, lamina maceration (bar = 50 μm). **E**. *Bactris acanthocarpoides*, stegmata, stem maceration (bar = 100 μm). **F**. *Tahina spectabilis*, stegmata with spherical silica bodies and lignified walls, lamina maceration (bar = 25 μm). **G**, **H**. *Sclerosperma mannii*, stegmata with hat-shaped silica bodies, lamina scrape, SV. **G**. (bar = 50 μm), **H**. (bar = 20 μm). **I**. *Bactris* sp., stem, mucilage canal with raphides, LS (bar = 50 μm). **J**. *Eremospatha* sp., stem, mucilage canal with raphides after cell wall breakdown, LS (bar = 200 μm). **K**. *Desmoncus oxycanthos*, lamina with isolated raphide sac, LS (bar = 50 μm). **L**. *Desmoncus oxycanthos*, young lamina with isolated raphide sacs, TS (bar = 50 μm). **M**. *Pseudophoenix lediniana*, wax on lamina abaxial epidermal groove, TS (bar = 50 μm). **N**. *Pseudophoenix sargentii*, wax filling lamina abaxial epidermal groove, TS (bar = 50 μm).

including both dicotyledons and monocotyledons, remains unexplained. Its limited appearance in a visible form in plants may be the result of incidental absorption in its soluble form by roots, subsequent distribution by the vascular system, and final sequestration in a solid polymerized but opalescent form. This makes its scattered appearance in unrelated groups of plants quite intriguing. Silicon in palms occurs in the form of silica bodies within small cells to which the term stegmata ('Deckzellen' or 'cover cells' in the German literature) is given (Mettenius 1864). Silica in monocotyledons, of which stegmata are but one example is almost (except for certain orchids) restricted to commelinid monocotyledons, of which palms are an early diverging lineage (Prychid et al. 2003). Stegmata each include only a single silica body and occur in continuous or discontinuous files next to vascular or non-vascular fibres, of leaf and stem, but uncommonly in roots. Silica may occur in cell walls of the stem epidermis of some rattans (Fig. 41H), a feature much in need of further investigation. Of particular interest is the variability with which stegmata are associated with the transverse veins of the lamina (Fig. 25B). To a large extent the presence of stegmata in this location correlates with abundant fibre sheathing cells of transverse veins.

The silica bodies of palm stegmata are easily recognized optically because they have a refractive index (~1.4) different from that of the mounting medium (~1.5) in which sections are usually made permanent (Fig. 23C, G, H). Although less easily seen in aqueous mounts, they are conspicuous in macerated material mounted in dilute glycerin (Fig. 23A, B, D, E). Silica is best removed from fixed material by means of HF before tissue is processed (Chapter 1). Otherwise silica bodies behave like sand particles in dulling microtome blades and tearing paraffin ribbons. The silica cell itself has an unequally thickened cell wall, with the basal wall (next to the attendant fibre) thickened (Fig. 23F), often lignified, and sometimes pitted, in contrast to the thin unpitted distal wall. The thickened part of the cell wall may retain an impression of the silica body after this has been dissolved.

Types of stegmata

Two cell types can be recognized according to the shape of the silica body.

1) **Spherical**, or druse-like and more or less symmetrical, but less so in smaller bodies, the surface spinulose (Fig. 23A–C).
2) **Hat-shaped**, or conical, asymmetric and with a flattened smooth base, the conical portion somewhat spinulose (Fig. 23E, G, H).

The size of stegmata varies considerably, even within the same leaf, because larger stegmata occur next to larger vascular or fibre bundles with smaller stegmata next to narrower bundles or even single non-vascular fibres; size variation in closely associated stegmata can be seen (Fig. 23A).

Taxonomic distribution

The systematic distribution of stegmata is such that not only individual species but most larger groups include only one type of silica body, as summarized in Table 2.

Table 2 Silica body types in palms

Subfamily	Spherical	Hat-shaped
Calamoideae	All	None
Coryphoideae	Most	Only Caryoteae
Nypoideae	None	All (1 sp.)
Ceroxyloideae	All	None
Arecoideae	Many	Few
	Podococceae	Iriarteeae
	Reinhardtieae?	Chamaedoreeae
	Roystoneeae	Sclerospermeae
	Cocoseae—some	Cocoseae—Bactridinae
	Attaleinae	Oranieae
	Elaeidinae	
	Manicarieae	
	Euterpeae	
	Geonomateae	
	Areceae	

In a phylogenetic context (Fig. 25A) one can suggest that spherical silica bodies are plesiomorphic for palms and that hat-shaped silica bodies are derived separately in several major groups. One seemingly consistent character is the presence or absence of stegmata in association with the transverse veins of the lamina. They are present in unrelated groups (Fig. 25B) always of the spherical type, as seen in comparison with Fig. 23A.

Developmental study

Incidental observation shows that silica cells in palms originate early and independently of each other, i.e. their cell files do not originate from a common mother cell, and silica appears early within them as initially small particles. The study of Schmitt et al. (1995) shows stegmata development in detail for the rattan *Calamus axillaris*. This involves the progressive thickening of the cell wall on the fibre side, deposition of silica within an enlarging vacuole that displaces the cell nucleus, and the ultimate degeneration of the cytoplasm as an unusual form of apoptosis.

Raphide sacs and mucilage canals

Raphide sacs (Fig. 23I–L) are single cells containing bundles of needle-like crystals of calcium oxalate or **raphides** (Frey 1929; Prychid and Rudall 1999) The crystals are usually embedded in mucilage-like material, which may stain intensely in microscopic preparations (Fig. 23J). Raphide sacs are usually enlarged and differentiate early within meristematic tissues but always initially as thin-walled cells, usually solitary (Fig. 23L), but sometimes forming continuous linear series (Fig. 23I) without marked cell extension. If walls between successive cells break down a continuous canal (**mucilage canal**) results (Fig. 23J). Mucilage canals occur most consistently in Calamoideae, especially in stem and root in which a cut surface can exude quantities of mucilage including refractive crystals that are visible with a hand lens. Otherwise where they are not obvious in aerial parts, mucilage canals can be common in the outer cortex of roots.

Raphide sacs as single cells are always colourless, i.e. without chloroplasts, and so are most easily recognized in the lamina mesophyll. They occur to a greater or lesser extent in all palms as ideoblasts in ground tissue, possibly universally in leaf and stem. They can become thick-walled in older roots, a transformation requiring detailed study. Additional fine crystalline material can occur, especially in the lamina epidermis, but we have not examined its distribution in detail. Raphide sacs are least easily seen in the ground tissue of leaf axis and stem, especially where the raphide crystals have been lost in sectioning. However, other cell contents, e.g. mucilage, may be useful in identifying them, especially as the mucilage may retain the outline of crystals lost in histological processing. As we have mentioned, in roots, they are most common external to the middle cortex but longitudinal sections are needed to establish if these enlarged cells are single raphide sacs or mucilage canals.

Raphide sacs have been reported for *Cocos* in the thin-walled cells of epidermal trichomes when they presumably can only be seen in immature material before distal shield cells are lost (Frey 1929). Raphide crystals have been reported as abundant in the styles and stigmas of Hyophorbeae (= Chamaedoreeae) and Ceroxyleae by Uhl and Moore (1973) and interpreted as a possible herbivore deterrent. In a survey of 148 taxa, but not including *Nypa*, Zona (2004) found raphides in embryos of Coryphoideae, Ceroxyloideae, and Arecoideae but not in Calamoideae.

Tannin

Tannins, i.e. polyphenolic substances, are conspicuous in most palm tissues as light to dark brown, or even black deposits within thin-walled but otherwise unspecialized cells. They normally fill the cell vacuole and fixation and sectioning may result in their displacement. Microscopically tannins show various aspects, from homogenous to foam-like, possibly when the vacuole is subdivided. The amount of tannin may increase with the age of the material; dried material of the lamina, which we have used extensively, usually becomes tannin-rich. Tanniniferous cells usually differentiate early and so are soon recognized in meristematic tissue. In stems in which secondary changes occur in ground tissues, the early-formed tannin cells retain their original size, indicating that they probably do not undergo metabolic or size changes with age (Fig. 16 cf. A and B). Where ground tissue become very lacunose, tannin cells become loci from which expansion occurs, seen as dark-coloured cells from which other enlarged cells radiate. In the root cortex radial plates of cells separating air lacunae can be marked by regularly distributed tannin cells (Fig. 84L). In phloem tissue, phloem parenchyma can sometimes be distinguished easily from companion cells by its tannin contents (Fig. 22H).

Systematically useful information on tannin distribution is lacking. In some Calaminae distinctive large tannin cells occur in the lamina, but not elsewhere, e.g. *Daemonorops* spp. as reported in Tomlinson (1961). In a number of genera in the tribe Areceae, tannin fills the lumen of the inner cortical fibres ('Kentia fibres') in a distinctive way (Fig. 88O).

Starch

Tomlinson (1961) concluded that starch grains are too variable in size and shape within palm species to offer any diagnostic value, although Wellendorf (1963) presents a somewhat more positive analysis, which we have not pursued. Starch grains are most abundant in ground tissue of leaf and stem but much less common in roots, and there mostly in the medulla. Starch granules are characteristically present in the guard cells of stomata (Fig. 3G). Starch is most abundant in the stems of hapaxanthic palms, i.e. those that flower distally after a long vegetative period (e.g. Caryoteae, *Corypha, Metroxylon, Raphia*). The starch functions as the massive carbohydrate reserve for the flowering process. It has considerable value as food for humans, as in *Metroxylon* (true sago) when harvested before flowering (Schuiling 2009).

CHAPTER 8 SPINES

Spines are here defined as any structure that ends in a sharp point without regard to its origin, i.e. avoiding definitions like prickle, thorn, and spine, which can refer to the method of origin of a sharp-pointed structure according to its homologies or other developmental details. The function of spines is obvious, because they are a defence mechanism, usually against larger herbivores. Spines in palms are restricted to certain taxonomic groups (hence one can distinguish armed from unarmed palms) and so have systematic or diagnostic significance, but within this broad distinction there is great diversity in the kinds of spines that palms produce, these kinds then often having restricted distributions. In principle, however, spines are sclerotic, in order to maintain their rigidity, without regard to original anatomical features. Spines may have a superficial origin, or develop from internal tissues by breakdown of surrounding less resistant tissue; root- and rootlet-spines have internal origins because endogenous development is a root character. The subject is dealt with in greater detail in Tomlinson (1961; 1962a) and Fisher (1981).

The following synoptic key provides a guide to palm spine morphology; it is followed by a listing of spine types in different taxonomic groups with some anatomical examples.

Synoptic key to spines

1. Spines originating from **superficial** structures. 2
1A. Spines originating from **internal** vascular tissues (**vascular spines**) or organs that arise endogenously (**root-** and **rootlet-spines**) 5
2. Spines resulting from a modification of an **existing aerial organ** (leaflet, rachilla) 3
2A. Spines originating as **emergences** from superficial tissue, i.e. not a modification of an existing organ; subcategories are determined by their location 4
3. **Leaflet spines**, as reduced leaflets that become sclerotic and sharp pointed (**acanthophylls**); these may be **basal** (*Phoenix*) or **distal**, as in the reflexed distal leaflets, widely spaced on the cirrus of climbing palms. . . . . *Desmoncus* (Fig. 24N) and Ancistrophyllinae
3A. **Rachilla-spine**, from the development of a sharp-tip to the inflorescence rachilla. . . . . (*Elaeis*)
3B. **Petiole spines,** as emergences on the margin of the petiole, where they can be triangular and resemble teeth (*Corypha*,

Fig. 24H), but may be long and straight with a bulbous base (*Pholidocarpus*), recurved (*Copernicia*, Fig. 24 L), or irregular . . . .(*Borassus*, Fig. 24J)
4. Spines on **vegetative organs**, e.g. surface of lamina, ribs or leaflet margins, leaf axis, stem. [The configuration of the spine in this category is very variable; they are commonly flattened or aggregated in various ways; universally they are erected by the expansion of a cushion of parenchyma at their insertion and can leave a shallow impression in the surface of the supporting organ (Fig. 24D, F).]. 7
4A. Spines on **reproductive organs**, e.g. bracts, inflorescence axis and even fruits 8
5. Spines as modified roots 6
5A. Spines resulting from persistent spine-tipped remains of leaf sheath **vascular** bundles particularly well developed in *Zombia* (Fig. 24A) . . . . *Rhapidophyllum*
6. **Root spines**, resulting from the modification of an adventitious root developing endogenously from the palm stem (Fig. 24P); these may be branched . . . .*Cryosophila*
6A. **Rootlet spines**, resulting from the modification of a second-order root on a first-order root as on the aerial stilt-roots of. . . . . *Socratea*
7. Spines on **margins** and **midribs** of leaflets . . . . e.g. *Raphia, Bactris*
7A. Spines on the margin of the **petiole**, where they are irregular (or triangular and resemble teeth, (e.g. *Corypha*; Fig. 24H), but may be long and straight . . . . *Pholidocarpus*
7B. Spines on the surface of the **leaf axis** (usually abaxially) . . . .*Bactris* spp.
7C. Spines on the **stem surface**, only revealed after leaf fall, ...*Astrocaryum*
7D. Spines aggregated on **petiole** in transverse linear series . . . . *Salacca* (Fig. 24D)
8. Spines on **bract** outer surface . . . . *Bactris*
8A. Spines on surface of **fruits** . . . . e.g. *Acrocomia*

Elaeis should appear twice in the key because the spines at the base of the rachis represent proximally the remains of large vascular bundles of the leaf sheath (internal origin) and distally the persistent midrib of the basal leaflets (essentially external origin), with a transition between the two (arrows in Fig. 24K).

Fig. 24 Spines. A. *Zombia antillarum*, stem with persistent, spiny leaf bases. **B**. *Rhapidophyllum hystrix*, apex of expanding leaf sheath with spines separating from sheath. **C**. *Rhapidophyllum hystrix*, spine, mostly fibres and reduced vbs, TS (bar = 500 μm). **D**. *Salacca magnifica*, petioles with spines. **E**. *Salacca magnifica*, spine, vbs and peripheral sclerenchyma, TS (bar = 200 μm). **F**. *Oncosperma tigillarium*, stem with spines. **G**. *Oncosperma tigillarium*, spine, vbs and peripheral sclerenchyma TS (bar = 200 μm). **H**. *Corypha utan*, petiole with marginal spines (teeth). **I**. *Corypha utan*, spine (tooth), vbs and peripheral eroded sclerenchyma, TS (bar = 500 μm). **J**. *Borassus madagascariensis*, petiole (on each side) with marginal spines (teeth). **K**. *Elaeis guineensis*, leaf with pinna midrib spines (above arrows) and fibrous sheath spines (below arrows). **L**. *Copernicia alba*, petiole spines (teeth). **M**. *Copernicia alba*, petiole spine (tooth) base, vbs and thick-walled sclerenchyma, TS (bar = 500 μm). **N**. *Desmoncus orthacanthos*, pinna pair reduced to recurved acanthophylls at lamina tip. **O**. *Desmoncus orthacanthos*, pinna hook, vbs with thick fibre layer, TS (bar = 500 μm). **P**. *Cryosophila* sp., stem with root spines. **Q**. *Cryosophila* sp., root spine with stele and peripheral sclerenchyma, TS (bar = 500 μm).

TAXONOMIC DISTRIBUTION

One can readily distinguish between groups with and without conspicuous spines (i.e. armed or unarmed palms).

Calamoideae: predominantly spiny, often fiercely so; in addition to their defence function there can be reflexed **hooks**, often aggregated to form **cat's-claw** spines, that function as grapnels; either on the leaf axis (abaxially) or on the extended terminal **cirrus** or **flagellum** of climbing palms. Rootlet spines occur in *Eugeissona*.

Exceptional taxa include Mauritiinae, which are almost totally unarmed (save for the fierce root spines present on the stems of *Mauritiella*); and *Pigafetta* and *Metroxylon* in which the 'spines' are soft, thread-like, and not dangerous (Fig. 29D).

Nypoideae: unarmed.

Coryphoideae:
Unarmed coryphoid groups: Sabaleae, Chuniophoeniceae, Caryoteae (except for incipient vascular-spines in *Arenga*).
Armed coryphoid groups: (see list). Many groups are significantly armed; mostly spines are on the petiole margin and can be conspicuous and bulbous-based (*Pholidocarpus*), recurved (*Copernicia* and *Hyphaene*), produced as irregular marginal teeth (*Corypha*), or occur as minute saw-like teeth (*Serenoa*). These should be considered emergences in a conventional sense, despite the statement in Dransfield et al. (2008b) that this condition does not exist in this subfamily, possibly based on mistaken homologies. Otherwise they are of leaf sheath origin and then vary from conspicuous, as in *Zombia,* or inconspicuous but needle-like, as in *Rhapidophyllum* with transitions to less formidable structures in e.g. *Guihaia* and *Trachycarpus*. Aerial root spines are distinctive in *Cryosophila*.
Borasseae: include armed and unarmed taxa. Petioles with prominent marginal teeth present in *Borassus, Hyphaene*, and *Medemia*, but not necessarily in all species. The distal petiole margins of *Latania* and *Satranala* are sometimes provided with small, rudimentary teeth.
Phoeniceae: basal leaflet-spines diagnostic.
Rhapidinae: unarmed in *Rhapis*, some *Trachycarpus*. Armed: leaf sheath vascular spines in *Guihaia, Maxburretia, Trachycarpus* spp., but especially well-developed in *Rhapidophyllum*; marginal petiolar spines well developed only in *Chamaerops* (present but minute in some *Trachycarpus* spp.; absent from other genera).
Trachycarpeae (exclusive of Rhapidinae): Marginal petiolar spines present in most genera (uniformly so in Livistoninae), but occasionally only the juvenile leaves have spines (e.g. *Saribus* spp., *Washingtonia*). Petiolar spines sometimes distinctive, as in the long, straight, bulbous-based spines of *Pholidocarpus* and the densely crowded, minute spines of *Serenoa* that collectively produce a saw blade-like edge. Most unplaced Trachycarpeae have marginal petiolar spines, but *Colpothrinax* and *Pritchardia* are wholly unarmed. Rootlet

spines are reported as widely distributed in Coryphoideae by Seubert (1997).

Arecoideae: most members of this large group are unarmed, with the following exceptions:

Iriarteeae: rootlet-spines on aerial roots, e.g. *Dictyocaryum, Socratea*; 'black bristles' reported for *Iriartella*.

Cocoseae:
Bactridinae: usually heavily armed with spines on stem, leaf axis, outer surface of bracts, leaf blade and even fruits, as in *Astrocaryum*.
Attaleinae: mostly unarmed but leaf sheath vascular-spines in some *Butia* and *Syagrus* spp.; *Polyandrococos* has spiny rachilla tips.
Elaeidinae: *Elaeis* with spiny region at base of petiole equalling a transition from vascular-spines to persistent midrib spines of basal leaflets; in addition rachilla-spines occur in the inflorescence and even the stigmas mature as recurved spiny structures.

Areceae:
Oncospermatinae: with spines in all genera on the leaf sheath and stem (Fig. 24F), sometimes only in juvenile stages (e.g. *Deckenia*); the stem spines in *Tectiphiala* with a bulbous base.
Verschaffeltiinae: spines on stem and/or leaf sheath in *Acanthophoenix, Nephrosperma, Phoenicophorium* and *Roscheria*, but usually only in juvenile stages; rootlet-spines on the aerial prop-roots of *Verschaffeltia*.

ANATOMY OF SPINES

Huard (1967) has made the most extensive study of the anatomy of spines, recognized as difficult objects to section because of their sclerotic texture. His objective was to identify putative palm fossils. Our own studies are limited, but we demonstrate that spines equivalent to a specific organ (leaflet, root) show its intrinsic anatomy to be modified so as to become highly sclerotic. This can include lignification and wall thickening in surface layers and ground tissues and the exaggeration of fibre sheaths of vascular bundles.

Our illustrations show as much as anything the technical difficulties of freehand sectioning these structures. They show the spine of *Rhapidophyllum* (Fig. 24B), a complex of narrow vascular bundles embedded in peripheral sclerenchyma (Fig. 24C), the reduced anatomy of an acanthophyll in *Desmoncus* (Fig. 24N) with extensive exaggeration and sclerification both of bundle sheath fibres and ground tissue (Fig. 24O), and the sclerenchyma development of outer cortex and stele in the root spine of *Cryosophila* (Fig. 24P, Q).

However, of unexpected significance is that spines that should be traditionally classed as 'emergences' develop vascular bundles in numbers proportional to the size of the spine. This is shown for the flat spines (Fig. 24D) of *Salacca* (Fig. 24E), the long stem spines (Fig. 24F) of *Oncosperma* (Fig. 24G), the sharp petiolar spines (Fig. 24H) of *Corypha* (Fig. 24I), and the recurved petiolar teeth (Fig. 24L) of *Copernicia* (Fig. 24M). The spines of *Elaeis* (Fig. 24K) have not been studied but are likely to be complex in view of their inherent morphology, and those of *Borassus* (Fig. 24J)

resisted all attempts to section them freehand, so we shed blood in vain! Many spines in palms are supported by a basal cushion of parenchyma. In spines that originate as structures appressed to the surface on which they originate this cushion can have the initial function of erecting the spine into its final orientation. Acanthophylls may rotate almost 180° in climbing palms (Fig. 24N).

CHAPTER 9 CLASSIFICATION, PHYLOGENY, AND ANATOMICAL EVOLUTION OF PALMS

INTRODUCTION

Wherever palms grow they are unmistakable. Their appreciable diversity, prominence in many tropical ecosystems, economic importance, and frequent aesthetic appeal attracted the attention of many early botanists, and has sustained interest in their biology to the present.

A major focus throughout has been the development of a system of classification for the family. The publication of a new edition of *Genera Palmarum* (Dransfield et al. 2008b), whose classification we adopt in this book, fundamentally realizes the goal of providing a classification that will be robust to the test of time and builds on the extensively innovative first edition (Uhl and Dransfield 1987). However, since the pioneering studies of Martius (1823–1850), whose classification of palms provides the foundation upon which successive classificatory schemes were built, there has been a shift in emphasis toward developing an evolutionary perspective of the family. Although evolutionary relationships can be expressed to a degree in the form of a classification, with the advent of modern phylogenetic approaches, the use of a branching tree diagram has become standard practice because it can explicitly depict the set of nested relationships among taxa. The classification outlined in *Genera Palmarum* (Dransfield et al. 2008b) divides the family into five subfamilies: Calamoideae, Nypoideae, Coryphoideae, Ceroxyloideae, and Arecoideae, but this convention is adopted as a convenience to communication and not because Dransfield and colleagues (or we) think that these subfamilies are evolutionary equivalents. The relationship among the five subfamilies is depicted in detail in Figs. 25–28, and in a more general way in the introductory figure for each subfamily in Part 2 of this book. We therefore wish to emphasize a 'tree-thinking' or phylogenetic approach to understanding evolution in the palm family, and use available phylogenetic data to reconstruct the evolutionary history of many anatomical characters. But systematic studies are always based on the work of predecessors, and the essential first step in any comparative study is the selection of material, which must be based on some system of classification.

Palms are generally challenging subjects for systematic studies because their frequent large proportions make the preparation of adequate herbarium specimens time-consuming and difficult. Even the best specimens can be problematic to interpret because of their fragmentary nature. Hence, field studies have been critical to the advancement of knowledge of the systematics of the family—perhaps more so than any other group of tropical angiosperms. Among the most energetic and resourceful field botanists of the 20th century was H.E. Moore, Jr., whose detailed field investigations of nearly every genus of palms then known made possible many important advances to palm taxonomy and classification because he was able to observe palms as whole organisms. An enthusiastic collaborator, Moore engaged colleagues from diverse botanical disciplines—such as anatomy, morphology, and cytology—to study palms, resulting in a broad base of information that could be used to establish a then unprecedentedly accurate systematic foundation. Indeed, the first comprehensive treatment in English (Tomlinson 1961) of the systematic anatomy of palms (but cf. Solereder and Meyer 1928) was facilitated by Moore's collegial support having been initiated under the guidance of C.R. Metcalfe at the Royal Botanic Gardens, Kew. The great synthesis of the research efforts of Moore and his colleagues is *Genera Palmarum* (Uhl and Dransfield 1987). One of the major advances that *Genera Palmarum* made beyond most prior classifications was adopting an evolutionary perspective, which was based on ideas of evolutionary trends in palms that were derived from comparative studies of palms and monocots as a whole (Moore 1973; Moore and Uhl 1978; Uhl and Moore 1982). Lacking a phylogenetic foundation, the polarity of character states—assessments of what is 'primitive' and 'derived'—was characterized a priori, and the resulting classificatory scheme justified by the goodness of fit of character data in the context of these trend-based transformation series. These ideas were developed from an 'intuitive' standpoint, based on the experience and interpretation of an observer.

The development of more objective methods of phylogeny reconstruction, as well as the use of DNA (deoxyribonucleic acid) sequence data as a source of character information to resolve evolutionary relationships, has led to many refinements in palm classification. All currently recognized groups are monophyletic: they contain all taxa descended from a presumed common ancestor. Further advances include adjustments to the circumscriptions of three of the five subfamilial clades, and the development of a robust phylogenetic hypothesis of the relationship among these groups (Asmussen et al. 2006; Baker et al. 2009). More detailed references to treatments of individual subgroups are provided in the later systematic discussion.

The phylogenetic placement of certain morphologically isolated groups of palms has occasioned surprise. Who would have thought that the fishtail palms and relatives (Caryoteae), with their very distinctive physiognomy and anatomy, would have a close relationship to the giant fan palms of Corypheae and Borasseae? Or, that the ivory nut palms and relatives (Phytelepheae, formerly subfamily Phytelephantoideae), would be the sister lineage to the Andean wax palms and their relatives (Ceroxyleae), and placed within subfamily Ceroxyloideae? These examples demonstrate how closely related groups can diverge enormously in morphology and anatomy from a common ancestor. Molecular phylogenetic data therefore can provide more than just the means to refine classifications. Evolutionary trees based on molecular data also provide an independent framework for the reconstruction of ancestral states of structural characters. Consequently, the importance of structural information in palm systematics is shifting from being a source of character data used for phylogeny reconstruction toward being a subject for analysis, in which phylogenetic knowledge is used to understand the evolution of morphological and anatomical traits.

A major goal of this book is therefore to reconstruct the evolutionary history of anatomical characters to which previous studies attributed systematic significance, as well to highlight the systematic significance of new characters we identified during the course of this investigation. The approach we stress throughout is the discovery of synapomorphies—derived character states shared by two or more taxa and their most recent common ancestor. Our objective therefore complements the approach of Horn et al. (2009), who addressed the evolutionary significance of patterns of recurrent evolution among suites of leaf anatomical characters. This aspect of character evolution—the recurrence of similar structure—is termed homoplasy. Indeed, most anatomical characters in palms are homoplasious, and frequently to a great degree. To accurately dissect and quantify structural homoplasy, molecular phylogenetic data are needed for ancestral state reconstructions, for reasons discussed earlier. A given character state that has multiple, independent origins can therefore, potentially, be interpreted as a synapomorphy for each clade in which it originates; but such an assessment is conditional upon the level of phylogenetic hierarchy of interest.

Next, we present and discuss the results of the reconstructions of eight major anatomical characters in the context of the most recent and comprehensive palm family phylogeny (Baker et al. 2009; Supertree).

METHODS USED IN THE PHYLOGENETIC ANALYSES

Phylogenetic data

The trees used in the ancestral state reconstructions of anatomical characters are based on the best phylogenetic data available at the time of writing. Different approaches for obtaining a best estimate of phylogeny were taken for different groups, as detailed later. In all instances, we used genera as the terminal taxa in our trees, collapsing clades when multiple species in a genus were sampled. Genera that had state variation for a given character were coded as polymorphic in a way that was consistent with what we observed.

Ancestral state reconstructions, which required a comprehensive family phylogeny, used a set of 5000 equally optimal supertrees derived from a weighted matrix representation with parsimony (MRP) analysis of the expanded input data set of Baker et al. (2009). These were generated using PAUP* (Swofford 2002) from the data matrix available at: http://eunops.org/content/files-store. The analytical strategy we employed therefore attempts to take into account existing topological uncertainty through a sensitivity analysis by conducting our ancestral state reconstructions on the full set of 5000 optimal trees. The Calamoideae character evolution analyses were carried out on both the topology of the weighted MRP supertree and the supermatrix tree from Baker et al. (2009). The topologies of these trees are fully resolved. For the character analyses of Coryphoideae, the availability of newly-released and excellent data that was not included in Baker et al. lead us to adopt a metatree approach (Funk and Specht 2007) to obtaining the best estimate of the phylogeny of this group. We grafted the strict consensus topology of Cryosophileae of Roncal et al. (2008) and the most likely tree topology of Trachycarpeae of Bacon et al. (in subm.) onto the strict consensus topology of Coryphoideae recovered in Asmussen et al. (2006). The analyses for Ceroxyloideae are based on the phylogeny of Trénel et al. (2007).

Metroxylon, Nypa, Arenga, Chelyocarpus, Sabal, Juania, Pseudophoenix, Iriartea, and *Hyophorbe* were used as outgroups in the subfamilial analyses, following the topology of Asmussen et al. (2006). The outgroups provided an adequate distribution of character state variation outside the subfamily of interest, and improved assessment of the ancestral states at the root node of each subfamily.

Ancestral state reconstructions

Ancestral state reconstructions using parsimony optimizations were conducted for each character in Mesquite version 2.72 (Maddison and Maddison 2009). The number of steps for each character analysed was collected from each of the 5000 trees on which optimizations were examined by assessing the value for a single tree (i.e. Current tree in Mesquite), then using the Step Through Trees command to collect the full set of values. For the analyses, maximal and minimal values of forward and reverse character state changes were gathered for the clade of Arecaceae using the Summarize Changes in Selected Clade command using all trees, with the maximum number of mappings sampled for each character on each tree set to 50. For each character examined, optimizations for each tree were summarized onto the strict consensus of all 5000 trees using the Trace Character Over Trees command.

ANATOMICAL EVOLUTION IN PALMS

Stegmata

The type of silica body included within the stegmata is among the few anatomical characters that gained wide currency in the formulation of 'intuitive' classifications of palms. Silica body type comprises two, usually distinct character states: spherical (Fig. 23A–C) and hat-shaped (Fig. 23D, E, G, H). Each consistently characterize lineages within palms long considered 'natural' groups based on other structural evidence. Parsimony reconstructions of the evolution of silica body type indicate that spherical silica bodies are plesiomorphic within palms, and hat-shaped silica bodies independently evolved in *Nypa*, Caryoteae, and Arecoideae (Fig. 25A). Within Arecoideae, hat-shaped silica bodies have 4–6 independent origins. Although these results corroborate the perspective that silica body type is systematically informative at shallow levels of hierarchy, the extensive homoplasy inferred in the evolution of hat-shaped silica bodies suggests the need for a reappraisal of the variation that exists within this character state.

Hat-shaped silica bodies are appreciably variable among different palm lineages. Those of *Nypa* strongly contrast with all other hat-shaped types. *Nypa* possesses among the smallest silica bodies of any palm, and apart from having a roughly thickened central region, are most similar to those of *Eugeissona*, which are also outstanding in their small size, appearing as minute flakes (unlike other Calamoideae, in which silica bodies are clearly spherical). The silica bodies of *Orania* and *Reinhardtia*, may be considered hat-shaped on account of often being about hemispherical in form, but have sometimes been interpreted as spherical because they lack the flattened, peripheral rim evident in other hat-shaped types. There is scope for further research to address these variants in greater detail, possibly with a view toward subdividing the hat-shaped state into multiple states that better reflect the evolutionary history of silica bodies within the family.

Although all palm leaves have stegmata in association with the fibres of the longitudinal veins, stegmata are also associated with the transverse veins in several genera. The distribution of this character state largely corresponds with the presence of a thick fibre sheath surrounding the transverse veins, although exceptions to this correspondence exist. The transverse veins of some taxa with well-developed fibre sheaths are without stegmata (e.g. many Calamoideae, *Kerriodoxa*), and a few groups with transverse veins sheathed by a single layer of short sclerenchyma cells have associated stegmata (e.g. *Manicaria*, Phytelepheae; Fig. 71J). Ancestral state reconstructions suggest considerable complexity in the evolution of this character, with transverse vein stegmata having independent origins in all subfamilial clades except Nypoideae (Fig. 25B). In Calamoideae, this character state evolved independently in several genera, but given the topology used for our analysis, is not systematically significant at deeper hierarchical levels. Transverse vein stegmata are most widespread within Coryphoideae, and although the evolution of this character state is complex within the subfamily, it appears to have great systematic significance. The origin of transverse vein stegmata is a synapomorphy for the clade of *Corypha* + Borasseae. Elsewhere in Coryphoideae, appreciable phylogenetic uncertainty prevents the unequivocal reconstruction of ancestral states at many nodes; however, our data indicate that transverse vein stegmata evolved one or more times at or near the base of the clade inclusive of *Sabal* and *Trachycarpus*. The evolution of this character warrants further examination. Transverse vein stegmata are present in few lineages within the clade of Ceroxyloideae + Arecoideae, but are synapomorphic for Phytelepheae (Ceroxyloideae) and a core clade of subtribe Attaleinae (Arecoideae: Cocoseae), exclusive of *Beccariophoenix* and, perhaps also, *Jubaeopsis*. The presence of transverse vein stegmata is one of the many leaf anatomical features that evolved in parallel between Attaleinae and Coryphoideae. Within Arecoideae, this character state is also curiously present in *Leopoldinia*, *Manicaria*, and *Pelagodoxa*, which are among the few genera of the subfamily with corky-warted fruits.

Root

Palm root anatomy is challenging to study because there exists substantial structural variation that is attributable to, on one hand, the age of the plant, and on the other hand, the age of the root (root order being equal). We must emphasize that our studies relate to first-order roots, which have no significant absorbing function. They are the pathways for transport from high-order rootlets. Studying roots collected from mature, reproductive individuals largely obviates the issue of plant age; however, roots of widely varying age will be encountered. Our observations indicate that palm roots show sustained primary growth, so 'mature' features of palm roots may not appear, in some instances, for many years. Hence, it is often necessary to examine many roots from a given individual to accurately observe certain character states. We therefore make judicious use of both our own and other available data (particularly Seubert 1996a, b, 1997, 1998a, b) on palm root structure for the purpose of evolutionary analysis.

Palm roots must survive for decades, even if they are replaceable to an extent. The success of the protective outer layers of the root in enabling it to endure potentially harsh environmental conditions is therefore critical to its longevity. Consequently, an important developmental phenomenon present in many palm roots is the initiation of a phellogen that produces a periderm, which provides a renewable means of protecting of the root from damage. Ancestral state reconstructions of this character reveal that roots of Calamoideae lack the capability to produce a periderm, fundamentally contrasting with other palms. Within the clade of Arecaceae exclusive of Calamoideae, notable independent losses of roots with a phellogen occur in Cryosophileae and Ceroxyloideae exclusive of *Pseudophoenix*. The roots of many genera within these clades are distinctive among all palms in having the thickest and most strongly sclerotic exodermal and outer cortical regions. Calamoideae also have mechanical tissues abundantly deployed at their root periphery. These results suggest that in the evolution of palm roots, a trade-off may exist between protection via diffuse, sustained thickening—in which the layers external to the periphery of the inner cortex are eventually displaced by periderm tissue—and protection via persistent and strongly developed tissue regions external to the inner cortex. In Arecoideae, an adequate interpretation of

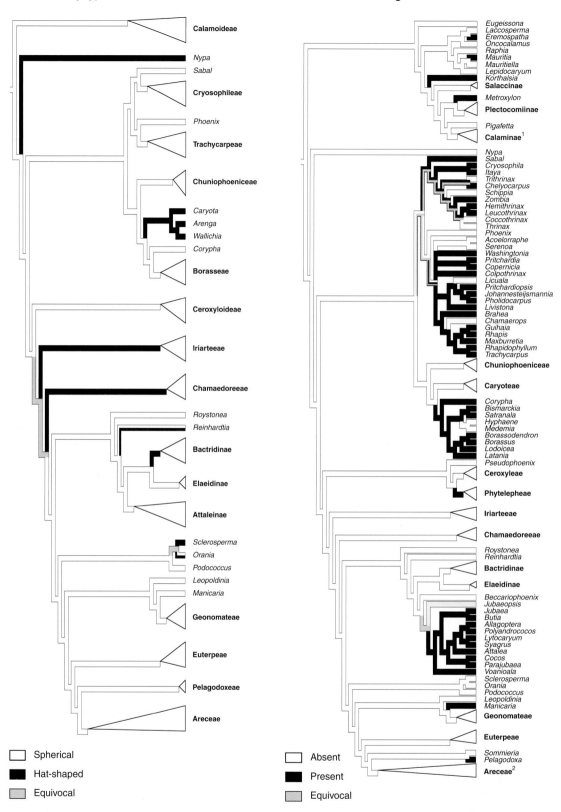

A. Silica body type

B. Association of stegmata with transverse veins

A. Silica body type (left tree terminals)

- Calamoideae
- *Nypa*
- *Sabal*
- Cryosophileae
- *Phoenix*
- Trachycarpeae
- Chuniophoeniceae
- *Caryota*
- *Arenga*
- *Wallichia*
- *Corypha*
- Borasseae
- Ceroxyloideae
- Iriarteeae
- Chamaedoreeae
- *Roystonea*
- *Reinhardtia*
- Bactridinae
- Elaeidinae
- Attaleinae
- *Sclerosperma*
- *Orania*
- *Podococcus*
- *Leopoldinia*
- *Manicaria*
- Geonomateae
- Euterpeae
- Pelagodoxeae
- Areceae

Spherical
Hat-shaped
Equivocal

B. Association of stegmata with transverse veins (right tree terminals)

- *Eugeissona*
- *Laccosperma*
- *Eremospatha*
- *Oncocalamus*
- *Raphia*
- *Mauritia*
- *Mauritiella*
- *Lepidocaryum*
- *Korthalsia*
- Salaccinae
- *Metroxylon*
- Plectocomiinae
- *Pigafetta*
- Calaminae[1]
- *Nypa*
- *Sabal*
- *Cryosophila*
- *Itaya*
- *Trithrinax*
- *Chelyocarpus*
- *Schippia*
- *Zombia*
- *Hemithrinax*
- *Leucothrinax*
- *Coccothrinax*
- *Thrinax*
- *Phoenix*
- *Acoelorraphe*
- *Serenoa*
- *Washingtonia*
- *Pritchardia*
- *Copernicia*
- *Colpothrinax*
- *Licuala*
- *Pritchardiopsis*
- *Johannesteijsmannia*
- *Pholidocarpus*
- *Livistona*
- *Brahea*
- *Chamaerops*
- *Guihaia*
- *Rhapis*
- *Maxburretia*
- *Rhapidophyllum*
- *Trachycarpus*
- Chuniophoeniceae
- Caryoteae
- *Corypha*
- *Bismarckia*
- *Satranala*
- *Hyphaene*
- *Medemia*
- *Borassodendron*
- *Borassus*
- *Lodoicea*
- *Latania*
- *Pseudophoenix*
- Ceroxyleae
- Phytelepheae
- Iriarteeae
- Chamaedoreeae
- *Roystonea*
- *Reinhardtia*
- Bactridinae
- Elaeidinae
- *Beccariophoenix*
- *Jubaeopsis*
- *Jubaea*
- *Butia*
- *Allagoptera*
- *Polyandrococos*
- *Lytocaryum*
- *Syagrus*
- *Attalea*
- *Cocos*
- *Parajubaea*
- *Voanioala*
- *Sclerosperma*
- *Orania*
- *Podococcus*
- *Leopoldinia*
- *Manicaria*
- Geonomateae
- Euterpeae
- *Sommieria*
- *Pelagodoxa*
- Areceae[2]

Absent
Present
Equivocal

Fig. 25 Summary of parsimony reconstructions of ancestral states for 5000 supertrees (weighted MRP analysis, expanded input tree) from Baker et al. (2009) displayed onto strict consensus for (A) silica body type and (B) association of stegmata with transverse veins of lamina. Major clades representing subfamilies and less inclusive ranks with uniform state coding are collapsed to their basal node and indicated by a triangle. Pale grey terminal branches indicate missing data. **A.** Silica body type optimizes onto all trees with 8 steps. Minimum and maximum number of forward and reverse changes: (0→1), min = 5, max = 8; (1→0), min = 0, max = 2. **B.** Association of stegmata with transverse veins of lamina optimizes onto 239 trees with 20 steps, 2067 trees with 21 steps, 2694 trees with 22 steps. Minimum and maximum number of forward and reverse changes: (0→1), min = 11, max = 19; (1→0), min = 3, max = 10. Superscript 1 to right of Calaminae: *Calamus* scored as polymorphic (0, 1). Superscript 2 to right of Areceae denotes independent gain of stegmata associated with transverse veins in *Nephrosperma*.

the complex systematic distribution of root phellogens is presently untenable. Further observations of mature root structure in this subfamily may significantly increase the number of taxa for which root phellogens are known. The phellogen present in the root of *Nypa* is unusual among all palms investigated as it produces few divisions that consist of minute cells (Seubert 1996b). The periderm produced is apparently such a thin layer that it does not significantly disrupt the tissues external to it. This is correlated with the root's permanent submersion.

The pattern of secondary cell wall thickening in the endodermis is among the few anatomical characters previous studies regarded as phylogenetically informative at deep levels of hierarchy within palms (Tomlinson 1961; Seubert 1996a, b, 1997, 1998a, b). Two basic character states exist: 1) unevenly thickened endodermal cell walls that appear U-shaped in TS, with the thickenings restricted to the radial and inner tangential walls (Fig. 38M), and 2) evenly thickened endodermal cell walls (sometimes scarcely so) that appear O-shaped in TS (Fig. 34J). Although the endodermal cell thickenings in many monocots are U-shaped (Kauff et al. 2000), our ancestral state reconstruction indicates that O-shaped thickenings are plesiomorphic for the family, with U-shaped thickening type synapomorphic for the clade of Coryphoideae + Ceroxyloideae + Arecoideae (Fig. 26B). *Eugeissona* and *Nypa* differ from most other palms in having endodermal cell walls that are scarcely thickened (Fig. 33I). In Calamoideae, U-shaped thickenings have independently evolved in *Lepidocaryum* and Calameae. A reversion back to the O-type thickening within Calameae characterizes the clade of *Pigafetta* and Calamineae. Within Coryphoideae, reversions back to O-shaped thickenings occur independently in Cryosophileae and the clade of Borasseae + *Corypha*.

In the first major survey of palm roots, Drabble (1904) defined two major types of cortical fibres—the 'Kentia-type' and 'Raphia-type'—each having a limited distribution within the family. The 'Kentia-type' fibre is distinguished by having relatively thin walls that are angular in TS outline; they are distributed within the cortex as solitary cells or are irregularly grouped in bundles (Fig. 86M). The systematic distribution of 'Kentia-type' fibres is largely confined the to Ceroxyloideae + Arecoideae clade, where they are present in most tribes. The absence of these fibres from certain tribes of Arecoideae, such as Oranieae and Sclerospermeae, has probable systematic significance; further investigation is warranted when a more robust phylogenetic hypothesis of this subfamily becomes available. 'Raphia fibres' characteristically develop thick secondary walls and are grouped in bundles that are circular in outline in TS (Fig. 55I). Stegmata, which are usually absent from palm roots, are always associated with these fibres. The systematic distribution of 'Raphia-type' fibres is wide and scattered, occurring in at least two, and perhaps three, subfamilies (Fig. 27A). Paradoxically, the root cortical fibres of *Raphia*, and indeed most other Calamoideae, may best be interpreted as 'Kentia-type' fibres because they lack stegmata and have relatively wide lumina (Seubert 1996a, 1997); 'Raphia-type' fibres are possibly present in only *Eugeissona* within this subfamily (Fig. 33H). Within Coryphoideae, 'Raphia-type' fibres are of outstanding systematic significance because their presence in the clade of Caryoteae + Borasseae + *Corypha* constitutes the only

structural synapomorphy presently known for this morphologically disparate group. In Arecoideae, these fibres independently evolved in Iriarteeae, *Manicaria*, together with subtribe Bactridinae (Cocoseae), in which they are present in (and possibly synapomorphic for) a group of three genera.

Stem

Stem anatomical features of palms have been little utilized for systematic purposes. As discussed in Chapter 5, this is chiefly because anatomical variation in palm stems must be understood in relation to both sustained primary growth, inherent to the *Rhapis* principle, and variation along a radius at a single level. Given the emerging knowledge of these complexities, the comparative anatomy of palm stems is an area ripe for much further inquiry. Of key interest in this regard is the evolution of habit within Arecaceae.

It is well-known that the scandent habit in palms has multiple origins within the family, and further, that some of these lineages have fundamental differences in stem structure (Baker et al. 2000c; e.g. *Calamus* vs. *Desmoncus*; Tomlinson et al. 2001; Tomlinson and Spangler 2002; Tomlinson and Zimmerman 2003). Less appreciated, however, is that the tall, arborescent palms likely also evolved independently many times within the family. Synthesizing these issues in a phylogenetic context suggests a fundamental contrast in stem evolution between Calamoideae, which contain the large majority of climbing palms, and the clade of Coryphoideae + Ceroxyloideae + Arecoideae (hereafter, the 'crown' clade of palms), which contains most arborescent palms. In keeping with this idea, our data suggests that there are general differences in stem structure between these two lineages, particularly with regard to the extent of sustained primary growth. *Nypa*, whose stem has many features that may be related to its unusual habit is difficult to interpret in this context. Stems of calamoid palms—admittedly best known from the rattan genera—are often highly homogenous, and, importantly, undergo little or no sustained primary growth (Fig. 18G, I). As the mechanical properties of such stems emphasize flexibility over rigidity, it may be unsurprising that many of the calamoid lineages that evolved the scandent growth form have the most pronounced speciation. Much of the flexural rigidity of calamoid rattan stems is, initially, conferred by persistent leaf bases that tightly envelop the stem (Isnard et al. 2005; Isnard and Rowe 2008a, 2008b), but subsequently by their inherent mature stem structure. Although there are a few outstanding examples of tree palms within Calamoideae (e.g. *Mauritia*), how these stems compare mechanically with those of tree palms of the crown clade Arecaceae remains to be known. Tree palms in the Calamoideae otherwise can be either hapaxanthic (e.g. *Metroxylon*, *Raphia*) or relatively short-lived (e.g. *Pigafetta*).

Contrasting with the Calamoideae, the stems of the crown clade show many structural features that enhance their physical and material properties to maximize success at tree building. Stems of this group frequently have many vascular bundles with thick fibre sheaths densely crowded toward the stem periphery, and in large tree palms a diversity of strategies of sustained primary growth are deployed (see Chapter 5; Waterhouse and Quinn 1978; Rich

A. Root periderm

B. Cell wall thickening pattern of endodermis

Absent

Present

Equivocal

O-shaped or unthickened

U-shaped

Equivocal

Fig. 26 Summary of parsimony reconstructions of ancestral states for 5000 supertrees (weighted MRP analysis, expanded input tree) from Baker et al. (2009) displayed onto strict consensus for (A) root periderm and (B) cell wall thickening pattern of endodermis. Major clades representing subfamilies and less inclusive ranks with uniform state coding are collapsed to their basal node and indicated by a triangle. Pale grey terminal branches indicate missing data. **A.** Root periderm optimizes onto all trees with 27 steps. Minimum and maximum number of forward and reverse changes: (0→1), min = 11, max = 14; (1→0), min = 13, max = 16. Superscript 1 to right of Chamaedoreeae denotes independent loss of periderm in *Wendlandiella*. **B.** Cell wall thickening pattern of endodermis optimizes onto all trees with 12 steps. Minimum and maximum number of forward and reverse changes: (0→1), min = 4, max = 5; (1→0), min = 7, max = 8. Superscript 1 to right of Lepidocaryeae denotes independent gain of U-shaped wall thickenings in *Lepidocaryum*.

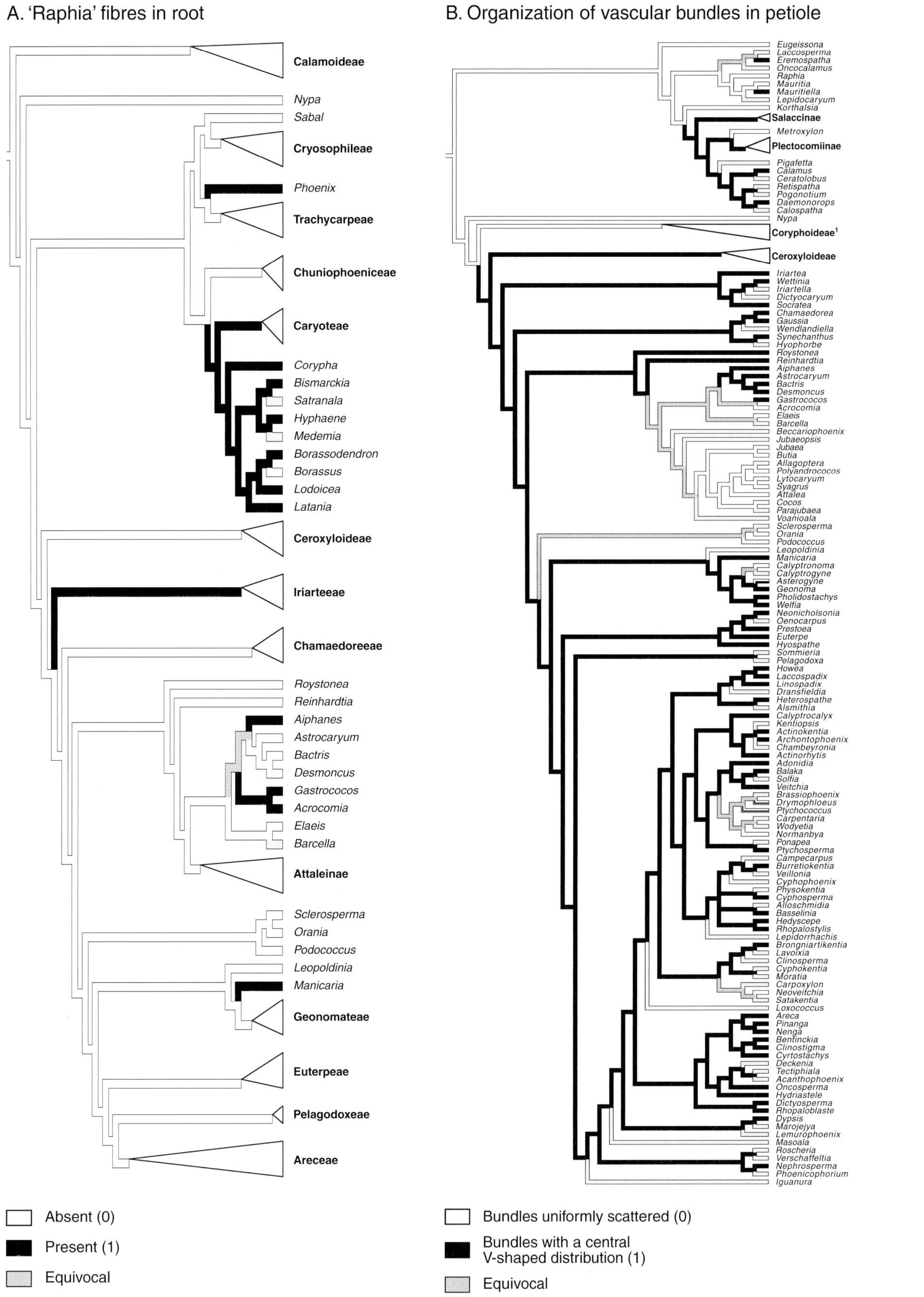

A. 'Raphia' fibres in root

Calamoideae
Nypa
Sabal
Cryosophileae
Phoenix
Trachycarpeae
Chuniophoeniceae
Caryoteae
Corypha
Bismarckia
Satranala
Hyphaene
Medemia
Borassodendron
Borassus
Lodoicea
Latania
Ceroxyloideae
Iriarteeae
Chamaedoreeae
Roystonea
Reinhardtia
Aiphanes
Astrocaryum
Bactris
Desmoncus
Gastrococos
Acrocomia
Elaeis
Barcella
Attaleinae
Sclerosperma
Orania
Podococcus
Leopoldinia
Manicaria
Geonomateae
Euterpeae
Pelagodoxeae
Areceae

B. Organization of vascular bundles in petiole

Eugeissona
Laccosperma
Eremospatha
Oncocalamus
Raphia
Mauritia
Mauritiella
Lepidocaryum
Korthalsia
Salaccinae
Metroxylon
Plectocomiinae
Pigafetta
Calamus
Ceratolobus
Retispatha
Pogonotium
Daemonorops
Calospatha
Nypa
Coryphoideae[1]
Ceroxyloideae
Iriartea
Wettinia
Iriartella
Dictyocaryum
Socratea
Chamaedorea
Gaussia
Wendlandiella
Synechanthus
Hyophorbe
Roystonea
Reinhardtia
Aiphanes
Astrocaryum
Bactris
Desmoncus
Gastrococos
Acrocomia
Elaeis
Barcella
Beccariophoenix
Jubaeopsis
Jubaea
Butia
Allagoptera
Polyandrococos
Lytocaryum
Syagrus
Attalea
Cocos
Parajubaea
Voanioala
Sclerosperma
Orania
Podococcus
Leopoldinia
Manicaria
Calyptronoma
Calyptrogyne
Asterogyne
Geonoma
Pholidostachys
Welfia
Neonicholsonia
Oenocarpus
Prestoea
Euterpe
Hyospathe
Sommieria
Pelagodoxa
Howea
Laccospadix
Linospadix
Dransfieldia
Heterospathe
Alsmithia
Calyptrocalyx
Kentiopsis
Actinokentia
Archontophoenix
Chambeyronia
Actinorhytis
Adonidia
Balaka
Solfia
Veitchia
Brassiophoenix
Drymophloeus
Ptychococcus
Carpentaria
Wodyetia
Normanbya
Ponapea
Ptychosperma
Campecarpus
Burretiokentia
Veillonia
Cyphophoenix
Physokentia
Cyphosperma
Alloschmidia
Basselinia
Hedyscepe
Rhopalostylis
Lepidorrhachis
Brongniartikentia
Lavoixia
Clinosperma
Cyphokentia
Moratia
Carpoxylon
Neoveitchia
Satakentia
Loxococcus
Areca
Pinanga
Nenga
Bentinckia
Clinostigma
Cyrtostachys
Deckenia
Tectiphiala
Acanthophoenix
Oncosperma
Hydriastele
Dictyosperma
Rhopaloblaste
Dypsis
Marojejya
Lemurophoenix
Masoala
Roscheria
Verschaffeltia
Nephrosperma
Phoenicophorium
Iguanura

Absent (0)

Present (1)

Equivocal

Bundles uniformly scattered (0)

Bundles with a central
V-shaped distribution (1)

Equivocal

Fig. 27 Summary of parsimony reconstructions of ancestral states for 5000 supertrees (weighted MRP analysis, expanded input tree) from Baker et al. (2009) displayed onto strict consensus for (A) 'Raphia-type' fibres on root and (B) organization of vascular bundles in petiole. Major clades representing subfamilies and less inclusive ranks with uniform state coding are collapsed to their basal node and indicated by a triangle. Pale grey terminal branches indicate missing data. **A.** 'Raphia-type' fibres in root optimizes onto all trees with 7 steps. Minimum and maximum number of forward and reverse changes: (0→1), min = 5, max = 7; (1→0), min = 1, max = 2. **B.** Organization of vascular bundles in petiole optimizes onto all trees with 17 steps. Minimum and maximum number of forward and reverse changes: (0→1), min = 5, max = 9; (1→0), min = 7, max = 12. Superscript 1 to right of Coryphoideae denotes independent gain of central V-shaped distribution of vascular bundles in *Rhapis*.

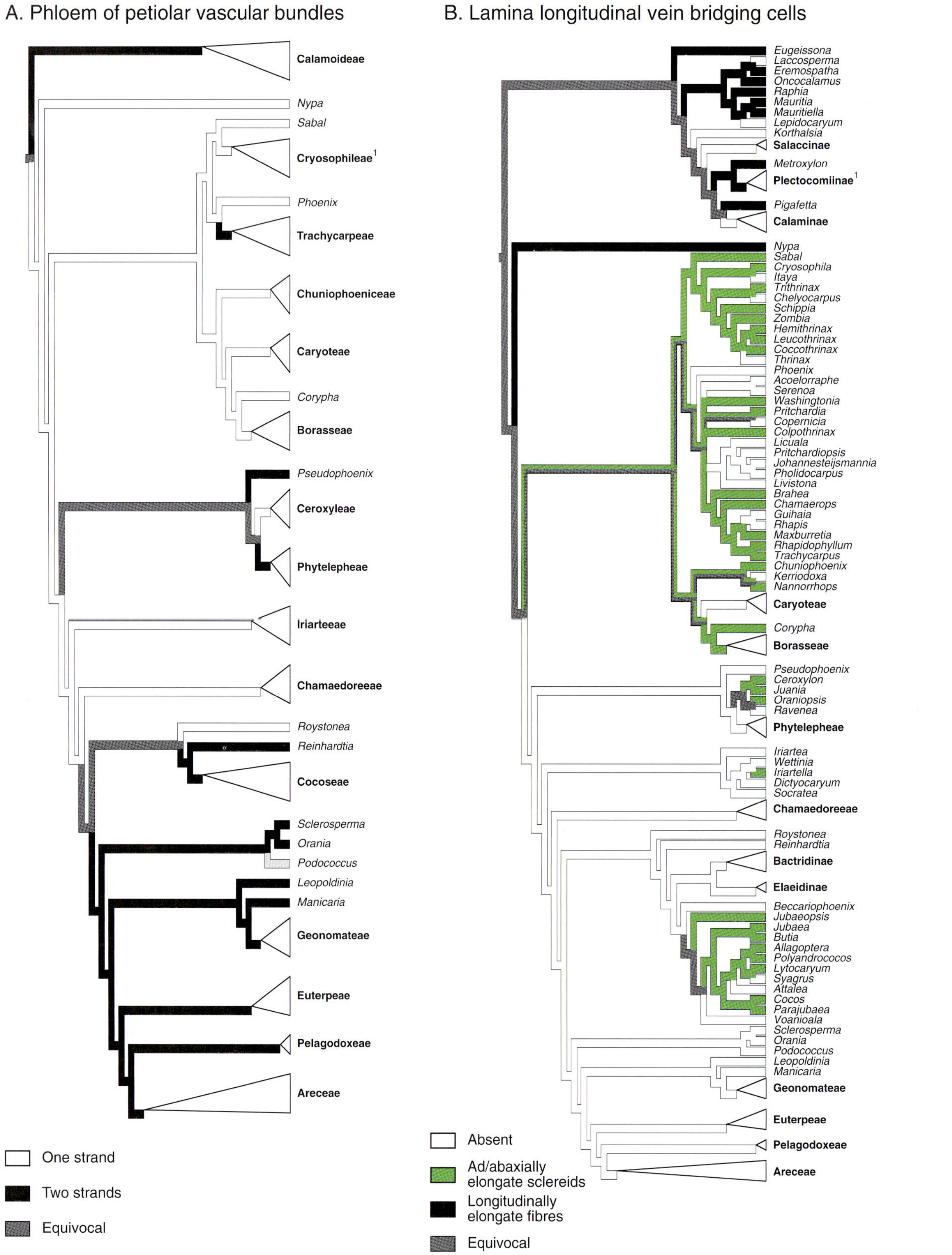

Fig. 28 Summary of parsimony reconstructions of ancestral states for 5000 supertrees (weighted MRP analysis, expanded input tree) from Baker et al. (2009) displayed onto strict consensus for (A) number of phloem strands in petiolar vascular bundles and (B) type of bridging cells of longitudinal veins of lamina. Major clades representing subfamilies and less inclusive ranks with uniform state coding are collapsed to their basal node and indicated by a triangle. Pale grey terminal branches indicate missing data. **A**. Phloem of petiolar vbs optimizes onto all trees with 7 steps. Minimum and maximum number of forward and reverse changes: (0→1), min = 4, max = 7; (1→0), min = 0, max = 3. Superscript 1 to right of Cryosophileae denotes independent gain of two phl strands in *Itaya*. **B**. Lamina longitudinal vein bridging cell type optimizes onto 675 trees with 28 steps and 4325 trees with 29 steps. Minimum and maximum number of forward and reverse changes: (0→1), min = 6, max = 14; (1→0), min = 7, max = 15; (0→2), min = 1, max = 5; (2→0), min = 2, max = 7; (1→2), min = 0, max = 0; (2→1), min = 0 max = 1. Superscript 1 to right of Plectocomiinae: *Plectocomiopsis* scored as polymorphic (0, 2). To review, superscripts must be added to Figs. 25, 26, and 28.

1987a, b; Tomlinson 1990). Fundamental to this is the capability for extreme cell longevity in the stems of many of the largest palms of this clade. As palm stems are without secondary thickening, the cells of palm stems must remain alive and metabolically active throughout the life of the plant. Probably unlike the arborescent calamoids, arborescent palms of the crown clade are known to reach ages of at least 200 years (e.g. *Roystonea regia*, Place des Palmiers, French Guiana; Tomlinson 2006), and many plants in cultivation throughout the tropics are more than a century old.

Comparative stem structure remains one of the final frontiers in palm biology. Our data on stem structure, which represent a significant increase in taxonomic coverage over that presented by Tomlinson (1961), require further synthesis. The described hypothesis of stem evolution in palms is indefinite in many respects, but in presenting it as such we wish to convey that the comparative anatomy of palms is far from being a closed book, and future work in this area will be critical to answering many basic questions in palm biology.

Petiole

Because much of the variation present in palm petioles corresponds with leaf size, there are few characters of systematic significance. Petiolar vascular organization is noteworthy among these characters, and our observations reveal that two major configurations exist within the family. Petiolar vascular bundles with a uniformly scattered distribution is the ancestral state at the root node of the family, and characterize many Calamoideae, *Nypa*, and almost all Coryphoideae (Figs. 12E; 27B; and 44A). The contrasting character state, in which petiolar vascular bundles have a distinct V-shaped distribution (Figs. 12A, C, D), is independently derived in the clade of Ceroxyloideae + Arecoideae and within Calaminae, but a few other relatively minor origins of the state are apparent as well (Fig. 27B). Existing character data for Arecoideae indicate that significant reversals back to a uniformly scattered state characterize Cocoseae, in part, and, perhaps a group of genera within Ptychospermatinae (Areceae), but missing data and potential phylogenetic uncertainty precludes a more definitive assessment at this time. Augmenting the above evolutionary perspective is the observation that differences in petiole vascular architecture are not correlated with leaf morphology (i.e. pinnate vs. palmate), which suggests that potentially differing leaf developmental programmes can produce similar appearing mature structure.

The only aspect of petiole anatomy to have gained currency in systematic use is the number of phloem strands present in the vascular bundles. As suggested by Parthasarathy (1968), petiolar vascular bundles in palms are divisible into two types based on this characteristic: 1) vascular bundles with a single, undivided strand (Fig. 12I, N); and 2) vascular bundles appearing to have two strands on account of an even division of the phloem by a median sclerotic partition (Fig. 12J–L). The range of diversity somewhat transcends this categorization, because in some taxa interpreted as having a single phloem strand, the tissue includes an irregular and complex network of sclerotic cells (e.g. *Nannorrhops*, *Phoenix*, *Roystonea*). Evolutionary interpretation of this character is partly clear. Although reconstructions of this

character are equivocal for the root node of the family, all other nodes along the subfamilial backbone of the phylogeny optimize with one phloem strand as the ancestral state (Fig. 28A). Within the clade of palms exclusive of Calamoideae, petiole vascular bundles with two phloem strands independently evolved in Trachycarpeae (Coryphoideae), Ceroxyloideae, and Arecoideae. Optimizations within the latter two subfamilies are equivocal. If the basic phylogenetic topology of Arecoideae used here is upheld by further research, the unusual petiolar phloem strands of *Roystonea* deserve further study to assess their homology.

Lamina

The leaf lamina provides the largest number of systematically important anatomical characters within Arecaceae. These characters are highly homoplasious across the family (Horn et al. 2009), but are often informative of phylogenetic relationships within each of the subfamilial clades. Although leaf anatomical character data largely do not corroborate the backbone relationship among the five subfamilies, the distribution of different types of specialized cells that serve as bundle sheath extensions is of evolutionary interest at this level of hierarchy (Fig. 28B). In Calamoideae and Nypoideae, longitudinal veins connect to the adaxial surface layers by a palisade of ad/abaxially elongate sclereids (Fig. 10F–H). In contrast, the clade of Coryphoideae, Ceroxyloideae, and Arecoideae possess a different type of vein bridging (when present) by longitudinally elongate fibres (Fig. 10D, E, I–L; Horn et al. 2009). Thus, the evolutionary history of this character is indicative of a novel shift in the anatomical design of the lamina within the family.

Character state optimizations at each subfamilial node are more definite, enabling the identification of synapomorphies. Horn et al. (2009) indicate (based on parsimony optimizations) that the presence of adaxial subepidermal fibres (Fig. 33A) and transverse veins sheathed by many layers of fibres (Fig. 11A) are synapomorphies for Calamoideae. The presence of large, inflated parenchyma cells adjacent to the protoxylem of the petiolar vascular bundles is unique to this subfamily (Fig. 32A–J). *Nypa* is characterized by a large number of autapomorphic states, which we detail in the Nypoideae chapter. Synapomorphies for Coryphoideae are: adaxial ribs including at least five, independent vascular bundles (Fig. 51C); transverse veins sheathed by many layers of fibres (Fig. 60A); the absence of non-vascular fibres free of the surface layers (Fig. 55A); and adaxial nonvascular fibres septate and with wide lumina (Fig. 52D, E). Ceroxyloideae are without any unambiguous synapomorphies in leaf anatomical character reconstructions presented by Horn et al. (2009); however, adaxial subepidermal fibres (Figs. 68A and 69A) are synapomorphic for this clade if data from Trénel et al. (2007) are adopted (which resolves *Ravenea* sister to other Ceroxyleae). Given current estimates of palm phylogeny, lack of resolution along the backbone of Arecoideae is the most significant factor contributing to the difficulty of discovering synapomorphies for this clade.

The richness of the systematic potential leaf lamina characters is best demonstrated within the subfamilies. We refer the reader to the subfamilial accounts in Part 2 for more complete examples and further discussion.

SYSTEMATIC ANATOMY OF PALMS (Part 2)

SUBFAMILY CALAMOIDEAE

INTRODUCTION

An early diverging group of palms recognized by the scale-bearing fruits (hence the earlier name lepidocaryoid palms) and frequent spiny armature (Fig. 29). They vary from tall trees to shrubby forms and include the great majority of climbing palms, notably the economically important rattans. They are most abundant in the Asian tropics, except for three palmate leaved genera constituting the subtribe Mauritiinae in South America, together with *Raphia taedigera* in Central America, the sole New World representative of an otherwise African/Madagascan genus. African calamoids are also represented by the three climbing genera of the subtribe Ancistrophyllinae and one species of *Calamus*. *Calamus* itself is the largest genus of palms (c. 350 spp.) and includes most commercial rattans.

Leaves are pinnately reduplicate except for the palmately reduplicate Mauritiinae, a distinctive condition otherwise found only in the anomalous *Guihaia* of the Coryphoideae. Leaf morphology in the climbing taxa can be specialized in the production of a terminal extension of the rachis (= cirrus) armed with either recurved distal leaflet spines (Ancistrophyllinae) or with cat's claw emergence spines as in many Calameae. Otherwise the climbing organ in most species of the subtribe Calaminae, including most *Calamus* spp. is a modified unbranched inflorescence (flagellum) also with cat's claw armature. The basic flower unit in the subfamily is a pair of flowers (dyad), which is also interpreted as ancestral. Modification within an archetypical pair of perfect flowers involves the establishment of unisexual flowers and some combination of flower types, although the two types within an individual do not differ much in size. Dioecy is common in the tribe Calameae, and clearly a derived condition. Complementary understanding of pollination mechanisms is lacking.

The group is ecologically diverse, usually within forest communities, but with a preference for wet habits. Rattans can be viewed as weeds, as by foresters, because they favour disturbed sites and establish best in tree-fall gaps. Although molecular evidence suggests that the Calamoideae is sister to all other palms, the extreme morphological diversity of many of its members illustrates their success in advancing well beyond any putative ancestral palm (Fisher and Dransfield 1977).

Long recognized as a natural group this subfamily shows much less diversity in anatomical features compared with the other large palm subfamilies. This relatively diminished diversity

is reflected in a constant suite of shared anatomical characters for the subfamily whereas subgroups are not usually clearly defined anatomically except for stem characters in climbing members, which clearly reflect features adaptive to the distinctive habit. The relationships among the taxa within the Calamoideae as based on molecular evidence shows three major clades (Fig. 30A), each of which corresponds to a tribe in a hierarchical system. *Eugeissonia* is monotypic and can be regarded as early diverging and thus sister to all other Calamoideae. The two remaining tribes are subdivided into 10 subtribes of some complexity. The largest is the Calaminae, which represents most extensively the climbing habit in palms.

Anatomical features of the subfamily Calamoideae

Leaf

LAMINA

1) Dorsiventral, the adaxial surface usually with few stomata or trichomes and **very uniform**.
2) Anticlinal walls of epidermal cells distinctly sinuous in SV; frequently with the sinuous effect exhibited in the cuticle which penetrates the anticlinal cell walls.
3) Epidermal cells **rectangular** in SV.
4) **Trichomes** (when present) commonly with thin-walled filamentous extension from a basal group of thick-walled pitted cells (Fig. 31B).
5) **Guard cells** frequently with the cell lumen constricted by wall thickenings except at the thin-walled poles (Fig. 31E).
6) **Fibres** always present in the hypodermal layers.
7) **Longitudinal veins** largely independent of the surface layers and situated in the mid- or abaxial mesophyll, at most few larger veins connected to the surface layers by sclerotic, often anticlinally-extended cells.
8) **Transverse veins** always running above the small longitudinal veins and with a well-developed fibrous sheath, the fibres sometimes extending freely into the mesophyll.

PETIOLE

1) **Protoxylem parenchyma** of vbs often extensively developed as a tissue of wide, empty parenchyma cells (Fig. 31D).
2) **Surface layers** including hypodermal fibres (cf. lamina).

Fig. 29 Calamoideae. **A**. Summary phylogeny of Arecaceae showing position of Calamoideae; with anatomical synapomorphies for subfamily.
B. *Eugeissona utilis*, Petiole vb, TS (Scale bar = 1 mm). **C**. *Raphia farinifera*, scaly fruits. **D**. *Pigafetta elata*, petioles and leaf sheaths. **E**. *Metroxylon sagu*, flowering plant viewed from above. **F**. *Raphia farinifera*, habit.

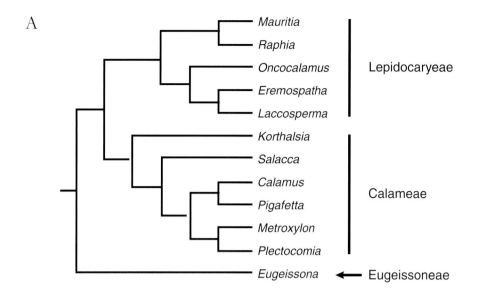

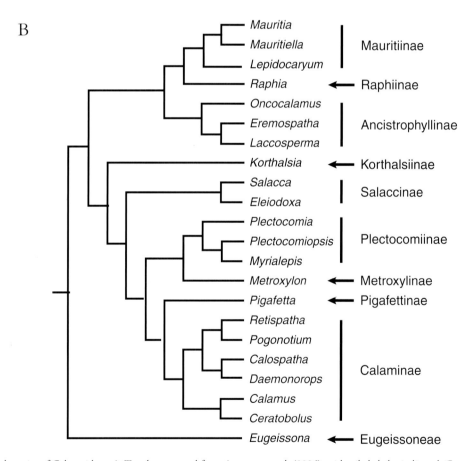

Fig. 30 Summary phylogenies of Calamoideae. **A**. Topology pruned from Asmussen et al. (2006), with tribal clades indicated. Generic sampling is incomplete, but all subtribes are represented by at least one exemplar. **B**. Topology pruned from the 'most congruent supertree' of a weighted matrix representation with parsimony (MRP) analysis from Baker et al. (2009), with subtribes indicated for Lepidocaryeae and Calameae. All constituent genera of Calamoideae are represented.

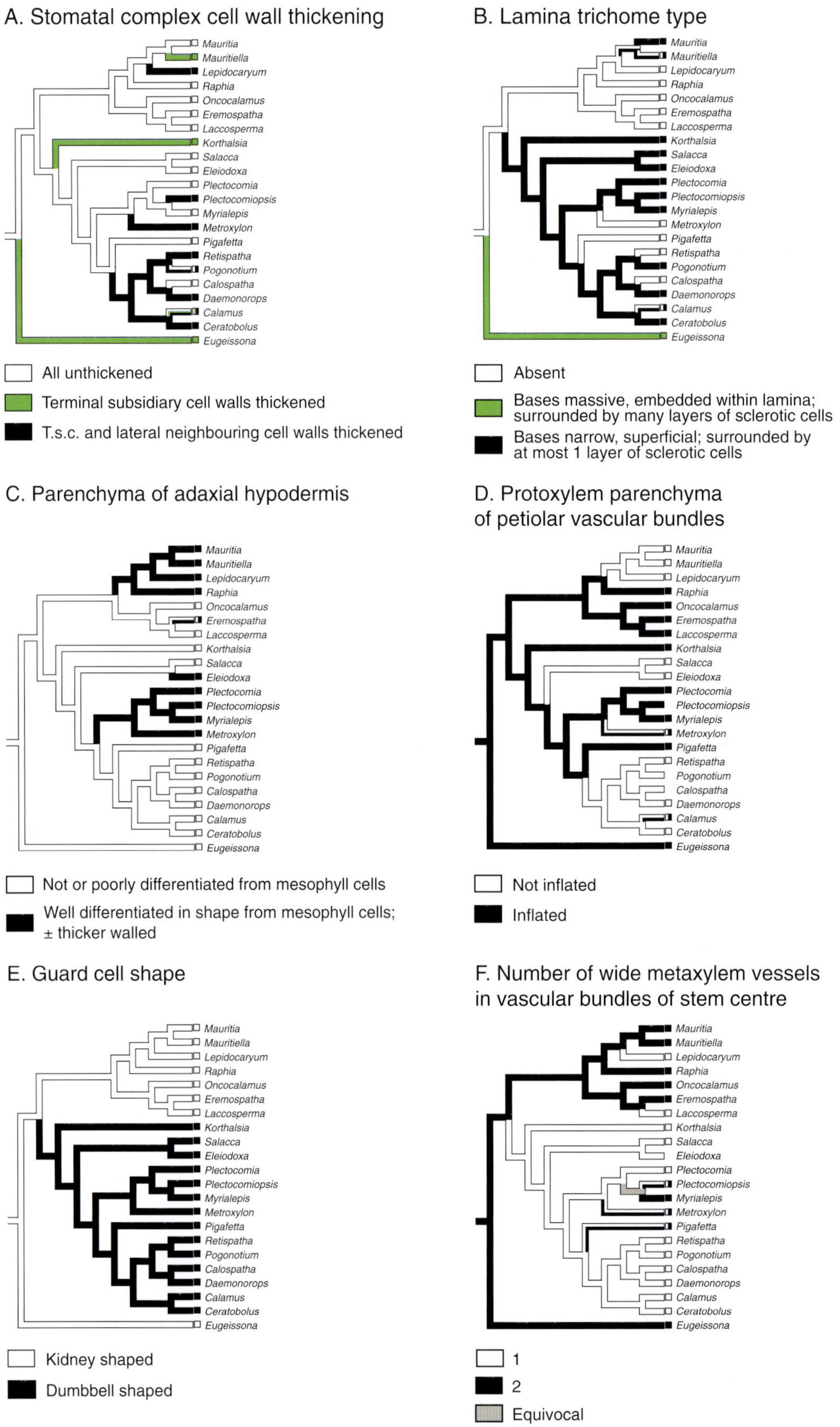

Fig. 31 Parsimony ancestral state reconstructions of selected anatomical characters within Calamoideae. All reconstructions shown are optimized on the tree indicated in Fig. 30B. Terminals without a coloured box adjacent to the branch tip have missing data. Number of steps reported for each optimization is the value within the subfamily. **A**. Stomatal complex cell wall thickening optimizes with 11 steps. **B**. Lamina trichome type optimizes with 9 steps. **C**. Differentiation of adaxial hypodermal parenchyma optimizes with 4 steps. **D**. Protoxylem parenchyma of petiolar vascular bundles optimizes with 5 steps. **E**. Guard cell shape optimizes with 1 step. **F**. Number of wide metaxylem vessels in the vbs of the stem centre optimizes with 7 steps.

3) **Vascular bundles** not usually with a conspicuous V-configuration in TS; each including a single wide mxy vessel and two phl strands.

4) **Peripheral vascular bundles** with well-developed fibre sheaths, but not confluent to form a continuous mechanical layer.

Stem

Central ground tissue sometimes becoming lacunose (e.g. *Metroxylon*) but not undergoing extensive late modification by cell expansion and division.

Root

1) **Cortex** rarely with fibre strands (*Raphia*).
2) **Endodermis** often with uniformly thickened walls, the lumen O-shaped in TS.

Vascular tissue

1) **Metaxylem vessels** in root and stem usually with simple perforation plates; in petiole with scalariform (rarely reticulate) perforation plates.
2) **Sieve tubes** in root with usually long, oblique compound sieve plates each with many small sieve areas; in stem and leaf with simple sieve plates on transverse or slightly oblique end walls. This arrangement is distinctly different from all other palms.

Note: whether these distinctive features of the Calamoideae could be seen as ancestral for all palms is doubtful, even though they are the sister group to all other palms. But they do point the way to features that might be looked for in palm fossils, especially the oldest.

This information on anatomical features is present as a preliminary guide to the later detailed systematic descriptions, but they also lead into a more comprehensive analysis of where the anatomical data can be used to support an estimated phylogeny.

Phylogeny and classification based on molecular evidence

Calamoideae, sister to all other Arecaceae, are the second most species-rich subfamily of palms, containing about 615 spp. within 21 genera (Dransfield et al. 2008b). The subfamily is distributed throughout the wet tropics, and includes the great majority of climbing palms, most notable of which are the economically important rattans. Also well represented within the group are self-supporting palms, which range in physiognomy from tall trees to shrub-like or acaulescent growth forms (Baker et al. 2000a; Dransfield et al. 2008b). Although Calamoideae vary substantially in habit and life history strategy (10 genera include hapaxanthic palms, Baker 2000a; Fig. 29E), the circumscription of the subfamily has remained stable since the earliest comprehensive classification of palms by Martius (1823–1850), because it possesses a highly diagnostic suite of morphological traits (Baker et al. 2000a; Dransfield et al. 2008b). The retrorse,

overlapping scales that cover the gynoecia and fruits of all taxa, are unique to the subfamily and constitute its most obvious defining feature (Fig. 29C). Other diagnostic character states that are characteristically present include spinescent foliar organs (spines occur variously on the sheath, leaf axis, pinna midrib, or pinna margin; organized into whorls in Calameae; Fig. 29D), a basic inflorescence unit of two flowers (dyad), and epitropous ovules. In concordance with the longstanding view that Calamoideae are an undoubtedly 'natural' lineage of palms, phylogenetic studies based on both morphological and molecular data uniformly resolve the subfamily as monophyletic with strong statistical support.

Comparative anatomical data are consistent with the above evidence. Calamoideae are the most anatomically cohesive subfamily of palms (setting aside the monotypic Nypoideae), particularly with respect to leaf and root anatomy (Tomlinson 1961; Seubert 1996a; Horn et al. 2009). The extent of leaf and root anatomical variation in Calamoideae is less than that present in other palm subfamilies, and a suite of character states, which are largely shared by all genera, distinguish the subfamily. From lamina anatomy, these are: rectangular epidermal cells with sinuous anticlinal walls (Fig. 7J, K); an adaxial series of non-vascular fbs in contact with the epidermis (Fig. 37D); orthogonally-oriented transverse veins situated adaxial to smaller longitudinal veins and sheathed by many layers of fibres (Figs. 11G and 35A, E); and, frequently, vertical sclereids that form a connecting girder between each of the larger longitudinal veins and the adaxial surface layers (Figs. 36D; 37A; and 38C; Tomlinson 1961; Horn et al. 2009). Of these character states, Horn et al. (2009) identified adaxial non-vascular fibres with a subepidermal distribution and transverse veins sheathed by many layers of fibres as synapomorphies for the subfamily (Fig. 29A). An additional leaf anatomical synapomorphy for the subfamily, albeit independently lost from three subtribal clades, is the occurrence of inflated parenchyma cells adjacent to the protoxylem within each petiolar vascular bundle (Figs. 31D and 32). This unusual feature is restricted to Calamoideae within palms, and is not known to be present in any putative sister lineages of monocots. The phloem of Calamoideae is also distinctive as the sieve tube elements of the stems and leaves almost always have simple sieve plates (Fig. 22G, H), unlike other palms (except *Nypa*), in which these organs characteristically have sieve tube elements with compound sieve plates.

Within Calamoideae, Baker et al. (2000a, b) established a basic phylogenetic hypothesis which is sustained by more recent analyses of global palm phylogeny (Asmussen et al. 2006; Baker et al. 2009, Supertree). These studies support the recognition of three tribal clades within the subfamily: 1) a monogeneric Eugeissoneae, sister to all other Calamoideae; 2) Lepidocaryeae, containing three subtribes; and 3) Calameae, containing six subtribes (Dransfield et al. 2008b; Fig. 30A, B). Competing hypotheses of deep phylogeny within Calamoideae exist. However, the above topology is the most frequently recovered in molecular phylogenetic studies employing multiple loci, and receives the strongest statistical support. Alternative topologies differ most importantly from the above in placing *Eugeissona* as variously sister to or nested within Lepidocaryeae, and, in one case, placing

Fig. 32 Petiole vbs of Calamoideae, with inflated pxy parenchyma, all TS. **A**. *Eugeissona insignis* (bar = 1 mm). **B**. *Oncocalamus* sp. (bar = 200 μm).
C. *Laccosperma* sp. (bar = 200 μm). **D**. *Raphia taedigera* (bar = 200 μm). **E**. *Korthalsia laciniosa* (bar = 200 μm). **F**. *Eleiodoxa conferta* (bar = 250 μm).
G. *Salacca affinis* (bar = 200 μm). **H**. *Metroxylon salomonense* (bar = 200 μm). **I**. *Myrialepis paradoxa* (bar = 200 μm). **J**. *Plectocomia muelleri* (bar = 200 μm).

Ancistrophyllinae sister to all other Calamoideae (Baker et al. 2000b, ITS; Baker et al. 2009, Supermatrix).

In our analyses of anatomical character evolution within Calamoideae, we optimized the total set of characters on each of the two topologies presented in Baker et al. (2009). To be judicious in our evolutionary interpretations, we present and discuss reconstructions of six of these characters (Fig. 31) that yield largely congruent optimizations on the two alternative topologies. Reflecting the results of previous morphological character analyses of Calamoideae (Baker et al. 2000a; Dransfield et al. 2008b), our analyses did not reveal any synapomorphies for Lepidocaryeae or the clade of Calameae + Lepidocaryeae. Therefore, we restrict the following discussion of our results to each tribal clade.

Eugeissoneae

Despite its key phylogenetic position as one of the earliest diverging lineages of palms, *Eugeissona* provides no important clues regarding ancestral anatomical states in palms. All fundamental aspects of its anatomy are like those of most other Calamoideae. Nevertheless, *Eugeissona* is one of the most anatomically distinctive genera of the subfamily, although the four autapomorphies that distinguish it seem to be rather slight features. Of this set of character states, stomata with sclerotic terminal subsidiary cells (Figs. 31A and 33D, E) and trichomes with massive bases included within the lamina mesophyll (Figs. 31B and 33F) have a wide distribution within palms, but evolved independently in *Eugeissona*. The two other autapomorphies are unique to the genus. First, the silica bodies of *Eugeissona* are minute and disciform, and so contrast with those of all other Calamoideae, which have spherical silica bodies. Among all palms, the silica bodies of *Nypa* most closely approximate those of *Eugeissona* in their unusually small size, but nevertheless differ in being distinctly hat-shaped. Second, *Eugeissona* is one of the few genera of palms to have fibre sclereids included within its lamina mesophyll (Fig. 33A; Tomlinson 1959b). These sclereids, present in most *Eugeissona* spp., are unlike those of other genera because they are short, usually vertically oriented within the leaf, and encrusted with stegmata.

Eugeissona is probably unique within the subfamily also in having wide, circular (in TS outline) bundles of thick-walled fibres present in the root cortex (Fig. 33H, I). It is unknown whether stegmata accompany these fibres; but if so, *Eugeissona* would, paradoxically, be the only genus of Calamoideae to possess fibres of the 'Raphia-type' (Seubert 1996a).

Lepidocaryeae

Lepidocaryeae contain three morphologically disparate subtribal lineages: Ancistrophyllinae, Mauritiinae, and Raphiinae (Fig. 30A, B). Ancistrophyllinae, consisting of three genera of African rattans, is resolved as sister to the other two tribes in many studies. The clade of Mauritiinae + Raphiinae contrasts with the latter subtribe in habit: it consists of solitary or clustering tree palms that may attain massive proportions (e.g. *Raphia*; Fig. 29F). Even this less inclusive clade has no obvious morphological synapomorphies, and several major morphological

characters sharply differentiate the two constituent subtribes. Most outstanding of these is leaf lamina shape—the three genera of Mauritiinae are unique within Calamoideae in having palmate leaves—but the two groups also strongly differ in flowering behaviour (*Raphia* is hapaxanthic) and inflorescence morphology. Comparative anatomical data complement existing morphological information because they corroborate hypotheses of relationship within Lepidocaryeae that were previously without supporting structural data.

The scandent habit in Ancistrophyllinae is noteworthy because its evolutionary origin is unequivocally independent from that of the Asian rattans (genera of Calameae; Baker et al. 2000a). Anatomical data have not been developed that might definitively address convergent evolution in the mechanics or hydraulics of the stems of Ancistrophyllinae, but one unique synapomorphy of the tribe would repay further investigation. The group is well characterized by a continuous hypodermal layer of stegmata in their stems (Fig. 34G). This character state can be viewed as potentially analogous to the silicified outer periclinal walls of the stem epidermis of the Asian rattans because such peripherally positioned silica deposits could contribute to stem flexural rigidity. However, the mechanical significance of silica deposits in stems of both groups remains to be investigated.

Despite the dissimilarity in leaf morphology between Mauritiinae and *Raphia*, leaf anatomical data do corroborate a hypothetical relationship between the two groups, and two character states optimize as synapomorphies for the clade: 1) the parenchyma of the adaxial hypodermis of all genera is well differentiated from the adjacent mesophyll parenchyma (Figs. 31C; 35A; and 36D), and 2) the pinna or segment margins include a stout marginal vein. Another similarity in the leaf anatomy of these four genera is that expansion tissue is only associated with ribs. All other genera of Calamoideae, except *Eugeissona* and *Pigafetta*, develop 'supernumerary' bands of expansion tissue in their pinnae that show no clear relationship to vasculature (e.g. Figs. 34A and 37C).

Calameae

This tribe contains 13 genera accommodated within six subtribes, and has a distribution almost wholly restricted to tropical Asia and the Pacific (only the West African *Calamus deerratus* occurs outside this range). The best available molecular phylogenetic evidence suggests the relationships among the subtribes indicated in Fig. 30A, B, although statistical support for much of this topology is not strong. From an evolutionary perspective, Calameae are among the most remarkable clades within Arecaceae because of the extreme lability in habit within the group and the strikingly unequal species richness among the constituent subtribal clades. For example, the rattan lineage Calamineae includes c. 80% of the species within the subfamily.

The monophyly of Calameae is strongly corroborated by comparative anatomical evidence, and four character states optimize as synapomorphies for the group. Most distinctive among these are stomata with dumbbell-shaped guard cells, which are present in all members of the clade and almost unique to the group within palms (Figs. 31E and 39C; only *Schippia*, in Coryphoideae, has similarly-shaped guard cells). Superficial trichomes with narrow bases are present on the leaves of most taxa, and although they are otherwise diverse in shape within the tribe, just *Mauritia* and *Mauritiella* within Calamoideae outside of Calameae possess comparable trichomes (Figs. 31B; 37E; 38E, G; and 40C, E). The vascular bundles of the stem centre have only a single, wide metaxylem vessel in most taxa, contrasting with most other Arecaceae outside of Arecoideae, which typically have two or more wide vessels in such bundles (Figs. 31F and 41D). Also optimizing as a synapomorphy for Calameae are endodermal cells that have a U-shaped wall thickening pattern (Figs. 26B; 38M; and 39K). The expression of this type of endodermis within Calameae shows distinct differences from its manifestation elsewhere in palms, because the inner periclinal wall is usually so strongly thickened that it occupies nearly the entire volume of the cell. A reversion back to an endodermal type in which all walls are equally thickened characterizes the clade of Pigafettinae + Calamineae.

SYNAPOMORPHIES FOR EACH OF SIX SUBTRIBAL CLADES OF CALAMEAE

- **Korthalsiinae:** 1) terminal subsidiary cells of stomata sclerotic; 2) peripheral cap of 'yellow cells' present adjacent to peripheral-most stem vascular bundles; 3) phloem of vascular bundles of stem centre with median partition; and 4) lamina with complete abaxial layer of epicuticular wax.
- **Salaccinae:** decrease in size and abundance of parenchyma associated with protoxylem of petiolar vascular bundles (i.e. no inflated protoxylem parenchyma present).
- **Metroxylinae:** 1) stomata with terminal subsidiary cells and lateral neighbouring cells sclerotic; and 2) terminal subsidiary cells unlobed.
- **Plectocomiinae:** peripheral cap of 'yellow cells' present adjacent to peripheral-most stem vascular bundles.
- **Pigafettinae:** expansion tissue of each pinna associated only with midrib.
- **Calaminae:** 1) decrease in size and abundance of parenchyma associated with protoxylem of petiolar vascular bundles; 2) stomata with terminal subsidiary cells and lateral neighbouring cells sclerotic; and 3) phloem of vascular bundles of stem centre with a median partition.

Note: Baker et al. (2000c) indicate the probable necessity of major generic recircumscriptions within subtribe Calamineae. As phylogenetic estimates of this group become more robust, the application of systematic anatomical data, particularly from leaf lamina, holds great promise for characterizing major clades of this subtribe.

TRIBE EUGEISSONEAE

This tribe includes only the genus *Eugeissona* of distinctive morphology and of systematic interest because molecular evidence places it as sister to all other Calamoideae. It includes 6 spp.

in the Malay Peninsula and Borneo (Dransfield 1970). Plants are extremely spiny, multiple-stemmed, low-growing or even acaulescent, but with disproportionately large leaves and large terminal inflorescences, (i.e. hapaxanthic). Branching is non-axillary in some spp. (e.g. *E. insignis, E. tristis*), and approaches a true dichotomy (Fisher et al. 1989). Perhaps *E. minor* is most distinctive because the axes, with the erect foliage leaves, are elevated to a height of 2–3 m on massive stilt roots up to 2.5 cm in diameter (Holbrook et al. 1985). Clonal separation of branch axes can produce vegetative spread. This unusual, essentially rhizomatous, habit recalls that of a few gingers. How the palm, which must germinate at ground level, achieves this aerial condition by some modified establishment growth, is unknown. The reported manufacture of 'fine walking sticks' from roots of this palm is an unusual application.

Note: as discussed in Dransfield et al. (2008b) the genus has a combination of specialized features within the Calamoideae, i.e. large flowers, numerous stamens, an endocarp, and method of germination. The distinctive dyad, each with a staminate and a hermaphrodite flower, is otherwise found within the subfamily only in *Metroxylon*.

Eugeissona (Fig. 33)
Leaf
LAMINA

Dorsiventral (Fig. 33A, B). **Trichomes** scattered on each surface, and with an elliptical deeply sunken base of sclerotic cells (Fig. 33F). **Epidermis:** cells with sinuous walls (Fig. 33C, E) usually thin-walled, but outer wall of epidermis thickened and wholly cutinized in *E. insignis*. **Stomata** with small g.c., often in linear series alternating with cells below hypodermal fibres; t.s.c. thick-walled (Fig. 33D, F). **Hypodermis** either absent or represented by a discontinuous irregular series of colourless cells alternating with fibres. **Mesophyll** with or without a ± distinct palisade layer. **Fibres** (Fig. 33A, B) solitary or in groups of 2–3 (but up to 15 in *E. insignis*) adjacent to each hypodermis, fewer abaxially, rarely free in abaxial mesophyll. Lignified **sclereids** (Fig. 33A) abundant in mesophyll of *E. minor* and *E. utilis* (few or absent from *E. insignis*) as short cylindrical cells throughout mesophyll, often extending vertically. **Veins** in abaxial mesophyll; only largest attached to each surface by sclerotic mesophyll cells. O.S. complete around smallest veins, the cells longitudinally extended and with sl. thickened wall; I.S. sclerotic and complete fibrous around larger veins. Phl of larger veins abundantly sclerotic. **Transverse veins** (Fig. 33G) frequent, regularly orthogonal, often extending above small longitudinal veins, with a well-developed fibrous sheath. **Stegmata** next to hypodermal fibres extremely small, easily overlooked.

Note: *Eugeissona* spp. differ from each other significantly in lamina anatomy making it difficult to characterize the genus. The following differences may be emphasized:

1. Lamina (Fig. 33A), ~7 cells deep, fibre-sclereids abundant in mesophyll, variously oriented, palisade absent, hypodermal fibre strands usually of 1–3 cells.... *E. minor, E. tristis*

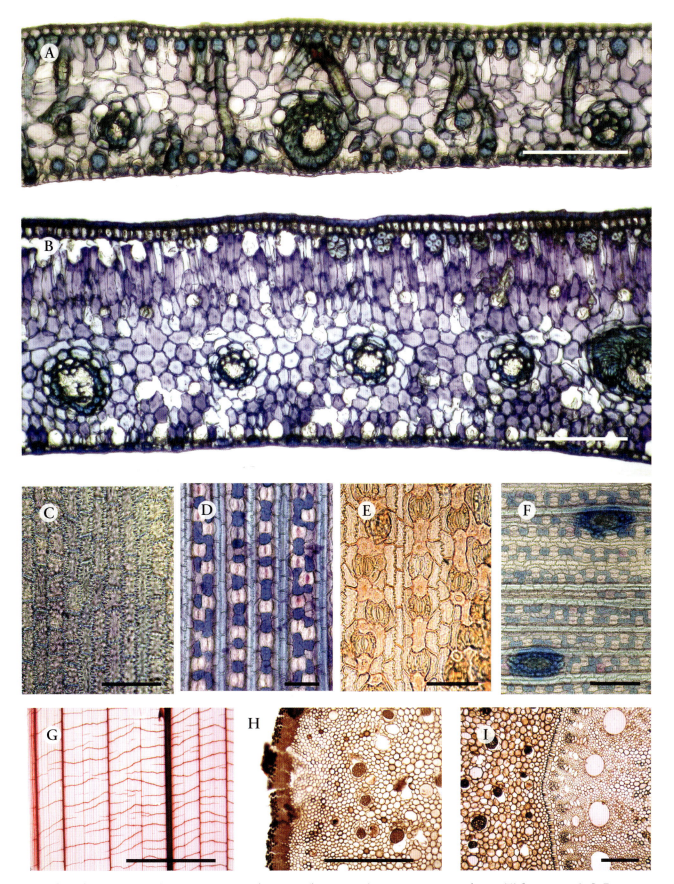

Fig. 33 Calamoideae (Eugeissoneae). **A**. *Eugeissona minor*, lamina, TS (bar = 100 μm). **B**. *Eugeissona insignis*, lamina, TS (bar = 100 μm). **C**. *Eugeissona insignis*, lamina adaxial epidermis, SV (bar = 50 μm). **D**. *Eugeissona insignis*, lamina abaxial epidermis, SV (bar = 50 μm). **E**. *Eugeissona insignis*, lamina abaxial epidermis, SV (bar = 50 μm). **F**. *Eugeissona insignis*, lamina abaxial epidermis with trichomes, SV (bar = 100 μm). **G**. *Eugeissona minor*, lamina cleared, SV (bar = 10 mm). **H**. *Eugeissona insignis*, root periphery, TS (bar = 500 μm). **I**. *Eugeissona insignis*, root stele with endodermis and inner cortex, TS (bar = 200 μm).

1A. Lamina (Fig. 33B), up to 12 cells deep; fibre-sclereids almost absent, small veins independent of surface layers; palisade well developed. Hypodermal fibre strands up to 15 cells thick . . . .*E. insignis*

PETIOLE

Vbs with 1 wide mxy vessel and a single phl strand; xyl parenchyma forming an enormously inflated parenchyma sheath delimited externally by thin-walled fibres (Figs. 29B and 32A). Fbs common in ground parenchyma, suspended within lacunose tissue. Stegmata not observed.

Stem (*E. utilis*)

Peripheral vbs with a single wide mxy vessel; central bundles with 2 wide mxy vessels; ground tissue including narrow fibre strands.

Root (Fig. 33H, I)

Cortex including wide to narrow fibre strands, but easily displaced in sectioning to leave an apparent space (Fig. 33H); middle cortex not lacunose; mucilaginous raphide canals as uniseriate series of elongated cells in outer cortex. **Endodermis** (Fig. 33I) with *outer* tangential wall thicker than inner wall in young roots; medullary vessels occasional.

Vascular elements

Xyl: many vessels with simple perforation plates in root, but scalariform in stem and petiole. **Phl:** simple sieve plates in leaf and stem but compound on oblique end walls in root.

Note: the apparent absence of stegmata from the lamina because of their minute size is particularly noteworthy.

TRIBE LEPIDOCARYEAE

A tribe of considerable diversity in tropical America, Africa, and Madagascar, subdivided into 3 subtribes that are geographically as well as morphologically contrasted. They may be distinguished as follows:

With pinnate leaves

1. Exclusively African; 3 genera, all climbing palms, the leaf with a terminal cirrus producing pairs of reduced leaflets (acanthophylls) that are reflexed (cf. *Desmoncus*) . . . . Ancistrophyllinae.
2. The single genus *Raphia*, usually single-stemmed palms with disproportionately long leaves; from tropical Africa to Central and South America (cf. *Elaeis*, the only other palm with a transatlantic distribution) . . . . Raphiinae

With palmate leaves

3. Exclusively America, widely distributed in Northern South America, mostly single-stemmed palms but with an appreciable size range . . . . Mauritiinae

SUBTRIBE ANCISTROPHYLLINAE ('WEST AFRICAN RATTANS') (Figs. 18C, F; 22A; 23J; and 34)

A group of 3 small genera with several poorly understood spp. restricted to tropical West Africa and distinguished in Africa from *Calamus deerratus* (Calaminae, a 'true rattan') in possessing a cirrus rather than a flagellum. The cirrus here has acanthophylls (widely-spaced reflexed spine-leaflets). Leaves are pinnately compound, the tubular leaf sheath with a persistent ochrea. *Eremospatha* (pleonanthic) and *Laccosperma* (hapaxanthic) are similar in their dyads of hermaphrodite flowers, a condition considered to be unspecialized among calamoid palms, whereas *Oncocalamus* (hapaxanthic) is unique within all palms in its ultimate cincinnate flower cluster which consists of a central female flower and two lateral mixed female and male flower complexes, each a separate sympodium.

Laccosperma (= Ancistrophyllum)

A small genus of c.7 spp. of multiple-stemmed climbing palms in West Africa, the aerial stems supported by leaves each with a long cirrus. Stem hapaxanthic as determined by a group of first-order inflorescence units each in the axil of 1 of the reduced distal leaves.

Eremospatha

Includes up to 12 spp. of climbing palms in tropical West Africa. The multiple stems each supported by leaves and their terminal cirri, these axes pleonanthic.

Oncocalamus

Five spp. of climbing palms in equatorial West Africa, the aerial stems multiple and hapaxanthic. Inflorescence as described above. For this reason the genus was at one time regarded as sufficiently isolated to constitute an independent subtribe, but its systematic association with other Ancistrophyllinae is now well established.

Anatomical features of the subtribe Ancistrophyllinae

Leaf

LAMINA

Dorsiventral (Fig. 34A, C). **Trichomes** absent. **Stomata** occasional to frequent adaxially (Fig. 34B). **Epidermis:** cells ± rectangular with markedly sinuous anticlinal walls, the cells narrow and shallow in costal regions, wider and deeper in intercostal regions (Fig. 34D). Outer epidermal wall moderately to extremely thick-walled. Abaxial epidermal cells somewhat smaller than adaxial. **Mesophyll:** palisade layers absent or somewhat differentiated (Fig. 34C). **Fibres** mostly restricted to hypodermis in large strands alternating with files of 1–2 ± cubical but poorly differentiated cells or colourless cells absent, other fibres may occur in the

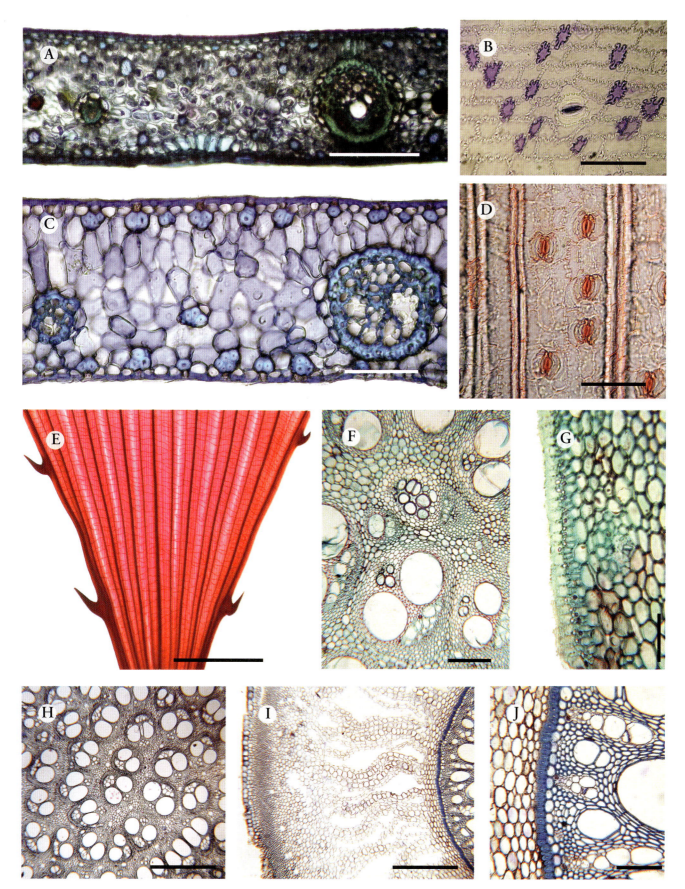

Fig. 34 Calamoideae (Lepidocaryeae: Ancistrophyllinae). **A**. *Oncocalamus macrospathus*, lamina, TS (bar = 100 μm). **B**. *Oncocalamus macrospathus*, lamina adaxial epidermis, SV (bar = 50 μm). **C**. *Laccosperma robustum,* lamina, TS (bar = 50 μm). **D**. *Laccosperma acutiflorum*, lamina abaxial epidermis, SV (bar = 50 μm). **E**. *Eremospatha wendlandiana*, base of cleared pinna, SV (bar = 10 mm). **F**. *Oncocalamus* sp., stem vbs and leaf trace, TS (bar = 200 μm). **G**. *Oncocalamus* sp., stem periphery showing silica bodies in hypodermis, TS (bar = 50 μm). **H**. *Eremospatha* sp., stem centre, TS (bar = 1 mm). **I**. *Eremospatha* sp., root periphery and cortex, TS (bar = 500 μm). **J**. *Eremospatha* sp., root stele near endodermis, TS (bar = 100 μm).

mesophyll (Fig. 34A) or alternate with veins (Fig. 34C). **Veins** diffuse, in abaxial mesophyll independent of surface layers except for few large veins attached to abaxial and especially adaxial surface by anticlinally-extended sclerotic cells. **Transverse veins** (Figs. 11G and 34E) wide, sheathed by fibres and mostly running above small veins to connect only the larger veins. **Midrib** projecting adaxially, vbs surrounded by a common sclerotic sheath.

The 3 genera may be distinguished by the following features of lamina anatomy:

1. Adaxial epidermis (Fig. 34B) including silicified and wholly cutinized short cells, sometimes paired, alternating along a cell file with 1–5 longer cells; hypodermal fibres mostly subepidermal to sub-hypodermal, mesophyll fibres common towards each surface (palisade scarcely differentiated) . . . . *Oncocalamus.*
1A. Epidermis without long/short cells in the same file (palisade differentiated) 2
2. Epidermal cells dimorphic, those above hypodermal fibre strands (costal cells) small, wholly cutinized, outer wall thin; cells not above hypodermal files (intercostal cells) larger, thick-walled . . . . *Laccosperma*
2A. Epidermal cells not dimorphic, outer wall thick, abaxially equal to depth of cell lumen, stomata sometimes sl. sunken, the g.c. overarched by adjacent cells . . . . *Eremospatha*

PETIOLE (Fig. 32B)
Vbs including a massive strand of inflated parenchyma cells as a pxy sheath and completed by a narrow sclerotic sheath (Fig. 32 B, C); single wide mxy vessel, phl with two strands.

Stem (Fig. 34F–H)

Stems in this subtribe show a few distinctive common features, but are considerably contrasted, as follows:

COMMON STEM FEATURES
Epidermis wholly cutinized with outer wall thickened. **Hypodermis** including a layer of **sub-hypodermal fibres** with associated small stegmata (Fig. 34G). **Cortex** narrow, few cells (c. 5–15) wide with a single series of narrow vbs or fbs. **Vbs** with 1 or 2 wide mxy vessels, the perforation plates simple, on transverse end walls; phl a single strand. **Raphide sacs** present, but only conspicuous in *Eremospatha* (Fig. 23J).

Note: the distinctive configuration of the phl in leaf traces as they depart from the stem (Fig. 18F) needs further study

DISTINCTIVE FEATURES OF STEM VBS IN THE 3 GENERA

- *Laccosperma* (Fig. 18C, F)—central vbs with 1 wide mxy vessel and a single phl strand (Fig. 18C), sieve plates simple, on transverse end walls, **transverse commissures** absent, phl configuration in departing leaf traces associated with apparent bridge bundles (Fig. 18F); small central vbs absent.
- *Eremospatha* (Fig. 34H)—vbs with 2 wide mxy vessels and a single phl strand (Fig. 34H), sieve plates long, oblique, compound, with narrow sieve areas. **Transverse commissures** common, each with a single narrow vessel and short irregular vessel elements, the sheathing cells few and thin-walled. No special configuration of phl cells in the departing

leaf traces. Narrow scattered central vbs suggest a stem vasculature without bridges and continuing axial bundles, thus differing from palms showing the '*Rhapis*-principle'.
- *Oncocalamus* (Fig. 34F)—vbs with 2 wide mxy vessels and a single phl strand (Fig. 34F); sieve plates not observed. **Transverse commissures** uncommon. Narrow scattered central vbs also suggesting the *Calamus*-type of vascular construction.

This elementary analysis shows that *Oncocalamus* and *Eremospatha* share a number of common features, with the possible interpretation that the 'discontinuous' vascular system described for *Calamus* could exist here but not in *Laccosperma*. Further study is needed to clarify this relationship, which points to considerable diversity within the subtribe.

Root

LACCOSPERMA
Epidermis thin-walled, large-celled. **Outer cortex** including a narrow zone of thin-walled cells and a wide zone of mixed thin- and thick-walled narrow cells, the innermost zone a 3–4 cell wide layer of thick-walled, pitted fibres delimiting the middle cortex. **Inner cortex** wide, the outer layers compact and including raphide canals, the wide inner layer aerenchymatous, with radially-extended air lacunae separated by mostly uniseriate plates of either wide or collapsed cells. **Fibres** absent. **Stele** wide. **Endodermis** with uniformly thickened cells, O-shaped in TS. Pericycle 1–2-layered. Vascular tissues embedded within thick-walled conjunctive tissue, the mxy vessels v. wide, those towards the centre separated from the more peripheral polyarch system as medullary vessels. **Medulla** thin-walled.

EREMOSPATHA (Figs. 22A and 34I, J)
Corresponds closely to *Laccosperma* but with a uniform layer of sclerenchyma in the outer cortex and more numerous raphide canals (Fig. 34 I); stele with O-shaped endodermal cells and uniseriate pericycle (Figs. 22A and 34J).

Vascular tissues

Mxy in stem and root with wide vessels, the perforation plates simple and transverse; in leaf axis with scalariform perforation plates on long oblique end walls. **Sieve tubes** with long compound sieve plates in the root and simple, transverse sieve plates in the stem, except in *Eremospatha* where the sieve plates in the stem, are on long oblique end walls unlike the simple sieve plates in the stem that otherwise characterizes the whole subfamily.

SUBTRIBE RAPHIINAE (Figs. 29F and 35)

Includes only the genus *Raphia,* with a transatlantic distribution from Madagascar and tropical Africa to Central and South America. Habitually it favours swampy soils. They are usually tree palms of distinctive appearance (Fig. 29F) because the trunk is disproportionately short in relation to the large leaves, in the extreme to 25 m long. Trunkless spp. without a visible above-ground stem also occur (e.g. *R. humilis*). Palms are hapaxanthic eventually producing a series of long pendulous inflorescence units each in the axils of the few distal leaves as the flowering phase is achieved. The inflorescence or the trunk can be tapped to make palm wine. Raffia fibre, used horticulturally as a binding material, is obtained by stripping off the surface layers of young leaves, presumably before they become much lignified. The condensed first-order branch units are diagnostically useful, with the male flowers aggregated distal to the female flowers and ultimately the fruits. Superficially the palms appear unarmed, but they have sharp spines on the leaflet margin and midrib.

Leaf

LAMINA

Dorsiventral (Fig. 35A). **Hairs** absent. **Epidermis** thin-walled except for the outer wall, especially adaxially; adaxial epidermis uniform, large-celled; abaxial epidermis smaller-celled and thinner-walled. Adaxial epidermis (Fig. 35B) without costal bands, cells above hypodermal fibre strands sl. wider than those elsewhere; anticlinal walls somewhat sinuous at lower focal levels, but markedly so at the level of the cuticle. Abaxial epidermis without sinuous walls (Fig. 35C, D) distinctly segregated into narrow costal bands of narrow, elongated cells and wider intercostal bands with 1–2 series of stomata, the cells shorter and less regular than the costal cells. **Stomata** congested, in irregular files, t.s.c. short and wide, the standard pattern with neighbouring trapezoid cells not readily observed. Guard cells small, thin-walled each with 2 equal cutinized ledges; the guard cells somewhat sunken because of the sl. enlarged adjacent cells. **Hypodermis** shallow, 1-layered below each surface and including irregular strands of fibres, adaxially alternating with files of small thin-walled colourless cells; abaxially with a more continuous thin-walled layer, the fibre strands narrower and more often separated from the epidermis. **Mesophyll** wide, with a well-developed adaxial palisade of 2–3 anticlinally extended cells; abaxial mesophyll of looser cells. **Fibres** mainly restricted to hypodermal layers, usually absent elsewhere except for occasional solitary fibres or narrow strands, especially at the level of the veins or the palisade. **Veins** in the abaxial mesophyll independent of the surface layers except for few large veins attached above and below to the hypodermis, the adaxial cells tall; O.S. complete around most smaller veins, but present only laterally around large veins, the sheath cells elongated parallel to the vein. I.S. of fibres continuous around large veins except for lateral sclerotic cells at the level of the xyl; usually present only below in the small veins. Phl of large veins divided into 2 separate strands by a strand of somewhat sclerotic but thin-walled cells. **Transverse veins** (Fig. 35A, E) either numerous, wide, extending above the small veins and mostly connecting only to the large veins, the fibrous sheath well developed and usually including a wide mxy vessel, the phl of separate narrow strands; or small longitudinal veins may be connected by smaller transverse veins. **Midrib** prominent adaxially, including a sclerotic cylinder with 1 large abaxial vb and smaller adaxial, often inverted vbs. **Expansion cells** in small strands abaxially on each side of midrib.

Note: *Raphia* spp. seem very uniform in their lamina anatomy and we have found only sl. quantitative variation in structural features, suggesting that it might be difficult to separate spp. using anatomical features.

PETIOLE

Peripheral **vbs** with well-developed fibrous bundle sheath, congested and forming a peripheral but not confluent sclerotic zone. **Central vbs** not forming a conspicuous V-shape, each usually with a single wide mxy vessel and 2 phl strands; the bundle sheath rather thin, especially surrounding the pxy region, which includes a wide band of empty enlarged and thin-walled cells. Ground parenchyma uniform but including narrow vbs. **Transverse commissures** frequent and usually included within irregular bands of narrow cells (Pykkö 1985).

Stem

Surface layers of suberized cells in old stems produced by etagen-divisions in outer cortical cells. **Cortex** narrow, including transient leaf traces and numerous narrow fibrous and vbs. **Central cylinder** abruptly delimited by outer layer of its lignified ground tissue. Outer vbs with massive fibrous phl sheath with a transition internally to more diffuse vbs, but still each with a massive sheath of fibres external to the phl (Fig. 35F). Central vbs somewhat more diffuse, wide, with 2 wide mxy vessels (Fig. 35F) and undivided phl; the fibre sheath rather narrow. **Ground parenchyma** uniform, undergoing limited late expansion to form a somewhat lacunose tissue with abundant starch.

Note: the stem anatomy is rather distinctive (a '2-vessel type'), with wide mxy vessels and well developed phl sheath fibres.

Root

Epidermis large-celled, thin-walled. **Cortex** with outer 3 layers of 1) thin-walled; 2) thick-walled; and 3) thin-walled cells; the last layer transitional to the wide middle and inner cortex and including raphide sacs (Fig. 35G). Cortical tissue with narrow radiating plates of thin-walled, often collapsed cells (Fig. 20F) within which are numerous fibre-strands possibly of the 'Raphia-type', i.e. short, wide-lumened and thin-walled cells aggregated in varying sizes (Fig. 35G). Innermost cortex a narrow, thin-walled layer 3–5 cells wide. **Stele** narrow, with a thin-walled medulla. **Endodermis** (Fig. 35H) with uniformly thickened cells, O-shaped in TS. **Pericycle** with 1(–2) layers of pitted short cells. Vascular tissues restricted to the peripheral conjunctive tissue. Aerial roots produced (Cardon 1978).

Vascular elements

Vessels in root with simple transverse perforation plates intermixed with elements with oblique scalariform perforation plates; elements in stem wide, mostly with transverse or oblique end walls with simple plates; elements in leaves with long oblique scalariform plates with many thickening bars. **Sieve tubes**, with simple transverse sieve plates in stem and leaf; roots not observed.

Cell inclusions

Stegmata present in stem and leaf; silica bodies spherical, spinulose, each enveloped by thickened basal cell wall. **Raphide sacs**

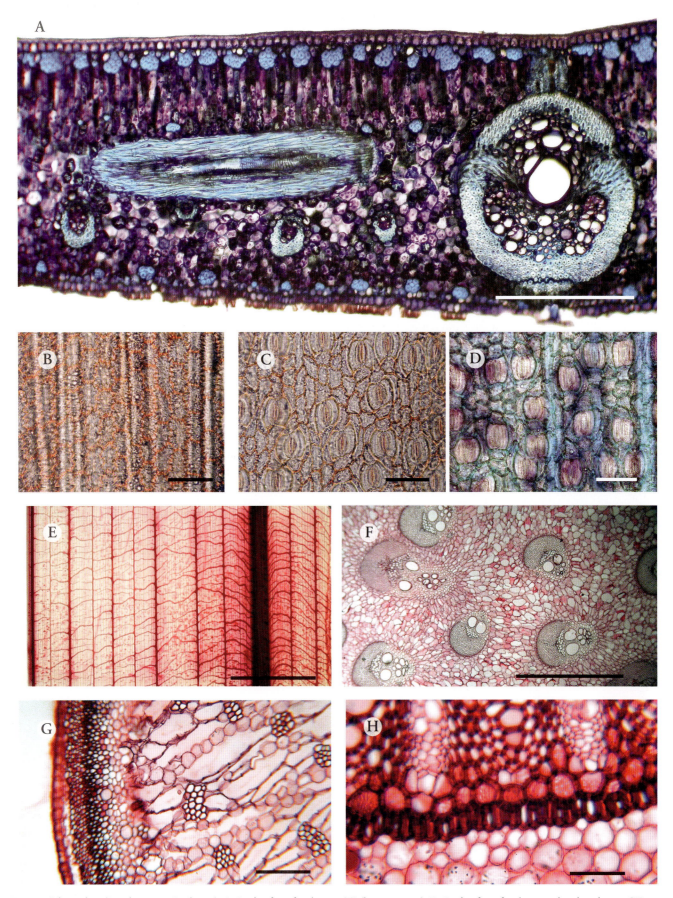

Fig. 35 Calamoideae (Lepidocaryeae: Raphiinae). **A**. *Raphia farinifera*, lamina, TS (bar = 200 μm). **B**. *Raphia farinifera*, lamina adaxial epidermis, SV (bar = 50 μm). **C**. *Raphia farinifera*, lamina abaxial epidermis, SV (bar = 50 μm). **D**. *Raphia farinifera*, lamina abaxial epidermis, SV (bar = 50 μm). **E**. *Raphia farinifera*, cleared lamina with midrib, SV (bar = 10 mm). **F**. *Raphia* sp., stem centre, TS (bar = 2 mm). **G**. *Raphia taedigera*, root periphery and cortical fbs, TS (bar = 200 μm) **H**. *Raphia taedigera*, root, outer stele with endodermis, TS (bar = 50 μm).

common in lamina mesophyll; narrow mucilage canals present in outer cortex of root.

Note: it is debatable if the 'Raphia-type' fibre strands first described by Drabble (1904), contrasted with the 'Kentia-type' fibres, actually occur in *Raphia*. The strands in *Raphia* (Fig. 35G) in some respects more resemble the typical 'Kentia-type' (Fig. 86M) and lack stegmata, which may be considered a defining feature of the 'Raphia-type', as illustrated in *Eugeissona* (Fig. 33H, I). Further comparative study is needed, and especially the examination of isolated fibre types from macerated material, to resolve the paradox.

SUBTRIBE MAURITIINAE (Fig. 36)

The subtribe includes 3 genera, the only reduplicate-leaved palms with palmate leaves (but cf. *Guihaia*), indicative of the developmental plasticity of the palm leaf, but never becoming as markedly costapalmate as in several Coryphoid genera.

Lepidocaryum (Fig. 36C)

A genus with few spp. of small palms in the forest understorey with clustering stems described as initially rhizomatous. It is unarmed except for short bristles on the margins and ribs of the narrow, widely separated leaf segments (Kahn and Mejia 1987).

Leaf

Reduplicate. Palmately compound, hastula not developed.

LAMINA

Dorsiventral (Fig. 36C). **Trichomes** absent. **Bristles** up to 5 mm long, each with a somewhat swollen base extending into a rigid multicellular sclerotic spine. **Epidermis** with narrowly rectangular cells with markedly sinuous and anticlinally thickened walls, the cell lumen much occluded. Epidermis of both surfaces differentiated into costal bands 2–3 cells wide above hypodermal fibres, the cells narrow, elongated, alternating with intercostal bands of somewhat wider, shorter cells, 1–2 cells wide to which the stomata are restricted. **Adaxial epidermis**: costal cells with a thick cutinized outer wall, the inner wall thin; intercostal cells with a thin outer cutinized layer and a thick uncutinized inner, cellulosic wall. **Abaxial epidermis** less markedly differentiated, with up to 3 stomatal series in each intercostal band. **Stomata** diffusely distributed and fewest on the adaxial surface, each with 2 short thin-walled l.s.c. deeper than normal epidermal cells; rhombohedral neighbour cells usually easily recognized. **Hypodermis** with regular bands of **fibres** alternating with short thin-walled and pitted hypodermal cells, the fibre bands 2–6 cells wide; fibre bands on the lower surface with fewer fibres. Additional small fbs or solitary fibres common in the mesophyll, often in 2 equidistant series towards each surface. Hypodermal cells around each substomatal chamber usually of 4 L-shaped or 2 C-shaped cells determined by a double or single cell file. **Mesophyll** (Fig. 36C) without palisade layers, cells irregular, somewhat lacunose toward s.st.ch.; permeated by extensive system of **fibre-sclereids**. **Veins**

narrow, smaller situated abaxially and independent of surface layers, larger connected to each surface by sclerotic cells; O.S. of conspicuous cubical cells with thickened pitted walls, incomplete below; I.S. almost always completely sclerotic, including narrow thick-walled fibres below, or above and below the largest veins, but of pitted, thick-walled elongated sclerotic cells elsewhere. **Vascular tissues** v. reduced, phl strand narrow, xyl obscure in all except larger veins. **Transverse veins** conspicuous, forming a well-developed system, the bundle sheath completely fibrous, with many fibres extending into the mesophyll and appearing like fibre-sclereids in TS, some extending to the epidermis. **Stegmata** common next to fbs and vb sheath fibres, the cells small with small irregularly spherical silica bodies.

Stem

Surface layers represented by a persistent layer of sclereids. **Cortex** narrow, with numerous fbs and few vbs. **Central cylinder** with outermost congested vbs, each with a well-developed fibre cap, maturing first in the outermost vbs. Peripheral and most central vbs with 1 wide mxy vessel (a '1-vessel' palm) but 2 mxy vessels only in outgoing leaf trace. Most vbs with a single phl strand. Ground parenchyma cells without late expansion.

Vascular elements

Stem mxy vessels with long oblique to v. oblique scalariform perforation plates, the vessel elements short. Phl sieve tubes with transverse to oblique simple sieve plates.

Mauritiella (Figs. 9B and 36D–K)

Moderate-sized, multiple-stemmed palms, including about 14 spp. in northern South America, the stems armed with spine roots exposed when the leaf sheaths ultimately abscise.

Leaf

Reduplicate, shortly costapalmate with a hood-shaped adaxial hastula, the leaf segments long and narrow. Bristle-like appendages, each with a bulbous base present on segment margins and adaxially on midrib, projecting apically. Ribs and adjacent surfaces with filamentous scales grading into microscopic hairs.

LAMINA

Dorsiventral (Fig. 36D). **Trichomes** present abaxially, with a filamentous thin-walled distal portion arising from a narrow basal group of thick-walled pitted cells, the distal portion usually projected apically and often bifid (Fig. 36F). **Epidermis: adaxial epidermis** v. uniform; cells square to shortly elongated or with slightly oblique end walls, anticlinal walls slightly sinuous; outer wall thick and uniformly cutinized. **Abaxial epidermis** (Fig. 36E) with somewhat sinuous walls, papillose, especially the cells surrounding the g.c., and including costal bands 1–2 cells wide alternating with intercostal bands with regular files of stomata. **Stomata** with a regular and uniform neighbour cells and short t.s.c. (Fig. 9B), those of each successive complex in the same file usually contiguous. G.c. slightly sunken; outer cavity partly occluded with wax deposits as seen in surface view. **Hypodermis**

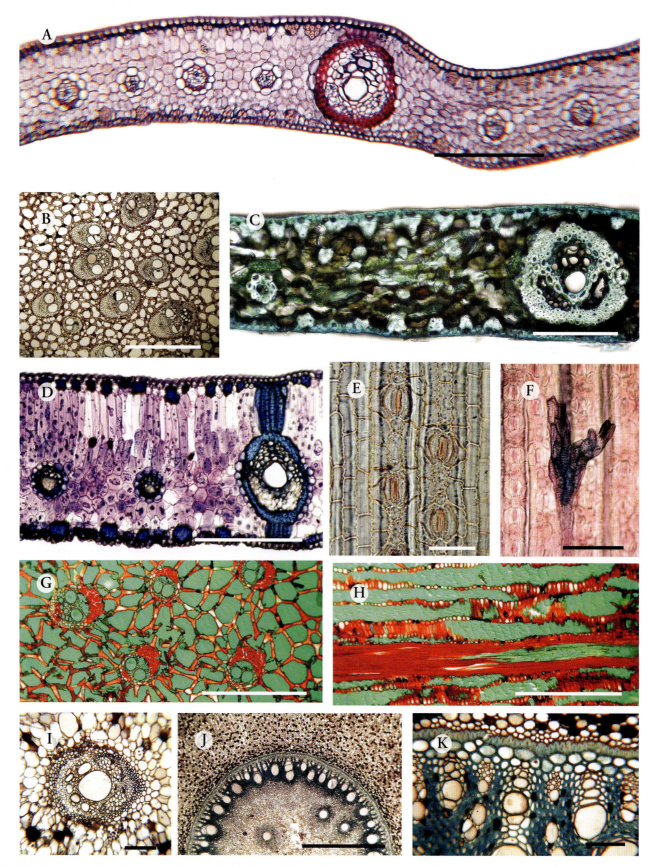

Fig. 36 Calamoideae (Lepidocaryeae: Mauritiinae). **A**. *Mauritia flexuosa*, lamina, TS (bar = 200 µm). **B**. *Mauritia* sp., stem, subperiphery vbs., TS (bar = 2 mm). **C**. *Lepidocaryum tenue*, lamina, TS (bar = 100 µm). **D**. *Mauritiella* sp., lamina TS (bar = 250 µm). **E**. *Mauritiella armata*, lamina abaxial epidermis, SV (bar = 50 µm). **F**. *Mauritiella* sp., lamina abaxial epidermis with trichome, SV (bar = 100 µm). **G**. *Mauritiella aculeata*, stem centre, air spaces filled with green-staining embedding matrix, TS (bar = 2 mm). **H**. *Mauritiella aculeata*, stem centre, air spaces filled with green-staining embedding matrix, LS (bar = 2 mm). **I**. *Mauritiella armata*, petiole vb, TS (bar = 200 µm). **J**. *Mauritiella armata*, root, stele and inner cortex, TS (bar = 2 mm). **K**. *Mauritiella armata*, root, stele at endodermis, TS (bar = 100 µm).

1-layered below each surface, adaxially including bands each of 2–5 slightly lignified fibres alternating with files 2(–3) cells wide of thin-walled short colourless cells. Abaxial hypodermis with fewer, smaller fbs or solitary fibres alternating with wider bands of thin-walled cells, interrupted to form s.st.chs. **Fibres** absent from mesophyll except for occasional strands (possibly free-vein endings) at same level as smaller veins. **Mesophyll** (Fig. 36D) including a single layer of tall palisade cells, the mesophyll cells otherwise isodiametric and compact. **Veins** narrow, situated toward the lower surface, mostly independent of surface layers except few larger veins with girder-like bundle sheath extensions of tall sclerotic cells adaxially and shallow fibrous extensions abaxially (Fig. 36D). O.S. of thin-walled pitted rectangular cells, incomplete below; I.S. of abaxial fibres completed adaxially by pitted sclerotic cells of the sheath extension. Vascular tissues of smaller veins reduced; only larger veins with abundant xyl and phl. **Transverse veins** few, not prominent, sheathed by abundant fibres, which may extend into the mesophyll as short extensions, sometimes to the hypodermis. **Stegmata** common in continuous files, silica bodies spherical.

PETIOLE (Fig. 36I)
Terete distally, with a conspicuous V-shaped vb arrangement. **Vbs**: outer not forming a pronounced sclerotic cylinder but bundle hierarchy very regular. Central vbs very uniform and regularly distributed. Large vbs with 1 wide mxy vessel; phl divided into 2 separate strands by a wide sclerotic partition (Fig. 36I). Bundle sheath fibres developed only outside phl, the outermost vbs with a somewhat more pronounced bundle sheath. **Ground tissue** without air spaces, including many small vbs with transition to narrow fbs.

Stem (Fig. 36G, H)
Well-developed lacunae (aerenchyma) between vbs. See comments under stem note.

Root (Fig. 36J, K)
Cortex uniform with many tannin cells. **Endodermis** with O-shaped cells. **Pericycle** 2-layered. Medullary vessels occasional.

Mauritia (Fig. 36A, B)

Tall imposing single-stemmed unarmed palms forming extensive stands and with many uses; because of their abundance they constitute a major resource.

Leaf

LAMINA (Fig. 36A)
Mauritia flexuosa (Fig. 36A) differs from *Lepidocaryum* as follows:

Adaxial epidermis uniform, without differentiation between costal and intercostal cells; palisade of short anticlinally-extended cells somewhat differentiated; fibres of transverse veins without distal extension into mesophyll as fibre-sclereids; fibre strands exclusively hypodermal.

PETIOLE
Without a distinct V-shaped arrangement of central vbs. Vbs with 1 mxy vessel and 2 widely separated phl strands.

The following commentary indicates some additional petiole features:

- *Mauritia flexuosa*. Vbs with v. narrow fibre sheath around phl only. Ground parenchyma loose, with well-developed intercellular spaces, including scattered narrow vbs and in LS including short cells interspersed among files of long cells.
- *Mauritia vinifera* has similar vbs, but lacks short cells in the ground parenchyma, which has wider air canals in part developed by cell expansion. Also the transverse commissures are unusual because of their well-developed fibre sheaths.
- *Mauritia* sp. HEM 9478. Vbs with well-developed fibre sheath around both xyl and phl. Ground parenchyma with or without short cells, including narrow fibre strands with few (3–6) cells but infrequent narrow vbs.

These apparent differences mostly likely illustrate the variation that can be encountered along a single large leaf axis.

Stem (Fig. 36B)
Peripheral vbs with 1 wide mxy vessel, 1 phl strand and massive fibre sheath, the fibres long delayed in their maturation. Outermost vbs with a well-developed 'yellow cap' of short pitted sclerenchyma cells (cf. *Korthalsia*). Central vbs with 2 wide mxy vessels and 1 phl strand. **Central ground parenchyma** becoming extensively lacunose, with a lattice-like network of horizontally-extended cells, often lobed or Y-shaped (Fig. 36B).

Note: the contrast of this axis with that of *Lepidocaryum* in part relates to the (10 or more times) difference in diameter of their stems.

Root
Surface layers becoming sclerotic with age. **Cortex** uniform, ground parenchyma loose and enclosing narrow air-canals through collapse of cells. Outer cortex including conspicuous raphide canals and scattered thick-walled elongated cells. Tannin abundant in longitudinal cell files. **Endodermis**: cells O-shaped, the lumen almost occluded by thick walls. **Pericycle** mostly 2-layered. **Medulla** wide, including few medullary mxy vessels and raphide canals.

Vascular elements
Mxy in root with simple transverse perforation plates, in stem with oblique scalariform perforation plates with numerous (10–20) thickening bars, in leaf with long oblique scalariform plates with many bars. **Sieve tubes** in root with long oblique, compound sieve plates, with numerous small sieve areas; in stem and leaf with simple sieve plates on ± transverse end walls. Stem mxy elements in *Mauritia* and *Mauritiella* wide to very wide (up to 350 µm wide) but much narrower in *Lepidocaryum* in relation to its smaller size.

Note: the 3 genera may be separated anatomically as follows:

Lamina

1. Lamina mesophyll with extension of bundle sheath fibres from transverse veins as discrete fibre-sclereids . . . . *Lepidocaryum*
1A. Lamina mesophyll with limited extension of ends of fibres from transverse veins; fibre-sclereids absent. . . . *Mauritiella*
1B. Lamina mesophyll without fibre extension from transverse veins or with fibre ends shortly intruding . . . . . . *Mauritia*

Summary for the Mauritiinae

The subtribe conforms to the general features of the subfamily in terms of epidermal structure, fibre distribution in the lamina and sinuous lamina epidermal walls. Trichomes are usually absent (but cf. Fig. 36F) and large lamina veins are connected to epidermis by short anticlinally extended sclerotic cells (e.g. in *Mauritiella* Fig. 36D) and usually associated with well-developed palisade layers. Roots are without cortical fibre strands.

Stem notes

Mauritella most closely resembles *Mauritia* in its stem anatomy as a 2-vessel palm with lacunose ground parenchyma:

1. Stem narrow (to 5 cm diameter), without a distinctive 'yellow cap' to outermost vbs of the central cylinder; ground parenchyma cells not inflated, compact. Central vbs mostly with 1 wide mxy vessel (~100 μm diameter) . . . . *Lepidocaryum*
1A. Stem wide (up to 70 cm diameter), with a distinctive 'yellow cap' to outermost vbs of the central cylinder; ground parenchyma cells enlarged to enclose wide air canals, the tissue markedly spongy. Central vbs mostly with 2 wide mxy vessels (up to 300 μm wide). . . . *Mauritia, Mauritiella*

A distinctive feature is the presence in all genera of mxy vessel elements with scalariform perforation plates in the stem vbs.

TRIBE CALAMEAE (see p. 87)

SUBTRIBE KORTHALSIINAE (Figs. 18A, B; 32E; 37B, C, E, H, I, M)

The genus *Korthalsia* of SE Asia, including almost 30 spp. of high-climbing palms is recognized easily by the praemorse and rhombohedral leaflets narrowed at the base to a short petiolule, but with distally divergent venation. The aerial stems can be branched (possibly dichotomously) and end in a terminal aggregate of lateral units subtended by reduced leaves and so can be described as hapaxanthic (cf. *Raphia*). The leaf axis is extended into a long cirrus armed only with recurved hooks (i.e. no acanthophylls) as the climbing organ. The genus is unusual in the subfamily because flowers are hermaphrodite and solitary in the axils of bracts on the rachilla, the usual dyads presumably lost by reduction. Many spp. are ant-plants ('ant rattans'), in which ants shelter in an inflated ochrea (the tubular extension of the leaf sheath above the insertion of the petiole). The genus extends from the Andamans to New Guinea and into Indochina. Spp. show a wide ecological range within forest communities suggestive of pioneering ability. Despite its abundance and association with other rattans it has limited commercial application, largely because the canes are unattractively patterned.

Leaf (Fig. 37B, C, E, H, I)

LAMINA

Dorsiventral (Fig. 37B, C). **Trichomes** (Fig. 37E) frequent, mostly on the abaxial surface, each with a sunken elliptical base constricted at its insertion with several irregular tiers of thick-walled cells extending into the often filamentous distal shield cells. **Epidermis** with outer wall somewhat thickened; adaxial epidermis (Fig. 37H) uniform with more or less cubical or sl. extended cells, the walls undulate to sinuous; abaxial epidermis (Fig. 37I) with narrow costal bands below hypodermal fibres and somewhat wider intercostals bands, costal cells smaller than those of adaxial surface, the cell walls often not sinuous; intercostal cells irregular, somewhat sinuous-walled. **Stomata** in intercostal regions, not in regular files; t.s.c. well differentiated, often short and with thickened cell walls, each with prominent lobes overarching g.c. somewhat (Fig. 37I). Guard cells small, the cell lumen somewhat constricted by thickened walls except at the thin-walled poles. **Hypodermis** of colourless cells absent, but including fibres or fibre strands (Fig. 37B, C). **Mesophyll** with or without a distinct palisade, the abaxial cells spherical, loose. **Fibres** frequent in the mesophyll and hypodermis, solitary or in wide strands of many cells; fibre strands may be conspicuous as a discontinuous adaxial hypodermal layer of narrow strands together with larger abaxial strands (Fig. 37C). Otherwise fibres infrequent adaxially but represented by wide abaxial strands, either at the same level as the veins or in the abaxial hypodermis (Fig. 37B). **Veins** in abaxial mesophyll, mostly independent of surface layers, only a few larger veins attached to each surface by sclerotic cells. O.S. usually incomplete below small veins and incomplete above and below large veins; cells ± cubical. I.S. usually completely fibrous or 1–2-layered below small veins and then completed by sclerotic parenchyma adaxially. **Transverse veins** frequent in adaxial mesophyll and usually running above small longitudinal veins, wide with an extensive fibrous sheath. **Ribs** prominent adaxially at each fold (no midrib) the sclerotic cylinder including 1 large abaxial vb and smaller, often inverted adaxial bundles. **Expansion cells** (Fig. 37C) scattered in bands on the abaxial surface.

PETIOLE (Fig. 32E)

Hypodermis sclerotic, including frequent fibrous strands. **Vbs** not forming a conspicuous V-shape. Bundle sheaths near abaxial surface of petiole confluent to form an almost continuous peripheral sclerotic zone, elsewhere separated by sclerotic ground parenchyma. Central vbs with the pxy region including a wide, thin-walled tissue of empty inflated cells (Fig. 32E); mxy with 1 wide vessel; phl with 2 strands.

LEAF SHEATH (Fig. 18A)

Tubular, prominently ligulate, the ligule ochreate, the ochrea sometimes inflated and ant-inhabited; hypodermis including a well-developed series of fbs transitional internally to small vbs.

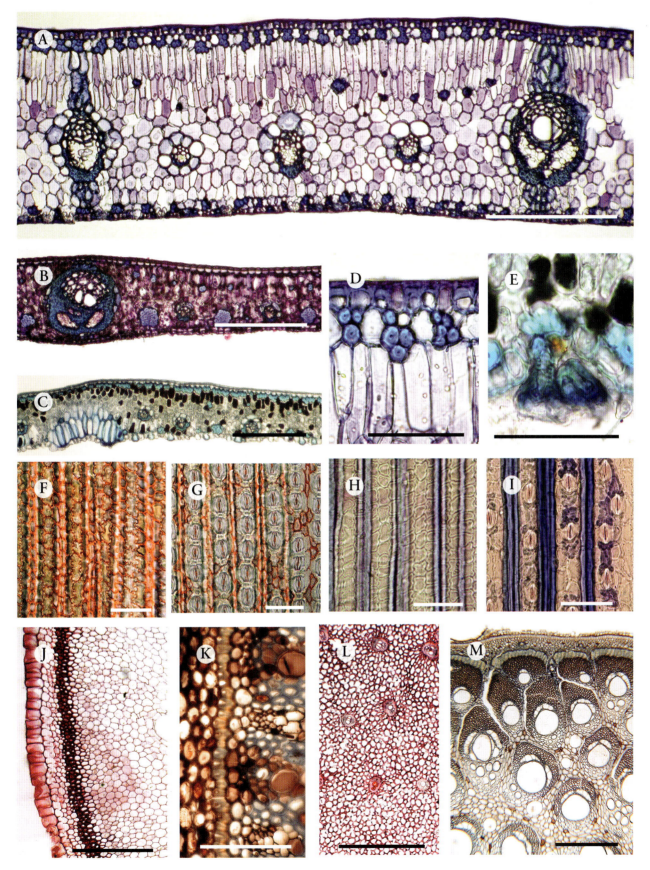

Fig. 37 Calamoideae (Calameae: Korthalsiinae & Metroxylinae). **A**. *Metroxylon vitiense*, lamina, TS (bar = 200 μm). **B**. *Korthalsia laciniosa*, lamina, TS (bar = 200 μm). **C**. *Korthalsia zippelii*, lamina, TS (bar = 200 μm). **D**. *Metroxylon vitiense*, lamina adaxial surface, TS (bar = 50 μm). **E**. *Korthalsia zippelii*, lamina abaxial surface with trichome, TS (bar = 50 μm). **F**. *Metroxylon vitiense*, lamina adaxial epidermis, SV (bar = 50 μm). **G**. *Metroxylon vitiense*, lamina abaxial epidermis, SV (bar = 50 μm). **H**. *Korthalsia cheb*, lamina adaxial epidermis, SV (bar = 50 μm). **I**. *Korthalsia cheb*, lamina abaxial epidermis, SV (bar = 50 μm). **J**. *Metroxylon salomonense*, root periphery, TS (bar = 250 μm). **K**. *Metroxylon sagu*, root outer stele with endodermis, TS (bar = 100 μm). **L**. *Metroxylon sagu*, stem centre, TS (bar = 5 mm). **M**. *Korthalsia echinometra*, stem periphery, TS (bar = 500 μm).

Main **vascular system** (Fig. 18A) of a single series of vbs, each with a massive bundle sheath, infilled with small vbs, thickened dorsal part of sheath with 2 vascular series. Pxy region not including enlarged parenchyma cells. **Ground parenchyma** of loose cells expanding radially from the large vbs. **Spines** prominent on outer surface of ligule, each including a sclerotic vasculated tip developing on a massive band of extended parenchyma cells.

Stem (Figs. 18A, B and 37M)

Cortex (Fig. 37M) v. narrow, only a few cells wide and with a single series of diffuse narrow bundles except near the leaf insertion. **Central cylinder** with v. uniformly congested vbs, the outermost narrow and with well-developed fibrous bundle sheaths forming a distinct sclerotic layer. Outermost fibres of the outermost vbs forming a 'yellow cap' (Fig. 37M), with little affinity for stain as seen in TS, i.e. a band of early maturing short thick-walled cells, the walls distinctly pitted as seen in LS. **Central vbs** (Fig. 18B) each with a narrow sheath, a single v. wide (up to 350 μm) mxy vessel, the phl including 2 separate strands situated in a position lateral to the mxy vessel ± as in the Calaminae; sieve tubes in a single series; ground parenchyma loose, the cells often lobed and forming wide intercellular spaces.

Note: although not examined 3-dimensionally, the stem of *Korthalsia* lacks the narrow axial bundles and transverse commissures that are characteristic of *Calamus* (q.v.); the leaf traces also appear to give off an axial bundle at their point of departure from the stem. Stems of *Korthalsia* are described as of limited value as rattans (Dransfield et al. 2008b: p. 169) because they are disfigured by nodal scars and adherent inner leaf sheath layers. However, more important disqualifying features are likely to be found in their contrasted vasculature, not of the *Calamus*-type and the limited lignification of the ground parenchyma, possibly together with the very wide raphide canals (Fig. 18B).

Root

Not studied.

Cell inclusions

Raphide sacs conspicuously wide in the aerial parts; forming continuous mucilage canals in petiole and stem. **Stegmata** with spherical silica bodies, including the transverse veins of the leaf.

Vascular tissues

Mxy vessels in stem with long elements and simple transverse perforation plates. **Sieve plates** in phl of petiole and stem simple and on transverse or sl. oblique end walls.

TRIBE SALACCINAE (Figs. 24D, E; 32G; 38A, B, D–G, K–M)

Including the 2 genera *Eleiodoxa* (hapaxanthic) and *Salacca* (pleionanthic) of SE Asia, from the Malay Peninsula to the Philippines. Both are low-growing or acaulescent palms favouring wet forest locations, but generally avoided because of their abundant sharp spines (Fig. 24D). Despite this *Salacca zalacca* (= *Salacca edulis*; in Malay, *salak utan*) is cultivated over wide areas for its fruit with a fleshy mesocarp. A morphological peculiarity is that the inflorescence unit emerges through a groove in the back of the leaf base. Fisher and Mogea (1980) showed that this is because the unit arises as a bud in the normal median plane but at the base of a narrow adaxial cleft. This is enlarged by radial expansion of the leaf base and encloses the developing inflorescence, which eventually emerges abaxially through tissue weakened by cell necrosis to form the groove. This occupation of the normal axillary position by a reproductive shoot would appear to preclude the development of a vegetative bud in that position. However, vegetative buds are developed at the same node but displaced circumferentially by an average of 130°. *Eleiodoxa*, being hapaxanthic and so developing no basal inflorescence units, has basal vegetative buds in the normal axillary position as potential suckers.

Salacca (Figs. 24D, E; 32G; 38A, D, E, K–M)

A moderate-sized genus (~15 spp.) of essentially acaulescent palms, but still incompletely collected and described. They range from Indochina through the Philippines to Borneo and Java.

Leaf

Often robust, to 7 m tall in the largest sp., reduplicate and pinnately compound or entire. The leaf axis usually armed abaxially with solitary or clustered long rigid spines (Fig. 24D). The account is based on *Salacca zalaccca*, with differences in other taxa outlined later.

LAMINA

Dorsiventral (Fig. 38A). **Trichomes** (Fig. 38E) frequent abaxially and usually situated below a fibrous strand; each with a base of few irregular sclerotic cells extending into a flattened filament of thin-walled cells, the filament directed towards the apex of the leaflet. **Epidermis** with occasional bands of small costal cells above and below large veins. Adaxial epidermis uniform, cells large, cubical or sl. longitudinally extended, with markedly sinuous anticlinal walls due to undulations of outermost cuticular deposit. Abaxial epidermis without distinct costal bands, cells similar to but less regular than those of the adaxial surface. **Stomata** (Fig. 38D) diffuse, uniformly scattered but not in distinct files; t.s.c. not differentiated from other epidermal cells; g.c. each with 2 prominent cutinized ledges, the cell lumen constricted except at the thin-walled poles by thickening of outer wall. **Hypodermis** (Fig. 38A) absent. **Mesophyll** without distinct palisade layers, the tissue loose except for the compact layer adjacent to the abaxial epidermis. **Fibres** solitary or in groups of up to 6 cells, mostly adjacent to each surface. **Veins** situated abaxially in mesophyll, mostly independent of surface layers except for larger veins attached to surface layers by small sclerotic cells. **Transverse veins** frequent in adaxial mesophyll, running above longitudinal veins; wide with sheathing fibres; vascular tissues often including 2 phl strands. **Midrib** prominent adaxially, with

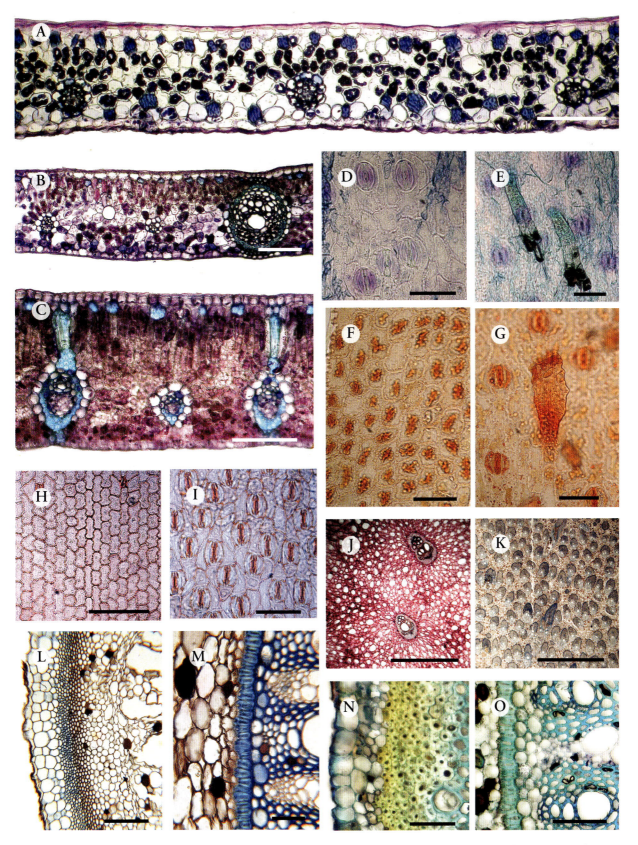

Fig. 38 Calamoideae (Calameae: Salaccinae & Pigafettinae). **A**. *Salacca zalacca*, lamina, TS (bar = 200 μm). **B**. *Eleiodoxa conferta*, lamina, TS (bar = 100 μm). **C**. *Pigafetta elata*, lamina, TS (bar = 100 μm). **D**. *Salacca zalacca*, lamina abaxial epidermis with stomata, SV (bar = 50 μm). **E**. *Salacca zalacca*, lamina abaxial epidermis with trichomes, SV (bar = 50 μm). **F**. *Eleiodoxa conferta*, lamina adaxial epidermis, SV (bar = 50 μm). **G**. *Eleiodoxa conferta*, lamina abaxial epidermis with trichome and stomata, SV (bar = 50 μm). **H**. *Pigafetta elata*, lamina adaxial epidermis, SV (bar = 100 μm). **I**. *Pigafetta elata*, lamina abaxial epidermis, SV (bar = 50 μm). **J**. *Pigafetta filaris*, stem centre, TS (bar = 2 mm). **K**. *Salacca zalacca*, stem centre to subperiphery, TS (bar = 2 mm). **L**. *Salacca zalacca*, root periphery with epidermis, TS (bar = 200 μm). **M**. *Salacca zalacca*, root outer stele and endodermis, TS (bar = 50 μm). **N**. *Pigafetta elata*, root periphery with epidermis, TS (bar = 100 μm). **O**. *Pigafetta elata*, root outer stele and endodermis, TS (bar = 100 μm).

hypodermal fibres, including a sclerotic cylinder enclosing a large abaxial and smaller, often inverted vbs. **Expansion cells** as indistinct abaxial bands on each side of the midrib and occasionally elsewhere.

PETIOLE

Hypodermis represented by a continuous series of fbs. Central **vbs** not in a conspicuous V-shape, but still forming 2 series with same orientation. Mxy with 1 wide vessel and 2 phl strands. Each vb with a fibrous phl sheath. **Ground parenchyma** as seen in LS differentiated into transverse bands of wide longitudinally-extended cells alternating with plates of 2–3 layers of small flattened cells, to which stored starch, if present, is restricted. Inflated pxy parenchyma cells absent or little developed (Fig. 32G).

Stem (Fig. 38K)

Short, narrow, mostly subterranean and with short internodes, the leaf traces thus inserted almost horizontally. **Surface layers** becoming suberized but not undergoing cell division. **Cortex** wide with v. numerous crowded and wide fbs. **Central cylinder** with a wide peripheral zone of congested bundles, the outermost narrowest (Fig. 38K); phl sheath well developed. Central bundles also relatively congested and similar to peripheral bundles.

Root (Fig. 38L, M)

Surface layers (Fig. 38L) including 4–5 layers of contrasted cells; collapsed epidermal cells, wide parenchyma cells, a narrow band of lignified (?) fibres with an abrupt transition to an unlignified layer of similar cells and then a gradual transition to the ground parenchyma of the inner cortex. **Cortex** lacunose, the lacunae radially extended to form a series often continuous across the cortex. Cortical fibres absent. **Endodermis** (Fig. 38M) with massive U-shaped wall thickenings almost occluding the cell lumen. Pericycle a single layer. Medulla sometimes including wide vessels.

Cell inclusions

Mucilage canals frequent in ground parenchyma of leaf and stem. **Stegmata** present in leaf and stem but absent from root.

Vascular elements

Vessels in root with long elements and simple transverse perforation plates; elements in stem shorter and with irregular end walls but with mostly simple perforation plates; elements in petiole with scalariform plates on long oblique end walls. **Sieve tube** elements in root with compound sieve plates on usually oblique end walls; elements in stem and petiole with simple sieve plates on ± transverse end walls.

Note: other spp. of *Salacca* correspond to this description in most respects, but suggested differences include:

- *S. magnifica*. Abaxial fibre strands in lamina mesophyll, wide, including up to 15 cells and not usually in contact with the epidermis; adaxial strands fewer and smaller.
- *S. wallichiana*. Adaxial epidermal cells large, with a thick cuticle.

Eleiodoxa (Figs. 32F and 38B, F, G)

One widely distributed sp., *E. conferta*, from Peninsula Malaysia to Thailand, Borneo, and Sumatra. It is a common understorey palm of freshwater swamp forests.

Leaf

Large, pinnately compound with whorls of long spines on the abaxial surface of the leaf axis.

LAMINA

This can be distinguished from *Salacca* most readily by the adaxial epidermal cells with less sinuous walls (Fig. 38F); the dissimilar hairs, with a broad distal extension (Fig. 38G); the thick-walled t.s.c. of the stomatal apparatus; the g.c. with the outer cuticular ledge larger than the inner; fbs almost exclusively in the hypodermis (Fig. 38B).

PETIOLE

Lacks the differentiation into long and short ground parenchyma cells as described for *Salacca*. Little inflation of pxy parenchyma in vbs (Fig. 32F).

Stem

Not studied

Root

Not studied.

SUBTRIBE METROXYLINAE

Including a single genus.

Metroxylon (Figs. 10F; 11A, E; 32H; 37A, D, F, G, J–L)

The genus *Metroxylon* (5 spp.) from SE Asia, westward from the upper Malay Peninsula to Samoa and northward to Micronesia. They are all tall, but either single- (*M. vitiensis*) or multiple- (*M. sagu*) stemmed palms, the lateral axes then somewhat rhizomatous initially (Schuiling 2009). They have a conspicuously hapaxanthic habit (Fig. 29E) resulting in a huge terminal panicle (Tomlinson 1971a), except for the pleonanthic *M. (Coelococcus) amicarum* in the Caroline Islands. They usually occupy swampy habitats where they can occur in large numbers. The palms are important as a source of starch (sago, typically from the soft inner tissues of the stem) which becomes most abundant just before flowering. The large pinnately reduplicate leaves also make good thatch. The palms are easily harvested although they are armed with long spines, which can be quite soft in many varieties.

The commercial value of the true sago palm (*Metroxylon sagu*) remains little emphasized although it is widely cultivated and used locally as a source of starch and thatch throughout SE Asia. Many named cultivars (= folk varieties) exist, as summarized in Schuiling (2009).

Note: in Uhl and Dransfield (1987) the genus was associated with *Korthalsia*, despite the habit contrast, because both have at least some flowers hermaphrodite, but the dyad condition is lost from *Korthalsia*. Molecular evidence as to its placement remains ambiguous as discussed in Dransfield et al. (2008b).

Leaf
LAMINA

Dorsiventral (Figs. 10F and 37A). **Trichomes** absent. **Adaxial epidermis** (Fig. 37F) with rectangular longitudinally-extended cells and sinuous anticlinal walls; outer wall thickened and cutinized. Cell files differentiated into longer, narrower deeper cells above hypodermal fibres and shorter, wider, thicker-walled cells above normal hypodermal cells (Fig. 37D). **Abaxial epidermis** (Fig. 37G) with narrow costal bands 1–2 cells wide below the hypodermal fibres alternating with wider intercostal bands. Costal cells either similar to those of adaxial surface or longer; intercostal cells short, wide. **Stomata** mainly abaxial with 1–2 series in each intercostal band, usually with only 1 short t.s.c. separating consecutive g.c. pairs of the same series (Fig. 37G); t.s.c. short, thick-walled and slightly overarching g.c. Guard cells small, each with 2 cutinized ledges, the cell lumen constricted somewhat by wall thickening except at the thin-walled poles. **Hypodermis** (Fig. 37D) shallow, forming a layer 1–3 cells deep of alternating fibre-strands and thin-walled cubical to somewhat elongated colourless cells. Abaxial hypodermis smaller-celled and interrupted by s.st.ch. between fibre strands. **Mesophyll** (Fig. 37A) with 2–3 adaxial palisade layers, the abaxial cells compact, larger and isodiametric in the mid-mesophyll. **Fibres** as described above and more or less restricted to the hypodermis, solitary or in few-celled strands; fibre strands in the mesophyll elsewhere either absent from *M. sagu* or little developed as in *M. vitiensis*. **Veins** (Fig. 37A) independent of surface layers except for a few large veins attached to each hypodermis by 1–4 series of columnar cells with somewhat thickened and lignified walls. O.S. conspicuous, usually complete around small veins but incomplete above and below large veins; sheath cells more or less cubical and somewhat thicker-walled than mesophyll cells. I.S. usually represented by abaxial fibres and sclerotic parenchyma elsewhere. Phl of larger veins divided into 2 separate strands by a median sclerotic partition. **Transverse veins** wide, with limited vascular tissues but sheathed by numerous fibres (Figs. 10F and 11A, E). **Midrib** with a sclerotic cylinder enclosing a large abaxial vb and smaller adaxial inverted vbs. **Expansion cells** conspicuous as abaxial bands on each side of the midrib, together with bands elsewhere in the lamina at infrequent intervals.

PETIOLE

Vbs congested towards periphery, each then with a well-developed fibrous sheath; occasionally confluent so as to produce a rigid outer mechanical zone. Central vbs lacking a clear V-arrangement; each with a narrow fibrous phl sheath and 1 wide mxy vessel and 2 phl strands. Pxy parenchyma cells extensively inflated (Fig. 32H). **Ground parenchyma** cells including transverse bands of longitudinally-extended cells alternating with bands of smaller cells.

Stem (Fig. 37L)
Surface layers producing a bark of suberized and tanniniferous cells derived from numerous etagen-like divisions in a peripheral phellogen. Central cylinder with a wide peripheral zone of congested vbs, each with a massive radially-extended fibrous phl cap. Central parenchyma spongy and lacunose by secondary transverse expansion of cells so as to enclose wide air lacunae, the central vbs thus widely separated (Fig. 37L). Starch abundant in the ground tissue. Central vbs mainly with 1 wide mxy vessel.

Root (Fig. 37J, K)
Surface layers (Fig. 37J). **Epidermis** initially of large thin-walled cells; hypodermal layers including a single layer of wide cells, 2–3 layers of small cells and a sclerotic zone transitional to the thin-walled cells of the middle cortex. **Phellogen** developing in older roots. Middle cortex including air spaces produced by collapse of radial files of cells, fibres absent. **Endodermis** (Fig. 37K) becoming thick-walled with O- or U-shaped cells in older roots. The early report by Drabble (1904) of fibre strands in the root cortex has not been confirmed in our material.

Cell inclusions
Stegmata with spherical silica bodies. **Mucilage canals**, as modified raphide sacs abundant in ground parenchyma of petiole and cortex of root.

Vascular tissues
Mxy vessels usually with simple perforation plates in the root, but a mixture of simple and scalariform perforation plates in the stem; in the petiole exclusively with scalariform plates on long oblique end walls. **Sieve tubes** in root with long oblique compound sieve plates; in stem and petiole mostly with simple transverse sieve plates.

SUBTRIBE PIGAFETTINAE

Including the genus *Pigafetta* (2 spp.), a stately dioecious palm of higher altitudes, from Sulawesi to Papua New Guinea with pioneering properties and characterized by the dense covering of soft spines (Fig. 29D). David Fairchild considered it his favourite palm.

Pigafetta (Figs. 29D and 38C, H–J, N, O)
Leaf
LAMINA

Dorsiventral, thick (Fig. 38C). **Trichomes** absent. **Epidermis**: adaxial epidermis large-celled, cells rectangular with coarsely sinuous anticlinal walls (Fig. 38H), without distinct costal bands; abaxial epidermis without costal bands. **Stomata** (Fig. 38I) congested, without distinct longitudinal series, the intervening cells irregular and without markedly sinuous walls, g.c. with short thick-walled t.s.c. **Hypodermis** absent. **Mesophyll** with a 2–3-layered palisade, the cells of the uppermost 2 layers markedly anticlinally extended. **Fibres** either in small lignified strands next to or within the adaxial hypodermis or larger unlignified strands scattered in the mesophyll and independent of the surface layers. **Veins** in abaxial mesophyll, the smaller independent of the surface layers and usually with a complete O.S., largest veins attached to each surface by sclerotic parenchyma above and below, the adaxial cells of sclerotic parenchyma including a single series of tall cells (Fig. 38C).

PETIOLE

Not studied.

Stem

Central ground tissue (Fig. 38J) lacunose, with uniform air canals developed by cell expansion, the vbs diffusely scattered.

Root

Surface layers (Fig. 38N) including wide epidermal cells with thickened outer walls, 1–2 thin-walled hypodermal layers and an abruptly delimited ring of sclerenchyma fairly abruptly transitional to the outer cortex. **Cortex** without fibres. **Endodermis** (Fig. 38O) with thick walls, O-shaped in TS. **Pericycle** 2–3-layered.

Vascular elements

Mxy in vbs of stem with long vessel elements, the simple perforation plates on transverse end walls.

SUBTRIBE PLECTOCOMIINAE

(Figs. 18J, K; 32I; and 39)

Three genera in SE Asia, all climbing palms, with the climbing organ a cirrus, (modified distal extension of the leaf axis) but without acanthophylls, i.e. with usually robust cat's-claw spines. Plants are uniformly hapaxanthic but also are dioecious, a unique situation in palms. Staminate flowers may be paired (*Plectocomia, Plectocomiopsis*) or solitary (*Myrialepis*); pistillate flowers are always solitary. There are frequent anomalies of branching, as with basal suckers either in non-axillary positions (*Myrialepis*) or bulb-like (*Plectocomia*) (Fisher and Dransfield 1979). The stems have limited use as rattans because the leaves separate incompletely from the stem without leaving a clean scar. Stem anatomy distinguishes the 3 genera from each other, but also markedly from that of the true rattans (Calaminae).

Myrialepis (Figs. 18K; 32I; and 39B, E–I)
Leaf

LAMINA (Fig. 39B, E, G)

Dorsiventral (Fig. 39B). **Trichomes** (Fig. 39E) above and below hypodermal fibrous strands on both surfaces, but least common adaxially. Each with a persistent sunken base of sclerotic pitted cells extended marginally into a flattened filament of ephemeral thin-walled cells. **Epidermis**: cell walls conspicuously sinuous with wide bands of costal cells above and below few large veins; adaxial epidermis (Fig. 39E) with large short cells, rectangular but not obviously distinct above hypodermal fibres. Abaxial epidermis with costal bands of elongated cells below the hypodermal fibres alternating with intercostal bands of cells with more pronounced sinuous walls. **Stomata** occasional adaxially, restricted to intercostal bands abaxially, often including only a single series; t.s.c. not well differentiated; l.s.c. with chloroplasts; g.c. with lumen constricted by wall thickenings except at the poles. **Hypodermis** 1-layered adaxially the cell files often replaced by fibres, absent abaxially but fibre strands equally numerous, hypodermal layer more compact than other mesophyll cells. **Fibres** ± restricted to hypodermal strands. **Veins** equidistant from each surface, small veins independent of surface layers but few large veins attached to each surface by narrow sclerotic cells. Cells of O.S. chlorenchymatous; I.S. of small veins often only 1-layered and only fibrous abaxially. **Transverse veins** diffuse (Fig. 39G), narrow, usually connecting only to larger veins and extending above smaller longitudinal veins, sclerotic parenchyma sheath often only 1-layered.

PETIOLE (Figs. 32I and 39F)

V-shaped arrangement of vbs not conspicuous. Peripheral vbs with massive fibrous bundle sheaths but never confluent as a continuous series; **central vbs** (Fig. 39F) with a well-developed fibrous sheath interrupted at the level of the mxy; mxy usually with 2 wide vessels, phl with 2 separate strands. Pxy parenchyma cells somewhat inflated (Fig. 32I). **Ground parenchyma** including numerous narrow vbs or fbs; parenchyma differentiated into plates of short cells alternating with wider plates of long cells.

SHEATH

Including a single series of large veins with massive fibrous sheaths alternating with mostly abaxial small veins and fibrous strands.

Stem (Figs. 18K and 39H, I)

Peripheral vbs (Fig. 39H) each with a well-developed fibrous sheath forming an almost continuous sclerotic cylinder, with an almost continuous 'yellow cap' of short, thick-walled pitted cells (cf. *Korthalsia*) but interrupted by exiting large leaf traces at the node (Fig. 39H). Central vbs diffuse, without a well-developed fibre sheath, each with 2 v. wide mxy vessels and a single phl strand with wide sieve tubes (Figs. 18K and 39I). **Ground parenchyma** cells remaining thin-walled and unexpanded, with a differentiation into long and short cells, in some samples including numerous wide fbs (Fig. 39I) or small vbs.

Vascular elements

Vessels in the leaf with scalariform oblique perforation plates, in the stem with simple transverse perforation plates. **Sieve tubes** in stem and leaf with simple sieve plates on ± transverse end walls.

Cell inclusions

Stegmata with spherical silica bodies; those in the leaf forming widely separated linear series, but not in the stem. Stegmata v. numerous to the outside of the 'yellow cap' in the stem. **Raphide canals** conspicuous as wide elements in leaf and stem.

Plectocomia and *Plectocomiopsis*

These resemble *Myrialepis* closely, but differ in some important details notably in the stem.

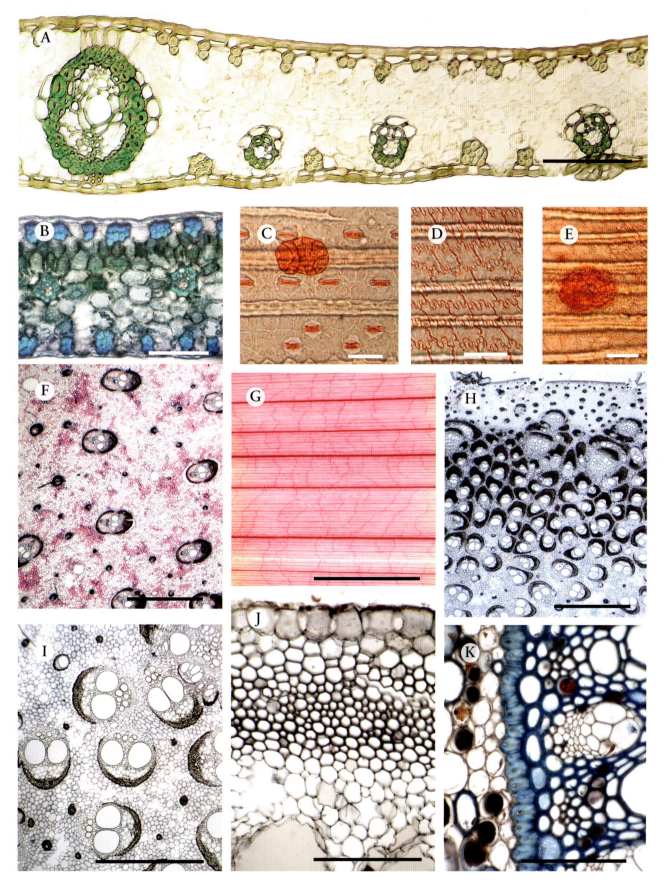

Fig. 39 Calamoideae (Calameae: Plectocomiinae). **A**. *Plectocomiopsis geminiflora*, lamina, TS (bar = 100 μm). **B**. *Myrialepis paradoxa*, lamina, TS (bar = 100 μm). **C**. *Plectocomiopsis geminiflora*, lamina abaxial epidermis with trichome, SV (bar = 50 μm). **D**. *Plectocomiopsis geminiflora*, lamina adaxial epidermis, SV (bar = 50 μm). **E**. *Myrialepis paradoxa*, lamina adaxial epidermis with trichome, SV (bar = 50 μm). **F**. *Myrialepis paradoxa*, petiole, TS (bar = 2 mm). **G**. *Myrialepis paradoxa*, lamina cleared, SV (bar = 10 mm). **H**. *Myrialepis paradoxa*, stem periphery, TS (bar = 2 mm). **I**. *Myrialepis paradoxa*, stem centre, TS (bar = 1 mm). **J**. *Plectocomiopsis geminiflora*, root periphery, TS (bar = 1 mm). **K**. *Plectocomia elongata*, root centre, endodermis and outer stele, TS (bar = 100 μm).

Plectocomia (Figs. 7J, K; 18J; 32J; and 39K)

Leaf

LAMINA

Dorsiventral. **Epidermis**: adaxial cells with sinuous walls (Fig 7J, K). **Hypodermis** 1 layered below each surface, with alternating bands of fibres and colourless thin-walled cells. Chlorenchyma with a well-developed 1–2-layered palisade. **Veins** in abaxial mesophyll.

PETIOLE

Vbs with 1 wide mxy vessel and 2 phl strands; parenchyma around pxy elements conspicuously inflated (Fig. 32J).

Stem

Vbs at periphery of central cylinder with a distinct 'yellow cap' (cf. *Korthalsia*). Central vbs with a single mxy vessel and phl strand (Fig. 18J). **Ground tissue** including numerous narrow fibrous strands, circular in TS and with locally present bands of short cells.

Root

Surface layers include **epidermis** of wide thin-walled cells, the outer wall somewhat thickened; a 2–3-layered thin-walled **exodermis**; a thick-walled 5–6 cell layer gradually transitional to compact outer cortex, much as in *Plectocomiopsis* (q.v.). Middle cortex with radiating usually uniseriate plates of cells enclosing radial air-canals. Fibres absent. **Endodermis** (Fig. 39K) with thick walls, O- to U-shaped in TS. **Pericycle** 1(–2)-layered. Isolated mxy vessels towards centre of **medulla** in larger roots.

Vascular tissues

Mxy vessels in root and stem with simple transverse perforation plates; **Sieve tubes** in root with long oblique sieve plates; in stem and petiole with simple transverse sieve plates.

Plectocomiopsis (Fig. 39A, C, D, J)

Leaf

LAMINA (Fig. 39A, C, D)

Dorsiventral (Fig. 39A). **Trichomes** occasional on abaxial surface with a somewhat sunken elliptical base of thick-walled cells (Fig. 39C). **Epidermis** with outer wall thickened. Adaxial epidermis: cells with markedly sinuous walls (Fig. 39D). Cells adjacent to stomata sl. thick-walled. **Hypodermis** (Fig. 39A) with numerous fibre strands but without colourless cells. **Veins** (Fig. 39A) mostly towards and sometimes in contact with the abaxial surface, the larger attached to both surfaces with sclerotic parenchyma cells. **Mesophyll** without a palisade. Non-vascular fibres exclusively hypodermal. **Transverse veins** numerous, with a well-developed fibrous sheath, the ends of the fibres sometimes extending freely into the mesophyll.

PETIOLE

Central vbs each with a single wide mxy vessel and phl strand; ground tissue with transverse plates of small cells.

Vascular tissues

Wide **vessels** in leaf, stem, and root with simple, transverse perforation plates; **sieve tubes** with long compound sieve plates in root, with mostly simple, transverse sieve plates in stem and leaf.

Cell inclusions

Raphide canals, common in all parts except outer cortex; in root wide and often with a surrounding layer of small cells resembling an epithelium. **Stegmata** absent from roots, in lamina often in continuous series, but dispersed elsewhere.

Summary

This subtribe is distinctive in a number of shared common features. **Epidermis** with sinuous walls, large-celled, without conspicuous differentiation of costal and intercostals cell files on adaxial surface. **Trichomes** often present below hypodermal fibre strands. G.c. of stomata with middle lumen constricted by wall thickening but thin-walled at the poles. **Stem** with peripheral 'yellow cap' of short sclereids; the outgoing leaf traces with abundant pxy; central vbs either with 2 wide mxy vessels (*Myrialepis*) or 1 wide mxy vessel (*Plectocomia* and *Plectocomiopsis*); including scattered small vbs and fibre strands.

Note: in the absence of 3-dimensional vascular analysis one cannot compare the vascular system of the stems of Plectocomiinae with the distinctive arrangement described for *Calamus* (q.v.). The construction of the vb in the two groups is different, in the former with 1 or 2 wide mxy vessel and a single phl strand (Fig. 18J, K), and in the latter always with 1 wide mxy vessel and 2 phl strands (Fig. 18L). In addition, stems have a marked contrast between central and peripheral bundle densities; this, combined with the absence of wall thickening and lignification of the ground parenchyma makes them of limited use as rattans.

SUBTRIBE CALAMINAE (THE 'TRUE RATTANS')

The largest group of climbing palms (over 350 spp.), including many spp. as a commercial source of canes (rattans) and extensively used in the industrial manufacture of furniture, but also with many local uses. The whip-like climbing organ may be a **flagellum** (a modified inflorescence unique to this group) or a **cirrus** (the distal extension of the leaf axis). The latter organ differs from that in *Desmoncus* (Bactridinae) and Ancistrophyllinae because the grapnels are hooked emergences rather than acanthopylls (reflexed spine-tipped leaflets). Rarely these organs are absent from the few non-climbing members of the group. A distinctive, but unfortunately not obvious, probable synapomorphy for the group is the presence of a discontinuous stem vascular system as demonstrated in *Calamus* by 3-dimensional analysis (Tomlinson et al. 2001; Tomlinson and Spangler 2002). The feature can be recognized initially but indirectly in single stem sections, but still requires more extensive confirmation.

The palms in this group range widely in the Old World, but largely as the genus *Calamus*. A distinctive floral feature is the presence of solitary staminate rather than paired flowers, representing a presumed modification of the floral dyad otherwise found in most Calamoideae.

The subfamily shows most of the synapomorphies of the Calamoideae but there is some variation, especially in lamina anatomy. However, this runs across genera, which therefore cannot be defined precisely in anatomical terms. Some features are:

1) Absence of a colourless **hypodermis**.
2) **Trichomes** with a terminal filamentous extension.
3) Frequent extension into the mesophyll of the **tips of the fibres** sheathing the transverse veins. (This seems transitional to the development of fibre-sclereids in the mesophyll of a few taxa, e.g. *Daemonorops* spp., and bears comparison with similar features in some Coryphoideae.)
4) **Hypodermal fibres** sometimes augmented by mesophyll fibre strands at the level of the small veins, as in some *Calamus* and *Daemonorops* spp.
5) Frequent development of **thick-walled cells** (t.s.c. and neighbour cells) around the stomata.

Calamus (Figs. 12A, B; 18D, E, G–I, L; 40A–G, K; and 41A–H, K, L)

Calamus is the largest genus of palms (~350 spp.) in SE Asia from India to Fiji, north to southern China and south to Queensland, with *Calamus deerratus* the sole representative in continental Africa. They are of necessity forest palms, but with a strong tendency to become weedy in disturbed sites. Our sampling is limited but based on a molecular phylogeny of the genus (Baker et al. 2000c), in an attempt to represent as much diversity as possible. The taxonomy of the genus is difficult to study in the absence of field work because of the spiny habit, and frequent morphological differences between juvenile and adult phases, quite apart from the tendency for lengthy canes to flower only terminally.

Leaf

Pinnate, reduplicately compound, the leaflets rarely of the fishtail type, with an erose apex (*C. caryotoides*).

LAMINA (Fig. 40A–G, K)

Dorsiventral (Fig. 40A, D, G, H). **Trichomes** present in some spp., usually restricted to abaxial surface, but there may be transitions between trichomes and spines, the latter typically marginal; each with a narrow sclerotic and often uniseriate base extending into an apically directed filament of thin-walled cells (Fig. 40B, C, E). **Epidermis** either with a thin cuticle, the outer wall usually thin, or the epidermis thicker-walled and wholly cutinized. Adaxial epidermis usually v. uniform, or with occasional stomata near midrib and above hypodermal expansion cells. Adaxial cells rectangular, but sometimes with sl. oblique end walls; anticlinal cell walls sinuous. Cell files often contrasted between narrow cells above hypodermal fibre strands (costal cells) and wider cells between fibre strands (intercostals cells), the difference sometimes further contrasted

in terms of wall thickness and differences in cuticularization. Abaxial epidermis (Fig. 40B, C, K) usually with smaller cells than on adaxial surface, with few wide costal bands below largest veins and narrow costal bands below hypodermal fibres, the intercostal bands with discontinuous files of stomata, the intervening cells somewhat irregular. **Stomata** with long or short t.s.c., these and the neighbouring cells distinctly thicker-walled than other cells (Fig. 40B, C); g.c. each with the cell lumen constricted by wall thickening except at the poles. **Hypodermis** of colourless cells absent, but including extensive fibre strands. **Mesophyll** infrequently with a well-developed palisade of 1–2 layers of cells, otherwise the adaxial mesophyll little differentiated from the rest of the mesophyll (Fig. 40A, D, G). **Fibres** most often exclusively as hypodermal strands of 1 to several cells next to the epidermis (Fig. 40F), and then usually most abundant adaxially or few abaxially in the mesophyll, but in many spp. with additional fibre strands, either at the same level as the small veins (Fig. 40A), or with few wide cells scattered throughout the mesophyll (Fig. 40G). **Veins** in the abaxial mesophyll and independent of the surface layers except the few largest attached to each surface by short sclerotic cells. O.S. either complete around small veins or incomplete above and below larger veins. I.S. well developed and fibrous except for lateral sclerotic cells; phl of largest veins divided into 2 separate strands by a median sclerotic partition (Fig. 40A). **Transverse veins** in adaxial mesophyll extending above small longitudinal veins, wide, sheathed by fibres that often extend at their tips into the adjacent mesophyll as incipient fibre-sclereids. **Midrib** prominent abaxially, hypodermal fibres frequent; sclerotic cylinder including a large abaxial vb and smaller inverted vbs.

PETIOLE (Fig. 12A, B)

Vbs only exceptionally forming a distinct V-arrangement (Fig. 12A); peripheral vbs with massive fibrous sheaths, but not confluent (Fig. 12B), the adjacent ground parenchyma sometimes sclerotic. Each vb with usually 1 wide mxy vessel and 2 phl strands, but lacking the inflated pxy seen in most other groups (cf. Fig. 32). Additional small vbs and fbs commonly present.

Stem (Figs. 18D, E, G–I, L and 41A–E, G, H)

Vbs uniformly distributed, without forming a marked peripheral mechanical layer (Figs. 18D, G, H and 41A). **Epidermis** typically with a silicified outer wall (Fig. 41H). **Cortex** narrow (Fig. 41B). **Central cylinder** without many narrow or congested peripheral vbs but including conspicuous leaf traces near the node (Fig. 41B, G). Central **wide vbs** (Figs. 18L and 41C, D) each with a single wide mxy vessel and 2 lateral phl strands; mxy vessel rarely with tyloses in functional stems (Fig. 41E). Additional scattered **small vbs** (Fig. 41D) of varying sizes indicating the basal beginnings of axial vbs that do not arise as branches from outgoing leaf traces (Tomlinson et al. 2001; Tomlinson and Spangler 2002). **Transverse commissures** as the only connections between larger axial vbs, identical in anatomy to the transverse commissures of the petiole. **Ground tissue** of somewhat thick-walled cells often somewhat lobed or stellate in TS and somewhat lacunose in ways that may be diagnostic (Weiner 1992).

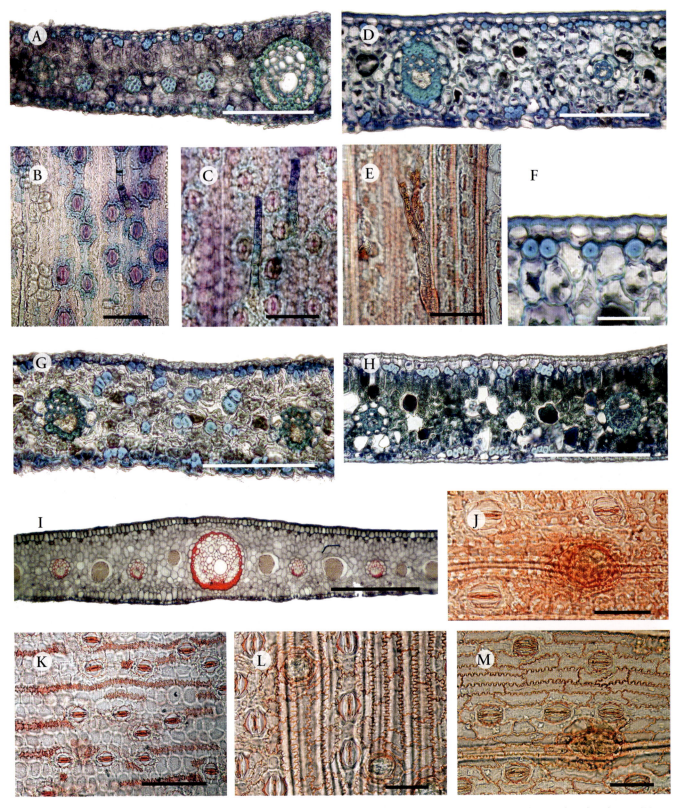

Fig. 40 Calamoideae (Calameae: Calaminae) lamina. **A**. *Calamus caesius*, lamina, TS (bar = 100 μm). **B**. *Calamus caesius*, lamina abaxial epidermis, SV (bar = 50 μm). **C**. *Calamus caesius*, lamina abaxial epidermis with trichomes, SV (bar = 50 μm). **D**. *Calamus thysanolepis*, lamina, TS. (bar = 100 μm) **E**. *Calamus thysanolepis*, lamina abaxial epidermis with trichome, SV (bar = 100 μm). **F**. *Calamus thysanolepis*, lamina adaxial epidermis, TS (bar = 25 μm). **G**. *Calamus castaneus*, lamina, TS (bar = 100 μm). **H**. *Ceratolobus pseudoconcolor*, lamina, TS (bar = 100 μm). **I**. *Daemonorops* sp., lamina with major vein, TS (bar = 200 μm). **J**. *Daemonorops didymophylla*, abaxial epidermis with trichome, SV (bar = 50 μm). **K**. *Calamus blumei*, abaxial epidermis, SV (bar = 100 μm). **L**. *Ceratolobus discolor*, abaxial epidermis with trichome, SV (bar = 50 μm). **M**. *Pogonotium ursinum*, abaxial epidermis with trichome, SV (bar = 50 μm).

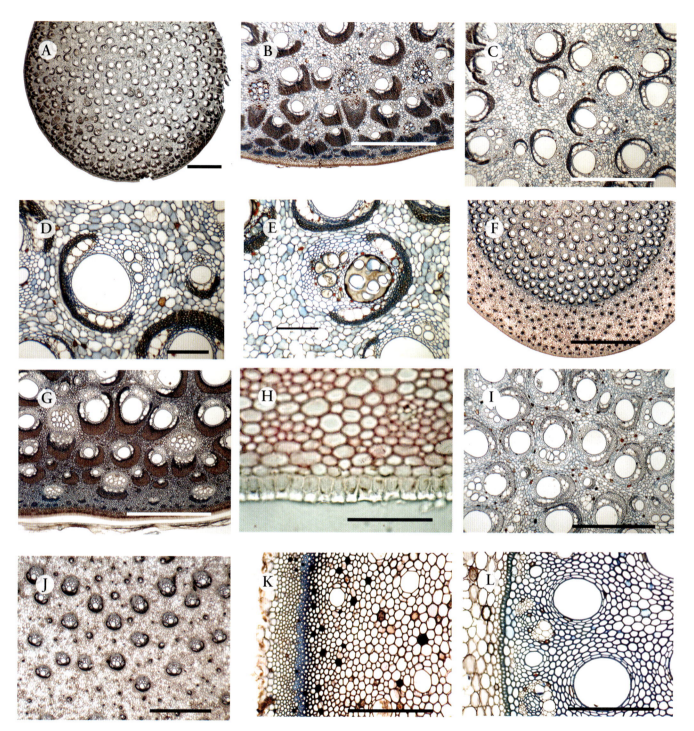

Fig. 41 Calamoideae (Calameae: Calaminae). **A.** *Calamus deerratus*, stem, TS (bar = 2 mm). **B.** *Calamus deerratus*, stem periphery, TS (bar = 1 mm). **C.** *Calamus deerratus*, stem centre, TS (bar = 1 mm). **D.** *Calamus deerratus*, stem central vb, TS (bar = 200 μm). **E.** *Calamus deerratus*, stem vb with tyloses, TS (bar = 200 μm). **F.** *Calamus deerratus,* rhizome, TS (bar = 2 mm). **G.** *Calamus* sp., stem periphery and leaf sheath, TS (bar = 1 mm). **H.** *Calamus gracilis*, Stem periphery with silicified epidermis, TS (bar = 100 μm). **I.** *Daemonorops angustifolia*, stem centre, TS (bar = 1 mm). **J.** *Daemonorops angustifolia*, petiole vbs, TS (bar = 2 mm). **K.** *Calamus deerratus*, root periphery, TS (bar = 200 μm). **L.** *Calamus deerratus*, root outer stele with endodermis, TS (bar = 200 μm).

RHIZOME (Fig. 41F)

Epidermis not silicified and little cutinized. **Cortex** wide, including few fbs but without conspicuous leaf traces. **Central cylinder** enclosed by a sclerotic layer of the inner cortex. Central **vbs** usually with a single phl strand. Central ground tissue with wide intercellular spaces; included small vbs not observed.

Note: a study to establish if the rhizome of rattans has the basic *Rhapis*-principle of vascular construction and so contrasts with that of the aerial stem has not been carried out but is much desired.

Root (Fig. 41K, L)

Surface layers (Fig. 41K) with 4–5 distinct layers radially transitional from one to the next. **Epidermis** thin-walled, large-celled, the outer wall often appearing conical in TS. **Hypodermis** of 2–3 thin-walled cells transitional to narrow thicker-walled cells completed internally by a narrow layer of thicker-walled sclerenchyma transitional in turn to inner cortex with wide mucilage canals. **Inner cortex** with radiating air canals, fibres absent. **Endodermis** with uniform wall thickenings, O-shaped in TS (Fig. 41L). **Medulla** including few medullary vessels.

Vascular elements

Vessel elements in root with simple transverse perforation plates; in stem wide and >200 um in diameter and the entire vessel >3 m long (Fisher et al. 2002) with simple transverse perforation plates; in leaf varying from simple transverse to long scalariform perforation plates on oblique end walls. This might be seen as a derived condition in relation to the hydraulic benefits of reduced resistance to water flow. **Sieve tube elements** in leaf and stem with simple sieve plates on transverse end walls; in root with compound sieve plates on oblique end walls.

OTHER GENERA OF CALAMINAE

The other genera of Calaminae fall within the considerable range of variation we have seen even within our still limited study of *Calamus* and so offer no distinguishing generic characters.

Daemonorops (Figs. 8C; 40I, J; and 41I, J)

As pointed out in Tomlinson (1961), *Daemonorops*, though morphologically distinct from *Calamus* shows a similar range of anatomical features so that the 2 genera cannot be consistently separated at the structural level. In view of the large size of both genera and the considerable variation in leaf morphology along there very extended stems which show a transition from juvenile to adult phases, any more precise claim would be misleading. A few features may be mentioned.

Leaf

LAMINA (Fig. 40I, J)

Dorsiventral (Fig. 40I). **Trichomes**, if present, with a sclerotic base extended into a distal filament (Fig. 8C). **Epidermis** with rectangular elongated cells and sinuous walls. **Stomata** (Fig. 40J) with ± dumbell-shaped g.c. **Hypodermis** represented only by fibres or fbs, additional fibre strands sometimes present in the mid-mesophyll at the level of the smaller veins (Fig. 40I). **Mesophyll** with or without palisade layers, sometimes including large **tannin cells. Transverse veins** with a well-developed fibrous sheath, the fibres seemingly transitional to mesophyll fibre-sclereids. **Fibre-sclereids** common in the mesophyll in some spp., and associated with the transverse veins.

PETIOLE (Fig. 41J)

Central vbs with 1 wide mxy vessel and 2 separate phl strands; ground tissue including small scattered fbs and vbs (Fig. 41J).

Stem (Fig. 41I)

Large central vbs with 1 wide mxy vessel and 2 lateral phl strands. Vascular construction as seen in single sections corresponding closely to the unusual anatomy described for *Calamus*, with small vbs of varying size interspersed among the large vbs, and frequent transverse commissures, implying a similar disconnected hydraulic system.

Root

As in *Calamus*, without cortical fibres; **endodermis** with O-type cells; **stele** with frequent medullary vessels.

Vascular elements

Distribution as in *Calamus*, but elements in leaf axis sometimes with uniformly reticulate perforation plates.

Ceratolobus (Fig. 40H, L)

LAMINA

Dorsiventral (Fig. 40 H). **Trichomes** present on both surfaces, most abundant abaxially; each with a basal series of thick-walled pitted cells, extending into a distal filament. **Epidermis** with rectangular unextended cells with sinuous anticlinal walls, outer wall thickened. Adaxial epidermis with occasional stomata. **Stomata** (Fig 40L) with t.s.c. and neighbour cells thick-walled, pitted; g.c. cells with 2 cutinized ledges, wall thickened except at the thin-walled poles. **Hypodermis** of colourless cells absent but including fibres. **Mesophyll** with or without a 1–2-layered palisade. **Fibres** almost restricted to hypodermal layers except for occasional cells in the adaxial mesophyll, sometimes forming an almost continuous adaxial layer, the individual fibres thick-walled and with a v. narrow lumen. **Veins** in mid- to abaxial mesophyll, mostly independent of surface layers except for few large veins sometimes continuous with surface via sclerotic cells. **Transverse veins** running above longitudinal veins, wide with a well-developed fibrous sheath, the ends of the fibres extending into the adjacent mesophyll.

Pogonotium (Fig. 40M)

LAMINA

Dorsiventral. **Trichomes** common on both surfaces, usually situated above or below a hypodermal fibre strand (Fig. 40M). **Epidermis** with sinuous anticlinal walls; thin-walled abaxially. **Stomata** with thick-walled t.s.c. and neighbour cells. **Mesophyll** with palisade layers including enlarged colourless or tannin cells; **fibres** exclusively hypodermal as small strands.

Retispatha

LAMINA

Dorsiventral. **Trichomes** absent. **Epidermis** with sinuous anticlinal walls. **Hypodermis** absent as a colourless layer, but with frequent fibre-bundles. **Mesophyll** without a palisade layer. **Transverse veins** with massive fibre sheaths, the tips of the fibres extending into the mesophyll. **Stegmata** distinctive, essentially spherical but with few long surface spicules.

SUBFAMILY NYPOIDEAE

INTRODUCTION

This monotypic subfamily is represented by the mangrove palm, *Nypa fruticans*, which is naturally distributed from Sri Lanka, throughout SE Asia to northern Australia, and many of the western Pacific Islands. It has been introduced to other parts of the tropics. *Nypa* (Fig. 42A) has a number of biological attributes which are unique within palms: a dichotomously branched massive rhizome (Tomlinson 1971b) submerged at high tide (Fig. 42B); leaves without a well-developed sheath; reorientation of each leaf in the phyllotactic spiral into an erect position; axillary inflorescences with lateral wholly male branches below a terminal female rachilla (Fig. 42C); possibly fly pollination; an incipiently viviparous seedling which seems inverted in its morphology because the plumule is exserted first (Fig. 42E). For details see Tomlinson (1986).

Nypa occupies estuarine mud in conditions where there is limited wave action and its clonal spread can produce extensive pure stands. It represents one extreme of the considerable ecological spectrum exhibited by palms. Where introduced, as in West Africa, it can behave as an invasive species. In the Asian tropics the palm has considerable economic value as a source of thatch; its sugary phloem sap is tapped from the inflorescence, and its developing endosperm is a jelly-like sweetmeat. It is the basis for a local wine industry in Malaysia and has been suggested as a possible source of biofuel (Fong 1986, 1987), hence gasoline from seawater!

The fossil record of *Nypa* is one of the oldest attributable to a modern angiosperm genus (Muller 1981). Based on occurrences of the fossil pollen genus *Spinizonocolpites* from the Maastrichtian of Central and South America, Africa, India, and Asia, *Nypa* attained a near global distribution by the Late Cretaceous (Muller 1981; Gee 2001; Dransfield et al. 2008b; Gomez-Navarro et al. 2009). Tertiary (notably Eocene) fossil pollen and fruits attributable to *Nypa* imply that it sustained, if not increased, its distribution in tropical and temperate paleoclimates worldwide. However, apparent absence of *Spinizonocolpites* throughout much of its former range in the Middle Miocene implies that the current distribution of *Nypa* is best explained by extinction throughout much of its range. It remains unknown to what extent the lineage, now represented by a single species, might have diversified during its history. *Nypa* represents one of the earliest-diverging lineages of palms, sister to a clade containing subfamilies Coryphoideae, Ceroxyloideae, and Arecoideae (Fig. 42D).

LEAF

Pinnately compound, up to 5 m long, petiole terete, sheath with a basal groove (or 2 grooves at the level of a stem dichotomy). Leaflets reduplicate, with large brown scales on the abaxial surface of the midrib, but no other visible indumentum.

Lamina (Fig. 43)

Dorsiventral (Fig. 43A, C). **Trichomes** (Fig. 43E, G), possible hydathodes, uniformly scattered on both surfaces, each with 2 small deeply sunken cutinized cells, often with a protruding thin-walled cell between them; distal cells not produced. **Epidermis** shallow, outer wall thick and cutinized (Fig. 43H). Adaxial cells somewhat transversely extended (Fig. 43D, E), costal regions not differentiated. Abaxial epidermis with narrow costal bands of rectangular, somewhat longitudinally-extended cells; intercostal bands wide, with irregular cells, especially around stomata, but similar to those of upper epidermis. **Stomata** (Fig. 43I) diffuse, cells in same files often short and wide; t.s.c. short, overarching g.c. somewhat; l.s.c. deep, extending well into hypodermal layers (Fig. 43J). G.c. large, sunken, each with a prominent inner and outer ledge, the slightly thickened wall adjacent to the stomatal pore bearing many smaller ledges, giving in TS a toothed appearance (Fig. 43J). **Hypodermis** 2-layered adaxially, 2–3-layered abaxially; outer adaxial cells hexagonal to spindle-shaped and transversely extended in surface view (Fig. 43E) and with thickened, lignified, pitted walls (Fig. 43C, E, H). Inner adaxial cells thin-walled and cubical. **Mesophyll** with a 1–2-layered adaxial palisade (Fig. 43C), the abaxial cells more isodiametric become looser near the s.st.ch. **Fibres** in the mesophyll forming a distinct adaxial and abaxial series between the larger veins, rarely in contact with the hypodermis; the fibre strands cylindrical or sl. anticlinally extended, those in the adaxial mesophyll usually the largest (Fig. 43C). **Veins** (Fig. 43A, C) divisible into 3 main types: 1) frequent small veins independent of surface layers in the mid-mesophyll, the O.S. of cubical cells usually complete; 2) medium-sized veins continuous to each hypodermis by colourless, slightly thick-walled vertically extended mesophyll cells

Fig. 42 *Nypa fruticans*. **A**. Habit of plant (cultivated at Allerton Garden, Hawaii). **B**. Habit of plant in fruit at high tide (Labutale, Papua New Guinea). **C**. Inflorescence with terminal pistillate and lateral staminate rachillae (cultivated at Allerton Garden, Hawaii). **D**. Summary phylogeny of Arecaceae showing position of Nypoideae; anatomical autapomorphies for Nypoideae indicated; formal optimization of character states not shown. **E**. Germination of a viviparous embryo, young embryo on left, young seedling in fruit on right.

113

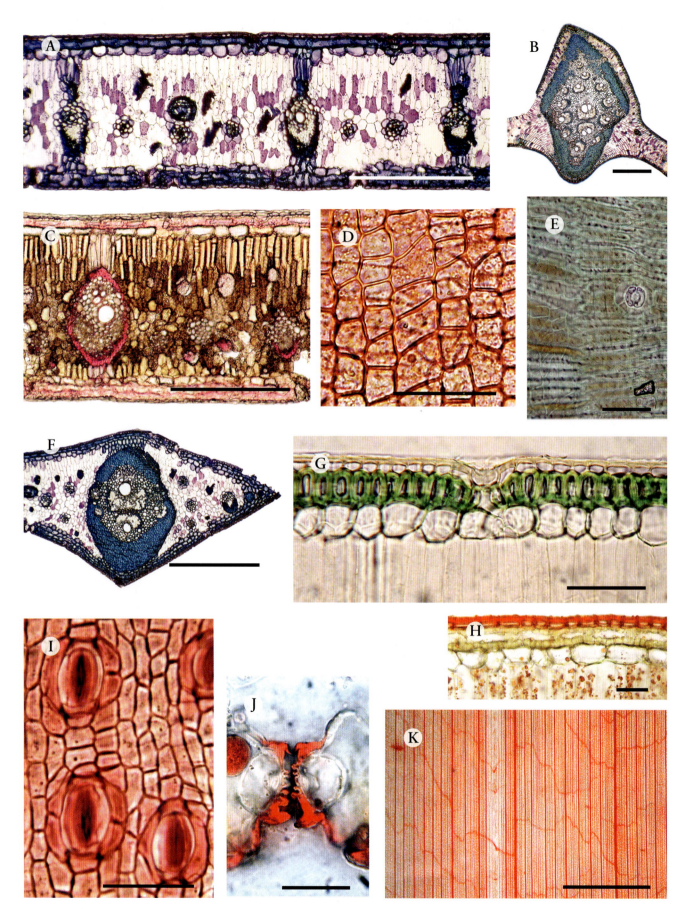

Fig. 43 *Nypa fruticans*, lamina. **A.** Lamina, TS (bar = 250 μm). **B.** Adaxial rib, TS (bar = 500 μm). **C.** Lamina, TS (bar = 250 μm). **D.** Epidermis, adaxial, SV (bar = 50 μm). **E.** Trichome ('hydathode') adaxial surface (epidermal cells in focus on left, hypodermis on right), SV (bar = 50 μm). **F.** Marginal vein, TS (bar = 500 μm). **G.** Adaxial surface with sunken trichome ('hydathode'), TS (bar = 50 μm). **H.** Adaxial surface without trichome, TS (bar = 20 μm). **I.** Stomata, abaxial epidermis, SV (bar = 20 μm). **J.** G.c. of stomata, TS (bar = 25 μm). **K.** Cleared lamina, SV (bar = 1 mm).

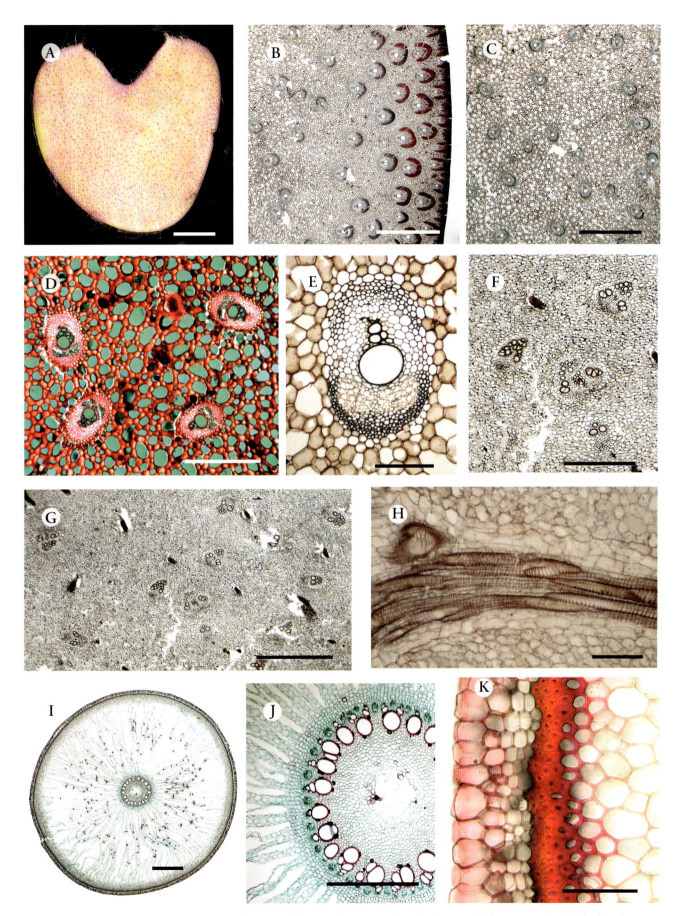

Fig. 44 *Nypa fruticans*. **A**. Petiole, TS (bar = 1 cm). **B**. Petiole periphery, TS (bar = 2 mm). **C**. Petiole centre, TS (bar = 2 mm). **D**. Petiole vb and aerenchyma, TS (bar = 1 mm). **E**. Petiole vb and aerenchyma matrix, TS (bar = 1 mm). **F**. Stem central vb, TS (bar = 1 mm). **G**. Stem central vbs, TS (bar = 2 mm). **H**. Stem central vb, LS (bar = 200 μm). **I**. Root, TS (bar = 1 mm). **J**. Root stele, TS (bar = 500 μm). **K**. Root epidermis and superficial layers, TS (bar = 100 μm).

(Fig. 43A); 3) few large veins with shallow mesophyll connections. I.S. only well developed in larger veins, with adaxial and abaxial fibres. **Phl** of larger veins not sclerotic, undivided. **Transverse veins** (Fig. 43K) diffuse, as wide as small longitudinal veins, each with a wide tracheary element and several phl strands; sheathed by sclerenchyma. **Midrib** (Fig. 43B) prominent adaxially with a sclerotic cylinder including small vbs and fibre strands. **Marginal rib** (Fig. 43F) also including smaller peripheral or inverted vbs. **Expansion cells** abaxial (Fig. 43B).

Petiole (Fig. 44A–F)

Petiole without a distinct V-shaped vb configuration (Fig. 44A). Hypodermis sclerotic (Fig. 44B). Surface layers sometimes suberized and divided as in an etagen meristem. Central **ground tissue** markedly lacunose in the basal portion of the petiole by enlargement of intercellular spaces, forming a reticulate arrangement of mostly uniseriate cell plates (Fig. 44C–E). Peripheral vbs congested and each with a massive sclerotic phl sheath (Fig. 44B). Central vbs diffuse, each with 1 wide mxy vessel and a single phl strand (Fig. 44F). **Transverse commissures** infrequent, fibre strands absent.

Leaf sheath

Vbs inverted adaxially, sclerotic tissue restricted to adaxial surface, ground tissue as in petiole. **Fibre strands** (cf. petiole) numerous, especially in surface layers. **Tannin** cells frequent, as in ground tissue and phl parenchyma.

STEM (Fig. 44G, H)

Rhizomatous, with v. irregular distribution of vbs (Fig. 44G). The '*Rhapis*-principle' of vascular construction has not been established. **Surface layers** with frequent etagen-type cell divisions producing a suberized but not lignified tissue. **Cortex** wide with a loose, lattice-like texture. Fbs frequent but vbs few; root traces often visible. **Central cylinder** (Fig. 44F–H) with diffuse instead of crowded vbs and with irregular axial orientation. **Vbs** narrow with a thin-walled sheath (Fig. 44F, H) except for somewhat thick-walled fibres next to phl. Mxy of narrow elements, usually 2 narrow vessels, but not v. regular; pxy extensive. Phl as a single strand. **Ground tissue** also including numerous wide fibre strands, circular in TS with up to 30 sl. thick-walled cells, together with small vbs. Both fbs and small vbs can appear as if freely suspended in the lacunose ground tissue. **Tannin** cells common. **Raphide sacs** not observed.

ROOT (Fig. 44I–K)

Up to c. 5 mm wide; never v. sclerotic, **stele narrow**, without vascular anomalies. **Surface layers** narrow. **Epidermis** (Fig. 44K) not papillate, thin-walled, the outer wall scarcely thickened. **Hypodermis** 3–4 layers, also thin-walled, the outermost layer tannin rich, outermost layer of cortex a 2–3-celled lignified thick-walled sclerotic layer with c. 10 cell layers within. **Middle cortex** (Fig. 44I) v. wide, lacunose with radiating mostly uniseriate plates of wide cells separated by plates of collapsed cells that form the intercellular space system. **Innermost cortex** with 3–4 layers of small compact cells. **Stele** narrow, with a narrow medulla of collapsed cells (Fig. 44J). **Endodermis** remaining thin-walled, but nevertheless lignified; pericycle also thin-walled. Mxy vessels narrow, continuous with short pxy poles; phl. strands narrow, circular in TS. Conjunctive tissue remaining thin-walled.

VASCULAR ELEMENTS

Xylem: mxy elements of root with scalariform perforation plates on long oblique end walls and numerous narrow thickening bars; those of stem, short and with short oblique scalariform perforation plates, which include few thickening bars (Fig. 44H); those of **petiole** with long v. oblique scalariform perforation plates with many bars. **Phl** of petiole and stem with simple sieve plates on transverse end walls, but unreported for the root.

CELL INCLUSIONS

Tannin cells common, e.g. in ground tissue of stem and root. **Raphide sacs** absent from root, but inconspicuous in stem and leaf. **Stegmata** abundant in discontinuous files next to leaf fibres, inconspicuous in stem and absent from root. Silica body **hat-shaped**, the basal wall of the silica cell scarcely thickened.

Notes: *Nypa* is sister to all other palms except Calamoideae (Fig. 42D). It is distinguished by many features that may relate to its unique habit, e.g. possible glandular trichomes or hydathodes, multiple cuticular ledges of the g.c. and well-developed colourless hypodermis with the outer layer sclerotic. Mechanical tissues are only developed extensively in the leaves, in relation to their erect orientation, with the distal portion of old leaves breaking at a quite standard level, the scar rounded and sealed. The leaf axis may function in gas exchange in much the same way as a pneumatophore, facilitated by its aerenchymatous ground tissue and development of lenticels.

SUBFAMILY CORYPHOIDEAE

INTRODUCTION

The Coryphoideae (Fig. 45) constitute the second largest sub-family of Arecaceae with respect to the number of genera (46) and the third with regard to number of species (Dransfield et al. 2008b). The distribution and diversity of this pan-tropical assemblage is so great that we confine its details to the description of individual groups. The genera are distributed among seven tribes of markedly unequal size; three contain only a single genus (Fig. 46). The current classification of the subfamily (Dransfield et al. 2005, 2008b) is largely based on the molecular phylogenetic study of Asmussen et al. (2006), and, with a few notable exceptions, is broadly congruent with previous classificatory schemes for the subfamily based on structural data alone (e.g. Uhl and Dransfield 1987). Beyond the current classification, the phylogenetic evidence allows for the recognition of four (or possibly just three) well-supported clades within the subfamily (Fig. 46): 1) New World thatch palm clade; 2) *Phoenix*; 3) Trachycarpeae [+ *Phoenix*]; and 4) syncarpous clade (Dransfield et al. 2008a, b).

The first of these clades contains the monogeneric Sabaleae and sister tribe Cryosophileae. A recent study by Roncal et al. (2008) makes substantial advances in clarifying the relationships among the genera of Cryosophileae, and provides the phylogenetic framework with which we discuss anatomical evolution within the group. The (possible) second major clade consists of the monogeneric Phoeniceae and sister tribe Trachycarpeae. This is the best supported hypothesis of relationship for *Phoenix* in any analysis to date (Asmussen et al. 2006; nearly complete sampling of the family at the generic level with four plastid loci). But as other studies do suggest alternative hypotheses, the relationship of *Phoenix* to other Coryphoideae is probably best considered to be an unsettled issue at this time. In less broadly sampled studies (Crisp et al. 2009; Bacon et al. in subm.) or else in those integrating morphological data (Baker et al. 2009), *Phoenix* is resolved as sister to the New World thatch palm clade with poor support; hence Dransfield et al. (2008b) treat *Phoenix* as a separate major clade of Coryphoideae. Bacon et al. (in subm.) have made great strides in providing a well-supported phylogeny of Trachycarpeae, and we employ their data for our analysis. The final of the three major clades of Coryphoideae is the syncarpous clade (Dransfield et al. 2008a, b), which contains tribes Chuniophoeniceae, Caryoteae, Corypheae, and Borasseae. Published estimates of the phylogeny of the syncarpous clade using plastid molecular data (Asmussen et al. 2006; Dransfield et al. 2008a) are congruent with a more detailed study by Bayton (2005).

Coryphoideae are perhaps the most structurally diverse of the five palm subfamilies, yet they possess a suite of leaf morphological features that clearly characterize the group. Nearly all Coryphoideae are readily distinguishable from other palms in possessing palmate leaves with induplicate segmentation (i.e. the single-folds V-shaped in cross section), and both these character states are probable synapomorphies for the subfamily. There are a limited number of exceptions to the above generality, occurring in taxa that are phylogenetically well nested within the group, and which always express one of the two synapomorphic character states. Pinnate leaves have independent origins within Coryphoideae in *Phoenix* and tribe Caryoteae (Horn et al. 2009), but these taxa still have induplicately folded pinnae. Likewise, a few genera of subtribe Rhapidinae (Trachycarpeae) have unusual types of leaf splitting but always have palmate leaves, with *Guihaia* unlike all other members of the subfamily in its reduplicately segmented leaves. Most palmate-leaved taxa also have an adaxial hastula present at the junction of the petiole and lamina. Although an assessment of the homology and evolution of the hastula remains a topic in need of further inquiry, current knowledge suggests it is unique to Coryphoideae (cf. Mauritiinae).

The most obvious anatomical synapomorphy for the subfamily is the structure of the *adaxial* rib, which in nearly all genera includes at least five large, independent vascular bundles. Other palms differ from this state in that the vascular bundles of the adaxial rib are enclosed within a complete cylinder of sheathing sclerenchyma. Although pinna or segment midrib structure has been previously adopted for comparative purposes, examining adaxial ribs across palm taxa is the logical choice from both the standpoint of homology (i.e. the structures are part of the same series) and from the biological consideration that reduplicate-leaved palms do not produce abaxial ribs. Many genera of Coryphoideae do express the characteristic 'Coryphoid' rib structure in the abaxial ribs of the unsegmented palman, but our investigation reveals a hitherto unappreciated wealth of diversity in abaxial rib structure within the subfamily that is not present in their adaxial ribs (Fig. 47A). Other character states reconstructed as synapomorphies for Coryphoideae, but with a more complex distribution both within and outside of the subfamily include:

Fig. 45 Coryphoideae. **A**. *Bismarckia nobilis*, young tree with waxy leaves. **B**. *Borassus aethiopum*, natural population on the Accra Plains, Ghana. **C**. *Corypha utan*, tree in fruit with most leaves fallen. **D**. *Caryota mitis*, twice compound leaf. **E**. Summary phylogeny of Arecaceae with anatomical synapomorphies for subfamily. **F**. *Trachycarpus fortunei*, solitary palm cultivated in Yorkshire, England. **G**. *Phoenix reclinata*, plant with multiple stems. **H**. *Hyphaene indica*, tree with dichotomous stem. **I**. *Arenga hookeriana*, understorey treelet.

Text in figure E:

Calamoideae
Nypoideae
Coryphoideae
Ceroxyloideae
Arecoideae

Adaxial rib including 5 or more independent vascular bundles

Nonvascular fibres always in contact with surface layers

Transverse veins sheathed by many layers of fibers

Adaxial fibes septate with wide lumina

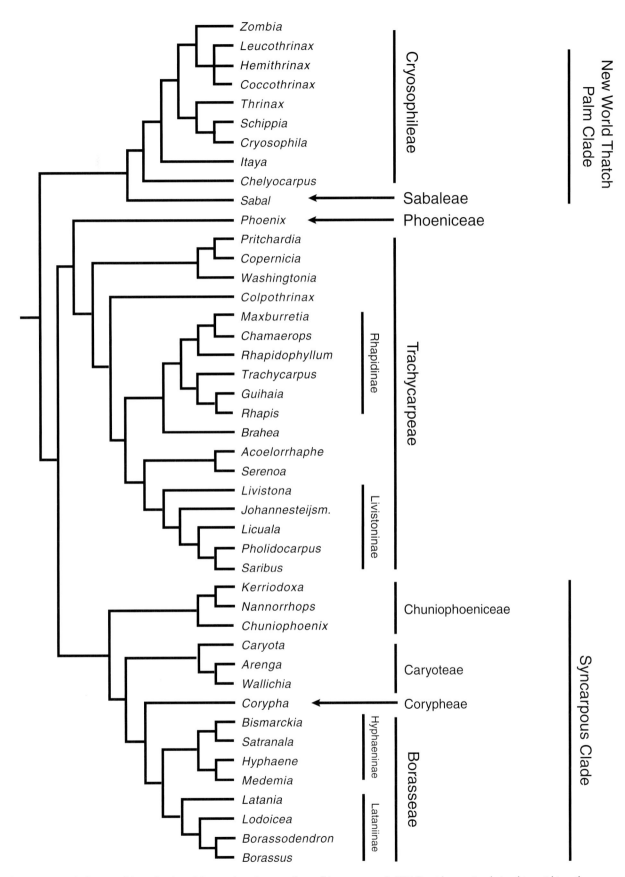

Fig. 46 Summary phylogeny of Coryphoideae. Metatree based on topology of Asmussen et al. (2006), with generic relationships within tribe Cryosophileae resolved according to Roncal et al. (2008), and those within tribe Trachycarpeae resolved according to Bacon et al. (in subm.). *Tahina* (Chuniophoeniceae) and *Trithrinax* (Cryosophileae) are not represented.

A. Abaxial rib structure

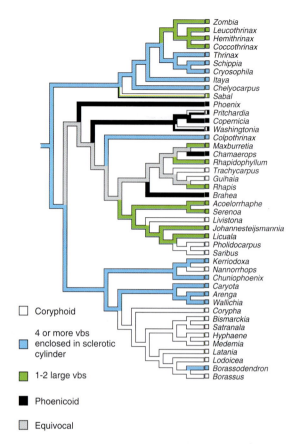

Coryphoid

4 or more vbs enclosed in sclerotic cylinder

1-2 large vbs

Phoenicoid

Equivocal

B. Attachment of longitudinal veins to adaxial surface layers of lamina by fibre girder

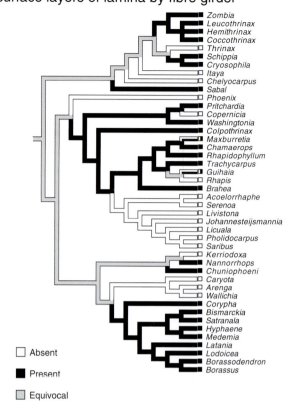

Absent

Present

Equivocal

C. Fibre type present adaxially in lamina

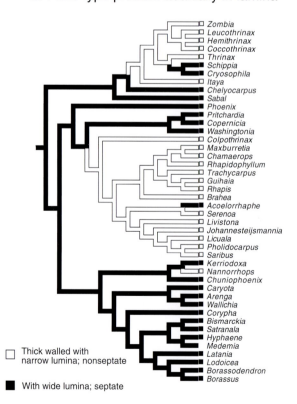

Thick walled with narrow lumina; nonseptate

With wide lumina; septate

D. Position of transverse veins in lamina

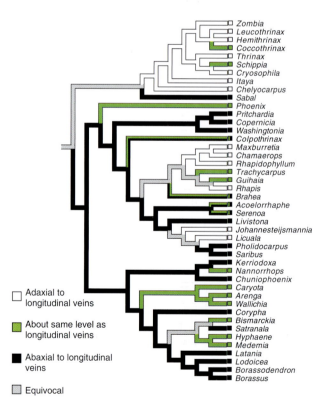

Adaxial to longitudinal veins

About same level as longitudinal veins

Abaxial to longitudinal veins

Equivocal

Fig. 47 Parsimony ancestral state reconstructions of anatomical characters within Coryphoideae. All reconstructions shown are optimized on the tree indicated in Fig. 46. Terminals without a coloured box adjacent to the branch tip have missing data. Number of steps reported for each optimization is the value within the subfamily. **A.** Abaxial rib structure optimizes with 18 steps. **B.** Attachment of longitudinal veins to adaxial surface layers by fibre girder optimizes with 11 steps. **C.** Fibre type optimizes with 5 steps. **D.** Position of transverse veins in lamina optimizes with 17 steps.

A. Vasculature of lamina segment margin

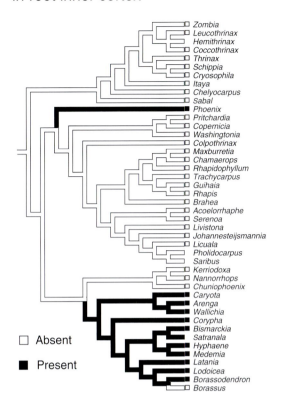

Zombia
Leucothrinax
Hemithrinax
Coccothrinax
Thrinax
Schippia
Cryosophila
Itaya
Chelyocarpus
Sabal
Phoenix
Pritchardia
Copernicia
Washingtonia
Colpothrinax
Maxburretia
Chamaerops
Rhapidophyllum
Trachycarpus
Guihaia
Rhapis
Brahea
Acoelorrhaphe
Serenoa
Livistona
Johannesteijsmannia
Licuala
Pholidocarpus
Saribus
Kerriodoxa
Nannorrhops
Chuniophoenix
Caryota
Arenga
Wallichia
Corypha
Bismarckia
Satranala
Hyphaene
Medemia
Latania
Lodoicea
Borassodendron
Borassus

□ Without stout marginal vein

■ Including stout marginal vein

▨ Equivocal

B. Presence of Raphia type fibre bundles in root inner cortex

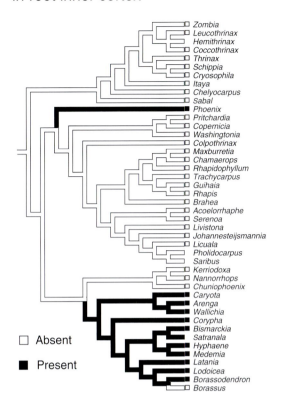

Zombia
Leucothrinax
Hemithrinax
Coccothrinax
Thrinax
Schippia
Cryosophila
Itaya
Chelyocarpus
Sabal
Phoenix
Pritchardia
Copernicia
Washingtonia
Colpothrinax
Maxburretia
Chamaerops
Rhapidophyllum
Trachycarpus
Guihaia
Rhapis
Brahea
Acoelorrhaphe
Serenoa
Livistona
Johannesteijsmannia
Licuala
Pholidocarpus
Saribus
Kerriodoxa
Nannorrhops
Chuniophoenix
Caryota
Arenga
Wallichia
Corypha
Bismarckia
Satranala
Hyphaene
Medemia
Latania
Lodoicea
Borassodendron
Borassus

□ Absent

■ Present

C. Cell wall thickening pattern of root endodermal cells

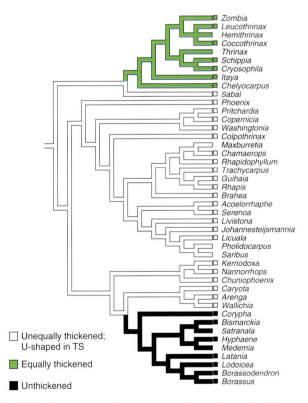

Zombia
Leucothrinax
Hemithrinax
Coccothrinax
Thrinax
Schippia
Cryosophila
Itaya
Chelyocarpus
Sabal
Phoenix
Pritchardia
Copernicia
Washingtonia
Colpothrinax
Maxburretia
Chamaerops
Rhapidophyllum
Trachycarpus
Guihaia
Rhapis
Brahea
Acoelorrhaphe
Serenoa
Livistona
Johannesteijsmannia
Licuala
Pholidocarpus
Saribus
Kerriodoxa
Nannorrhops
Chuniophoenix
Caryota
Arenga
Wallichia
Corypha
Bismarckia
Satranala
Hyphaene
Medemia
Latania
Lodoicea
Borassodendron
Borassus

□ Unequally thickened; U-shaped in TS

▨ Equally thickened

■ Unthickened

D. Ground tissue type in root stele

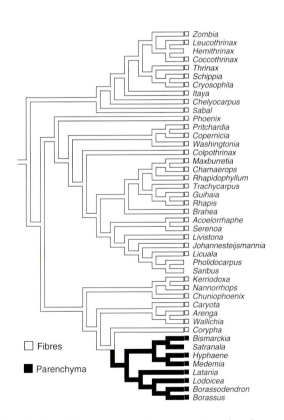

Zombia
Leucothrinax
Hemithrinax
Coccothrinax
Thrinax
Schippia
Cryosophila
Itaya
Chelyocarpus
Sabal
Phoenix
Pritchardia
Copernicia
Washingtonia
Colpothrinax
Maxburretia
Chamaerops
Rhapidophyllum
Trachycarpus
Guihaia
Rhapis
Brahea
Acoelorrhaphe
Serenoa
Livistona
Johannesteijsmannia
Licuala
Pholidocarpus
Saribus
Kerriodoxa
Nannorrhops
Chuniophoenix
Caryota
Arenga
Wallichia
Corypha
Bismarckia
Satranala
Hyphaene
Medemia
Latania
Lodoicea
Borassodendron
Borassus

□ Fibres

■ Parenchyma

Fig. 48 Parsimony ancestral state reconstructions of anatomical characters within Coryphoideae. All reconstructions shown are optimized on the tree indicated in Fig. 46. Terminals without a coloured box adjacent to the branch tip have missing data. Number of steps reported for each optimization is the value within the subfamily. **A**. Vasculature of lamina segment margin optimizes with 8 steps. **B**. 'Raphia fibres' in root inner cortex optimizes with 3 steps. **C**. Cell wall thickening pattern of endodermis optimizes with 2 steps. **D**. Ground tissue type in root stele optimizes with 1 step.

1) the presence of wide-lumened, septate fibres (Fig. 47C); 2) transverse veins sheathed by many layers of fibres; and 3) the absence of non-vascular fibres 'free' in the mesophyll, without connection to the surface layers (Horn et al. 2009).

Within Coryphoideae, anatomical data is of greatest systematic significance among the tribal clades and within less inclusive levels of hierarchy. We found no unequivocal anatomical synapomorphies that support the relationships among the major clades. Analyses of most characters reveal substantially homoplasious histories, but nevertheless they allow for the discovery of synapomorphies for many clades. In the following discussion, we highlight the characters of greatest systematic significance within each major clade. This discussion is illuminated by the cladograms of Figs. 47–50 and can be supplemented with the illustrations in the later detailed diagnoses.

Phylogenetic analysis of Coryphoideae clades

Clade 1: *New World thatch palm clade* (Fig. 46)

The most outstanding feature that characterizes this clade is that the lamina of most genera is deeply divided by a central abaxial split to the apex of the leaf axis (Dransfield et al. 2008b). Leaf anatomical data are not strongly informative of the relationship between the two tribal subclades, as most shared character states are clearly plesiomorphies (Figs. 47–50; Horn et al. 2009).

The polarity of most character states that collectively distinguish *Sabal* from other Coryphoideae remains ambiguous. From lamina anatomy, these states include the presence of a fibre girder that connects the longitudinal veins to the adaxial surface layers (Fig. 47B), transverse veins situated abaxial to the longitudinal veins (Fig. 47D), a substomatal chamber delimited by four L-shaped cells (Fig. 49D), and isolateral histology (Horn et al. 2009). The Coryphoid abaxial rib structure (sometimes including 1–2 large vascular bundles; Zona 1990) in *Sabal* evolved independently from other Coryphoideae, and optimizes as an autapomorphy for the genus (Fig. 47A). Perhaps unique to *Sabal* are sunken stomata with guard cells that have markedly unequally thickened walls (Tomlinson 1961). The absence of trichomes on the lamina is also noteworthy.

In contrast to *Sabal*, Cryosophileae is distinguished by several anatomical synapomorphies. The first of these is the loss of all stomata from the adaxial surface of the lamina, which characterizes all genera except for the isolateral *Trithrinax* (not sampled in the phylogeny). The lamina of many dorsiventral taxa of Coryphoideae has adaxial stomata in lower density than on the abaxial surface (Fig. 50A). Another lamina anatomy character that is synapomorphic for Cryosophileae is the gain of non-vascular fibres in contact with the adaxial epidermis (Fig. 50D). This character state is independently derived elsewhere in Coryphoideae, but it is found in few other genera. From root anatomical data (Seubert 1997), a third synapomorphy for Cryosophileae is the presence of endodermal cells with equally (and often strongly) thickened walls (Fig. 48C), which are unlike those of other Coryphoideae.

The position of the transverse veins in the lamina is also distinctive for Cryosophileae, as they are typically adaxial to the longitudinal veins. This state is reconstructed as ancestral for the tribe, but its polarity remains equivocal (Fig. 47D).

Within Cryosophileae, the presence of a large marginal vein in the lamina segment is a synapomorphy for the clade inclusive of *Zombia* and *Cryosophila* (Fig. 48A); but the feature is absent from *Cryosophila*. Non-septate fibres with thick walls and narrow lumina are present in the clade encompassing *Zombia* and *Itaya* (Fig. 47C). A clade of Caribbean endemics including *Zombia* and *Coccothrinax* is characterized by the reduction in vascular and mechanical tissue in the abaxial rib of the leaf (Fig. 47A). This is a surprising result in light of the studies of Uhl (1972) and Evans (1995), who investigated the lamina anatomy of *Chelyocarpus*, *Cryosophila*, *Itaya*, and *Schippia*, which current evidence indicates are early-divergent groups within the tribe. These authors regarded the abaxial ribs of the above genera, which have a sclerotic cylinder that encloses usually at least four vascular bundles, as being anatomically unusual within Coryphoideae. Our analysis indicates that this type of abaxial rib structure is plesiomorphic within the subfamily (Fig. 47A). That *Thrinax* s.s. possesses an abaxial rib structure like that of the above four genera is congruent with current molecular phylogenetic hypotheses of its relationship within the tribe (Lewis and Zona 2008; Roncal et al. 2008). Thus, the removal of the satellite genus *Hemithrinax* and the recognition of *Leucothrinax* as distinct from *Thrinax* s.s. is justified in light of both structural and molecular data.

Clade 2: Phoenix (Fig. 46)

Phoenix possesses a distinctive combination of character states in its lamina anatomy that make it among the most recognizable of Coryphoid genera. The abaxial rib hardly protrudes below the pinna and is almost devoid of vascular tissue so that the body of the rib is largely occupied by expansion tissue and few fibre bundles (Fig. 55B). This 'phoenicoid' rib otherwise is found in Coryphoideae in five genera of Trachycarpeae, but most notably in a clade including *Pritchardia*, *Copernicia*, and *Washingtonia*, which is sister to the rest of the tribe. Although the results of character analysis for abaxial rib structure allow for multiple evolutionary interpretations, a majority of the most parsimonious reconstructions optimize the phoenicoid rib as a synapomorphy for the clade of *Phoenix* + Trachycarpeae (Fig. 47A). The second lamina anatomy character of systematic significance in *Phoenix* is the architecture of its substomatal chamber, which is unusual in that 3–9 cells interlock to form up to six irregularly shaped apertures (Fig. 9J). Three genera of Rhapidinae (Trachycarpeae) have a substomatal chamber like that of *Phoenix*, but our results yield an equivocal interpretation of the evolution of this character state (Fig. 49D). In other palms the substomatal chamber is most often uniaperturate and delimited by either two C-shaped or four L-shaped cells. The most conspicuous exceptions are found in Borasseae.

In addition, *Phoenix* has many distinctive character states that have a complex distribution elsewhere in Arecaceae. Among these is the presence of short, sinuous transverse lamina veins each sheathed by 1–2 layers of sclerotic parenchyma (Horn et al. 2009).

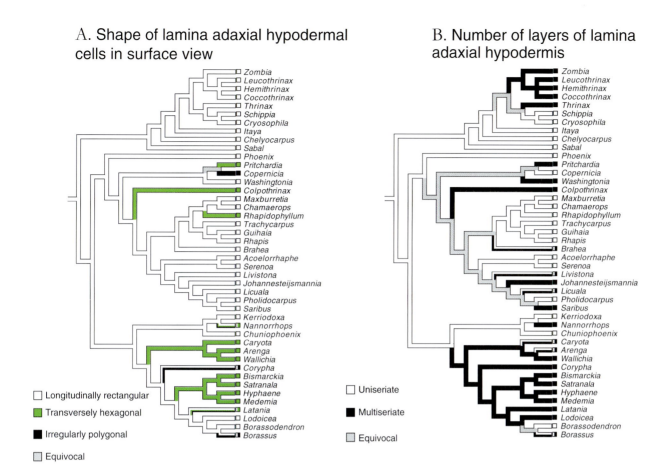

A. Shape of lamina adaxial hypodermal cells in surface view

- ☐ Longitudinally rectangular
- ▣ (green) Transversely hexagonal
- ▣ (black) Irregularly polygonal
- ▣ (grey) Equivocal

B. Number of layers of lamina adaxial hypodermis

- ☐ Uniseriate
- ▣ (black) Multiseriate
- ▣ (grey) Equivocal

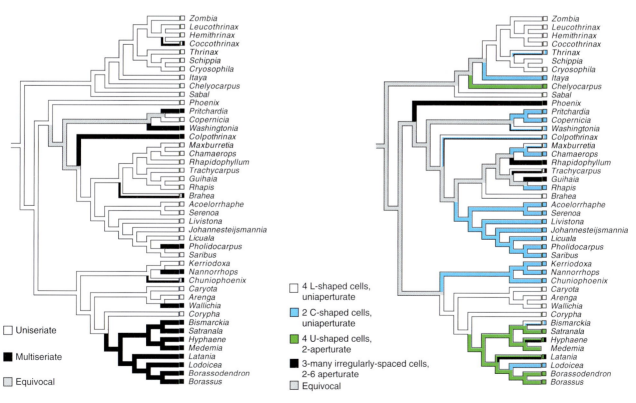

C. Number of layers of lamina abaxial hypodermis

- ☐ Uniseriate
- ▣ (black) Multiseriate
- ▣ (grey) Equivocal

D. Lamina substomatal chamber architecture

- ☐ 4 L-shaped cells, uniaperturate
- ▣ (blue) 2 C-shaped cells, uniaperturate
- ▣ (green) 4 U-shaped cells, 2-aperture
- ▣ (black) 3-many irregularly-spaced cells, 2-6 aperture
- ▣ (grey) Equivocal

Fig. 49 Parsimony ancestral state reconstructions of anatomical characters within Coryphoideae. All reconstructions shown are optimized on the tree indicated in Fig. 46. Terminals without a coloured box adjacent to the branch tip have missing data. Number of steps reported for each optimization is the value within the subfamily. **A.** Shape of adaxial hypodermal cells of lamina optimizes with 10 steps. **B.** Number of layers of adaxial hypodermis of lamina optimizes with 16 steps. **C.** Number of layers of abaxial hypodermis of lamina optimizes with 10 steps. **D.** Lamina s.st.ch. architecture optimizes with 20 steps.

A. Distribution of stomata on lamina

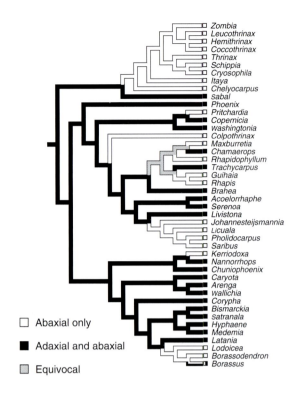

□ Abaxial only

■ Adaxial and abaxial

▨ Equivocal

B. Contour of anticlinal walls of adaxial lamina epidermal cells

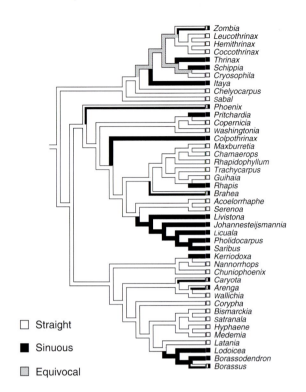

□ Straight

■ Sinuous

▨ Equivocal

C. Differentiation of lamina adaxial epidermal cells

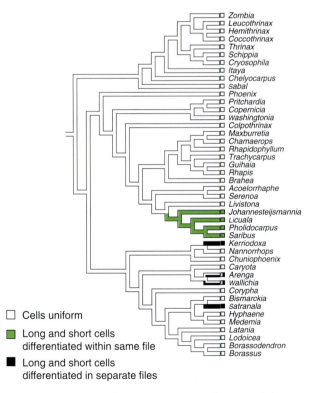

□ Cells uniform

■ Long and short cells differentiated within same file

■ Long and short cells differentiated in separate files

D. Distribution of nonvascular fibres associated with lamina adaxial surface layers

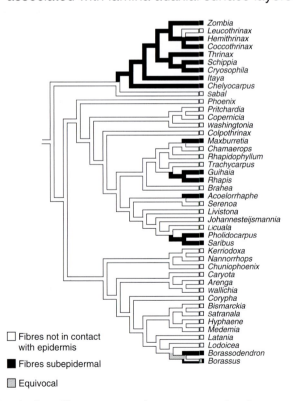

□ Fibres not in contact with epidermis

■ Fibres subepidermal

▨ Equivocal

Fig. 50 Parsimony ancestral state reconstructions of anatomical characters within Coryphoideae. All reconstructions shown are optimized on the tree indicated in Fig. 46. Number of steps reported for each optimization is the value within the subfamily. **A.** Distribution of stomata on lamina optimizes with 10 steps. **B.** Contour of anticlinal walls of adaxial lamina epidermal cells optimizes with 15 steps. **C.** Differentiation of types of adaxial epidermal cells optimizes with 5 steps. **D.** Distribution of non-vascular fibres associated with lamina adaxial surface layers optimizes with 8 steps.

The relative position of these veins, situated at the same level as the longitudinal veins, is an autapomorphic character state for the genus (Fig. 47D). The 'Raphia-type' fibre bundles in the inner cortex of the root of *Phoenix* have appeared independently from those present in taxa of the syncarpous clade (Fig. 48B).

Clade 3: Trachycarpeae (Fig. 46)

Trachycarpeae is the largest tribe of Coryphoideae, containing 18 genera and about 270 species (Dransfield et al. 2008b). The clade is characterized anatomically without exception by petiolar vascular bundles with the phloem strand divided in half by a median sclerotic partition.

Within Trachycarpeae, several large clades are defined by leaf anatomical synapomorphies. A clade consisting of almost the whole of the tribe, exclusive of *Copernicia*, *Pritchardia*, and *Washingtonia*, have non-vascular fibres that are thick-walled with narrow, non-septate lumina (Fig. 47C). A reversal back to septate fibres with wide lumina occurs only in *Acoelorrhaphe*. The clade of *Acoelorrhaphe*, *Serenoa*, and Livistoninae is characterized by a synapomorphic loss of fibre girders that connect longitudinal veins to the adaxial surface layers (Fig. 47B).

Subtribe Livistoninae is characterized by the sinuous anticlinal walls of the adaxial epidermal cells (Fig. 50B). Here the recircumscription of *Livistona* and resurrection of *Saribus* (Bacon et al. in subm.) makes eminent sense anatomically, as *Saribus* contains all species formerly included in *Livistona* that have adaxial epidermal cells differentiated into long and short cells within the same cell file (see Livistoninae for further discussion). With *Livistona* s.s. resolved as sister to all other Livistoninae, the presence of this unique type of differentiated adaxial epidermis optimizes as a synapomorphy for the clade of Livistoninae exclusive of *Livistona* (Fig. 50C). An additional synapomorphy for this clade is the loss of adaxial stomata (Fig. 50A).

Clade 4: Syncarpous clade (Fig. 46)

This clade of four tribes is extremely diverse in both morphology and anatomy. As the informal clade name suggests, all genera have a syncarpous gynoecium with a common pollen tube transmitting tract present within the ovary (Dransfield et al. 2008b). Shared anatomical features among the four tribes are all plesiomorphies or character states of ambiguous polarity (Figs. 47–50; Horn et al. 2009).

Tribe Chuniophoeniceae, sister to the other tribes, is anatomically heterogenous, with each of the constituent genera well defined by autapomorphies (Dransfield et al. 2008a; Fig. 59). The complex mosaic of anatomical character states present among the genera is matched by their contrasted gross morphology. Structural evolution within the group is likely correlated with their strongly contrasted habitats. The tribe may be loosely characterized on the basis of lamina anatomy, as three of the four genera have abaxial ribs with the vascular bundles included within a sclerotic cylinder whereas *Nannorrhops* has a Coryphoid abaxial rib. All genera have a substomatal chamber delimited by two C-shaped cells. Additionally, Horn et al. (2009) identify the presence of non-vascular fibres free in the mesophyll as a synapomorphy for the tribe.

The clade containing tribes Caryoteae, Corypheae, and Borasseae is among the most morphologically incongruous groupings revealed by molecular phylogenetic analyses of Arecaceae. Caryoteae, with pinnate to bipinnate leaves that contain little mechanical tissue, are strongly contrasted with *Corypha* and Borasseae, which have large, often massive, costapalmate leaves replete with fibres. Many other structural differences exist (Dransfield et al. 2008b), but anatomical data is, at least, partly congruent with this intriguing phylogenetic discovery. The presence of 'Raphia-type' fibre bundles in the inner cortex of the roots of all genera investigated (with the exception of *Borassus*) provides a convincing synapomorphy for this clade because cortical fibres of this type are otherwise rare within the family (Fig. 48B; Seubert 1997). Another synapomorphy for this clade is the presence of a multilayered adaxial hypodermis (Fig. 49B).

The three genera of Caryoteae are structurally disparate from their close relatives and so form a morphologically and anatomically well-defined clade. On the basis of lamina anatomy, Tomlinson (1961) suggested a relationship between Caryoteae and various reduplicate-leaved groups he informally recognized, which are now circumscribed within Arecoideae. Horn et al. (2009) explicitly addressed the convergence in lamina anatomy between Caryoteae and the clade of Arecoideae + Ceroxyloideae, showing that the shared similarities evolved independently. Thus, the character states that impart the 'Arecoid' *Gestalt* to Caryoteae are all synapomorphies for the tribe, and include: 1) spindle-shaped adaxial epidermal cells; 2) transversely extended, hexagonal adaxial hypodermal cells in the lamina (Fig. 49A); 3) transverse veins sheathed by sclerotic parenchyma; and 4) the loss of a regularly repeating series of non-vascular fibre bundles attached to the surface layers. Additional synapomorphies for the tribe are the hat-shaped silica bodies, whereas other Coryphoideae have spherical silica bodies, and the presence of transverse striations on the cuticle of the outer periclinal wall of each guard cell. The latter character state is unique within the palm family.

The petiole structure of Caryoteae, because it is terete distally, is also distinctive within Coryphoideae. Within the terete portion of the petiole, the vascular bundles near the periphery have massive phloem fibre sheaths that are confluent to form an essentially complete peripheral sclerotic cylinder.

The clade of *Corypha* + Borasseae, containing many of the most massive fan palms, is characterized by two anatomical synapomorphies. The first of these involves an independent origin of the Coryphoid abaxial lamina rib structure in this clade, presumably by the loss of the outer sclerotic sheath (Fig. 47A). Seubert (1997) linked *Corypha* and Borasseae together because both taxa have root endodermal cells with essentially unthickened walls. The latter character state is unique within Coryphoideae, which otherwise have root endodermal cells with thickened walls (Fig. 48C).

The lamina anatomy of *Corypha* has few distinctive apomorphies, despite the fact that morphologically it is among the most readily recognizable genera of the subfamily. A combination of character states that are largely plesiomorphies characterizes the genus, which include: 1) all longitudinal veins bridged to the adaxial surface layer (Fig. 47B; Horn et al. 2009); 2) long, orthogonally-oriented transverse veins sheathed by many layers of fibres and situated abaxial to the longitudinal veins (Horn et al. 2009); and 3) a hypodermis that is two-layered adaxially (Fig. 49B) and one-layered abaxially (Fig. 49C), with cells that are irregularly polygonal in surface view (Fig. 49A). The stomata are distinctive in that the outer cuticular ledges of the guard cells are prominent and strongly cutinized, together with lateral subsidiary cells that appear to be fused to form a unit.

Tribe Borasseae has long been recognized as a natural group within palms, as the constituent genera share many distinctive floral, inflorescence, and fruit characters (Dransfield et al. 2008b). From an anatomical perspective, Tomlinson (1961) concurred with this idea, pointing out that the genera share a suite of lamina anatomy features that include the attachment of longitudinal veins to the adaxial surface layers by fibre girders and the association of stegmata with the transverse veins. He conceded, however, that the group is not sharply distinct from other Coryphoid fan palms, a viewpoint sustained by evolutionary analyses of the above two characters. Both characters have a complex distribution within the subfamily, with fibre girders plesiomorphic for Borasseae (Fig. 47B), and stegmata associated with the transverse veins having an equivocal optimization for the tribe.

Our character analyses identify three synapomorphies for Borasseae. The first and most distinctive of these is the presence of substomatal chamber architecture in which four U-shaped cells, arranged in opposing pairs, delimit two apertures. This unusual configuration of cells is apparently derived from the usual substomatal chamber arrangement of four L-shaped cells that delimit a single aperture (Fig. 49D), which Tomlinson (1961) considered to be the most widespread type within palms. A multilayered abaxial hypodermis also characterizes the tribe (Fig. 49C). The root anatomy of Borasseae is unlike all other Coryphoideae in that the ground tissue of the stele is parenchymatous, and not fibrous (Fig. 48D; Seubert 1997). For a more detailed discussion of the evolution of lamina symmetry in this clade, see Horn et al. (2009).

Conclusions

The above discussion of character evolution in Coryphoideae highlights the extensive range of major characters and character states that are of importance at the broadest levels of systematic hierarchy. But the richness of the diversity of lamina anatomy within the subfamily is also exemplified at less inclusive levels, so that all 46 genera of the subfamily are probably distinguishable on the basis of lamina anatomy. Although it is beyond the scope of the present treatment to analyse this wealth of information in a phylogenetic context, the keys and descriptions below will serve as a starting point for future research.

Key to tribes of Coryphoideae based on leaf and anatomical characters

1. Leaves induplicately imparipinnate (or bipinnate); inner cortex of roots with fbs accompanied by stegmata 2

1A. Leaves palmate to costapalmate; inner cortex of roots with or, more often, without fibres 3

2. Margins of pinnae (or pinnules) distally praemorse; midrib of pinnae with many large vbs enclosed within a complete cylinder of sclerenchyma; stegmata with hat-shaped silica bodies; adaxial epidermis of lamina with spindle-shaped cells; outermost adaxial hypodermal layer of lamina of transversely extended hexagonal cells; s.st.ch irregularly defined by 2–5 cells that form a single aperture; g.c. with transverse cuticular striations . . . . Caryoteae

2A. Margins of pinnae entire throughout; midrib of pinnae with no or at most few small vbs; stegmata with spherical silica bodies; adaxial epidermis of lamina of ± square to longitudinally elongate rectangular cells; adaxial hypodermis of lamina of longitudinally extended rectangular cells; s.st.ch defined by 4–6 cells, interlocking such that they form 2–4 apertures; g.c. with an even surface, lacking cuticular striations . . . . Phoeniceae

3. Petiolar vbs with phl strand consistently divided into equal halves by a median sclerotic partition; medulla of first-order roots consistently including vessels (or rarely stele dissected); abaxial ribs of lamina variable but sometimes like that of *Phoenix*; leaf sheath rarely cleft below petiole (*Washingtonia*) . . . . Trachycarpeae

3A. Petiolar vbs with phl strand lacking sclerotic cells or with sclerotic cells irregularly distributed (rarely with a median sclerotic sinus, but not consistently within a given petiole, except in *Itaya* [Cryosophileae]); medulla of first-order roots with or without vessels; abaxial ribs of lamina variable, but never phoenicoid; leaf sheath commonly cleft below petiole 4

4. Endodermal cells of first-order roots with walls strongly and unevenly thickened, the outer periclinal walls remaining relatively thin, such that the cell lumen appears distinctly U-shaped in TS 5

4A. Endodermal cells of first-order roots with walls unthickened or evenly thickened such that the cell lumen appears elliptical in TS 6

5. S.st.ch of lamina defined by 4 L-shaped cells . . . . Sabaleae

5A. S.st.ch of lamina defined by 2 C-shaped cell . . . . .Chuniophoeniceae

6. Epidermis of roots 1-layered; inner cortex of roots without fibres; transverse veins of lamina adaxial to longitudinal veins if extending across many large veins; abaxial ribs including 1–3 large, independent vbs, or if including at least 5 vbs, then these included within a common fibre sheath . . . . Cryosophileae

6A. Epidermis of roots many-layered; inner cortex of roots commonly with fbs accompanied by stegmata; transverse veins of lamina abaxial to longitudinal veins if extending across many large veins; abaxial ribs including many, large independent vbs 7

7. Vessels of first-order roots embedded in parenchyma . . . . .Borasseae

7A. Vessels of first-order roots embedded in fibres . . . . .Corypheae

TRIBE SABALEAE

Sabal (Figs. 9I and 51)

An isolated, monotypic tribe within the Coryphoideae (16 spp.; Zona 1990; Quero 1991). It is a New World genus with a somewhat circum-Caribbean distribution, extending to the Pacific coast of Central America and Mexico but also widely distributed in the southeastern United States as far north as North Carolina. *Sabal palmetto* is the state tree of both Florida and South Carolina. The genus is therefore unusual because it extends from the tropics well into warm temperate regions. Habit in the genus is diverse, but without branched vegetative axes. Some small species are trunkless, the seedling axis thickening but remaining underground, as in *S. etonia* and *S. minor*. These acaulescent species might be described as persistent juveniles. Otherwise palms are single-stemmed, and often of imposing stature (e.g. *S. causiarum*). All are unarmed, pleonanthic palms, with much branched inflorescences that bear flowers singly along the rachillae; the gynoecium is fully syncarpous.

Leaf

Strongly costapalmate (weakly so in *S. minor*). The lamina is divided into halves by an abaxial split distal to the apex of the costa. Otherwise segmented into induplicate, single folds by adaxial splits. Adaxial hastula a narrowly V-shaped crest, the apex acute. Abaxial hastula sometimes present as a low ridge along the margins of the costa. Petiole unarmed. Leaf sheath cleft below the petiole, often persistent.

LAMINA (Figs. 9I and 51A–C, F–I)
Isolateral (Fig. 51A, B) except in *S. mauritiiformis*. **Trichomes** absent. **Epidermis** wholly cutinized, outer wall much thickened. Costal regions narrow and slightly more numerous adaxially than adaxially; alternating with somewhat wider intercostal regions. Costal cells ± polygonal to rectangular and sl. longitudinally extended, the walls never markedly sinuous, intercostal cells less regular, always short. **Stomata** (Figs. 9I and 51F, G–I) rather diffuse, not in regular files but sunken, the stomatal depression often occluded by wax deposits; t.s.c. cells short, overarching the guard cells; l.s.c. sunken to the outer hypodermal level, the outer stomatal cavity commonly wax-filled; g.c. distinctive in TS, each with an asymmetric cell lumen constricted in the middle, the wall beneath the outer cutinized ledge thickened, the thin back wall protruding into the l.s.c. **Hypodermis** 1(–2)-layered beneath each surface; cells small, thin-walled and ± cubical or sl. elongated in costal regions; usually 4 L-shaped cells surrounding each s.st.ch. **Mesophyll** palisade-like below each surface, either equally so, or the abaxial palisade less well differentiated; cells papillose, especially near s.st.ch. Central mesophyll cells large, compact ± spherical but with few chloroplasts. **Fibres** (Fig. 51A) septate, next to hypodermis either as non-vascular fibrous strands or as fibrous bundle sheath extensions; connecting all veins to adaxial hypodermis, or less commonly connecting larger veins additionally to abaxial hypodermis. Fibrous strands usually anticlinally extended, the abaxial strands commonly opposed to a small adaxial vein. **Veins**: O.S. always incomplete above small veins and present lateral to large veins; cells cubical or sl. anticlinally extended. I.S. of well-developed lignified fibres around large veins, becoming more like sclerotic parenchyma around small veins; phl of larger veins often sclerotic, commonly divided into 2 or more separate strands, or a major and minor lateral strand. **Transverse veins** with well-developed fibrous sheaths, commonly enclosing more than 1 phl strand; equidistant from both surfaces, running below the smaller adaxial veins where they do not connect to longitudinal veins. **Ribs** (Fig. 51C) most prominent abaxially, but ribs of both surfaces similar, sometimes quite tall, each with 2–many independent vbs, the surface layers resembling those of the lamina. Shape of midrib in TS varying from square to triangular and asymmetric, largely as determined by leaf size. **Expansion cells** in a single band within each fold opposite the rib, and usually including fibrous strands. **Stegmata** frequent, in long fairly continuous files adjacent to both vascular and non-vascular fibres; also scattered along sheathing fibres of transverse veins. **Tannin cells**: their distribution said to be diagnostically useful by Zona (1990).

PETIOLE (Fig. 51D, E)
Non-vascular fbs concentrated toward the periphery, becoming scarce to absent centrally (Fig. 51E). Vbs ± uniformly scattered, with decreasing density toward the centre. Larger vbs (Fig. 51D) with predominantly 2 large mxy vessels (infrequently up to 5 of moderate size) and a single large phl strand collateral to the mxy. Peripheral to the principal strand of vascular tissue within each bundle are 2 small phl strands, positioned 1 to each side of the mxy, representing the phl of inserted transverse veins.

Stem (Fig. 51L)

Surface layers sclerotic, but with age sloughing off or replaced by a **periderm** of sometimes impressive thickness (e.g. *S. causiarum*). **Cortex** relatively narrow, including many large vbs, each with a massive sheath of phl fibres, and numerous, smaller non-vascular fbs. **Central cylinder** (Fig. 51L) with densely parenchymatous ground tissue; peripheral ± radial files of these cells becoming somewhat tangentially expanded with age. Vbs including 2 wide mxy vessels and an unpartitioned phl strand; sheath of phl fibres well developed, but much less so in comparison to the cortical vbs. Non-vascular fbs widely scattered.

Root (Fig. 20H and 51J, K)

S. palmetto. **Epidermis** uniseriate, with cells wholly lignified, persistent; walls unevenly thickened, with outer periclinal and anticlinal walls v. strongly thickened relative to inner periclinal wall (Fig. 20H). **Exodermis** not appearing differentiated from outer cortex in TS (Fig. 51K), although Seubert (1997) reports a clearly differentiated exodermis. **Outer cortex** with cells always lignified; differentiated into 2 zones: a peripheral zone of densely packed, sclerotic, colourless parenchyma of 9–14 layers; and an internal zone 5–12 layers thick of v. thick-walled fibres. **Inner cortex** differentiated into 3 zones, the middle aerenchymatous (Fig. 51J); without mechanical tissue; cells uniformly unlignified (except below), tannins uncommonly or not accumulated.

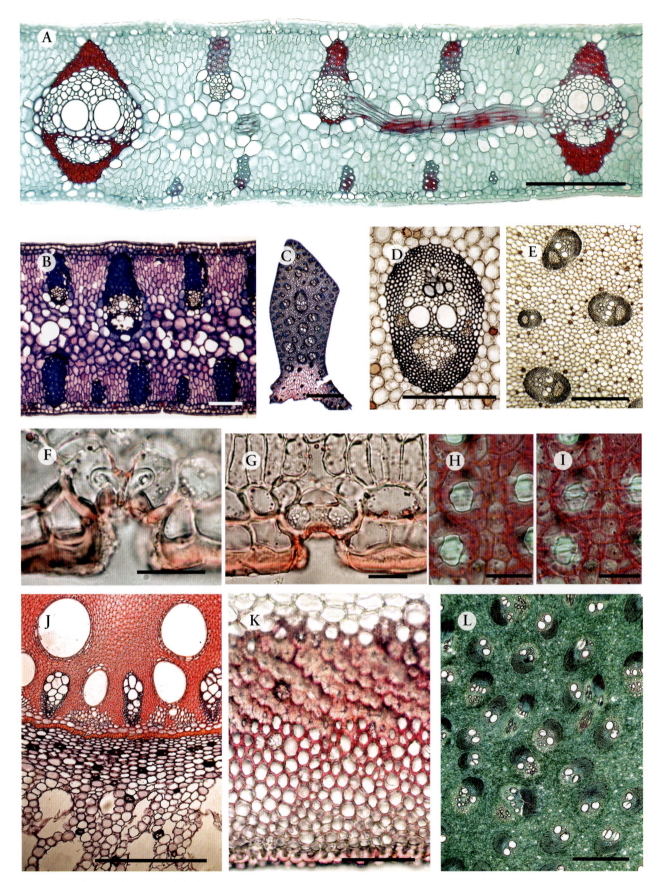

Fig. 51 Coryphoideae (Sabaleae). **A**. *Sabal palmetto*, lamina, TS (bar = 100 μm). **B**. *Sabal causiarum*, lamina, TS (bar = 100 μm). **C**. *Sabal causiarum*, lamina, adaxial rib, TS (bar = 1 mm). **D**. *Sabal etonia*, petiole vb, TS (bar = 200 μm). **E**. *Sabal etonia*, petiole centre, TS (bar = 500 μm). **F**. *Sabal causiarum*, stomata, TS (bar = 20 μm). **G**. *Sabal causiarum*, stomata, LS (bar = 20 μm). **H**. *Sabal etonia*, abaxial epidermis focused on hypodermis, below g.c., SV (bar = 25 μm). **I**. *Sabal etonia*, abaxial epidermis focused on stomata, above hypodermis, SV (bar = 25 μm) **J**. *Sabal etonia*, root stele and inner cortex, TS (bar = 500 μm). **K**. *Sabal palmetto*, root periphery, TS (bar = 100 μm). **L**. *Sabal palmetto*, stem centre, TS (bar = 1 mm).

128

Peripheral zone containing mucilage canals bordered by differentiated parenchyma; the mucilage canal complexes become lignified with age. **Endodermis** wholly lignified, walls unevenly thickened; cells shallowly U-shaped in TS. Pericycle 1–3-layered, lignified. **Stele** with vessels embedded in ground tissue of lignified fibres. **Medulla** a narrow cylinder of unlignified parenchyma. Without medullary vessels in small roots; including occasional medullary vessels in large roots.

Note: Zona (1990) provides illustrations and a thorough analysis of the distribution of some anatomical features of the lamina of 15 species in the form of a data matrix for cladistic purposes. His investigation is innovative because he described characters that have been little used by previous workers. Among these characters are: length and course of the individual transverse veins, extent of the palisade, and distribution of tannin cells in different lamina tissues.

TRIBE CRYOSOPHILEAE
(Figs. 21C and 52–54)

A group of New World palms, containing 10 genera with induplicate, palmate leaves. Like its sister taxon *Sabal*, but unlike other palmate-leaved Coryphoideae, the lamina is sometimes divided into halves by an abaxial split distal to the apex of the leaf axis. Cryosophileae are always pleonanthic, and bear (usually) perfect flowers with 1–3 (–6) free carpels. There are two main centres of distribution, one in the Caribbean, including South Florida, the other in South America as far south as Argentina. Most are single-stemmed palms of modest stature. Many Caribbean species are adapted to dry limestone country.

Anatomical features of the tribe Cryosophileae

Leaf
Palmate to shortly costapalmate. Lamina frequently deeply divided into halves by an abaxial split distal to the petiole, or without such a split in *Thrinax* and genera of the clade inclusive of *Zombia* and *Coccothrinax*. All other divisions are by adaxial splits into multifold or single-fold segments; the multifold segments often with further segmentation to a more shallow depth. Adaxial hastula usually deltoid, sometimes inrolled. Abaxial hastula less prominent, rounded or deltoid. Petioles always unarmed. Sheath sometimes cleft below petiole, and usually disintegrating into 2 contrarotating helical series of vbs, these often with massive fibre sheaths. Both series of bundles are continuous distally into persistent spines in *Trithrinax* and *Zombia*.

LAMINA (Figs. 52; 53; and 54A–G)
Dorsiventral (Figs. 52D, E and 53A–F) but isolateral in *Trithrinax* (Fig. 52F). **Trichomes** of 2 basic types: few-celled, minute, trichomes (*Chelyocarpus, Cryosophila, Itaya* (Fig. 54 E), *Zombia*) and large, knob-like or peltate, multicellular trichomes, usually sunken within the surface (*Hemithrinax* (Fig. 54H), *Leucothrinax, Schippia, Thrinax, Trithrinax*). **Epidermis** (Figs. 53I, J and 54A–G): cells rectangular, the anticlinal walls sinuous (Fig. 54C), or more often, straight. **Stomata** exclusively abaxial, or also with an adaxial distribution in *Trithrinax*; superficial to deeply sunken. Subsidiary cells sometimes papillose (*Itaya, Zombia*; Fig. 53J). **Hypodermis** 1–2-layered adaxially, mostly 1-layered abaxially; usually including a discontinuous to continuous series of septate or non-septate fibres. Hypodermal parenchyma usually rectangular, sometimes lignified. S.st.ch. usually delimited by 4 L-shaped cells, but other configurations exist (see genera). **Fibres** with a distribution adjacent to each surface layer; frequently grouped in bundles distributed in a regularly repeating series (but not or weakly so in *Chelyocarpus, Itaya, Schippia*); fbs sometimes massive (*Coccothrinax* spp. (Fig. 53D–F), *Hemithrinax* spp.). **Veins** in the central, or more often, abaxial mesophyll; the larger often in contact with the abaxial surface layers. Larger vbs either free of adaxial surface layers (*Chelyocarpus, Itaya*) or more often attached by a fibre girder; vbs with abaxial fibre girder in *Coccothrinax* spp. and *Trithrinax*. Large vbs including 1–few large mxy vessels; phl strand often including a network of sclerotic cells but divided in *Itaya*. Vein sheath variable. Segment margin often including a stout marginal vein. **Ribs**: adaxial rib usually including 4 or more independent vbs; devoid of vascular tissue in *Itaya*. Abaxial rib (Fig. 52A–C) variable, see key and generic descriptions. **Transverse veins** (Fig. 53G, H) mostly situated adaxial to the longitudinal vbs; course straight or sometimes sinuous (*Zombia*) and extending between major vbs or from rib to rib; sheathed by fibres sometimes extending into the adjacent mesophyll (*Coccothrinax* spp., *Thrinax* (Fig. 53B), *Zombia*). Transverse veins accompanied by stegmata in several genera. **Expansion tissue** mostly in a single band opposite the body of each rib (Fig. 52A, C), but 2 bands adjacent to the abaxial ribs in *Cryosophila, Schippia* (Fig. 52B), and *Trithrinax*.

PETIOLE (Figs. 21C and 54I)
Hypodermal layers including numerous fibre strands. **Vbs** with a uniformly scattered distribution, but crowded at the abaxial periphery with relatively strongly developed fibrous sheaths. Each vb (Figs. 21C and 54I) with 2 wide mxy vessels or with 2 groups of smaller vessels. Phl a single, unpartitioned strand (divided in half by a central sclerotic partition in *Itaya*); not or somewhat sclerotic. **Tannin cells** abundant, uniformly scattered in the ground tissue.

Stem (Fig. 54M, N)
Trichomes sometimes frequent and with multiseriate bases. **Cortex** (Fig. 54M) narrow, including abundant fbs (sometimes massive as in *Coccothrinax*) and often also small vbs; leaf traces with well-developed bundle sheaths around the phl. **Central cylinder** (Fig. 54 N) with fairly uniform, initially congested vbs, those of the outer cortex each with a well-developed fibrous sheath next to the phl, the xylem associated with sclerotic parenchyma. Mxy including several narrow vessels (2 in *Trithrinax*), pxy abundantly developed. Ground tissue in mature stems with parenchyma remaining highly compact (*Coccothrinax, Thrinax, Trithrinax*) or sometimes secondarily expanded radially around the vbs to create large intercellular spaces that secondarily may become occupied by inflated cells, often with tannin contents, *Zombia* (Fig. 54N).

129

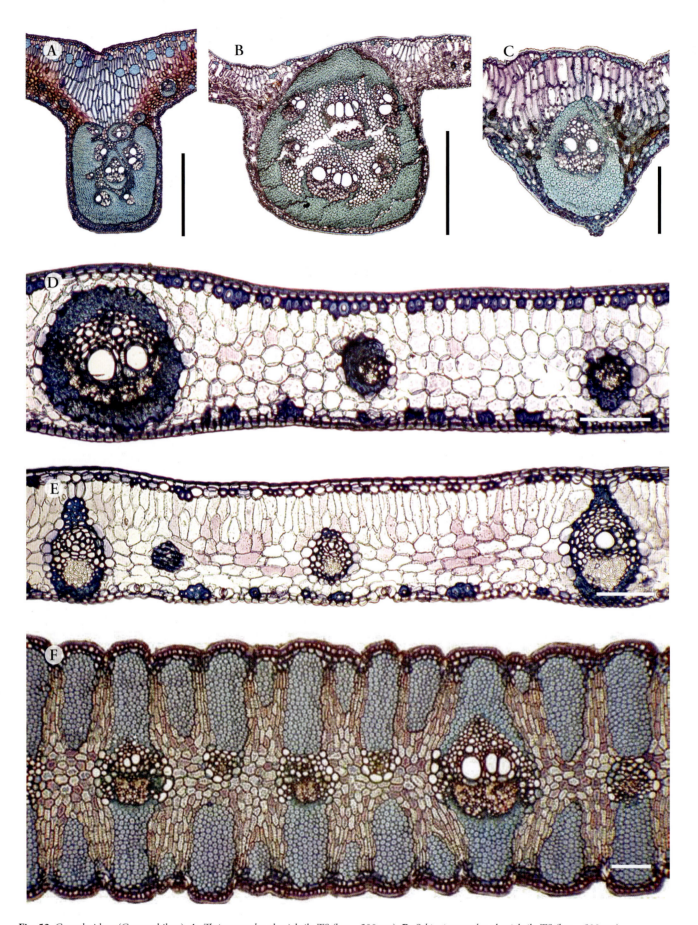

Fig. 52 Coryphoideae (Cryosophileae). **A**. *Thrinax excelsa*, abaxial rib, TS (bar = 500 μm). **B**. *Schippia concolor*, abaxial rib, TS (bar = 500 μm). **C**. *Coccothrinax miraguama*, abaxial rib, TS (bar = 200 μm). **D**. *Chelyocarpus chuco*, lamina, TS (bar = 100 μm). **E**. *Cryosophila warscewiczii*, lamina, TS (bar = 100 μm). **F**. *Trithrinax campestris*, lamina, TS (bar = 100 μm).

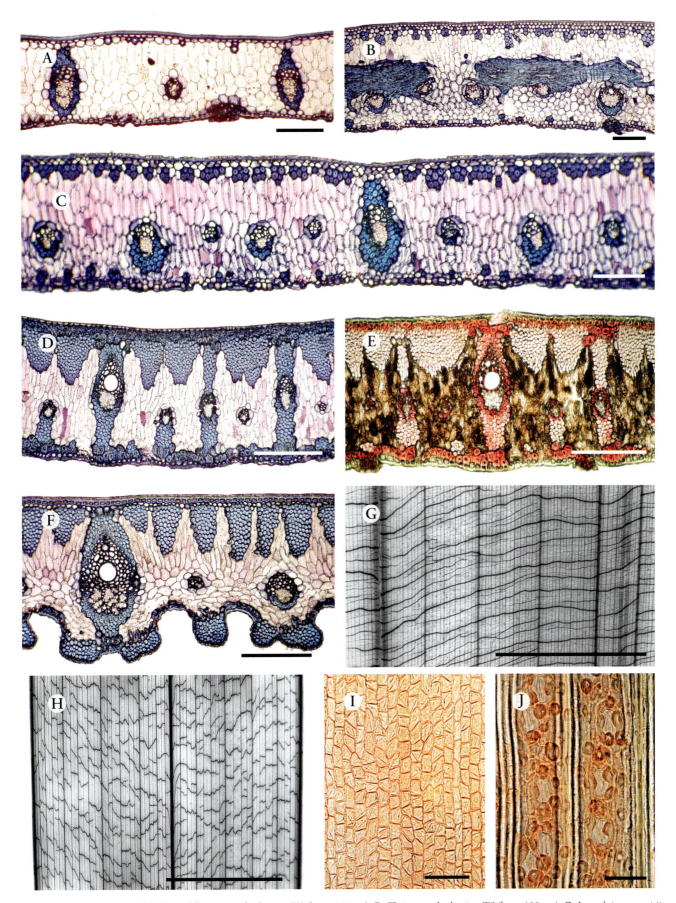

Fig. 53 Coryphoideae (Cryosophileae). **A**. *Schippia concolor*, lamina, TS (bar = 100 μm). **B**. *Thrinax excelsa*, lamina, TS (bar = 100 μm). **C**. *Leucothrinax morrisii*, lamina, TS (bar = 100 μm). **D**. *Coccothrinax ekmanii*, lamina, toluidine blue,TS (bar = 200 μm). **E**. *Coccothrinax ekmanii*, lamina, phloroglucinol + HCl, TS (bar = 200 μm). **F**. *Coccothrinax crinita*, lamina, TS (bar = 200 μm). **G**. *Itaya amicorum*, leaf clearing, SV (bar = 10 mm). **H**. *Zombia antillarum*, leaf clearing, SV (bar = 10 mm). **I**. *Zombia antillarum*, lamina, adaxial epidermis, SV (bar = 50 μm). **J**. *Zombia antillarum*, lamina, abaxial epidermis, SV (bar = 50 μm).

131

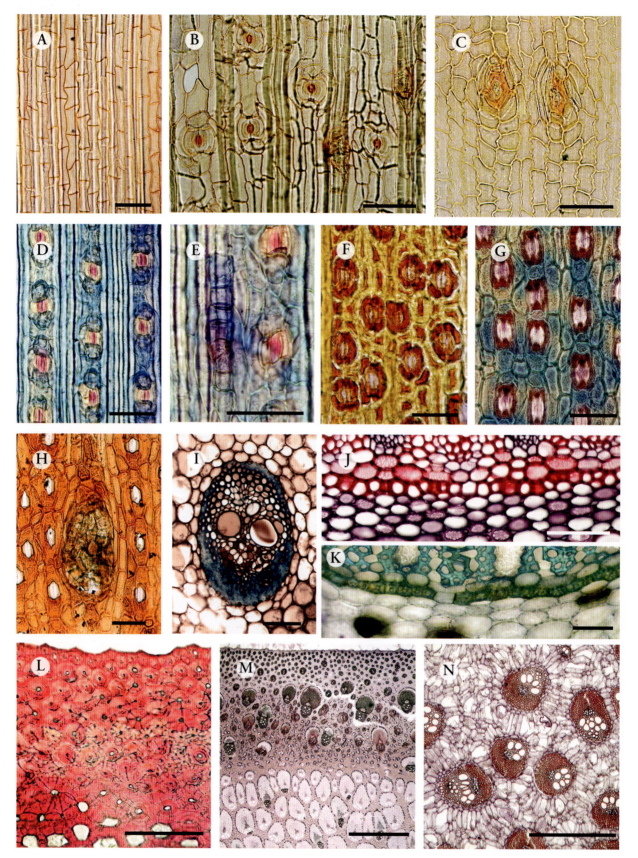

Fig. 54 Coryphoideae (Cryosophileae). **A**. *Chelyocarpus chuco*, lamina, adaxial epidermis, SV (bar = 50 μm). **B**. *Chelyocarpus chuco*, lamina, abaxial epidermis, SV (bar = 50 μm). **C**. *Itaya amicorum*, lamina, adaxial epidermis, SV (bar = 50 μm). **D**. *Itaya amicorum*, lamina, abaxial epidermis, stomata, SV (bar = 50 μm). **E**. *Itaya amicorum*, lamina, abaxial epidermis, trichome, SV (bar = 50 μm). **F**. *Hemithrinax rivularis*, lamina, isolated abaxial hypodermis, SV (bar = 50 μm). **G**. *Hemithrinax rivularis*, lamina, abaxial epidermis, focus on epidermis, SV (bar = 50 μm). **H**. *Hemithrinax rivularis*, lamina, abaxial epidermis, trichome, SV (bar = 50 μm). **I**. *Chelyocarpus* sp., petiole vb, TS (bar = 100 μm). **J**. *Cryosophila* sp., root, endodermis, TS (bar = 100 μm). **K**. *Zombia antillarum*, root, endodermis, TS (bar = 50 μm). **L**. *Zombia antillarum*, root, periphery, exodermis, phloroglucinol + HCl, TS (bar = 100 μm). **M**. *Zombia antillarum*, stem periphery, TS (bar = 10 mm). **N**. *Zombia antillarum*, stem centre, TS (bar = 10 mm).

Root (Fig. 54J, K, L)

Root system usually dense, much branched; bearing lateral root spines (except *Itaya*, with root tubercles; Seubert 1997). Periderm not produced. **Epidermis** 1-layered; cells with weakly to strongly thickened outer walls; of varying persistence. **Exodermis** occasionally not differentiated from the outer cortex, but frequently differentiated as 1–many layers of fibres that often accumulate tannins (Seubert 1997). In TS alone, distinction between exodermis and outer cortex often obscure. **Outer cortex** commonly of wholly lignified cells, sometimes homogenous (Fig. 54L), but usually separable into 2–3 zones based on cell wall thickness (Seubert 1997). Exodermis and outer cortex collectively forming a persistent, many-layered protective unit, in *Leucothrinax* up to 35 layers thick. Sclerotic mucilage canals occasional within the outer cortex. **Inner cortex** of unlignified parenchyma usually strongly contrasted with the outer cortex; with 3 zones; uniformly parenchymatous, apparently uniformly unlignified in young roots, but with few, scattered cells developing slightly thickened, lignified walls in old roots. Aerenchyma in the middle zone of the inner cortex sometimes weakly developed (in *Trithrinax* not developed and the inner cortex essentially homogenous). **Endodermis** (Fig. 54J, K) of sl. tangentially elongate cells, these elliptical in TS; walls evenly thickened, lignified. Pericycle 1–2-layered, sclerotic, lignified. **Stele** with numerous xylem poles and narrow, radially extended phl strands embedded in ground tissue of lignified, thick-walled fibres; without structural anomalies. Medulla with narrow, central parenchymatous cylinder; a few scattered cells within becoming sclerotic with age.

Key to genera of Cryosophileae based on anatomical characters of the lamina

1. Abaxial rib including at least 5 vbs enclosed within a common fibre sheath, either as a complete cylinder or the cylinder discontinuous adaxially. If rib with 3 or fewer discrete vbs (*Chelyocarpus repens*), then all longitudinal vbs of the lamina free of adaxial surface layers and always in contact with the abaxial surface layers 2
1A. Abaxial rib including just 1 or 2 large vbs that are discrete, not enclosed within a common fibre sheath. If the rib includes 3 discrete bundles (rare in *Coccothrinax*, *Leucothrinax*), then some longitudinal vbs bridged to adaxial surface layers by a fibre girder 6
2. Adaxial non-vascular fibres with v. narrow cell lumen and the thick wall at least as wide as the diameter of the lumen 3
2A. Adaxial non-vascular fibres with wide cell lumen and thin walls narrower than the diameter of the lumen 4
3. Adaxial hypodermis mostly 1-layered, ± exclusively of fibres; longest transverse veins extending from midrib to segment margin; abaxial trichomes of mature laminae simple and of fewer than 10 cells, distributed costally, with superficial unicellular bases . . . . *Itaya*
3A. Adaxial hypodermis usually 2-layered, with a discontinuous series of fibres (single and grouped in bundles) interspersed among parenchyma cells; longest transverse veins extending

much less than half the span of the segment rib to its margin; abaxial trichomes of mature lamina persisting as stout elliptical bases of more than 15 cells, included within the lamina tissue to the level of the mesophyll . . . . . . *Thrinax*
4. Fibre sheath surrounding bundles of abaxial rib with an adaxial discontinuity at the level of the expansion tissue (or absent from *C. repens*); expansion tissue in a single broad band opposite the vascular tissue of the abaxial rib; all longitudinal veins free from adaxial surface layers. . . . . *Chelyocarpus*
4A. Fibre sheath surrounding bundles of abaxial rib a complete cylinder without discontinuity; expansion tissue opposite vascular tissue of abaxial rib in 2 bands, 1 to either side of the sheathing fibre cylinder; moderately large longitudinal veins bridged to adaxial surface layers by fibre girder (except *Cryosophila cookii*). 5
5. Non-vascular fbs adjacent to abaxial surface layer absent or nearly so . . . . *Schippia*
5A. Non-vascular fbs adjacent to abaxial surface layers present in a regularly repeating series . . . .*Cryosophila*
6. Lamina histology isolateral; stomata present on both surfaces in about equal density . . . . *Trithrinax*
6A. Lamina histology dorsiventral; stomata exclusively abaxial 7
7. T.s.c. and lateral neighbouring cells of stomata with papillose protrusions; mesophyll with several, v. irregular series of non-vascular fbs without connection to surface layers . . . . *Zombia*
7A. T.s.c. and lateral neighbouring cells of stomata without papillae; mesophyll either without any free non-vascular fbs, or, rarely present, multiple series not evident between longitudinal vbs 8
8. Stomata not sunken (though sometimes included within depressed furrows), or if sunken then the largest adaxial fibre girders of the longitudinal veins at least 10 cell layers deep; trichome bases not sunken, usually opposite bundles of non-vascular fibres . . . . *Coccothrinax*
8A. Stomata sunken (lamina topography always planar); most extensively developed adaxial fibre girders of the longitudinal veins at most 6 cell layers deep; trichome bases within the lamina tissue and extending to the outermost layer of the spongy mesophyll, appearing deeply sunken 9
9. Smaller longitudinal vbs typically with a complete O.S. of inflated parenchyma cells . . . .*Hemithrinax*
9A. Smaller longitudinal vbs typically with an incomplete O.S. of inflated parenchyma cells; vein sheath completed abaxially by thick-walled fibres of the I.S. . . . . *Leucothrinax*

Schippia (Figs. 52B and 53A)

Monotypic, small to moderate, solitary palms endemic to Belize. *Schippia* is distinguished by its long-pedicellate flowers that have a well-developed perianth and unicarpellate gynoecium, which matures into a relatively large, spherical, white berry. Within its restricted distribution, *Schippia concolor* inhabits a variety of forest types from open, pine-dominated woodlands to moist subtropical forest (Balick and Johnson 1994).

Leaf

Lamina with a central abaxial split to apex of petiole, otherwise deeply divided into induplicate, single-fold segments.

LAMINA

Dorsiventral (Fig. 53A), costal regions differentiated on abaxial surface only. **Trichomes** restricted to the intercostal regions of the abaxial surface; peltate, sl. sunken with the limited expanse of thin-walled shield cells flush with the lamina surface, the shield narrowly elliptical, about 6 cells wide, with 2–3 layers of the base included within the lamina tissue. **Epidermis** of rectangular cells with mostly straight anticlinal walls and strongly thickened outer periclinal walls. **Stomata** sunken, the g.c. dumbbell-shaped, with the outer cuticular ledges more prominent than the inner. **Hypodermis** differentiated adjacent to adaxial surface only, of a single layer of rectangular cells interrupted by subepidermal fibres. S.st.ch. not clearly differentiated. **Fibres** large, with wide, septate lumina; present only in contact with the adaxial surface, singly or in groups of up to 4. **Veins** in central to abaxial mesophyll, the smaller free, the larger in contact with both surface layers and forming a small adaxial girder of 3–5 layers of thick-walled fibres. Segment margin including a large marginal vein. **Abaxial rib** (Fig. 52B) with about 7 large vbs included within a complete sclerotic cylinder. **Expansion tissue** in 2 bands, 1 to either side of the sclerotic cylinder. **Transverse veins** short, sheathed by 1–2 layers of sclerotic parenchyma.

Trithrinax (Fig. 52F)

A genus of about 3 spp. distributed in open and frequently arid environments of warm temperate regions of SE South America (S Bolivia, Paraguay, Argentina, SE Brazil, Uruguay). *Trithrinax* species are moderate, clustering or solitary palms with persistent leaf sheaths that include many series of stout fbs that extend distally as spines. Flowers are trimerous, with an apocarpous gynoecium; stamen filaments greatly exceed the length of the other floral organs at anthesis.

Leaf

Lamina moderately to deeply divided by central split, and thence into single-fold segments.

LAMINA

Isolateral with well-differentiated costal bands (Fig. 52 F). **Trichomes** weakly peltate, costally distributed on both surfaces; the persistent bases circular to irregularly elliptical in surface outline. Bases 2–6 cells wide, lignified. **Epidermis** of rectangular, or less commonly irregularly polygonal, cells with straight anticlinal walls and strongly thickened outer periclinal walls. **Stomata** sunken, with irregular thickenings of epicuticular wax surrounding the aperture, and often also extending over the aperture so as to partly occlude it. **Hypodermis** mostly 2-layered below each surface, cells rectangular, the intercostal cells of similar size and shape to those of the epidermis, the costal about twice as wide as the epidermal cells, S.st.ch. formed by 4 L-shaped cells. **Mesophyll** with palisade layers adjacent to both surfaces and a central zone of larger, compact ± ellipsoidal somewhat thick-walled cells, with conspicuous pitting. **Fibres** thick-walled, with

narrow lumina, non-septate. Fibres absent or nearly so from the mesophyll, but present in stout bundles attached to both surfaces so as to form a series irregularly alternating with the fibre girders of the vbs. **Veins** always attached to both surface layers, except for the smallest, by massive fibre girders (small vbs sometimes free in *T. schizophylla*). O.S. interrupted by the fibre girder; of thick-walled, inflated parenchyma cells; I.S. relatively weakly developed, of 1–few layers of sclerotic parenchyma. Large vbs including 1–few large mxy vessels; phl strand including an extensive but irregular network of sclerotic cells. Segment margin truncate in TS outline, without a stout marginal vein. Abaxial **rib** including a single vb of moderate size. **Expansion** tissue poorly differentiated from mesophyll, in 2 bands, 1 to either side of the vb. **Transverse veins** short, sheathed by 2–4 layers of fibres, without accompanying stegmata.

Zombia (Figs. 11F; 20G, K; 24A; 53H–J; and 54K–N)

One sp. (*Z. antillarum*), endemic to Hispaniola. A moderate-sized, clustering fan palm (stems to ~8 m) of open, dry vegetation that is easily recognized by its persistent, spiny leaf bases. The mouth of each leaf sheath produces a distal series of erect spines that become deflected downwards as the sheath expands (Fig. 24A). Flowers with a rudimentary perianth and unicarpellate gynoecium that matures into a globose, white berry; the seed is strongly lobed.

Leaf

Palmate, divided into single-fold segments; without a central split. Adaxial hastula prominently 3-lobed.

LAMINA (Figs. 11F and 53H–J)

Dorsiventral. **Trichomes** mostly on abaxial surface in costal regions, each with a short uniseriate base of 1 or more sclerotic cells ending in a longer filament of thin-walled cells. **Epidermis**: adaxially with little differentiation of costal regions, v. uniform; cells not or little elongated with transverse to sl. oblique end walls (Fig. 53I), the cell nucleus large in relation to overall cell size. Cells shallow, wholly cutinized, with thickened outer wall. Abaxial epidermis differentiated into narrow costal bands 1–3 cells wide below fibrous strands and somewhat wider intercostal bands each with 1–2 irregular files of stomata. Broader bands below larger veins with rectangular somewhat elongated cells. Stomatal files papillate, the papillae largely developed on the t.s.c. and neighbouring cells of each stomatal complex, each papilla with a cavity continuous with the cell lumen. **Stomata** (Fig. 53 J) sunken with outer cavity largely formed by papillae, the g.c. cells partly enclosed by thin-walled l.s.c., t.s.c. short; g.c. with 2 cutinized ledges. **Hypodermis** 2-layered adaxially, the outermost layer a continuous series of cells interrupted by numerous narrow fibrous strands, inner layer of somewhat larger cells, with further fibre strands. Abaxial hypodermis 1-layered, the cells thin-walled, the files sometimes replaced by fibres but much less abundantly than on the adaxial surface. **Fibres** lignified, including additional small strands in the mesophyll. **Mesophyll** compact, the adaxial-most layer forming an indistinct palisade, intercellular spaces only enlarged close to substomatal cavities. **Veins** in abaxial mesophyll, mostly independent of but toward the abaxial surface, larger veins adaxially continuous with hypodermal layers via fibrous bundle sheath extensions. O.S. complete to incomplete around smaller veins interrupted somewhat above and below larger veins. I.S. fibrous or with sclerotic pitted parenchyma cells laterally. Adaxial rib most prominent, including 5 large, independent vbs. Abaxial rib including a solitary large vb. **Transverse veins** frequent, wide, sometimes extending above smaller longitudinal veins; rather short, with an irregularly jagged course (Figs. 11F and 53H). Fibre sheath accompanied by stegmata; fibre ends commonly intruding conspicuously into the mesophyll. Stegmata infrequent adaxially; somewhat more frequent abaxially, associated with both vascular and non-vascular fibres.

Stem
As earlier (Fig. 54M, N).

Root
As earlier (Figs. 20G, K and 54K, L).

Coccothrinax (Figs. 52C and 53D–F)
The largest genus of the tribe, with about 50 spp. widely distributed around the Caribbean, but with a clear centre of species diversity in Cuba. All species inhabit seasonally dry, high light environments, and many are narrow endemics to areas with peculiar or otherwise extreme edaphic conditions. They are small to moderate, solitary or clustering, often slender palms with unsplit leaf sheaths that are highly variable in form among the spp.; sheaths occasionally persistent, and then with the distal fbs modified into short spines or large wefts of soft fibres. The unicarpellate gynoecium matures into a usually purplish-black coloured fruit; seed deeply grooved.

Leaf
Palmate, divided into induplicate single-fold segments; silvery-white abaxially on account of epicuticular waxes. Sheath entire and disintegrating into fibres, but otherwise diverse.

LAMINA (Figs. 52C and 53D–F)
Dorsiventral (Fig. 53D–F), with series of ridges and furrows on abaxial surface of *C. argentea*, *C. borhidiana*, *C. crinita* (Fig. 53 F), *C. proctorii*, and *C. salvatoris*. Abaxial surface with a complete layer of long, narrow, thin strands of epicuticular wax. **Trichomes** costally distributed on abaxial surface; superficial, even when large; variable in size among the spp., ranging from minute, irregularly-shaped aggregations of 3–5 cells (e.g. *C. alba*, *C. barbadensis*, *C. spissa*) to massive, elliptical, knob-like processes of up to 40 sclerotic cells (e.g. *C. cupularis*, *C. miraguama*, *C. proctorii*). **Epidermis**: costal regions conspicuously differentiated only on abaxial surface. Adaxial epidermis with rectangular cells and straight anticlinal walls, usually longitudinally elongate; abaxial epidermis with irregularly polygonal cells in intercostal regions, longitudinally rectangular costally. **Stomata** mostly in irregular files, with 1–4 series present intercostally.

Among the 18 spp. examined, stomata are sunken in just *C. ekmanii*, *C. miraguama*, *C. proctorii*, and *C. salvatoris*. In these taxa the t.s.c. strongly overarching the g.c., and together with the neighbouring cells, forming an aperture that is lined with wax deposits. **Hypodermis** irregularly 2(–3)-layered, adaxial outermost layer of uniformly rectangular cells, usually with only slightly thickened walls, but the cells sclerotic in *C. borhidiana* and *C. ekmanii*. More internal parenchymatous layers intermixed with the adaxial non-vascular fibres, the cells sometimes ± anticlinally extended. Abaxial hypodermis 1(–2)-layered, including the abaxial series of fbs; parenchyma cells usually sclerotic (but not in, e.g. *C. alba*, *C. barbadensis*, *C. readii*, *C. spissa*). S.st.ch. formed by mostly 4 L-shaped cells, always with thin walls. **Fibres** narrow, thick-walled, not or poorly lignified, non-septate; mostly associated with the surface layers; groups of fibres free in the mesophyll rarely present and without any regular distribution. Adaxial series of non-vascular fibres with cells grouped in moderate to massive bundles elliptical to broadly wedge-shaped in TS outline, 5–15 cells deep, the groupings sometimes so broad adaxially as to be nearly confluent. Abaxial series of non-vascular fibres in bundles of 2–20 cells (or even larger in taxa with fibrous ridges, e.g. *C. crinita* (Fig. 53F); usually with a regular but sparser distribution than the adaxial series. Hypodermal and mesophyll fibres vary in degree of lignification (Fig. 53D, E). **Veins** situated in the abaxial mesophyll, with few to most vbs bridged to adaxial surface by narrow, but usually well-developed fibre girder with few to many cell layers. Among all Coryphoideae with dorsiventral laminae, several spp. unique in prominent abaxial as well as adaxial fibre girders. *Coccothrinax ekmanii* noteworthy in the small vbs attached to the abaxial surface; in other taxa small bundles free of the surfaces. Large vbs with 1 large mxy vessel and the phl strand with few to many sclerotic cells. Segment margin with a large marginal vein. **Ribs**: adaxially most prominent, with at least 5 independent vbs, abaxial rib (Fig. 52C) with 1–2(3) independent vbs. **Expansion tissue** in single wide band opposite the body of the rib (Fig. 52C). **Transverse veins** short, with a 2–5-layered sheath of fibres or sclerotic parenchyma; peripheral fibres sometimes protruding into adjacent mesophyll.

Note: hitherto unappreciated, the lamina anatomy of *Coccothrinax* is the most variable of any genus of Coryphoideae, and is among the most varied within the family, so justifying the detailed description.

Hemithrinax (Fig. 54F–H)
A Cuban endemic genus of 3 spp., each with a narrow distribution, occurring on exposed slopes of limestone hills or areas with ultramafic soils. *Hemithrinax* species are small to moderate, solitary palms, sometimes bearing leaves with very short petioles, which impart an unusual, congested appearance to the crown. The flowers have stamens with short filaments and extrorse anthers, and a unicarpellate gynoecium that matures into a small, white fruit.

Leaf
Lamina divided into single-fold, induplicate segments; without a central split. Sheath cleft below petiole.

LAMINA (Fig. 54F–H)

Dorsiventral. **Trichomes** (Fig. 54 H) restricted to abaxial surface, or, in *H. compacta*, also present in low density on adaxial surface. Trichomes knob-like, circular to elliptical, up to 9 cells wide; each with a base of strongly sclerotic cells sunken within the lamina. Costal regions differentiated on both surfaces, but much more numerous abaxially. **Epidermis** (Fig. 54G, H) adaxial cells rectangular, with straight anticlinal walls (or frequently isodiametric to transversely extended in SV in *H. ekmanniana*); abaxial cells more irregularly shaped. **Stomata** sunken. T.s.c. elliptical to broadly Y-shaped, strongly overarching the g.c.; forming a distinct aperture together with the neighbouring cells. **Hypodermis** irregularly 2-layered adaxially, the outer layer of thin-walled rectangular parenchyma cells occasionally interrupted by fibres. Inner layer strongly discontinuous, consisting of uniseriate files of anticlinally-extended parenchyma cells (but sclerotic in *H. ekmanniana*) delimiting large groups of fibres. Abaxial hypodermis (Fig. 54F) of rectangular and irregularly polygonal cells, sclerotic in *H. ekmanniana* and *H. rivularis*, but thin-walled in *H. compacta*. S.st.ch of 4 L-shaped cells. **Fibres** thick-walled and non-septate, mostly associated with surface; less commonly free in mesophyll. Adaxial fibres grouped in irregular bundles so densely distributed as to be nearly confluent; forming a zone of fibres 5–9 layers deep in *H. compacta* and *H. rivularis*, or to 20 layers in *H. ekmanniana*. Abaxial non-vascular fibres grouped in bundles of up to 10 cells but with a much less dense distribution than adaxially. **Veins** situated in the central mesophyll, the larger in contact with the adaxial surface by a broad fibre girder. Large vbs with massive I.S. adjacent to ad- and abaxial surfaces; including 1 large mxy vessel and single phl strand with few sclerotic cells. Smaller vbs independent of surface and usually with a complete O.S. Segment margin including a large marginal vein with a massive fibre sheath. Abaxial **rib** including 1–2 large, independent vbs. **Transverse veins** short, with a sheath of 3–4 layers of fibres.

Note: the 3 spp. of this genus are well characterized by differences in lamina anatomy.

Leucothrinax (Fig. 53C)

Containing 1 sp., *L. morrisii*, formerly included within *Thrinax*. Molecular phylogenetic data resolve *Leucothrinax* as sister to *Hemithrinax*. Distributed in S Florida, W Cuba, Bahamas, and the Lesser Antilles, where it grows in open, coastal vegetation on skeletal, limestone soils. It is a solitary, unarmed palm with a slender axis.

Leaf

Lamina divided into single-fold, induplicate segments; without a central split. Abaxial surface whitish on account of dots of wax associated with the stomata. Sheath cleft below petiole.

LAMINA

Dorsiventral (Fig. 53C). **Trichomes** restricted to abaxial surface and mostly with an intercostal distribution; knob-like, elliptical, 4–6 cells wide, the base sunken. **Epidermis** with costal regions differentiated on both surfaces; much more frequent abaxially.

Epidermal cells rectangular, with straight anticlinal walls. **Stomata** sunken, in irregular, intercostal files of 1–3 series; t.s.c. trapezoidal, overarching the g.c.; together with the neighbouring cells forming a distinctive rectangular aperture. **Hypodermis** (Fig. 53C) 2-layered, adaxially, the outer a continuous layer of rectangular parenchyma cells; the internal layer of many files of shallow, rectangular, parenchyma cells including a series of colourless parenchyma cells that are further complemented by 1–2 (–4) layers of fibres. Abaxial hypodermis like the inner adaxial layer, but with fewer fibres; s.st.ch. defined by 4 L-shaped cells. **Fibres** thick-walled, non-septate, exclusively at surface; adaxially forming a dense zone of irregular bundles. **Veins** in central to abaxial mesophyll, the larger attached to adaxial surface by a fibre girder up to 5 cells deep as described above (Fig. 53C). Large vbs with 1 mxy vessel; phl with few sclerotic cells. Many smaller vbs with an O.S. that is discontinuous abaxially, but the vein sheathing completed by the well-developed I.S. fibres. Segment margin with large marginal vein. **Ribs** most prominent adaxially; abaxial rib with 2 independent vbs. **Transverse veins** short, extending at most between large vbs but adaxial to smaller vbs; sheath of 2–4 fibre layers.

Thrinax (Fig. 52A)

A genus of 3 spp., with 1 sp. widely distributed throughout much of the Caribbean basin, and 2 endemic to Jamaica. Like most Caribbean members of the tribe, its flowers bear a reduced perianth and are unicarpellate. *Thrinax* is differentiated from these similar appearing genera by its leaf sheath that is cleft below the petiole and in its flowers with well developed pedicels.

Leaf

Lamina divided into single-fold, induplicate segments; without a central split (Read 1975).

LAMINA (Fig. 52A)

Dorsiventral. **Trichomes** restricted to the abaxial surface, mostly opposite the groups of non-vascular fibres; persistent sclerotic bases ± peltate, narrowly elliptical deeply sunken. **Epidermis** of rectangular cells with sinuous anticlinal walls; costal regions differentiated on both surfaces, but more frequent abaxially. **Stomata** with Y-shaped t.s.c., the lobes overarching the l.s.c. **Hypodermis** 2-layered adaxially, including both rectangular parenchyma cells and interspersed fibres in small groups. Abaxial hypodermis of 1 layer of rectangular parenchymatous cells that include a series of fibres that are solitary, or grouped in bundles of up to 10 cells; s.st.ch. defined by 2–4 C- or L-shaped cells. **Mesophyll** with a well-differentiated palisade of 2–4 layers; tannin cells frequent and scattered. **Fibres** narrow and thick-walled, exclusively associated with the surface layers, adaxially forming an irregular series of numerous bundles often continuous with hypodermal fibres. Abaxial series with fewer bundles, including up to 10 cells. **Veins** in central mesophyll free of the surface, except for the largest in contact with both surfaces and then without fibre girders. Large vbs with 1 large mxy vessel; phl strand with an irregular network of sclerotic cells. O.S. often incomplete abaxially; I.S. well developed around large vbs. Segment margin including a conspicuous marginal vein. **Ribs** adaxially with at least 10 independent vbs; abaxial rib (Fig. 52A) with at least 5 large and few smaller vbs collectively enclosed within an 8–15-layered fibre sheath that is discontinuous adaxially below the expansion tissue. **Expansion tissue** in a single, wide band opposite the body of each rib. **Transverse veins** short, the largest mostly extending between adjacent large vbs; but adaxial to smaller vbs. Vein sheath of several layers of fibres accompanied by stegmata; the peripheral fibres loosely sheathing the vein and shortly protruding into the adjacent mesophyll.

Chelyocarpus (Figs. 52D and 54A, B, I)

All 4 spp. are small to moderate, understorey rainforest palms that occur at low elevations in W Colombia and Western Amazonia, from Ecuador south to Bolivia. Numbers of floral parts unusually variable, but all floral parts free, or nearly so; stamens have thickened, fleshy filaments. The pericarp of the fruit is sometimes corky and deeply fissured.

Leaf

The lamina is typically deeply split into several multifold segments, which are each shallowly split into single-fold segments; among the prominent divisions, a central split to the petiole is characteristic. Sheath entire, with a tubular weft of fibres.

LAMINA (Figs. 52D and 54A, B)
Dorsiventral (Fig. 52D). **Trichomes** restricted to costal regions of abaxial surface; minute, elliptical or scale-like, consisting of about 3–15 thick-walled cells. **Epidermis** (Fig. 54A, B) of rectangular cells; anticlinal walls straight or broadly undulate (Fig. 54A). Costal regions differentiated only on abaxial surface below small strands of subepidermal fibres. **Stomata** with elongated t.s.c. but overarching the g.c. (Fig. 54B). **Hypodermis** 1-layered below each surface, the parenchymatous cells rectangular and interspersed among groups of subepidermal fibres. S.st.ch. commonly delimited by 4 U-shaped cells in opposing pairs delimiting 2 apertures. **Fibres** adjacent to surface (Fig. 52 D); the adaxial series septate, with wide lumina, and in an exclusively subepidermal position; abaxial series subepidermal or in small bundles to 3 layers deep, lignified, and with narrow lumina. **Veins** in the abaxial mesophyll, mostly free of the surface, the largest in contact abaxially. Larger vbs with 2 wide mxy vessels and phl strand that is much-dissected by an irregular network of sclerotic cells. O.S. usually complete. I.S. well developed, of 1–5 layers of fibres, in larger vbs of uniform thickness throughout. **Ribs**: adaxially and abaxially semielliptical in TS, the adaxial largest. Adaxial ribs including at least 5 large, independent vbs. **Expansion tissue** in a single band opposite the body of the ribs, and also occasionally in bands within the lamina, but not associated with any vascular tissue. **Transverse veins** straight, sheathed by 3–5 layers of fibres accompanied by stegmata, adaxial to smaller vbs, otherwise extending between few large vbs (*C. chuco*), or from rib to rib (other spp.).

PETIOLE (Fig. 54I)
Central vbs with 2 wide mxy vessels and a single phl strand.

Cryosophila (Figs. 20N, O; 21C; 52E; and 54J)

A genus of 10 spp. ranging from SW Mexico to N Colombia. All are solitary palms of moderate stature, distinctive in having stems beset with root spines (McArthur and Steeves 1969). *Cryosophila* is among the most ecologically diverse of neotropical palm genera, occurring in habitats ranging from dry, open forest to the understorey of lowland rainforest.

Leaf

The lamina is strongly discolourous and divided into halves by a deep central split to the petiole, with each half further split (less deeply) into multifold segments, which are often more finely divided distally. Sheath entire.

LAMINA (Fig. 52E)
Dorsiventral, the abaxial surface with a series of shallow ridges and furrows in *C. guagara*. Epicuticular **waxes** developed as large flakes, forming a complete layer on the abaxial surface. **Trichomes** restricted to the abaxial intercostal regions; simple and few-celled. **Epidermis** with costal regions differentiated only abaxially. Adaxial epidermal cells mostly rectangular, sometimes irregularly so, or in *C. warscewiczii* often hexagonal and transversely extended; anticlinal walls ± straight. Abaxial epidermis with intercostal cells irregularly polygonal; costal cells regularly rectangular, with sinuous anticlinal walls. **Stomata** with the t.s.c. somewhat Y-shaped, the lobes somewhat papillose and overarching the l.s.c. and g.c. **Hypodermis** 1-layered below each surface, including files of rectangular parenchyma cells and a discontinuous series of subepidermal fibres. S.st.ch. delimited by 4 L-shaped cells. **Mesophyll** with a distinct palisade (Fig. 52E). **Fibres** septate, mostly subepidermal with wide lumina, unlignified (but abaxially thick-walled and lignified in *C. warscewiczii*); abaxially either subepidermal, or sometimes in 2-layered groups of up to 10 cells. **Veins** in central to abaxial mesophyll, with some vbs connected to adaxial surface by a narrow fibre girder up to 10 cells deep (exceptionally all vbs free of adaxial surface in *C. cookii*). Large vbs sometimes also in contact with abaxial surface; with 1 large mxy vessel; phl strand with few sclerotic cells. Segment margin without a large marginal vein. **Ribs**: adaxial ribs with many, independent vbs; abaxial rib with at least 5, large vbs enclosed within a complete cylinder of fibres. **Transverse veins** short, sheath of 1–5 layers of sclerotic parenchyma; stegmata absent.

Root (Fig. 54J)

Spines are determinate adventitious, aerial roots (Fig. 24P, Q). Inner cortex aerenchymatous (Fig. 20O). Endodermal cells at first O-shaped, becoming U-shaped (Fig. 54J). Medulla lacunose (Fig. 20N).

Itaya (Fig. 54C–E)

Itaya amicorum, the sole sp. of the genus, has a limited distribution in the western Amazonian regions of Brazil, Colombia, Ecuador, and Peru. It is a solitary palm that attains moderate stature with age, and grows in the understorey of lowland rainforest. It is distinct from both of the latter genera on the basis of floral structure, possessing an androecium with stamens more numerous (18–24) than the number of perianth parts and a unicarpellate gynoecium.

Leaf

The lamina of *Itaya* is similar in form to those of *Chelyocarpus* and *Cryosophila*. Sheath cleft below petiole.

LAMINA (Fig. 54C–E)

Dorsiventral, costal regions differentiated on abaxial surface only. **Trichomes** present on both surfaces, and of 2 types. Adaxial trichomes elliptical, with 1–2 central, thin-walled cells surrounded by 3–6 distinctly thicker-walled cells. Abaxial trichomes costally distributed; filamentous and simple (Fig. 54E). **Epidermis** (Fig. 54C, D) with sinuous anticlinal walls; adaxial cells mostly rectangular; abaxial intercostal cells irregular, but cells long and rectangular in the broad costal regions. **Stomata** intercostal, in uniseriate files (Fig. 54 D); t.s.c. dilated, inflated and ± lobed adjacent to stomata, each forming a papilla strongly overarching the g.c. **Hypodermis** largely without parenchymatous cells (except abaxially), otherwise consisting of a continuous, mostly single layer of thick-walled fibres with narrow lumina; those adjacent to the abaxial surface lignified; s.st.ch. delimited by 2 C-shaped or 3 arcuate cells. **Veins** in the abaxial mesophyll, mostly free from the surface layer, except the largest connected abaxially. Large vbs with 1–4 mxy vessels of intermediate size, and with the phl strand divided in half by sclerenchyma. Segment margin obtuse in TS outline, without a large marginal vein. **Ribs** strongly protruding abaxially, including 15–25 vbs enclosed within a common fibre sheath except for an adaxial discontinuity below the single band of **expansion tissue**. Adaxial rib hardly protruding and essentially devoid of vascular tissue; largely consisting of a single strand of expansion tissue with few fbs. **Transverse veins** adaxial to the smaller vbs, the longest extending from rib to rib; sheath of 2–5 layers of fibres often diverging from vein and ramifying freely in the mesophyll; accompanied by stegmata.

TRIBE PHOENICEAE

Phoenix (Figs. 12M and 55A–I)

The isolated dioecious genus *Phoenix* is easily recognizable, by its proximal leaflets modified as long sharp spines. It ranges from Africa continuously as far as Taiwan and the northern Philippines, but the estimates vary from 14–20 spp. because it hybridizes freely in cultivation (Barrow 1998). The most important commercial species is the date palm, *P. dactylifera*, but there are several widely planted ornamentals, notably the massive Canary Island date palm, *P. canariensis*, which represents the extreme of a range of size down to the diminutive shrubby *P. acaulis*. The most widely distributed species is *P. reclinata* which occurs throughout Africa. Palms are solitary or clustering. Although most spp. are thought of as occupying dry habitats, in deserts they are associated with access to ground water, e.g. oases, whereas *P. paludosa* occurs as a back-mangrove species in Africa. *Phoenix theophrasti*

on Crete, if not introduced, makes a second European species of native palms (cf. *Chamaerops*).

Leaf

Imparapinnate, with numerous single-fold, induplicate pinnae that are regularly arranged or grouped along the leaf axis. Proximal pinnae modified as stout, sharp spines. Unexpanded foliage with a continuous, indument-like layer of cellular debris present adaxially (technically referred to as a **haut** = German *Haut*, lit. 'skin'). Petioles usually short. Sheath remaining entire, densely fibrous at maturity, the dorsal stub persisting.

LAMINA (Figs. 9J and 55A–D)

Isolateral (Fig. 55A) or dorsiventral (*P. roebelenii*). **Trichomes** present costally on abaxial surface of *P. roebelenii*; isolateral taxa glabrous. Trichomes peltate, each with an elliptical base to 10 cells wide; the basalmost cells sclerotic and sunken; shield cells thin-walled. **Epidermis** (Fig. 55C, D) with costal regions differentiated on both surfaces of isolateral taxa, or only abaxially in *P. roebelenii*. Cells rectangular and longitudinally extended; anticlinal walls straight or sinuous (*P. roebelenii*); outer periclinal walls somewhat thickened. **Stomata** sunken or not (*P. roebelenii*); in irregular files, with 1–3 series intercostal; t.s.c. ± elliptical to rectangular, sl. longitudinally or transversely extended, overarching the g.c. Lateral neighbouring cells narrowly crescentic. **Hypodermis** 1-layered on each surface; uniformly parenchymatous with rectangular cells becoming irregularly polygonal in intercostal regions; s.st.ch. delimited by a network of 4–9, irregularly shaped cells (Figs. 9J and 55D) that delimit (1)2–6 apertures. **Mesophyll** with a 3–5-layered palisade but variously distinct either adaxially, or below both surfaces in isolateral spp.; cells frequently accumulating tannins. **Fibres** septate, in bundles distributed in a regularly repeating series both ad-and abaxially; bundles 2–10 cells deep (Fig. 55A). **Veins** in central mesophyll and free of surface layers; the larger vbs often bracketed by an extra layer of hypodermal parenchyma both ad- and abaxially. Small vbs with a complete O.S.; I.S. usually complete, of 1 layer of fibres or sclerotic parenchyma. Large vbs with O.S. discontinuous toward the surface layers; I.S. of many layers of fibres, best developed towards the surface. Large vbs with 1 large mxy vessel; phl strand with few sclerotic cells (*P. roebelenii*), or strand strongly dissected by an extensive sclerotic network. Pinnae margin without a large marginal vein. **Rib** (Fig. 55B) hardly protruding abaxially; essentially devoid of vascular tissue and consisting of a single band of expansion tissue that includes few fbs (**phoenicoid rib**). **Transverse veins** v. short, positioned at same level as vbs; sheathed by 1–2 layers of sclerotic parenchyma and without stegmata. Raphide sacs absent.

PETIOLE (Fig. 55E)

Surface layers including a 1-layered hypodermis of colourless cells. **Vbs** with a uniformly scattered central distribution; each usually with 2 or more wide mxy vessels (Figs. 12M and 55E); phl strand 1, but including many irregularly distributed groups of sclerotic cells. Within a given petiole, the sclerotic cells may occasionally partition the phl strand into equal halves. Non-vascular fbs congested at the periphery, widely scattered centrally. Tannin cells frequent throughout.

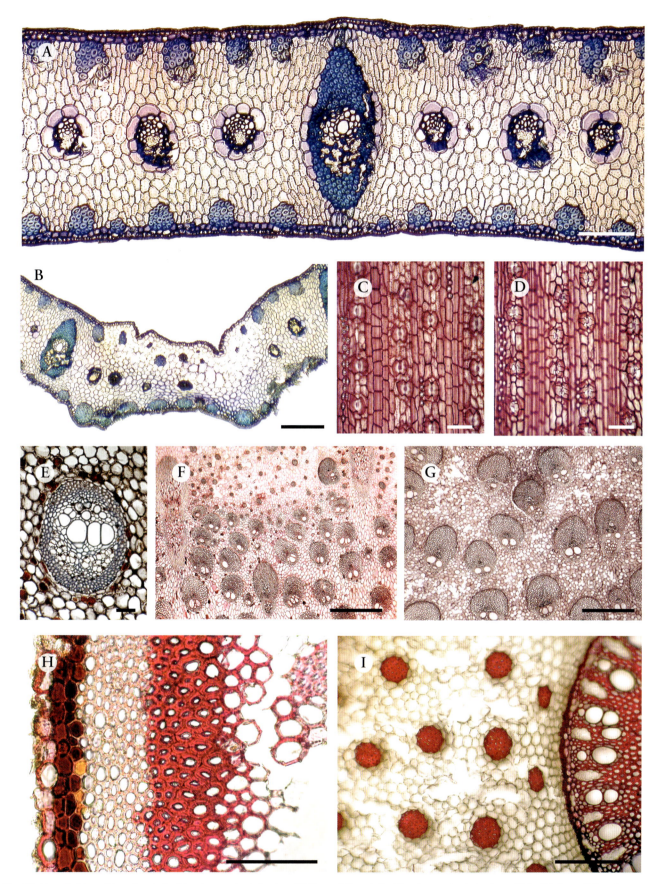

Fig. 55 Coryphoideae (Phoeniceae). **A**. *Phoenix canariensis*, lamina, TS (bar = 100 μm). **B**. *Phoenix canariensis*, lamina midrib, TS (bar = 200 μm). **C**. *Phoenix dactylifera*, lamina abaxial epidermis, focused on epidermal surface, SV (bar = 50 μm). **D**. *Phoenix dactylifera*, lamina abaxial epidermis, focused on hypodermis, SV (bar = 50 μm). **E**, *Phoenix* sp., petiole vb, TS (bar = 100 μm). **F**. *Phoenix dactylifera*, stem periphery, TS (bar = 1 mm). **G**. *Phoenix dactylifera*, stem centre, TS (bar = 1 mm). **H**. *Phoenix dactylifera*, root periphery, TS (bar = 100 μm). **I**. *Phoenix dactylifera*, root cortex and stele, TS (bar = 200 μm).

Stem (Fig. 55F, G)

Surface layers persistent, suberized, apparently not becoming replaced by a periderm with age. **Cortex** (Fig. 55 F) narrow, with many leaf traces orientated perpendicular to the axis, and numerous fbs. **Central cylinder** said by Tomlinson (1961) to be delimited from cortex by a complete cylinder of fused vbs, or the demarcation less abrupt (Fig. 55F). Central vbs (Fig. 55G) with a well-developed sheath of phl fibres; wide mxy vessels usually 2; phl strand undivided, including few sclerotic cells. **Ground parenchyma** undergoing various secondary expansion, showing cell enlargement, more intercellular spaces, and increased lignification within a stem as it ages.

Root (Fig. 55H, I)

Forming a periderm with age (Seubert 1997). **Epidermis** (Fig. 55H) 1-layered, of lignified cells with outer periclinal wall somewhat thickened. **Exodermis** 2–many layers of tannin-accumulating cells with relatively thin walls. **Outer cortex** with 3 zones of cells, the middle zone of cells with distinctly thicker walls than the inner or outer zones (but middle zone said by Seubert (1997) to have cells with relatively thinner walls). Inner cortex 3-zoned, the middle zone with well developed aerenchyma. **Inner cortex** (Fig. 55I) with bundles of strongly lignified **fibres** ('Raphia-type') accompanied by stegmata. **Endodermis** of cells with U-shaped thickenings, the outer periclinal walls remaining relatively thin. **Pericycle** 1–2-layered. **Stele** with vessels embedded in a ground tissue of fibres. **Medulla** parenchymatous, without vessels.

TRIBE TRACHYCARPEAE (Figs. 56–58)

This is the largest tribe of the subfamily, containing 18 genera and over 240 spp. The group is absent from Africa, and has centres of distribution in North America and the Caribbean, and temperate to tropical E Asia. The flowers are distinctive as the stamen filaments are often at least partly connate and the carpels essentially free, but with the styles appearing postgenitally fused in a majority of the genera. The infratribal classification is based largely on molecular phylogenetic data (Asmussen et al. 2006) that is loosely congruent with structural evidence (Dransfield et al. 2008b). Two subtribal clades—Livistoninae and Rhapidinae—are recognized, and are perhaps the most biogeographically and structurally cohesive clades within the tribe. Seven genera are unranked below the tribal level. Bacon et al. (in subm.) provide a new phylogenetic hypothesis for Trachycarpeae that resolves the relationships among the subtribes and unplaced genera with often appreciable support (see *Saribus* for details).

Leaf

Lamina palmate to moderately costapalmate (rarely strongly so); patterns of splitting variable, but usually with induplicate, single-folds evident, at least distally; central split not produced.

LAMINA (Figs. 56; 57A–I; and 58A–C, E, F)

Dorsiventral (Figs. 56A–D; 57A–C; and 58B, C) or isolateral (Fig. 58A). **Trichomes** if present, peltate (Fig. 56I, J) or knob-like (Figs. 56H and 58E). **Epidermis** (Fig. 56F) usually of longitudinally extended rectangular cells; anticlinal walls straight or less commonly sinuous. **Hypodermis** mostly 1–2-layered adaxially and 1-layered abaxially; cells mostly rectangular, exceptionally hexagonal and transversely extended; s.st.ch. architecturally diverse. **Fibres** almost always present as a series of ad- and abaxially distributed bundles that are attached to the surface layers; less commonly also present free of the surface layers. **Veins** mostly distributed in the central to adaxial mesophyll and frequently free of the surface layers, but if in contact with the surface without a fibre girder (except in a significant minority of genera). **Rib** (Fig. 57D, E) structure diverse, but significantly with the abaxial ribs of a few genera like that of *Phoenix* (e.g. *Brahea*, *Chamaerops*, *Copernicia*, *Washingtonia*). **Transverse veins** (Figs. 56K and 58F) mostly situated below or at the same level as the vbs, usually with a straight course; short, and extending just between large vbs, or extending from rib to rib (or margin).

PETIOLE (Figs. 57J, K and 58D)

Usually armed with a ± regularly repeating series of spines along the margins; these occasionally evident only on juveniles. **Trichomes** usually present, peltate, with sunken, suberolignified bases. **Epidermis** with strongly cutinized outer periclinal walls and anticlinal walls. **Hypodermis** 1–3-layered, usually not or weakly lignified. **Vbs** with a uniformly scattered distribution, but most densely aggregated toward the abaxial periphery. *Rhapis* exceptional in the wide V-shaped pattern of central vbs, otherwise absent (Fig. 58 D). Central vbs (Fig. 57J, K) with 1–2 large mxy vessels and always with the phl strand equally divided by a central sclerotic partition, imparting the appearance of 2 widely separate strands. Non-vascular fbs and small vbs concentrated peripherally, sometimes widely scatted centrally.

LEAF SHEATH

Often remaining entire, and not cleft below the petiole (tardily cleft below the petiole in *Washingtonia*), sometimes extended above the level of petiole insertion as a ligule (e.g. *Livistona*); sheath disintegrating into a tightly woven mass of coarse fibres at maturity, and sometimes also extending from the mouth of the sheath as a series of blunt spines.

Stem (Figs. 57L, M and 58G, H)

Surface layers becoming sclerotic, with conspicuous subepidermal sclereids in *Rhapis*. Periderm formation characteristic of many tree palms without persistent sheaths (*Brahea*, *Livistona*) but absent from palms with slender axes (*Acoelorrhaphe*, *Rhapis*). **Cortex** narrow, with few fbs (*Acoelorrhaphe*, *Rhapis*) or with numerous fbs and vbs in tree palms (*Brahea*, *Livistona*, *Trachycarpus*). **Central cylinder** with congested outer vbs (*Acoelorrhaphe*, *Copernicia*), or with the boundary between the cortex and central cylinder less sharply demarcated. Peripheral bundles with the phl sheath always well developed, sometimes massive. **Xylem** in central bundles each with 1 (*Licuala*, *Rhapis*), 2 (*Acoelorrhaphe*, *Brahea*, *Chamaerops*, *Copernicia*, *Livistona*, *Washingtonia*), or 4–6 (*Trachycarpus*) wide mxy vessels. Phl in a single strand, the sieve tubes relatively wide and conspicuous, with long compound sieve plates. **Ground tissue** cells remain unmodified with age in smaller stems (*Licuala*, *Rhapis*), but, in arborescent palms,

Fig. 56 Coryphoideae (Trachycarpeae: Rhapidinae). **A**. *Guihaia argyreta*, lamina, TS (bar = 200 μm). **B**. *Trachycarpus fortunei*, lamina, TS (bar = 200 μm). **C**. *Rhapidophyllum hystrix*, lamina, TS (bar = 100 μm). **D**. *Chamaerops humilis*, lamina with polarized light, TS (bar = 100 μm). **E**. *Chamaerops humilis*, lamina, abaxial epidermis with stomata, TS (bar = 50 μm). **F**. *Trachycarpus fortunei*, lamina, abaxial epidermis, SV (bar = 50 μm). **G**. *Trachycarpus fortunei*, lamina, abaxial epidermis and stomata, TS (bar = 100 μm). **H**. *Trachycarpus fortunei*, lamina, abaxial epidermis and trichome, LS (bar = 50 μm). **I**. *Guihaia argyreta*, lamina trichome, SV (bar = 500 μm). **J**. *Guihaia argyreta*, lamina trichome, LS (bar = 100 μm). **K**. *Rhapis excelsa*, cleared lamina, SV (bar = 10 mm).

with 2 distinct patterns of secondary expansion. In the more conspicuous pattern at the periphery of the central cylinder, files of cells with an irregularly radial orientation become tangentially dilated, and also with limited radial cell division imparting a sutured appearance to these tissue zones. Internally the above pattern of tissue expansion may occur less distinctly (*Livistona*), or else a second type of tissue expansion may be expressed, in which cells become radially dilated around each vb (*Acoelorrhaphe, Brahea, Chamaerops, Copernicia*). In *Acoelorrhaphe* large lacunae can form, but with age, they are filled by single enlarged cells via localized and sustained mitotic activity (Fig. 58 H). Narrow **fibrous strands** (or occasional small vbs) scattered in central ground tissue of many genera (*Acoelorrhaphe, Brahea, Chamaerops, Copernicia, Washingtonia*), sometimes restricted to the stem base. **Starch** often abundant. **Tannin** often conspicuous, especially in larger cells of ground tissue. **Stegmata**, with spherical silica bodies, common in long files.

Root (Fig. 58I–K)

Often with many, slender lateral roots. Sometimes bearing root spines (Rhapidinae: *Guihaia, Rhapidophyllum, Rhapis*; Seubert 1997). Periderm developing in many genera (Seubert 1997). **Epidermis** 1-layered, of cells with outer periclinal walls relatively unthickened to strongly thickened. **Exodermis** 1- to many-layered, of cells with equally, or uncommonly, unequally thickened walls; cells often accumulating tannins. **Outer cortex** highly variable, usually separable into 2 or 3 zones based on relative cell wall thickness, less commonly appearing homogenous. **Inner cortex** 3-zoned, with the aerenchyma of the middle zone typically well developed. **Stone cells** common in many genera, particularly in the inner zone, where they may be aggregated into a cylinder that surrounds the stele. In *Copernicia, Licuala,* and *Livistona*, fibres ('Kentia-like') without accompanying stegmata occur singly or, in *Copernicia* alone, in irregular bundles of up to 8 cells. **Endodermis** with unequally thickened cell walls, appearing U-shaped in TS, the outer periclinal walls remaining relatively thin. **Stele** with vessels embedded in ground tissue of fibres (Fig. 58I, J). **Medulla** regularly including few to many vessels. Medullary vessels surrounded by a thick sheath of fibres; when medullary vessels numerous, medullary parenchyma scanty (Fig. 58I) or absent. The vascular cylinder of *Licuala* is unique within Coryphoideae in being strongly dissected.

Key to subtribes and unplaced genera of Trachycarpeae based on lamina anatomy

1. Adaxial epidermis **2-celled type** within a given file: 1) *long cells* that are v. shallow and at least 5× longer than wide, with only the outer periclinal walls sometimes thickened; and (2) *short cells* that protrude abaxially between adjacent hypodermal cells, typically at most 3×longer than wide, and with the outer periclinal wall strongly thickened and the anticlinal and inner periclinal walls conspicuously and irregularly thickened; cells of intercostal areas with sinuous anticlinal walls . . . . Livistoninae except *Livistona*

1A. Adaxial epidermal **1-celled type**; cells of intercostal areas with sinuous or straight anticlinal walls. 2

2. Abaxial surface with strongly peltate trichomes (often eccentrically so); adaxial hypodermis 1-layered . . . . . . Rhapidinae

2A. Abaxial surface without strongly peltate trichomes (but sometimes with ± knob-like trichomes), or if trichomes strongly peltate, then adaxial hypodermis at least 2-layered . . . .(Unplaced genera) 3

3. Adaxial hypodermis at least 2-layered, with the outermost layer uniformly of transversely-extended, hexagonal cells in precise longitudinal files 4

3A. Adaxial hypodermis 1- or 2-layered with the outermost layer including many longitudinally-extended, rectangular cells, at least adjacent to the costae; if including hexagonal cells, then 1-layered 5

4. Largest transverse veins extending from abaxial rib to adaxial rib, or from midrib to segment margin. . . . . *Pritchardia*

4A. Largest transverse veins confined between adjacent major longitudinal veins . . . . *Colpothrinax*

5. Midrib (abaxial, at standard level) strongly protruding and including at least 5 vbs larger than any in the lamina; s.st.ch. consisting of a single aperture defined by 2 cells . . . *Livistona*

5A. Midrib (abaxial, at standard level) principally including 1–3 vbs larger than any in the lamina, or with an approximately phoenicoid construction: i.e. with little vasculature (at most 1 small vb); if midrib includes 5 or more large vbs, then the s.st.ch. sometimes 2–3-aperturate and commonly defined by 4 cells. 6

6. Largest transverse veins crossing many major longitudinal veins spanning at least 1/3 of the distance from abaxial rib to adaxial rib, or from midrib to segment margin . . . .*Copernicia*

6A. Largest transverse veins confined between adjacent major longitudinal veins 7

7. Adaxial non-vascular fibres septate and with relatively thin secondary walls and wide lumina 8

7A. Adaxial non-vascular fibres non-septate and with thick secondary walls and very narrow lumina 9

8. Hypodermis 2-layered on both surfaces. Stomatal density on both lamina surfaces about equal. S.st.ch. commonly defined by 3–4 hypodermal cells. . . . . *Washingtonia*

8A. Hypodermis 1-layered on both surfaces. Stomatal density on abaxial surface about twice that of adaxial surface. S.st. ch. defined by 2 hypodermal cells. . . . . *Acoelorrhaphe*

9. All vbs entirely free of surface layers . . . .*Serenoa*

9A. Many large longitudinal veins bridged to adaxial surface layers by fibre girders . . . . *Brahea*

SUBTRIBE RHAPIDINAE (Fig. 56)

The subtribe has an exclusively Laurasian distribution, and, further, is noteworthy in containing a large proportion of the palm genera occurring outside the tropics. All are small palms, often described as shrubby, i.e. multiple-stemmed and with short trunks. *Rhapis* provides a rare example in palms of the sympodial rhizomatous habit with dimorphic axes: horizontal with scale leaves and vertical with foliage leaves. There is a strong tendency to dioecism. The consistent presence of a gynoecium with essentially free carpels characterizes the clade, but is probably a secondarily derived condition.

Anatomical features of the subtribe Rhapidinae

Leaf

Several genera exhibit unusual types of leaf splitting.

LAMINA

Dorsiventral or isolateral (*Chamaerops* only). **Trichomes** peltate, restricted to the abaxial surface, or present on both surfaces in *Chamaerops*. **Epidermis**: adaxial cells longitudinally extended and rectangular (but in *Guihaia* often hexagonal and transversely extended); the outer periclinal wall sl. to strongly thickened; anticlinal walls straight or sinuous. **Stomata** sometimes sl. sunken; present abaxially in irregular intercostal files, and also adaxially near the segment margin (or frequent throughout in *Chamaerops*). T.s.c. longitudinally rectangular to semielliptical; lateral neighbouring cells conspicuous. **Hypodermis** 1-layered adjacent to each surface, with the parenchymatous cells forming a complete layer, or sometimes including a discontinuous series of subepidermal fibres; adaxial cells longitudinally extended and rectangular (or transversely extended and hexagonal in *Rhapidophyllum*); s.st. ch. defined by 2–6 cells variously configured. **Mesophyll** with weakly to moderately developed adaxial palisade; abaxial mesophyll usually with thickened, strongly pitted walls. **Fibres** thick-walled, non-septate, not or poorly lignified; exclusively associated with surface layers and usually distributed both ad- and abaxially as regularly repeating series of bundles (exceptions occur in *Rhapis* and *Rhapidophyllum*). **Veins**: larger vbs attached to adaxial surface by girder of fibres like the non-vascular fibres but free of the abaxial surface (or both free of surfaces in *Guihaia argyreta* and *Rhapis*). O.S. usually complete around small vbs; large vbs with depauperate O.S. and well developed I.S. of many layers of lignified fibres usually discontinuous with the O.S. Large vbs with 1–2 large mxy vessels; phl strand divided by a median partition of sclerotic cells and sometimes also with an irregular network. **Rib** vasculature various; expansion tissue in a single, wide band opposite the body of each rib. **Transverse veins** mostly short, the longest extending between large vbs, and above the smaller vbs; *Rhapis* is exceptional. Raphide sacs infrequent.

Key to genera of Rhapidinae based on lamina anatomy

1. Adaxial hypodermis uniformly parenchymatous, without subepidermal fibres 2
1A. Adaxial hypodermis including a discontinuous series of subepidermal fibres intercalated among the parenchyma cells 4
2. Non-vascular fbs associated only with adaxial surface layers, abaxial series absent or nearly so . . . . *Rhapidophyllum*

2A. Non-vascular fbs present in association with both ad- and abaxial surface layers 3
3. Stomata present with ± equivalent density on each epidermis. G.c. strongly elliptical in outline in TS, eccentrically positioned with the outer wall strongly thickened and occluding much of the cell lumen. Midrib depressed-rectangular in outline in TS, weakly protruding abaxially and with 1–2 large vbs, but these not larger than those in the lamina . . . . *Chamaerops*
3A. Stomata mostly abaxial. G.c. circular in outline in TS, walls uniformly thin or with the walls adjacent to the stoma sl. thickened. Midrib ovate in outline in TS, strongly protruding abaxially; including at least 5 vbs larger than those in the lamina . . . . *Trachycarpus*
4. Regularly repeating series of non-vascular fbs associated with surface layers absent. Anticlinal longitudinal walls of adaxial epidermal cells often regularly sinuous. Stegmata associated with transverse veins including silica bodies distinctly larger than those associated with the longitudinal veins and fibres . . . . *Rhapis*
4A. Regularly repeating series of non-vascular fbs in contact with surface layers well developed, at least adaxially. Anticlinal walls of adaxial epidermal cells not sinuous. Stegmata not associated with transverse veins, or with silica bodies equivalent in size to those of the longitudinal veins and fibres 5
5. Outer periclinal wall of adaxial epidermal cells at least 5× thicker than anticlinal walls. Shield cells of abaxial peltate trichomes collectively covering entire lamina surface. Trichome stalk positioned centrally relative to outline shape of shield cells. Segment margin including a massive marginal vein with a fibre sheath to at least 20 layers thick . . . . *Guihaia*
5A. Outer periclinal wall of adaxial epidermal cells equivalent in thickness to those of the anticlinal walls. Expanse of shield cells of abaxial peltate trichomes collectively very incompletely covering the lamina surface. Trichome stalk positioned eccentrically relative to outline shape of shield cells. Segment margin without a differentiated marginal vein of large size . . . . *Maxburretia*

Chamaerops (Figs. 8I and 56D, E)

A single sp., *C. humilis*, in the western Mediterranean (Spain, Sicily) and adjacent North Africa and the only palm native to continental Europe (cf. *Phoenix*). A small multiple-stemmed palm, acaulescent when depauperate, as often in the wild, but in cultivation with short, erect trunks 2–3 m long. A peculiar condition of pollinating insect attraction mediated by aromatic compounds originating from leaf tissue at the time of flower maturity has been claimed (Caissard et al. 2004).

Leaf

Palmate, always with induplicate, single-fold segments. Petiole diamond-shaped in TS; margins with series of spines. Sheath fibrous.

LAMINA (Fig. 56D, E)

Dorsiventral (Fig. 56D). **Trichomes** (Fig. 8I) costally distributed, eccentrically peltate and narrowly spindle-shaped; to 9 cells wide, with the short bases of suberolignified cells sunken within the lamina tissue. Thin-walled shield cells scanty or absent at maturity. **Epidermis** with costal regions frequent and alike on both surfaces except for narrower adaxial intercostal regions. Cells with straight anticlinal walls. **Stomata** (Fig. 56E) sl. sunken, the g.c. distinctive with outer cuticular ledge most prominent. **Hypodermis** uniformly parenchymatous; s.st.ch. delimited by 2 C-shaped cells to form 1 aperture. **Mesophyll** with palisade not or v. weakly differentiated. Ad- and abaxial series of non-vascular **fbs** both well-developed, the adaxial series with larger bundles (Fig. 56D). **Adaxial rib** including 3–4 independent vbs and a peripherally-oriented series of large fbs. **Abaxial rib** phoenicoid. **Transverse veins** short and with an irregularly sinuous course; sheathed by 1–2 layers of sclerotic parenchyma.

Guihaia (Fig. 56A, I, J)

Containing 2 distinct spp. endemic to SE China and N Vietnam, which grow on the slopes of karst limestone hills. They are small, clustering palms that are ± acaulescent or with short stems. *Guihaia* is the only genus of Coryphoideae in which the segmented portion of the lamina is divided precisely into *reduplicate* single folds.

Leaf

Palmate. Petiole unarmed. Sheath coarsely fibrous, the fibre ends projecting as blunt spines.

LAMINA (Fig. 56A, I, J)

Dorsiventral (Fig. 56 A); the abaxial surface with a series of low, rounded ridges opposite the large vbs, and otherwise irregularly undulate. **Trichomes** (Fig. 56 I, J) with a narrow stalk region up to 8 cell layers deep. Stalk cells and those central in the shield region sclerotic and suberolignified. Peripheral shield region a large, star-shaped expanse of many contorted, thin-walled cells fused in a single layer, but without lignin or suberin. **Epidermis** with outer wall much thickened. **Hypodermis** adaxially including a discontinuous series of subepidermal fibres; s.st.ch. consisting of a network of many, irregularly shaped cells that delimit up to 6 small apertures. **Fibres** adaxially grouped in irregularly-shaped bundles of 2–15 cells. Abaxial series of fbs with up to 4 cells, sparse and with an irregular distribution. **Veins** always free of the surface (*G. argyreta*; Fig. 56A) or with the largest connected to the adaxial surface by a shallow fibre girder (*G. grossefibrosa*). Larger vbs with a thick and complete I.S. of fibres. Segment margin with a massive marginal vein. **Adaxial rib** including a single v. large vb. **Abaxial rib** as in the palman of lamina, with at least 5 independent vbs and many peripherally-oriented fbs.

Trachycarpus (Figs. 7B, C and 56B, F–H)

A genus of about nine spp. distributed from the Himalayas of N India eastward to SE China, N Vietnam, and N Thailand, often occurring in montane forests and meadows to an eleva-

tion of 2400 m. *Trachycarpus* spp. are small to moderate, mostly solitary palms, with a lamina that is segmented into induplicate, single folds. The fruit usually develops from only one of the three carpels and is distinctively lobed or kidney-shaped. *Trachycarpus fortunei* is among the most cold-tolerant of all palms, being cultivated (with protection) as far north as Scotland in the UK and the city of New York in the USA (Fig. 45F).

Leaf

Palmate. Petiole unarmed or armed with fine teeth. Sheath coarsely fibrous.

LAMINA (Figs. 7B, C and 56B, F, G, H)

Dorsiventral (Fig. 56B). **Trichomes** (Fig. 56H) centrally peltate, elliptical; the shallow cells base and central boss of subero-lignified cells, up to 10 cells wide; distal expanse of thin-walled calls scanty or absent at maturity. **Epidermis** (Figs. 7B; 56 F) with straight anticlinal walls. **Hypodermis** 1-layered, uniformly parenchymatous (Fig. 7C); s.st.ch. formed by a network of 4–6 irregularly shaped cells that delimit 2–4 small apertures. Adaxial series of non-vascular **fibres** in large bundles 7–11 cells deep. Abaxial series of fbs sparse and sometimes not in contact with the surface. **Veins** often with 2 wide mxy vessels. Segment margin without a large marginal vein. **Abaxial rib** as described in previous key.

Rhapidophyllum (Figs. 20C; 24B, C; and 56C)

A monotypic genus endemic to the coastal plain of the SE USA, as an element of the understorey of riparian forests and wet hammocks, often where the soil is underlain by limestone. *Rhapidophyllum hystrix* is locally known as the needle palm, on account of the very long spines produced distally on the persistent leaf sheaths (Fig. 24B, C). It is a small, clustering, ± acaulescent palm, distinctive in that its lamina is divided into segments between the folds.

Leaf

Petiole margin unarmed.

LAMINA

Dorsiventral (Fig. 56C). **Trichomes** eccentrically peltate; expanse of thin-walled shield cells largely absent from mature lamina. **Epidermis**: cells with straight anticlinal walls. **Hypodermis** uniformly parenchymatous; s.st.ch. a network of 4–6 cells that delimit (1)2–4 apertures. Adaxial series of non-vascular **fibres** in bundles 3–7 layers deep. Abaxial series absent or nearly so. Segment margin with a large marginal vein. **Adaxial rib** including 1 large vb. **Abaxial rib** including 1 to few large vbs.

Root

Middle cortex with well-developed aerenchyma (Fig. 20C).

Maxburretia

The 3 spp. comprising this genus are each narrowly distributed within Peninsular Thailand and West Malaysia, all confined to the exposed slopes of limestone hills. They are small, clustering palms with relatively stout stems of short stature that are densely cloaked by persistent, spine-tipped leaf sheaths.

Leaf

Palmate, divided into induplicate, single-fold segments. Petiole essentially unarmed.

LAMINA

Dorsiventral. **Trichomes** eccentrically peltate, costally distributed, spindle-shaped; thin-walled shield cells scanty in mature lamina. **Epidermis**: cells with straight anticlinal walls. **Hypodermis** including discontinuous series of non-vascular fibres that are integrated into the fbs both ad- and abaxially; s.st.ch. delimited by 2 C- to 4 L-shaped cells that always define 1 aperture. **Fibres** adaxially with a dense distribution of narrow bundles 4–8 layers deep alternating with additional narrow fbs 1–2 cells deep. Abaxial non-vascular fibres in bundles of 2–7 cells, much less densely distributed. Segment margin without a marginal vein. **Abaxial rib** including 1–2 large and few small, independent vbs.

Rhapis (Figs. 7D, E; 21B; and 56K).

Containing about 8 spp. indigenous to SE Asia, from S China to N Sumatra in the forest understorey. *Rhapis* spp. are small, unarmed, clustering palms with v. narrow, reed- or canelike erect stems arising from a sympodial rhizome system. As in *Rhapidophyllum*, the lamina is segmented between the folds. *Rhapis excelsa* is popular as an ornamental, and its size, habit, and ease of cultivation has made it a tractable model system for the investigation of many fundamental questions in palm biology. Indeed, it is structurally the best-known palm (Tomlinson 1990).

Leaf

Palmate. Petiole unarmed. Sheath fibrous (Hastings 2003).

LAMINA (Fig. 56K)

Dorsiventral. **Trichomes** not sunken, mostly costally distributed, minute; bases of 1–3 sclerotic cells extending into a flattened, ± appressed filament of few thinner-walled cells. **Epidermis** with costal regions differentiated abaxially as uniseriate files below the series of subepidermal fibres; cells with sinuous anticlinal walls (Fig. 7C, D). **Hypodermis**: adaxially and abaxially including a discontinuous but densely distributed series of subepidermal fibres (less dense abaxially); s.st.ch. formed by 2 C-shaped cells delimiting 1 aperture. **Fibres**: ad- and abaxial series of non-vascular fibres occurring as irregularly distributed solitary or small groups of cells that accompany the subepidermal fibres. **Veins** uniformly free of the surface. Segment margin without a large marginal vein. **Ad- and abaxial ribs** each including only 1 large vb. Largest **transverse veins** (Fig. 56K) spanning about half the distance between adjacent ribs and adaxial to the smaller vbs. Each with a thick fibre sheath up to c. 6 deep accompanied by large **stegmata**.

PETIOLE

Vbs arranged in a shallow V-shape (Fig. 12C).

Stem

Vbs with 1 wide mxy vessel and a single phl strand (Fig. 21B).

SUBTRIBE LIVISTONINAE (Fig. 57)

A largely Asian group of induplicate-leaved fan palms, but with *Livistona carinensis* in the Horn of Africa and *Saribus* extending into New Caledonia. They are mostly moderate-sized, single-trunked palms. *Johannesteijsmannia* is distinctive in its large, diamond-shaped undivided leaves arising from a short stem. The trunk is robust in *Livistona* and *Pholidocarpus*. *Licuala* spp. are usually small and single- or multiple-trunked. Generic limits have been changed since Dransfield et al. (2008b), based on molecular studies of Bacon et al. (in subm.). See *Saribus* for details.

Anatomical features of the subtribe Livistoninae

Leaf

Palmate but varying in the degree of lamina segmentation, undivided in a few spp. (*Johannesteijsmannia*).

LAMINA

Dorsiventral (or weakly isolateral in *Livistona carinensis*). **Trichomes** absent, or present in some Australian *Livistona* spp. and in *Licuala*. Costal regions differentiated on abaxial surface, and also well differentiated adaxially in *Livistona*. **Epidermis**: cells rectangular with strongly sinuous anticlinal walls. In genera other than *Livistona*, epidermal cells of 2 distinct types (**long** and **short cells**) may be present within the same cell file. Long cells typically longitudinally elongate, v. shallow with narrow lumina, and uniformly thick walls. Short cells no more than 3× longer than wide, with wider lumina than the long cells and inner periclinal walls irregularly thickened. **Stomata** restricted to the abaxial surface, but also adaxial in *Livistona*; t.s.c. usually transversely elliptical, or less commonly also sl. longitudinally extended. Lateral neighbouring cells narrowly crescentic. **Hypodermis** 1–2-layered adjacent to each surface, uniformly of colourless parenchyma in *Johannesteijsmannia*, *Licuala*, and most *Livistona* spp., but otherwise including a discontinuous series of subepidermal fibres. Hypodermal cells adjacent to adaxial surface mostly longitudinally rectangular, but sometimes also transversely extended. Abaxial hypodermis of intercostal regions of longitudinally rectangular or sometimes mostly irregularly polygonal cells; s.st.ch. usually formed by 2 longitudinally oriented C-shaped cells, but in *Pholidocarpus* and *Saribus* cells often transversely oriented. **Fibres** thick-walled and non-septate; distributed in both ad- and abaxial series of small bundles of 2–10 cells that are in contact with the surface (but sometimes free of surface and with up to 18 cells in *Pholidocarpus*). **Fibres** also with a distribution free of the surface in *Johannesteijsmannia*, *Livistona* spp., and *Pholidocarpus*. **Veins** in central mesophyll (but abaxial in *Johannesteijsmannia* and *Licuala*) and usually always free of the surface except with adaxial contact in some *Livistona* spp.; fibre girders never developed. O.S. of smaller vbs complete (but see *Johannesteijsmannia*); usually complete around all vbs in *Licuala*, *Saribus*, and *Pholidocarpus*. I.S. of fibres and sclerotic parenchyma always complete; 1–2-layered in small vbs,

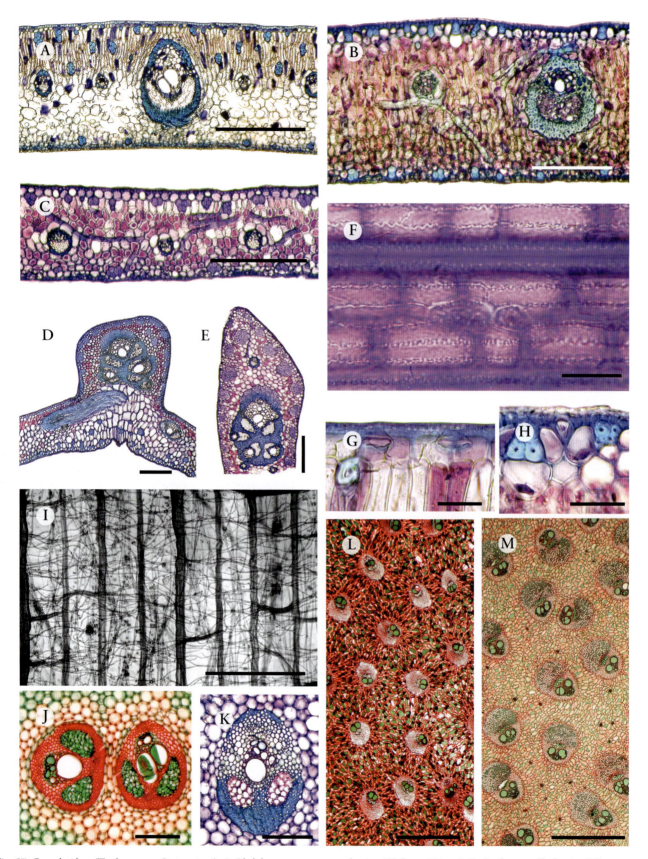

Fig. 57 Coryphoideae (Trachycarpeae: Livistoninae). **A**. *Pholidocarpus macrocarpus*, lamina, TS (bar = 200 μm). **B**. *Saribus merrillii*, lamina, TS (bar = 100 μm). **C**. *Licuala spinosa*, lamina, TS (bar = 200 μm). **D**. *Johannesteijsmannia altifrons*, adaxial lamina rib, TS (bar = 200 μm). **E**. *Licuala spinosa*, adaxial lamina rib, TS (bar = 200 μm). **F**. *Saribus merrillii*, adaxial epidermis, SV (bar = 25 μm). **G**. *Saribus merrillii*, adaxial epidermis with two short cells, LS (bar = 25 μm). **H**. *Saribus merrillii*, adaxial epidermis with short cell, TS (bar = 25 μm). **I**. *Saribus rotundifolius,* lamina venation and mesophyll fibres, clearing, SV (bar = 500 μm). **J**. *Licuala ramsayi*, petiole vbs, TS (bar = 100 μm). **K**. *Livistona chinensis*, petiole vb, TS (bar = 200 μm). **L**. *Licuala ramsayi*, stem centre, TS (bar = 2 mm). **M**. *Livistona chinensis*, stem centre, TS (bar = 2 mm).

2–4(–7)-layered in large veins. Large vbs including 1 large mxy vessel; phl strand evenly divided by a median sclerotic partition. Segment margin without a large marginal vein. Ad- and abaxial ribs with a diversity of vascular patterns. **Expansion tissue** in a single, wide band opposite the body of each rib. **Transverse veins** usually prominent, often extending from rib to rib or rib to margin (but see *Licuala*, *Livistona*); either abaxial to vbs, or in *Johannesteijsmannia* and *Licuala*, adaxial to vbs. Fibre sheath well-developed, with many fibres divergent from vein contour and ramifying freely in the mesophyll in *Licuala* and *Saribus* spp.; transverse veins accompanied by **stegmata** in all genera except *Licuala*. Raphide sacs occasional, but rather frequent in *Licuala*.

Key to genera of Livistoninae based on leaf anatomy

1. Adaxial epidermal cells of intercostal regions of 1 type (long cells only); costal regions well differentiated adaxially . . . . *Livistona*
1A. Adaxial epidermal cells of intercostal regions differentiated into 2 types within the same file, long and short cells, costal regions not or poorly differentiated adaxially 2
2. S.st.ch. defined by 2 lignified hypodermal cells; transverse veins always situated adaxial to the longitudinal veins, the longest regularly extending between adjacent ribs, and with associated fibres never becoming divergent or ramifying throughout the mesophyll. . . . . *Johannesteijsmannia*
2A. S.st.ch. defined by 2 unlignified hypodermal cells; transverse veins situated abaxial to the longitudinal veins, or if adaxial, then these veins rarely extending beyond 2 large veins; associated fibres becoming strongly divergent from vein contour and ramifying freely throughout the mesophyll 3
3. Trichomes present on abaxial surface of mature lamina as persistent, ± circular sunken bases; abaxial rib including 0–2 vbs larger than those within the lamina; transverse veins situated adaxial to longitudinal veins, the longest rarely extending the distance between 2 large veins and never from rib to rib, with associated fibres becoming strongly divergent from vein and appearing as fibre-sclereids in mesophyll . . . . *Licuala*
3A. Trichomes absent from mature lamina; abaxial rib including at least 5 vbs larger than those within the lamina; transverse veins situated abaxial to longitudinal veins, the largest regularly extending between ribs or from midrib to segment margin, with associated fibres tightly sheathing or divergent 4
4. Palisade mesophyll including many large fbs (up to 18 cells) free from the surface layers (Fig. 57A); transverse vein fibres always closely sheathing the vein. . . . . *Pholidocarpus*
4A. Palisade mesophyll without fbs free from surface layers, all adaxially distributed non-vascular fbs in contact with surface layers; transverse vein fibres sometimes becoming strongly divergent from vein and ramifying freely throughout mesophyll (Fig. 57B, I). . . . . (incl. *Pritchardiopsis*) *Saribus*

Livistona (Fig. 57K, M)

Livistona contains about 28 spp. of mostly moderate to tall, solitary palms, distributed from NE India and S Japan south throughout mainland SE Asia and Indonesia west of Wallace's line; also with a major, disjunct radiation in Australia and 1 sp. (*L. carinensis* = *Wissmannia carinensis*) in the Horn of Africa and S Arabian Peninsula. *Livistona* is ecologically diverse with spp. occupying habitats varying from wet, montane forest to desert oases. It is distinct from other genera of the subtribe by the combination of the following features: the lamina typically deeply divided into single-fold segments (or multifold by adaxial splits), the prophyll of each inflorescence subtending only 1 axis, and fruits that ripen green to blue.

Leaf
Palmate to costapalmate. Petiole armed with spines.

LAMINA

Dorsiventral. **Trichomes** present in the Australian *L. fulva* and costally distributed on the abaxial surface; elliptical and to 10 cells wide. **Epidermis** not including short cells. **Stomata** present on both surfaces, with few (*L. australis*) to many (*L. chinensis*) present adaxially, or the surfaces alike in *L. carinensis*. **Hypodermis** 1–2-layered adaxially, 1-layered abaxially; uniformly parenchymatous, or including subepidermal fibres in *L. australis*. **Fibres** usually most prominent adaxially. Irregularly repeating fbs without surface attachment in *L. australis* (and occasionally alternating with vbs) and *L. saribus* (in palisade). **Veins** situated in central to adaxial mesophyll; the larger vbs in contact with the adaxial surface, or continuous with both surfaces via a layer of hypodermal parenchyma in *L. carinensis*. O.S. discontinuous with a thickened region of the I.S. adaxially or adjacent to both surfaces. **Abaxial rib** including 5–20 large, independent vbs. **Adaxial rib** less prominent, narrowly elliptical in TS outline and with relatively fewer vbs. **Transverse veins** variable in length: short and extended just between larger vbs in *L. carinensis* to extending from rib to rib or rib to margin (e.g. *L. chinensis*, *L. saribus*); fibre sheath 3–6 layers thick.

PETIOLE (Fig. 57K)
Vbs with 1 mxy vessel and 2 phl strands.

Stem (Fig. 57M)
Central ground tissue without marked sustained cell expansion, central vbs with 2 mxy vessels, fibrous phl sheath well developed but maturing slowly.

Licuala (Figs. 12L and 57C, E, J, L)

Licuala with over 134 spp. is the largest genus of the subfamily. It is distributed from mainland SE Asia to Queensland, Australia, and Vanuatu, with most spp. occupying the understorey of rainforests. *Licuala* spp. are typically small to moderate, solitary or clustering palms. Most share a highly distinctive lamina morphology, in which the lamina is deeply divided by abaxial splits into multifold wedge-shaped segments, which are shortly segmented apically into induplicate single-folds.

Leaf

Palmate, as above or sometimes essentially entire. Petiole armed with spines.

LAMINA

Dorsiventral (Fig. 57C, E). **Trichomes** present on both surfaces, abaxially intercostal; circular to elliptical, with 1 (–2) layers of thin-walled cells encircling a central group of 4–10 sunken thicker-walled, tanniniferous cells; 2–3 layers thick in TS, but not interrupting the adjacent hypodermal layer. **Epidermis** including short cells adaxially. **Hypodermis** mostly 1-layered below each surface but sometimes 2-layered adaxially; uniformly parenchymatous. **Mesophyll** without a differentiated palisade. **Fibres** distributed in small bundles of 4–12 cells, in contact with the surface but without a series free of the surface. **Abaxial rib** including 1 large vb. **Adaxial rib** (Fig. 57E) including a rosette of 5–8, large, fused vbs; sometimes further accompanied by 1–2 independent vbs. **Transverse veins** short with an unlignified 2–3-layered fibre sheath, the fibres often extending, from the vein into the mesophyll (Fig. 57C).

PETIOLE

Central vbs with 1 wide mxy vessel and 2 phl strands (Figs. 12L and 57J).

Stem

Central ground parenchyma with limited cell expansion to create cell series radiating from vbs, the tissue somewhat lacunose (Fig. 57L).

Johannesteijsmannia (Fig. 57D)

A genus of 4 spp. in the Malay Peninsula, N Sumatra, and W Borneo, occupying the understorey of wet, primary rainforest. They are moderate, solitary palms with very short stems, and large, essentially undivided, diamond-shaped leaves. Fruits are corky-warted.

Leaf

Strongly costapalmate. Petiole short, with spines.

LAMINA

Dorsiventral. **Epidermis** adaxially including short cells. **Hypodermis** 2-layered adaxially, 1-layered abaxially; s.st.ch. distinctive, as described in key. **Mesophyll** without a distinct palisade. **Fibres** adjacent to surface as a series of small bundles of (1)2–5 cells. Other fbs as an irregular series in the adaxial mesophyll. **Veins** largely with the O.S. abaxially incomplete. **Abaxial rib** including 1 large vb. **Adaxial rib** (Fig. 57D) including a rosette arrangement of 6 or more vbs. **Transverse veins** as described in key, with a 5–7-layered fibre sheath.

Pholidocarpus (Fig. 57A)

A genus containing about 6 spp. indigenous to the S Malay Peninsula and much of Indonesia. They are massive, solitary palms, among the tallest of the subfamily (to c. 45 m), occurring in wet, lowland areas. The fruits are the largest of any genus of the subtribe, and are usually with a corky, strongly fissured pericarp.

Leaf

Costapalmate, major divisions multifold. Petiole armed with very large, stout spines.

LAMINA

Dorsiventral (Fig. 57A). **Epidermis** of both surfaces including short cells. **Hypodermis** 1-layered adaxially, 2-layered abaxially. Adaxial hypodermis with a discontinuous series of subepidermal fibres apparently independent of the adjacent fbs. Cells delimiting s.st.ch. typically transversely oriented. **Mesophyll** with well-developed adaxial palisade. **Fibres** distributed in bundles adjacent to the surface, together with an irregular series of large (up to 18-cell) bundles situated in the palisade mesophyll. **Abaxial rib** large and ± ovate in TS outline; including 20–25 large, mostly independent vbs, with the fibre sheaths of the peripheral bundles incompletely confluent. **Transverse veins** extending from rib to rib or rib to margin, with a 3–6-layered fibre sheath.

Saribus (Fig. 57B, F, G–I)

Since the publication of *Genera Palmarum* (Dransfield et al. 2008b) and the monograph of *Livistona* by Dowe (2009), a new phylogenetic study of tribe Trachycarpeae (Bacon et al. in subm.) establishes the need for revised generic circumscriptions within subtribe Livistoninae. In keeping with the conclusions of Bacon et al. (in subm.) we therefore amend the classification of Dransfield and colleagues by recognizing *Saribus*, which includes a morphologically distinct group of about 8 spp. formerly contained in *Livistona*, as well as the monotypic *Pritchardiopsis*. Material examined of *Saribus* includes the following taxa:

- *Pritchardiopsis jeanneneyi* (= *S. jeanneneyi*, comb. ined.).
- *Livistona rotundifolia* (= *S. rotundifolius*).
- *Livistona merrillii* (= *S. merrillii*, comb. ined.).
- *Livistona robinsoniana*, which is now considered a synonym of *L. rotundifolia* (= *S. rotundifolius*).
- *Livistona woodfordii* (= *S. woodfordii*, comb. ined.).

A genus containing 9 spp., with a distribution that includes the Philippines, Indonesia east of Wallace's line, thence east to the Solomon Islands and New Caledonia. *Saribus* spp. are all solitary rainforest palms of tree stature that bear 3 inflorescence axes subtended by the prophyll; fruits ripen orange to red. Circumscription of genus based on Bacon et al. (in subm.); containing *Pritchardiopsis* and spp. formerly included within *Livistona*.

Leaf

Costapalmate. Petioles typically armed with spines.

LAMINA (Fig. 57B, F–I)

Dorsiventral (Fig. 57B). **Epidermis** (Fig. 57G, H) including short cells with strongly thickened outer periclinal walls. **Hypodermis** 2-layered adaxially, 1-layered abaxially; each layer including a discontinuous series of subepidermal fibres that are incorporated into the overall fb series. **Mesophyll** with differentiated adaxial palisade. **Fibres** in bundles always in contact with surface layers. S.st.ch. occasionally delimited by 2 transversely extended cells. **Abaxial rib** ± ovate in TS, including 7–15, large independent vbs. **Transverse veins** (Fig. 57I) extending from rib to rib or rib to margin, with a 6–8-layered fibre sheath. In *S. merrillii* and *S. rotundifolius* peripheral fibres of transverse veins divergent from vein contour and ramifying freely in the mesophyll (Fig. 57B, I).

UNPLACED MEMBERS OF TRACHYCARPEAE

Acoelorrhaphe (Fig. 58G, H)

One sp. (*A. wrightii*) in S Florida and Central America. A multiple-stemmed fan palm of wet habitats, up to 10 m tall with stems c. 9 cm diameter.

Leaf

Palmate. Petiole long, armed with curved marginal spines. Sheath fibrous and persistent.

LAMINA

Dorsiventral. **Wax** layer covering cuticle and persistent; wax occupying outer stomatal chamber. **Trichomes** absent. **Epidermis** of each surface very similar; outer epidermal walls thick and uniformly cutinized, inner walls thin and not cutinized. Epidermis adaxially with longitudinally extended narrow rectangular cells; the end walls sometimes sl. oblique; costal regions not markedly differentiated, but diffuse stomata defining intercostal regions. Abaxial epidermis with somewhat thinner cell walls than those of adaxial epidermis, differentiated into narrow costal regions containing 6–8 files of shallow, narrow elongated and rectangular cells and wider intercostal regions with shorter, wider, and less regular cells. **Stomata** sunken in fairly regular longitudinal files, t.s.c. short; thin-walled l.s.c. overarched somewhat by neighbouring cells. **Hypodermis** 1-layered beneath each surface, but with abaxial files often replaced by fibres; cells cubical or somewhat longitudinally extended; s.st. ch. enclosed by 2 small C-shaped cells parallel to the stomatal pore. **Mesophyll** compact, the adaxial 1–2 layers forming an indistinct palisade, the central mesophyll of larger ± isodiametric cells, becoming smaller nearer the stomata. **Fibres** sl. lignified, septate, in strands of 2–20 cells forming irregular clusters beneath each surface, rarely solitary. Adaxial strands usually separated from the epidermis by a single colourless hypodermal layer; abaxial strands usually next to the epidermis. **Veins** equidistant from and always independent of the surface layers; widest veins often only 1 mesophyll layer from the hypodermis, O.S. of ± colourless cubical or vertically-extended cells, usually complete around the smallest veins, incomplete above and below the largest veins but incomplete and replaced by sclerotic parenchyma above smaller veins. Phl of largest veins divided into 2 separate strands by sclerotic parenchyma. **Transverse veins** few, narrow but sheathed by short fibres; at the same level as the longitudinal veins. **Adaxial rib** with about 5 large, independent vbs and irregular series of peripheral fbs. **Abaxial rib** with 1–2 large vbs. **Expansion cells** in single bands, including fibrous strands, within each fold of the lamina.

Stem (Fig. 58G, H)

Outer region of central cylinder forming a hard cylinder of sclerenchyma composed of large vb fibre caps and ground parenchyma (Fig. 58G). Stem centre less dense with enlarged ground parenchyma cells, infilling space between radiating cell series (Fig. 58 H).

Serenoa

Serenoa repens, the sole sp. of the genus, is a characteristic component of pine flatwoods and scrub vegetation on the coastal plain of the SE USA, from S South Carolina to Florida, and west to E Louisiana. The stems of *Serenoa* are decumbent and usually subterranean, enhancing its ability to survive the frequent wildfires that maintain the open, savannah-like aspect of many of the vegetation types in which it is often a dominant element. The stems of *Serenoa* are unusual among palms in producing axillary vegetative suckers, which facilitate the ability of this sp. to form large, clonal patches (Fisher and Tomlinson 1973).

Leaf

Palmate. Petiole margins armed with a series of fine spines resembling the teeth of a hacksaw blade; hence the common name of saw palmetto.

LAMINA

Isolateral. Thick layer of epicuticular wax present in some populations with silver coloured leaves. **Trichomes** absent. Lamina surfaces alike, with frequent, narrow costal bands. **Epidermis**: cells longitudinally rectangular with straight anticlinal walls; outer periclinal walls sl. thickened. **Stomata** sunken, the t.s.c. semielliptical, somewhat overarching the g.c. **Hypodermis** 1-layered below the surface, uniformly parenchymatous; the cells mostly longitudinally rectangular; s.st.ch. delimited by 2 C-shaped cells. **Mesophyll** with an ad- and abaxial palisade of 2–3 layers. **Fibres** narrow, thick-walled, non-septate; distributed in an ad- and abaxial series of bundles mostly with 8–15 cells, always in contact with the surface of. **Veins** in central mesophyll, always free of the surface. Smaller vbs with O.S. complete and I.S. mostly 1-layered. Larger vbs with O.S. discontinuous ad- and abaxially, the vein sheath completed by well-developed I.S.; including 1 large mxy vessel and phl strand divided by a broad median sclerotic partition. Segment margin truncate in TS outline, including a marginal vein. **Adaxial rib** most prominent, including 5–7 vbs and a peripheral series of fbs. **Abaxial rib** essentially phoenicoid, but sometimes with up to 2 larger vbs crowded at the periphery. **Transverse veins** short, at the same level as the vbs; sheathed by 2–5 layers of fibres.

Root

Differences in anatomy related to root order (= diameter) are described by Fisher and Jayachandran (1999).

Brahea (Fig. 58I–K)

A genus of about 10 spp., ranging from N Mexico to Guatemala. They are typically solitary palms of dry, open vegetation on rock outcrops.

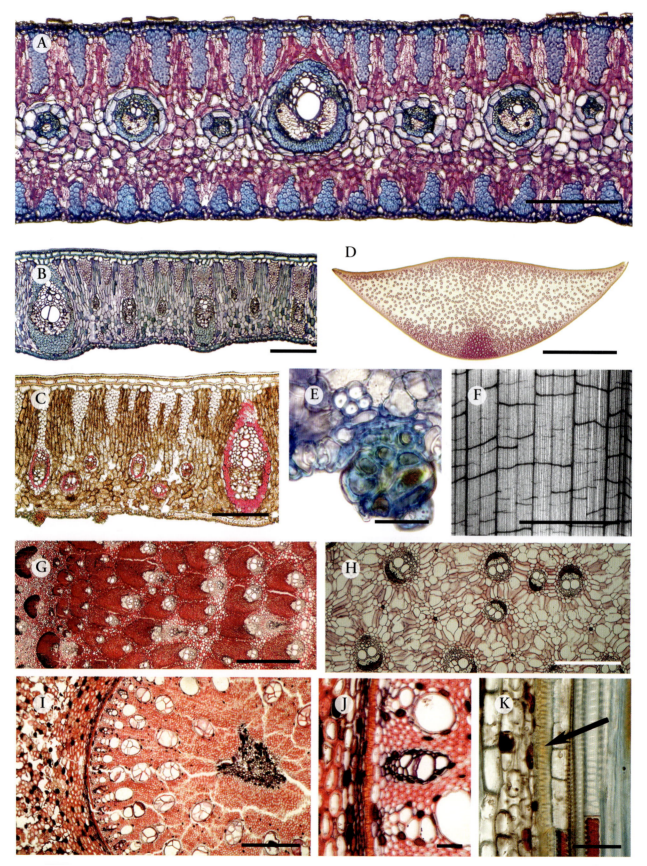

Fig. 58 Coryphoideae (Trachycarpeae: Unplaced Genera). **A.** *Copernicia baileyana*, lamina, TS (bar = 200 μm). **B.** *Colpothrinax wrightii*, lamina, toluidine blue, TS (bar = 200 μm). **C.** *Colpothrinax wrightii*, lamina, lignified walls stained pink with phloroglucinol, TS (bar = 200 μm). **D.** *Washingtonia robusta*, petiole, TS (bar = 10 mm). **E.** *Colpothrinax wrightii*, abaxial epidermal stomata and trichome, TS (bar = 25 μm). **F.** *Copernicia alba*, lamina clearing, SV (bar = 10 mm). **G.** *Acoelorrhaphe wrightii*, stem periphery, TS (bar = 1 mm). **H.** *Acoelorrhaphe wrightii*, stem centre, TS (bar = 1 mm). **I.** *Brahea* sp., root centre, TS (bar = 500 μm). **J.** *Brahea* sp., root endodermis and stele, TS (bar = 50 μm). **K.** *Brahea* sp., root endodermis (arrow), LS (bar = 50 μm).

Leaf

Shortly costapalmate. Petiole armed with short spines or unarmed.

LAMINA

Isolateral to dorsiventral (*B. brandegeei*, *B. salvadorensis*). **Trichomes** on intercostal abaxial surface of *B. salvadorensis* as large patches of weakly sclerotic cells; sclerotic cells otherwise absent. Epidermis with costal regions differentiated on both surfaces, or abaxial only in the dorsiventral spp. **Epidermis** of longitudinally rectangular cells; the anticlinal walls either straight or sinuous in *B. brandegeei*. **Stomata** on both surfaces of isolateral taxa, but restricted to abaxial surface of dorsiventral spp.; sunken, with the longitudinally semielliptical t.s.c. not overarching g.c.; aperture strongly occluded by canopy of epicuticular waxes in *B. armata*. **Hypodermis** 1–2-layered on both surfaces, the cells longitudinally rectangular; s.st. ch. delimited by 4 L-shaped cells. **Mesophyll** with palisade moderately developed adaxially and weakly developed abaxially. **Fibres** non-septate, thick-walled, with v. narrow lumina distributed in an ad- and abaxial series of bundles that are 5–11 layers deep and attached to the surface (or abaxial series ± absent in *B. salvadorensis*). Veins in central mesophyll. Larger vbs attached to adaxial surface by a broad fibre girder resembling adaxial fbs; the few largest occupying the full depth of the mesophyll; including 1 large mxy vessel and phl with an extensive network of sclerotic cells that are sometimes organized into a median partition. O.S. of small vbs ± complete; I.S. of 1–2 layers of fibres. Larger vbs with O.S. divided by fibre girder; I.S. thickened abaxially. Segment margin including a moderately developed vein. **Abaxial rib** phoenicoid. **Transverse veins** short, positioned at about the same level as the vbs; sheathed by 2–4 layers of fibres.

Root (Fig. 58I–K)

Cortex including irregular strands of wide-lumened fibres, forming an almost continuous cylinder in innermost cortex (Fig. 58I). **Endodermis** (Fig. 58J) with U-shaped wall thickening almost occluding cell lumen, the wall abundantly pitted (Fig. 58K). **Stele** (Fig. 58I) wide. **Medulla** almost completely fibrous, including numerous mxy vessels.

Colpothrinax (Fig. 58B, C, E)

Containing 3 spp. of solitary palms, with one endemic to savannahs of W Cuba and the other 2 in wet premontane to montane forests in Central America. The Cuban *C. wrightii* is known as the belly palm on account of the conspicuous bulge produced at midheight on the stems of many mature individuals. The flowers of *Colpothrinax* are similar to those of *Pritchardia*, but are with a more persistent corolla.

Leaf

Shortly costapalmate. Petiole unarmed. Sheath coarsely fibrous, long persistent.

LAMINA (Fig. 58B, C, E)

Dorsiventral (Fig. 58B, C). Costal regions differentiated only on abaxial surface; abaxial surface undulate in TS. **Trichomes** (Fig. 58E) present intercostally on abaxial surface only; ± peltate, spindle-shaped, 2–6 cells in diameter; expanse of shield cells not persisting at maturity. **Epidermis** of narrow, shallow, longitudinally rectangular cells with sinuous anticlinal walls; outer periclinal walls of *C. cookii* strongly thickened. **Stomata** restricted to abaxial surface; sl. sunken (*C. wrightii*) or not (*C. aphanopetala*); t.s.c. longitudinally semielliptical. **Hypodermis** 2-layered below each surface (Fig. 58B, C), uniformly parenchymatous, the cells strongly transversely extended and hexagonal; the peripheral layers v. thick walled; s.st.ch. commonly defined by 2 C-shaped cells (uncommonly to 4) with thinner walls than the surrounding hypodermis. **Mesophyll** with well-developed adaxial palisade of many layers, occupying ± half of the mesophyll (Fig. 58B, C). **Fibres** thick-walled, but not or poorly lignified; with narrow, non-septate lumina; fibres in bundles, with the ad- and abaxial series usually in contact with the surface layers (adaxial series sometimes free of surface in *C. cookii*). Adaxial series well developed (Fig. 58B, C), the bundles narrowly wedge-shaped in TS and 10–14 layers deep in *C. wrightii*, but less well developed in the other 2 spp. Abaxial series ± absent from *C. cookii*. **Veins** in central to abaxial mesophyll, with many large vbs in contact with the adaxial surface by a narrow fibre girder (most frequently so in *C. wrightii*). In *C. wrightii*, many large vbs also in contact with the abaxial surface by a girder-like extension of the I.S. Smaller vbs with complete O.S. and an I.S. of 1–2 layers of fibres; in *C. wrightii* these vbs at somewhat different levels within the mesophyll. Large vbs with 1 large mxy vessel and the phl strand with an extensive network of sclerotic cells sometimes organized into a median partition. Segment margin without a large marginal vein. **Adaxial rib** including 5–7 large, independent vbs. **Abaxial rib** most prominent, including 10–15 large vbs enclosed within a thick sclerotic adaxially discontinuous cylinder. **Transverse veins** short, situated at or below the level of the vbs, with a somewhat sinuous course; sheathed by 2 layers of sclerotic parenchyma.

Copernicia (Figs. 24L, M and 58A, F)

A genus of 21 spp. of solitary palms in NE South America and the Caribbean, with many endemic to Cuba. The Cuban radiation within the genus is spectacular, and contains several palms of bizarre or striking aspect. Nevertheless, the genus is well marked in having highly branched inflorescences with the ultimate order of branches (rachillae) usually bearing completely tubular bracts, and in having seeds with ruminate endosperm. *Copernicia prunifera* is the source of Carnauba wax, which is obtained from the hard, epicuticular wax layer on the leaves of this palm.

Leaf

Palmate to shortly costapalmate; adaxial hastula large in several Cuban spp. Petiole v. short to long, usually armed with stout marginal teeth (Fig. 24L, M).

LAMINA (Fig. 58A, F)

Dorsiventral (Fig. 58A), or sometimes weakly isolateral; the surfaces then ± alike. Thick layer of epicuticular wax present adaxially (Fig. 58A). **Trichomes** present costally on both surfaces; knob-like, elliptical and to 12 cells wide. Trichome bases sunken within the lamina tissue up to the level of adjacent fb and so interrupting the hypodermis. **Epidermis**: cells longitudinally rectangular costally and irregularly polygonal intercostally; anticlinal walls straight (cells mostly rectangular but with sinuous walls in *C. berteroana*). **Stomata** not or sometimes sl. sunken; t.s.c. small, circular to transversely elliptical. **Hypodermis** 1-layered next to each surface (but 2-layered adaxially in *C. berteroana* and *C. ekmanii*). Intercostal cells irregularly rectangular or polygonal; rectangular and longitudinally extended costally. S.st. ch. delimited by 2 C-shaped cells. **Mesophyll** with palisade of 4–6 layers differentiated at each surface; the cells commonly accumulating tannins. **Fibres** often with moderately wide, septate lumina, or in many Cuban spp. v. thick walled with narrow lumina; ad- and abaxial series of large and often narrow bundles attached to the surface (Fig. 58A). **Veins** in central to adaxial mesophyll, with either all vbs free of the surface (Fig. 58A), or in some spp. with the vbs attached to the adaxial surface without a specialized girder. Smaller vbs with a complete O.S. of thick-walled, inflated parenchyma cells and an I.S. of 1–2 layers of fibres. Large vbs with O.S. incomplete adaxially and sometimes also adaxially; I.S. of relatively even thickness, 2–5-layered, fibrous. Large vbs with 1 large mxy vessel and phl strand divided into equal halves by an often wide median sclerotic partition that may include a small region of phl. Segment margin without large marginal vein. **Adaxial rib** v. prominent, ridge-like, narrowly elliptical in TS; including about 10 large, independent, superimposed vbs. **Abaxial rib** phoenicoid. **Transverse veins** (Fig. 58F) of 2 ± distinct types: 1) small veins with 2–4-layered sheaths, at the level of and connecting the vbs; 2) large veins with 7–12-layered sheaths abaxial to the vbs, and extending below many large vbs (see key); accompanied by stegmata.

Note: the Cuban *C. rigida* differs from the above in having fbs free in the adaxial mesophyll and, especially, in having 2–3 additional, irregular series of small vbs that are scattered within the mesophyll.

Pritchardia

A genus indigenous to the South Pacific containing about 27 spp., most of which are narrow endemics of the Hawaiian Islands. They are handsome, solitary tree palms. The genus is florally distinctive, because the corolla abscises as a cap at anthesis.

Leaf

Palmate to shortly costapalmate. Petiole unarmed. Ribs of abaxial lamina surface, petiole, and sheath often with a dense, floccose indument.

LAMINA

Dorsiventral. Costal regions differentiated only on abaxial surface. **Trichomes** costally distributed on abaxial surface, peltate (or just a minute patch of sclerotic cells in *P. pacifica*), the base elliptical with thin-walled shield cells either scanty (e.g. *P. napaliensis*, *P. schattaueri*, *P. thurstonii*), or with a short, narrow stalk and large shield cells (e.g. *P. hardyi* and *P. martii*).

Epidermis of rectangular cells with sinuous anticlinal walls. **Stomata** restricted to abaxial surface, at most sl. sunken; t.s.c. circular to elliptical. **Hypodermis** 2–4(–5)-layered adaxially, the cells deep and appearing inflated in TS; surface layer of transversely-extended hexagonal cells. Abaxial hypodermis 2(3)-layered; cells transversely extended in intercostal regions; s.st.ch delimited by 2 C-shaped cells with strongly invaginated ends. **Mesophyll** with adaxial palisade of 3–4 layers. **Fibres** with wide lumina and septate; grouped in bundles of 2–10 cells. Adaxial series sometimes uniseriate and attached below hypodermis (*P. pacifica*), or included within the hypodermis (*P. minor*, *P. munroi*); more frequently multiseriate, with the outermost series of bundles included within the hypodermal layers, and innermost series sometimes free of the surface layers. Abaxial bundles always in 1 series attached to the surface. **Veins** in central to adaxial mesophyll; O.S. typically incomplete abaxially, the vein sheath then completed by a thickened region of large phl fibres of the I.S. Large vbs attached to the adaxial surface, sometimes by a modestly developed fibre girder; including 1 large mxy vessel and an equally divided phl strand. Segment margin without a large marginal vein. **Adaxial rib** equal to or more prominent than the abaxial, including at least 5 independent vbs. **Abaxial rib** sometimes ± phoenicoid, with a deep band of expansion tissue and few vbs crowded at the periphery, or much like the adaxial rib. **Transverse veins** extending from rib to rib or rib to margin; fibre sheath 2–4-layered, and accompanied by stegmata.

Washingtonia (Fig. 58D)

Containing 2 rather finely differentiated spp. of tall, solitary palms of the arid SW of North America, where they are restricted to oases or areas with more or less permanent water. The long inflorescences axes are distinctive in bearing long, sword-shaped bracts that are pendent on account of their thin texture.

Leaf

Costapalmate. Petioles of adult foliage armed with small spines (or appearing nearly unarmed), but developed proximally. Sheath cleft below petiole. Leaves conspicuously marcescent to the extent of being a fire hazard.

LAMINA

Dorsiventral, but with the surfaces ± alike, with similarly developed costal regions. **Epidermis**: cells longitudinally rectangular with straight anticlinal walls. **Stomata** sl. sunken, in 3–6 intercostal series of irregular files; t.s.c. semielliptical. **Hypodermis** 2-layered adjacent to each surface, uniformly parenchymatous; cells rectangular and transversely to longitudinally extended; s.st.ch. delimited by 2–4 C- to L-shaped cells. **Mesophyll** with moderately differentiated, 5-layered adaxial palisade (*W. filifera*), or the palisade also abaxially weakly differentiated (*W. robusta*). **Fibres** with wide, septate lumina; distributed in ad- and abaxial series in bundles of 15–20 cells in contact with the surface. **Veins** in central mesophyll and strongly differentiated into 2 size classes. Smaller vbs with complete O.S., I.S. 1-layered. Large vbs, all in contact with adaxial surface, and many also in contact with the abaxial hypodermis. These vbs exclusively attached adaxially with a broad fibre girder, otherwise stout, and with well developed I.S. Large vbs including 1 mxy vessel and a single phl strand divided into equal halves by a narrow sclerotic partition; each half further subdivided by network of thinner-walled sclerotic cells. **Adaxial rib** with 1 v. large vb. **Abaxial rib** ± phoenicoid, but with few vbs; rib mainly including expansion tissue. **Transverse veins**, short, straight, abaxial to vbs; fibre sheath accompanied by stegmata.

PETIOLE (Fig. 58D)

Without a distinct V-shaped arrangement of central vbs. Peripheral vbs, especially abaxially, with well-developed fibrous sheaths, forming an extensive sclerotic area.

TRIBE CHUNIOPHOENICEAE
(Figs. 59 and 60F, I)

Four genera with considerable geographic separation and morphological and ecological diversity, *Nannorrhops* in Arabia, *Kerriodoxa* in Thailand, *Chuniophoenix* in Vietnam, China, and Hainan, and the recently discovered *Tahina* in Madagascar. *Nannorrhops* has been described as the only true desert palm in its considerable drought resistance. The tribe is heterogenous, without obvious distinctive features except for some similarity of flower form within condensed cincinni with tubular bracteoles; carpels are basally fused. *Kerriodoxa* is the only strictly dioecious genus. The previous hypothesis (Uhl and Dransfield 1987) of a close relation with *Corypha* is not supported by molecular evidence. The genera are readily distinguished from each other by leaf anatomy, with *Chuniophoenix* itself somewhat heterogenous. This information provides another example of a small group of putatively related palms with evident anatomical diversity that seems highly correlated with ecological differences, in extreme contrasting *Nannorrhops*, of dry open habitats, with forest understorey palms. *Tahina* grows at the edge of limestone outcrops in low xerophytic woodlands with a long dry season.

Anatomical features of the tribe Chuniophoeniceae

Leaf

Lamina palmate or strongly costapalmate, usually divided into single or multifold segments. Adaxial hastula sometimes lacking. Petiole unarmed. Leaf sheath not disintegrating into fibres, typically split opposite the petiole, and also cleft below the petiole in all taxa except *Kerriodoxa* and *Chuniophoenix nana*.

LAMINA (Figs. 59A–H and 60F–L)

The lamina anatomy of this small tribe is at once perhaps the most diverse and anatomically disparate within Coryphoideae. The key below emphasizes these differences. Shared features include a s.t.ch. delimited by 2 C-shaped cells

PETIOLE

Crescentic in TS outline, or with a deep adaxial groove in *Chuniophoenix*. **Hypodermis** present in all taxa, but only well-developed in *Nannorrhops*, in which it consists of 4–6 layers of sclerotic cells. **Vbs** always uniformly scattered. Wide mxy vessels of central vbs: 1 in *Chuniophoenix*, 1–2 in *Kerriodoxa* and *Tahina*, 2 (–4) in *Nannorrhops*. *Nannorrhops* is notable for the extensive amount of pxy produced within each vb. Phl of central vbs always with a single unpartitioned strand, but sometimes including few (*Kerriodoxa*) or many (*Nannorrhops*) irregularly distributed sclerotic cells. *Chuniophoenix* is unlike the other genera in that the phl strand is completely sheathed by 3–5 layers of thick-walled fibres. Non-vascular fbs frequent throughout the ground tissue in *Kerriodoxa* and *Tahina*.

Stem

Observations only from *Nannorrhops*. **Cortex** with numerous fbs and many vbs. **Central cylinder** demarcated from cortex by dense region of small vbs each with a strongly developed sheath of phl fibres, the inner vbs much more widely spaced. Central vbs each with 2 widely spaced, large mxy vessels, and often including an intermediate range of tracheary elements graduating in size to the pxy; phl a single strand. **Ground tissue** of central cylinder without included fbs, but with many tannin cells. **Sustained primary growth** evident as parenchyma cells surrounding each vb of the cortex and central cylinder periphery become markedly dilated radially around each vb.

Root

Periderm produced to a limited extent in *Nannorrhops*. **Epidermis** uniseriate, of cells with ± unthickened walls or with the outer periclinal walls convex and thickened (*Kerriodoxa*). **Exodermis** of 2 to many layers of tanniniferous cells with walls moderately to strongly and equally thickened. **Outer cortex** 2-zoned, the inner zone of cells with much thicker walls than the outer zone. **Inner cortex** 3-zoned, with the aerenchyma of the middle zone poorly developed in *Nannorrhops*; sometimes including scattered stone cells concentrated in the innermost zone adjacent to the stele (*Kerriodoxa*, *Nannorrhops*). **Endodermis** of cells with unevenly thickened walls, appearing U-shaped in TS, the outer periclinal walls remaining relatively thin; cells accumulating tannins in *Kerriodoxa*. **Stele** with vessels embedded within ground tissue of fibres. **Medulla** parenchymatous, including vessels.

Key to genera of Chuniophoeniceae based on lamina anatomy

1. Outer cuticular ledge of each guard cell at least 2× larger than inner. . . . . *Chuniophoenix* (Fig. 59A)
1A. Outer cuticular ledge of each guard cell ± equivalent in size to the inner. 2
2. Lamina isolateral. Trichomes absent. Stomata deeply sunken. Transverse veins short, spanning a distance no greater than 3 longitudinal veins. Anticlinal walls of adaxial epidermal cells straight. Abaxial rib (midrib) without a peripheral fibre sheath; rib vbs independent of each other . . . . *Nannorrhops* (Fig. 59C)

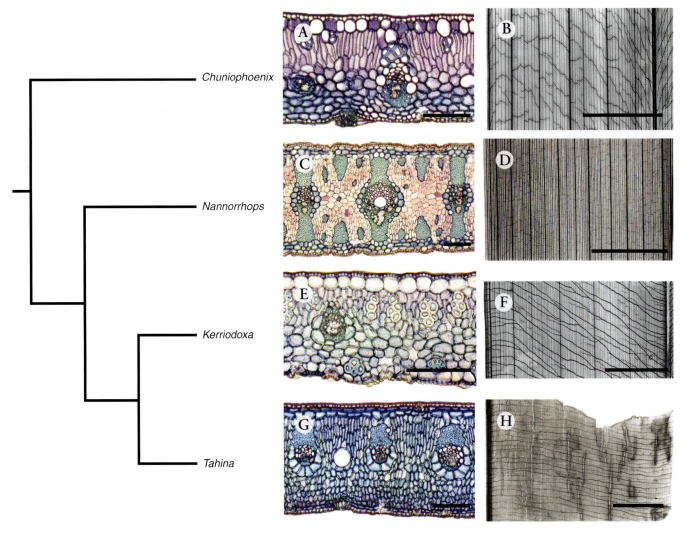

Fig. 59 Coryphoideae (Chuniophoeniceae). Phylogeny of tribe (Dransfield et al., 2008a) with images showing the diversity of lamina anatomy in TS and venation architecture, SV. **A**, **B**. *Chuniophoenix hainanensis*. **C**, **D**. *Nannorrhops ritchiana*. **E**, **F**. *Kerriodoxa elegans*. **G**, **H**. *Tahina spectabilis* (Bars for all lamina TS = 100 μm; bars for all lamina clearings = 10 mm).

2A. Lamina dorsiventral. Trichomes present on abaxial surface. Stomata not or scarcely sunken. Transverse veins long, the largest extending from midrib to segment margin. Anticlinal walls of adaxial epidermal cells regularly sinuous. Abaxial rib (midrib) with a peripheral fibre sheath of many layers that includes all vbs. 3

3. Veins of lamina not in contact with surface layers. Persistent trichome bases not sunken, 2–3 cells in diameter. Adaxial epidermis with both long and short cells present in different cell files; stomata absent. Free non-vascular fibres unlignified, septate, in bundles in the adaxial mesophyll . . . .*Kerriodoxa* (Fig. 59E)

3A. Veins of lamina each bridged to adaxial surface layers by a buttress of lignified fibres 3–6(7) layers deep. Persistent trichome bases sunken, about 8 cells in diameter. Adaxial epidermis with cells of uniform type; stomata occasional. Free non-vascular fibres non-septate, lignified, few mostly solitary in the abaxial mesophyll . . . . . . *Tahina* (Fig. 59G)

Chuniophoenix (Figs. 59A, B and 60G, H, J)

Two or 3 spp. in Indochina and Hainan, all small understorey palms of varying morphology; leaves palmate, without hastulae, the blade variously divided into segments with 1 or more folds.

Leaf

Palmate. Petiole with a deep adaxial groove. Sheath persistent and cleft below the petiole (*C. hainanensis*) or entire (*C. nana*); with a conspicuous floccose indumentum.

LAMINA (Figs. 59A, B and 60G, H, J)

Dorsiventral (Fig. 59A). **Trichomes**, if present (Fig. 60G), diffuse and mainly below veins, each with a basal group of sclerotic cells, the distal thin-walled cells somewhat persistent. **Epidermis** thin-walled but outer wall somewhat thickened; cell files not differentiated adaxially. **Stomata** (Fig. 60H) with enlarged outer ledge of g.c. **Hypodermis** 1-layered below each surface. **Mesophyll** with a well-developed adaxial palisade (Fig. 59A). **Fibres**

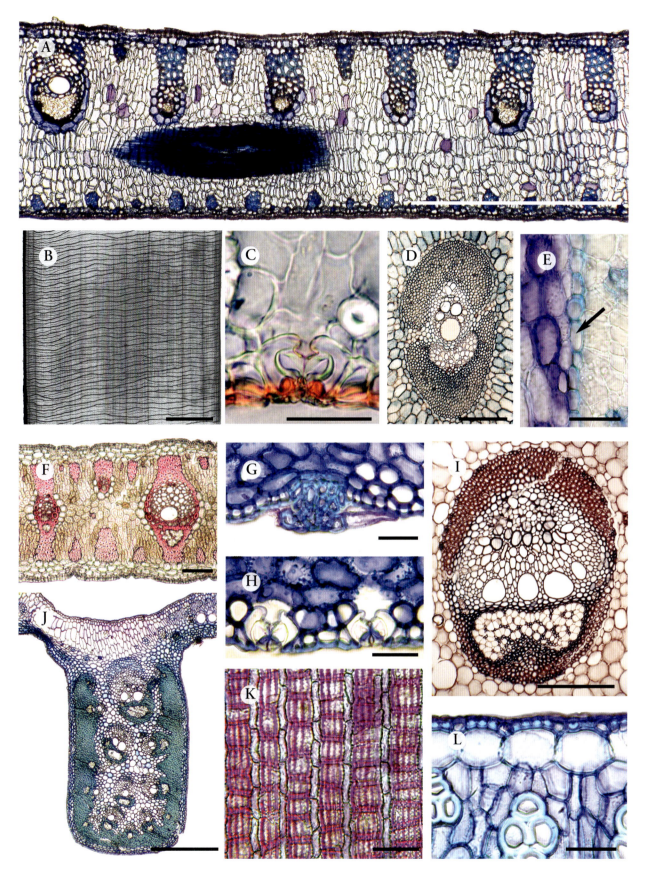

Fig. 60 Coryphoideae (Corypheae and Chuniophoeniceae). **A.** *Corypha umbraculifera*, lamina, TS (bar = 500 μm). **B.** *Corypha utan*, cleared lamina, SV (bar = 10 mm). **C.** *Corypha umbraculifera*, stomata, TS (bar = 25 μm). **D.** *Corypha* sp., petiole vb, TS (bar = 25 μm). **E.** *Corypha utan*, root endodermis (arrow), TS (bar = 50 μm). **F.** *Nannorrhops ritchiana*, lamina, phloroglucinol, TS (bar = 100 μm). **G.** *Chuniophoenix hainanensis*, lamina trichome in LS (bar = 250 μm). **H.** *Chuniophoenix hainanensis*, stomata in TS (bar = 25 μm). **I.** *Nannorrhops ritchiana*, petiole vb, TS (bar = 200 μm). **J.** *Chuniophoenix hainanensis*, lamina midrib, TS (bar = 500 μm). **K.** *Kerriodoxa elegans*, lamina adaxial epidermis, SV (bar = 50 μm). **L.** *Kerriodoxa elegans*, lamina adaxial epidermis, TS (bar = 25 μm).

155

few, with wide septate lumina, but in strands up to 15 cells wide, usually in palisade, absent or infrequent abaxially. **Veins** either all in contact with adaxial hypodermis (*C. hainanensis*) or free of hypodermis but in adaxial palisade (*C. nana*). **Transverse veins** (Fig. 59B) developed in varying degrees with a narrow to wide sclerotic bundle sheath, commonly extending below longitudinal veins. Raphide sacs conspicuous. **Midrib** (Fig. 60J) with a well-developed sclerotic sheath enclosing numerous small peripheral and central vbs, the sheath interrupted adaxially at the level of a large vb. **Expansion cells** forming a single adaxial series.

Nannorrhops (Figs. 20D, E; 59C, D; and 60F, I)

A single sp., *N. ritchiana*, in the drier montane parts of the Middle East. In the wild it is a low shrubby plant, without dependence on adjacent water and so described as a true desert plant. In cultivation it becomes erect and more tree-like. The habit overall is unique among palms because stems branch in 2 contrasted ways: basally via presumed normal axillary suckers; aerially via a true dichotomy in which 1 daughter axis soon produces a terminal inflorescence (i.e. is hapaxanthic) the other continuing vegetatively for some distance before repeating the dichotomy. This produces the effect of a low tree with a succession of terminal panicles.

Leaf

Costapalmate, without an adaxial hastula. Base cleft below the petiole; the mouth of the sheath with a copious brown woolly indumentum.

LAMINA (Figs. 59C, D and 60F)
Isolateral (Figs. 59C and 60F). **Trichomes** absent. **Epidermis**: without sinuous walls but with contrasted costal and intercostal bands. **Stomata** sunken and with short overarching t.s.c. **Hypodermis** 3–4-layered below each surface; s.st.ch surrounded by an irregular arrangement of cells. **Mesophyll** with distinct palisade layers towards each surface; central cells large, compact, and isodiametric. **Fibres** (Figs. 59C and 60F) numerous in anticlinally-extended strands, mostly in the mesophyll near each surface, central strands few, narrow or as solitary fibres. **Veins** mostly attached to each hypodermis by narrow fibrous girders completed by parenchyma cells, with additional small veins attached to either adaxial or abaxial hypodermis only, but then usually with an opposed independent fibre strand (Fig. 60F). **Transverse veins** (Fig. 59D) v. numerous, short, with a well-developed sclerotic sheath.

PETIOLE (Fig. 60I)
Vbs with several wide mxy vessels and abundant pxy, a single phl strand with an incomplete median constriction.

Root

Epidermis thin-walled (Fig. 20E). Outer cortex with thinner and thicker-walled layers. Stele with wide vessels in the medulla (Fig. 20D).

Kerriodoxa (Figs. 59E, F and 60K–L)

A single sp., *K. elegans*, of limited distribution in dry evergreen forests in Thailand. It has a short stem but its short stature and

leaf pattern makes it of considerable horticultural value; also unusual within the tribe in being dioecious.

Leaf

Palmate, with an adaxial hastula. Lamina strongly discolourous, with silvery undersurface representing persistent thin-walled trichome cells. Sheath entire.

LAMINA (Fig. 59E, F)
Dorsiventral (Fig. 59E). **Trichomes** restricted to costal regions of abaxial surface; minute (2–3 cells in diameter), distal cells persistent. **Epidermis** (Fig. 60K, L) with narrow, rectangular cells; adaxially with single files of somewhat wide deeper cells situated above the anticlinal walls of the hypodermal cells, the cell lumen large and the *outer* wall thickened, these alternating with 2–5 files of shallower cells with an *inner* thickened wall. Abaxial epidermis without differentiated cell files, but narrow costal bands below fibrous strands; intercostal bands wide. **Stomata** diffuse, at most sl. sunken with overarching, ± papillose t.s.c. and l.s.c.; g.c. with 2 equal cutinized ledges. **Hypodermis** 1-layered, thin-walled below each surface, abaxial cells smaller than adaxial (which appear inflated in TS); s.st.ch. surrounded by 2 longitudinally extended, C-shaped cells. **Mesophyll** uniform, but adaxial cells small, somewhat anticlinally extended as an incipient palisade. **Fibres** wide-lumened and septate, in strands of 2–6 cells in adaxial mesophyll and often next to hypodermis. Abaxial strands with fewer cells in smaller strands, always next to hypodermis. **Veins** in mid- to adaxial-mesophyll, independent of surface layers. O.S. incomplete; I.S. continuously sclerotic. **Transverse veins** (Fig. 59F) densely arranged with massive sclerotic sheath; running below most longitudinal veins often for considerable distances.

Tahina (Figs. 6A, B; 23F; and 59G, H)

A monotypic genus, *T. spectabilis*. A very rare palm of imposing stature but surprisingly only discovered in Madagascar in 2007 (Dransfield et al. 2008a). Surprising, because it rivals *Corypha* in size and is also hapaxanthic.

Leaf

Costapalmate, divided into multifold segments, each further divided into single-fold segments; adaxial hastula present. Abaxial ribs differentiated into a hierarchy of different sizes. Sheath cleft below petiole.

LAMINA (Figs. 6A, B; 23F; and 59G, H)
Dorsiventral (Figs. 6A and 59G). **Trichomes** restricted to abaxial surface, the large persistent base sunken and with numerous thick-walled cells. **Epidermis** wholly cutinized; adaxial cells rectangular with sinuous walls, abaxial cells shorter, less regular and without distinct costal bands. **Stomata** occasional adaxially; abaxial stomata with 2 additional cuticular ledges on the l.s.c., parallel to those on the g.c. T.s.c. short, overarching g.c. somewhat. **Hypodermis** up to 3-layered adaxially, the outer layer of somewhat transversely extended cells with thickened walls, the inner cells thin-walled; abaxial hypodermis 1-layered but 2-layered below sunken hairs. **Mesophyll** (Fig. 59G) with well-developed palisade up to 4 cells deep, abaxial mesophyll of up to 8

cell layers and less palisade-like. **Fibres** few and abaxial; solitary, lignified, and often next to the hypodermis; vascular fibres developed as extensions of the I.S. (Figs. 6A and 59G). **Veins** adaxial, all attached to hypodermis and with little size variation beyond the few large veins. O.S. complete only below, I.S. of sclerotic cells extended adaxially as a girder of lignified fibres completed by colourless cubical cells. **Transverse veins** (Figs. 6B and 59H) numerous, running below longitudinal veins, with a massive sclerenchyma sheath; often extending from the midrib to the leaf margin. **Stegmata** present adjacent to both longitudinal bundle sheath fibres and those of the transverse veins, the basal wall thickened and enclosing the spherical silica body (Fig. 23F). **Abaxial rib** with numerous vbs, the outer more or less confluent, the central bundles largest. Further illustrations of anatomy given in Dransfield et al. (2008a).

TRIBE CARYOTEAE
(Figs. 7H, I; 8F; 12E, F; and 61)

The tribe contains 3 genera indigenous to mainland SE Asia, with 2 of the genera more widely distributed in S India and throughout Indonesia, south to NE Australia and Vanuatu. Within each of the 3 genera, the habit varies considerably: all have representative species that are small, clustering palms, as well as solitary palms of sometimes massive tree stature. Most spp. occur in everhumid environments, and those of small stature are typically elements of forest undergrowth. Caryoteae are among the most structurally distinctive clades of palms. They are usually hapaxanthic, and are unique among palms exhibiting this type of flowering behaviour in producing inflorescences in a basipetal rather than acropetal sequence. Clearly synapomorphic for the tribe are the pinnate to bipinnate leaves that bear pinnae or pinnules with praemorse margins (Figs. 45I and 61G–I. Plants are typically monoecious, with the basic inflorescence unit a triad of a central female flower and 2 lateral male flowers (or derivative thereof). With the exception of the induplicate folding of the pinnae and the presence of a terminal pinna on each leaf (i.e. the leaves are imparipinnate), these palms are strikingly convergent upon a suite of character states otherwise best exemplified in Arecoideae (particularly tribe Iriarteeae). The arecoid-like aspect of these palms is also reflected in their anatomy, as the midrib structure, epidermal cell shape, and hypodermal cell shape of Caryoteae leaves are like those of many Arecoideae.

Anatomical features of the tribe Caryoteae
Leaf
Imparipinnate to bipinnate, with a terminal pinna. Sheath disintegrating into a complete cylinder of coarse fibres (the kittul of commerce, from *Caryota urens*); sometimes also including a conical ligule, which in young leaves forms a cap over the younger organs. Petiole unarmed, becoming terete distally (Fig. 12E, F).

LAMINA (Figs. 7H, I and 61A–I)
Dorsiventral. **Wax** layer of flake-like or hair-like cuticular proliferations covering the complete abaxial surface of the lamina, or lamina without specialized epicuticular waxes. **Trichomes** (Fig. 8F) when present abaxial, or sometimes also sparsely adaxial, of a characteristic and diagnostic rosette structure, consisting of 4–6 strongly sclerotic cells that surround 2 central, thinner-walled cells. These trichomes partly sunken into the lamina, but not interrupting the adjacent hypodermal layer. **Epidermis** with costal regions differentiated abaxially; of spindle-shaped cells on both surfaces (Figs. 7H and 61E). Adaxially with strongly thickened outer periclinal walls (many *Arenga* spp., *Wallichia*) or with all walls ± equally thick (most *Caryota* spp.); arranged in longitudinal to oblique files. Abaxial epidermal cells often in sinuous files that conform to the contour of the stomata. **Stomata** not (many *Caryota* spp.) or often sl. sunken (*Arenga*, *Wallichia*), the g.c. distinctive in always bearing a series of transverse **cuticular striations** (Fig. 61F) on their outer surface; often also with a limited adaxial distribution; t.s.c. sl. to distinctly overarching the g.c.; unlobed, or each with a pair of often papillose lobes adjacent to the stoma. Lateral neighbouring cells well differentiated, paralleling the l.s.c. **Hypodermis** mostly 1-layered beneath each surface, sometimes 2-layered adaxially (some spp. in all genera). Adaxial hypodermis of transversely extended cells hexagonal in distinct longitudinal files (Fig. 7I); frequently appearing somewhat inflated in TS. Abaxial hypodermal cells shallower but of similar shape. S.st.ch. delimited by 2–4(5) arcuate cells around a single aperture. **Mesophyll** scarcely to moderately differentiated into palisade and spongy layers; palisade 1–3-layered when present. Centrally positioned mesophyll cells often with tanniniferous contents. **Raphide sacs** common in mesophyll; ellipsoid, with sl. thickened walls. **Fibres** absent from many spp. (esp. *Caryota*). When present (spp. in all genera), septate with wide lumina and thin walls; often scattered in the mesophyll but independent of surface layers. **Veins** often entirely free of the surface layers, most commonly situated in the abaxial mesophyll, so that many are in contact with the abaxial hypodermis (but not by a bundle sheath extension); the largest vbs with 1 large mxy vessel and 1 phl strand that sometimes includes groups of sclerotic cells. Vbs with O.S. usually present laterally and interrupted by a thickened region of the I.S., especially on adaxial side. I.S. of 1 to few layers of sclerenchyma, fibrous adjacent to the phl. **Abaxial rib** strongly protruding abaxially and ± ovate in TS outline, containing many vbs enclosed within a complete sclerotic cylinder (*Arenga*, *Wallichia*). **Expansion tissue** in 2 bands, 1 to either side of sclerotic cylinder, but sometimes ± confluent adaxial to the cylinder. **Transverse veins** appearing v. slender, with an irregularly jagged and strongly sinuous course and at the same level as the longitudinal veins and nearly always in contact with them. Transverse veins narrowly sheathed by 1–2 layers of sclerotic parenchyma, and not accompanied by stegmata. **Stegmata** with hat-shaped silica bodies, present in discontinuous files along sheathing fibres of longitudinal vbs and, if present, also the non-vascular fibres.

PETIOLE (Fig. 12E, F)
In distal terete portion with 1–3 layers of colourless hypodermal cells and few layers of chlorenchyma, including raphide sacs. Peripheral vbs all with the phl facing outward, and with v. thick, confluent phl fibre sheaths, such that these series of bundles form

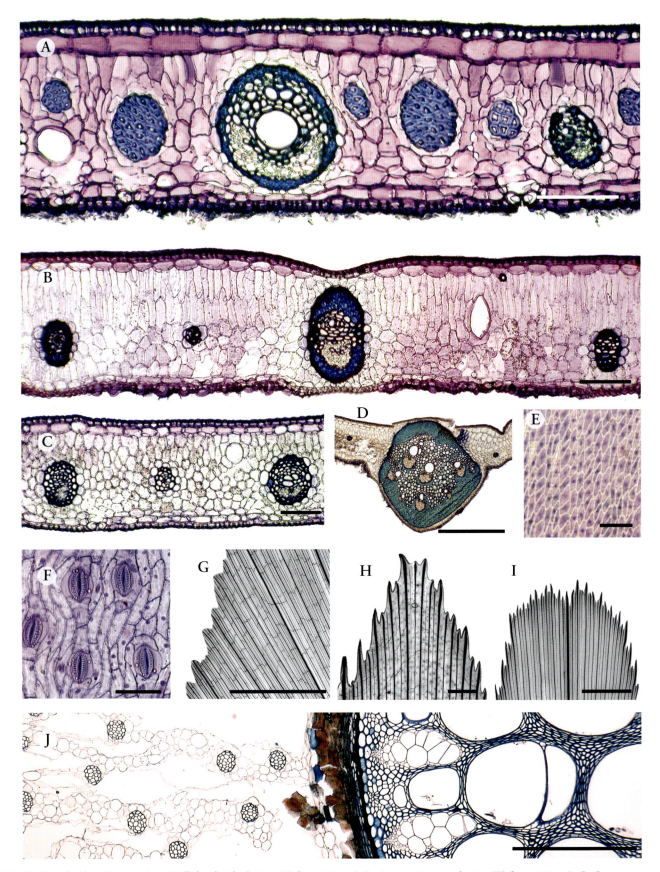

Fig. 61 Coryphoideae (Caryoteae). **A**. *Wallichia disticha*, lamina, TS (bar = 100 μm). **B**. *Arenga microcarpa*, lamina, TS (bar = 100 μm). **C**. *Caryota ophiopellis*, lamina, TS (bar = 100 μm). **D**. *Arenga microcarpa*, lamina midrib, TS (bar = 500 μm). **E**. *Arenga microcarpa*, lamina, adaxial epidermis, SV (bar = 50 μm). **F**. *Caryota obtusifolia*, lamina, abaxial epidermis, SV (bar = 50 μm). **G**. *Caryota ophiopellis*, lamina, clearing of praemorse margin of pinnule, SV (bar = 10 mm). **H**. *Arenga microcarpa*, lamina, clearing of pinna apex, SV (bar = 2 mm). **I**. *Wallichia disticha*, lamina, clearing of pinna apex, SV (bar = 10 mm). **J**. *Arenga pinnata*, root, inner cortex and outer stele, TS (bar = 500 μm).

a ± continuous sclerotic cylinder (Fig. 12F). Internally, vbs uniformly scattered, with increasingly variable orientation toward the centre. **Central vbs** with 1 large mxy vessel and 1 phl strand, usually not including sclerotic cells. **Ground tissue** sometimes including scattered non-vascular fbs.

Stem

Surface layers becoming strongly subero-lignified and sometimes producing a periderm. **Cortex** narrow, including many fbs. **Central cylinder** demarcated from cortex by a peripheral zone of densely crowded vbs with thick sheath of phl fibres. Central vbs with 1 (*Caryota*) or 2 (*Arenga*, *Wallichia*) large mxy vessels; phl with 1 undivided strand. **Ground tissue** including fbs in *Arenga pinnata* and *Wallichia disticha*. Expansion of ground tissues conspicuous in older stems of *Caryota*, the cells surrounding each vb radially dilated, creating large intercellular cavities. Tissue expansion in *Arenga* and *Wallichia* less pronounced, cells dilated without clear topographic reference to vbs or stem periphery, creating few intercellular spaces.

Root (Fig. 61J)

Periderm developed in older roots. **Epidermis** ± persistent, with thickened outer periclinal walls. **Exodermis** 1-layered, weakly or sometimes not differentiated from the 2 zones of the **outer cortex;** the outer zone 4–6-layered, of tanniferous cells with moderately thickened walls; the inner 8–12-layered, of cells lacking tannins and mostly with v. thick walls. Mucilage canals sometimes present. **Inner cortex** (Fig. 61J) with 3 zones, including large fbs concentrated in the middle, aerenchymatous zone, circular in TS outline and with accompanying stegmata. In older roots, innermost cortex with stone cells (Fig. 61J). **Endodermis:** cells with unevenly thickened walls, U-shaped in TS, the outer periclinal wall remaining thin. **Pericycle** 1-layered. **Stele** with vessels embedded in fibrous ground tissue extensively developed in outer stele and sometimes completely occupying the whole of the medulla (Fig. 20A, B). **Medulla** otherwise parenchymatous, without vessels.

Key to genera of Caryoteae based on lamina anatomy

1. Outermost hypodermal cells of both leaf surfaces with inclusions of crystal sand (calcium oxalate) in the form of minute styloids and prismatic crystals. Trichomes rarely present. Entire margin of pinnules not including a strongly differentiated marginal vein. Longitudinal veins of pinnules of even width throughout, not dilated and spathulate distally at praemorse margin (Fig. 61G) . . . . *Caryota*

1A. Outermost hypodermal cells without crystal inclusions. Trichomes frequent on abaxial lamina surface; consisting of a rosette of 4–6 sclerotic cells surrounding 2 central, thinner-walled cells. Entire margin of pinnae including a large marginal vein with a thick fibre sheath and little vascular tissue. Longitudinal veins of pinnae distally flaring into a terminal tracheid complex at praemorse margin (Fig. 61H, I) 2

2. Abaxial lamina surface wholly covered by flake-like cuticular proliferations arranged around each stomatal complex in a rosette, imparting an overall frosted appearance macroscopically . . . . *Arenga*

2A. Abaxial lamina surface without flakes of epicuticular wax, but sometimes cuticle appearing drawn out as fine hairs on each epidermal cell . . . . *Wallichia*

Caryota (Figs. 9G and 61C, F, G)

A genus of 13 spp., immediately recognizable by its bipinnate leaf. *Caryota* has a distribution about equivalent to that to the whole tribe. Species are known as fishtail palms, as the pinnules are obliquely wedge-shaped, each with a conspicuously praemorse distal margin and include many prominent longitudinal veins, but no distinct midrib.

Leaf

Mature foliage bipinnately compound, with a terminal pinna. Pinnae themselves with a terminal pinnule.

LAMINA (Fig. 61C, F, G)

Dorsiventral (Fig. 61C). Surfaces without epicuticular wax deposits. **Epidermis**: adaxial cells in longitudinal files and with straight or somewhat sinuous anticlinal walls. **Stomata** (Figs. 9G and 61F): t.s.c. not or only sl. lobed in *C. obtusa*, transversely elliptical to longitudinally semielliptical, never papillose. **Hypodermis**: layers on both surfaces including crystal sand in the form of minute styloids and prismatic crystals. **Fibres** wholly absent (Fig. 61C), except in *C. monostachya*, then single or sometimes paired in a ± regular series adjacent to both surface layers. Large, rib-like longitudinal **veins** of pinnules with a thick fibre sheath and a single, major phl strand accompanied by 2–4 much smaller phl strands, these individually isolated within the main sheath of phl fibres. **Marginal venation** (Fig. 61G): without terminal enlargement. **Transverse veins** as for tribe.

   Note: leaf anatomy within *Caryota* is rather uniform. *Caryota obtusa* and *C. ophiopellis* have a 2-layered adaxial hypodermis. Stomata are sunken only in *C. obtusa*.

Arenga (Figs. 8F; 9E; 20A, B; and 61B, D, E, H, J)

Arenga (20 spp.) is not only the largest genus of Caryoteae, but also the most varied morphologically and anatomically. As with *Caryota* the geographic distribution of the genus is about equivalent to that of the whole tribe.

Leaf

Imparipinnate, rarely undivided (e.g. some forms of *A. hookeriana*). The long, stout fibres that cloak the stems of many spp. are a product of the disintegration of the ligule, and not necessarily the basal portion of the fibrous leaf sheath.

LAMINA (Figs. 9E and 61B, D, E, H)

Dorsiventral (Fig. 61B). Abaxial epicuticular waxes present (see key). **Trichomes** present intercostally on both surfaces but most common abaxially (Fig. 8F). **Epidermis** (Fig. 61E): cells obliquely extended without sinuous anticlinal walls (except

A. undulatifolia); adaxially cells often in oblique files, but in distinct longitudinal files in *A. engleri*, *A. porphyrocarpa*, and *A. undulatifolia*. In *A. porphyrocarpa* and *A. undulatifolia*, uniseriate files of short cells with wide lumina occur above the longitudinal anticlinal walls of the underlying hypodermal cells, and so delimit a wide region of 4–6 files of long cells with narrow lumina. **Stomata** sl. sunken; t.s.c. bilobed adjacent to the g.c., the lobes papillose and overarching the aperture (Fig. 9E). **Hypodermis** usually 1-layered below each surface, but 2-layered adaxially in *A. undulatifolia*. **Fibres** sometimes absent (Fig. 61B) but even when present never abundant; solitary in the mesophyll in *A. hookeriana* and *A. undulatifolia*; in an irregular adaxial series of small bundles in *A. microcarpa*; series of small bundles adjacent to both surfaces in *A. tremula*. **Marginal venation** (Fig. 61H) with inflated vein ends. **Abaxial rib** (Fig. 61D) and **transverse veins** as for the tribe.

Root (Figs. 24A, B and 61J)

Representative of the whole tribe. The condition illustrated in Fig. 24 A, B, in which the medulla is almost wholly occupied by wide vessels, is in need of further investigation in light of the similar condition reported for *Corypha* (Tomlinson, 1961). Although not present or rare in Borasseae, abundant medullary vessels are uncommon in Coryphoideae outside of *Corypha* and Caryoteae, and constitute possible further structural evidence to support the close relationship between these 2 largely dissimilar groups.

Wallichia (Fig. 61A, I)

Containing 9 spp. restricted to NE India and mainland SE Asia. *Wallichia* is quite like *Arenga* morphologically, but is mostly clearly differentiated from the latter genus in having male flowers with sepals connate into a tube.

Leaf

Like that of *Arenga* but *Wallichia disticha* is distinctive in its 2-ranked crown of leaves.

LAMINA (Fig. 61A, I)

Dorsiventral (Fig. 61A). Specialized, trichome-like epicuticular **wax** elaborations present abaxially in *W. oblongifolia*. **Trichomes** present on both surfaces, most common abaxially; strongly sunken in *W. disticha*. Costal regions differentiated on both surfaces. Adaxial epidermis of *W. oblongifolia* with short and long cells differentiated like those of *Arenga* spp.; *W. disticha* with the cells arranged in oblique files. **Stomata** sunken, strongly so in *W. disticha*; t.s.c. bilobed, and in *W. disticha*, strongly papillose adjacent to g.c., overarching the aperture. **Hypodermis** 2-layered adaxially, the walls of the outer cells thickened and strongly pitted (Fig. 61A). **Fibres** distributed in bundles. In *W. oblongifolia*, bundles of up to 20 cells interdigitated between the vbs (Fig. 61A); in *W. disticha*, small bundles of up to 4 cells in irregular series both ad- and abaxially. **Veins**: larger vbs in contact with both surfaces. **Marginal venation**: vein ends inflated (Fig. 61I). Abaxial **rib** as for tribe. **Transverse veins** as for tribe, but those of *W. disticha* essentially without any sheathing cells.

TRIBE CORYPHEAE

Corypha (Figs. 13E–H, K; 20I; 24H, I; and 60A–E)

A genus of about 6 spp. from S India to N Australia, distinguished by its regularly spiny petiole margins, massive proportions, and hapaxanthic habit resulting in the single stem producing a gigantic terminal inflorescence. It is not a particularly long-lived palm because it flowers 40–60 years after seedling establishment. For inflorescence morphology see Tomlinson and Soderholm (1975) and Fisher et al. (1987).

Leaf

Costapalmate, the blade to 6 m diameter; hastula present on both surfaces; blade segmented to about half its radius, with abaxial ribs most prominent. Petiole to 4 m, the margins armed with teeth (Fig. 24H, I). Sheath cleft dorsally below petiole (Fig. 13E–H, K); often somewhat persistent and not disintegrating into a weft of fibres.

LAMINA (Fig. 60A–C)

Dorsiventral (Fig. 60A), but 2 surfaces v. similar; adaxial surface with fewer and wider costal regions. **Trichomes** apparently absent from mature lamina, but present above and below main ribs producing an ephemeral woolly indumentum. **Epidermis**: costal cells narrow, rectangular, with sl. sinuous walls; intercostal cells wider, shorter, less regular. **Stomata** (Fig. 60C) restricted to intercostal bands, but not in regular files, short, sunken in *C. utan* and *C. taliera*, but not in *C. umbraculifera*; t.s.c. short, wide and broadly crescentic to kidney-shaped, not overarching g.c. Outer cuticular ledge of g.c. prominent; inner ledge scarcely developed (Fig. 60C). **Hypodermis** 2-layered adaxially, 1-layered abaxially, the abaxial cells largest. S.st.ch. surrounded by 3–4 thin-walled, L-shaped cells. **Mesophyll** somewhat palisade-like beneath the adaxial surface, but scarcely so abaxially; central cells largest and more compact (Fig. 60A). **Fibres** septate, in bundles associated with each surface; adaxial bundles in 2 series mostly present as anticlinally-extended continuations of the vb sheath and in contact with the hypodermis, but with occasional small strands or single fibres alternating with the veins but still in hypodermal contact. Abaxial fbs smaller, rarely as solitary fibres, always in contact with the hypodermis or epidermis but not associated with vbs. **Veins** (Fig. 60A) restricted to adaxial mesophyll and each in contact with the adaxial hypodermis via an anticlinally-extended bundle sheath; the abaxial bundle sheath of narrow fibres less conspicuous (Fig. 60A). O.S. complete only on abaxial side of vein, not continuous with the hypodermis. Phl not sclerotic. **Transverse veins** (Fig. 60B) frequent, massive, extending below the longitudinal veins (Fig. 60A); fibrous sheath uniformly well developed; the phl sometimes surrounding the central xyl elements. **Raphide sacs** large, conspicuous within the central mesophyll. **Stegmata** infrequent, in files next to fibrous strands and bundle sheath fibres; silica body small and irregularly spherical.

PETIOLE (Fig. 60D)

Hypodermis 3–5 layered and of small, isodiametric, sclerotic cells. Non-vascular fbs concentrated near the periphery, but also common in lower density throughout the ground tissue. **Vbs** (Fig. 60D) with a scattered distribution, the central vbs with 1 large mxy vessel and 1 phl strand but usually without included sclerotic cells. Massive fibre sheath of vbs replaced by thick-walled parenchyma cells at level of mxy.

Stem

Cortex narrow, with many fbs and vbs. **Central cylinder** with peripheral zone of highly congested vbs, each with, a thick and often radially extended sheath of phl fibres. Central vbs each with 2 wide mxy vessels and a single phl strand with few sieve tubes; sheath of phl fibres modest in size relative to other Coryphoideae. **Ground tissue** of central cylinder little modified with age but including many non-vascular fbs; internally with abundant starch.

Root (Figs. 20I and 60E)

Periderm becoming thickened with age. **Epidermis** apparently of many layers of cells with ± unthickened walls and so apparently multiple; root hairs present when young (Fig. 20I). **Exodermis** many-layered, of thick-walled cells that accumulate tannins. **Outer cortex** of 1 layer of thick-walled cells. **Inner cortex** 3-zoned; including large fbs accompanied by stegmata ('Raphia-type'). **Endodermis** (Fig. 60E): cells with unthickened walls; outline elliptical in outline in TS. **Vessels** embedded within ground tissue of fibres. Medulla including vessels; Tomlinson (1961) reports the medulla is wholly occupied by vessels, each surrounded by a sclerotic sheath; see notes under *Arenga*.

TRIBE BORASSEAE (Figs. 62–65)

A distinctive group of 8 dioecious, pleonanthic genera with palmately compound, induplicate leaves, usually single-stemmed (exceptionally multiple-stemmed in some *Hyphaene* spp.) and with massive trunks. Several spp. of *Hyphaene* have trunks that also branch dichotomously, producing trees of distinctive habit. The tribe is exclusively Old World, ranging from West Africa to New Guinea. Inflorescences typically have thick, little-branched axes with the male flowers usually sunken in cincinnate clusters in rachilla pits whereas the female flowers are often very large. Fruits are large fibrous drupes which can have 3 separate seeds (pyrenes) surrounded individually by an endocarp.

Anatomical features of the tribe Borasseae

Leaf

Costapalmate, often strongly so. Adaxial hastula present or absent. Petioles armed or not. Leaf base always cleft below the petiole; often somewhat persistent to form a woody lattice work, and not disintegrating into fibres.

LAMINA

Isolateral or infrequently dorsiventral. Adaxial surface often with a thick layer of epicuticular wax. Abaxial surface sometimes with wax deposits distinctive for individual genera. **Trichomes** cos-tally distributed when present, abaxial only, or on both surfaces when strongly isolateral; usually peltate, sometimes massive. **Epidermis**: adaxially of longitudinally-extended rectangular cells, anticlinal walls straight or sinuous; outer periclinal wall at most moderately thickened. **Stomata** with lateral neighbouring cells producing an extra division parallel to the l.s.c. (uncommonly so in *Borassodendron* and *Borassus heineanus*), the g.c. sunken or less commonly not. **Hypodermis** mostly 2–3-layered both ad- and abaxially (1-layered adaxially in *Borassodendron* and *B. heineanus*); s.st.ch. often defined by 4 U-shaped cells in 2 opposing pairs delimiting 2 apertures. **Mesophyll** thin-walled, compact; always with well differentiated palisade either adaxial in dorsiventral lamina or also abaxial when distinctly isolateral. **Veins** central to adaxial in the mesophyll and always attached to adaxial surface by a fibrous bundle sheath extension (fibre girder) continuous with cells of the I.S. Taxa with isolateral leaves have few to many vbs with both ad- and abaxial fibre girders. O.S. of often inflated parenchyma cells discontinuous with the fibre buttresses. I.S. often of thick-walled, non-septate fibres. **Transverse veins** extending between few large vbs, in *Borassus* supported by complete sclerotic partitions. **Ad-** and **abaxial** ribs with at least 7 large independent vbs in mature foliage; the adaxial ribs with the bundles usually vertically superimposed. **Expansion tissue** in a single, wide band opposite the ribs; including few, large non-vascular fbs near the periphery. **Stegmata** in continuous files adjacent to all longitudinal veins and fibres, and also accompanying the transverse veins of all taxa except *Hyphaene* and *Medemia*. Stegmata of transverse veins irregularly distributed and sometimes much larger than those of vbs, with a thickened and lignified basal cell wall. Silica bodies spherical and spinulose, but in *Lodoicea* reported as hat-shaped in Tomlinson (1961).

PETIOLE

Epidermis provided with trichomes like those of the lamina, or sometimes with distinctive trichomes when the lamina is glabrous. *Bismarckia* provides an example of the latter condition, as its petiole has moderate indument of large, scurfy, dendritic trichomes (Fig. 62E). **Hypodermis** well-developed, of 1–4 layers of strongly lignified cells. **Central vbs** with a uniform scattered distribution, each with (1–)2 large mxy vessels and 1 phl strand, including few or no sclerotic cells. Fibre sheath of vbs well developed, but discontinuous lateral to the mxy, with fibres replaced by sclerotic parenchyma. **Fibres** abundant and congested at the petiole periphery, absent or widely scattered centrally.

Stem

Producing a periderm with age. **Cortex** with numerous large fbs and many small vbs. Peripheral region of **central cylinder** a zone crowded with vbs that have thick sheaths of phl fibres (crowding notably dense in *Hyphaene*). Central **vbs** with mostly 1 wide mxy vessel in *Borassus* and 2 wide mxy vessels in *Hyphaene* and *Bismarckia*. Phl strand always single. **Ground tissue** without fbs and densely parenchymatous except in *Bismarckia*, the only genus of Coryphoideae with pronounced aerenchymatous ground tissue. Expansion of ground tissue cells evident in the cortex of *Hyphaene*. Slight increases in cell size around vbs in stem centre evident in both *Hyphaene* and *Borassus*.

Root

Periderm development reported by Seubert (1997) as present in *Borassodendron* and *Latania*; investigation of older roots of other genera needed to ascertain the wider occurrence of this trait within the tribe. **Epidermis** many-layered, apparently multiple; of cells with evenly thin walls. **Exodermis** many-layered and strongly sclerotic. **Outer cortex** homogenous and of 2–4 layers of cells with thinner walls than the exodermis, or 2-zoned with the outer cells with distinctly thinner walls than in the inner zone (*Latania*). **Inner cortex** 3-zoned, but with aerenchyma of the middle zone poorly developed. Inner cortex including large circular bundles of fibres, in TS outline, accompanied by stegmata ('Raphia-type' fibres). *Borassus* and our material of *Borassodendron* distinctive within the tribe in being without any inner cortical fibres (the latter with such fibres according to Seubert 1997). Inner cortex of *Borassodendron* also said by Seubert (1997) to include solitary fibres of the 'Kentia-type'. **Endodermis**: cells with essentially unthickened walls, tangentially elliptical in TS. **Vascular tissue** of the stele embedded in parenchyma. Medullary vessels absent.

SUBTRIBE HYPHAENINAE (Fig. 62)

Key to genera of Hyphaeninae based on lamina anatomy

1. Mature lamina with large knob-like or peltate trichomes present costally on both surfaces; stegmata absent from transverse veins. [Isolateral] 2
1A. Mature lamina glabrous; stegmata present adjacent to transverse veins. [Dorsiventral or isolateral] 3
2. Non-vascular fbs regularly present without connection to surface layers and in 3 series: adaxial, central, and abaxial . . . . *Medemia*
2A. Non-vascular fbs irregularly present without connection to surface layers and with a predominantly central distribution . . . . *Hyphaene*
3. Lamina histology with **isolateral** symmetry. Longest transverse veins evidently short, extending a distance no greater than the span of 5 longitudinal veins . . . . *Bismarckia*.
3A. Lamina histology with **dorsiventral** symmetry. Longest transverse veins extending from midrib to near lamina margin . . . . *Satranala*

Bismarckia (Figs. 3 and 62E, K, L)

A monotypic genus endemic to Madagascar, where it is a common component of savannah vegetation on the seasonally dry, western part of the island. *Bismarckia nobilis* is a strikingly handsome palm, and forms that produce abundant epicuticular wax are prized as ornamentals for their silvery foliage.

Leaf

Strongly costapalmate; adaxial hastula asymmetric. Petiole unarmed.

LAMINA

Isolateral (Fig. 3). **Trichomes** absent from mature lamina. Surfaces alike, with epidermal cells rectangular or irregularly polygonal. **Stomata** (Fig. 3 F–H) sunken, the t.s.c. longitudinally semielliptical to transversely elliptical; l.s.c forming a flask-shaped receptacle (in TS), which entirely include the g.c. **Hypodermis** 3-layered adjacent to each surface, of hexagonal cells transversely extended in the intercostal regions; s.st.ch. delimited by 2 C-shaped to 4 L-shaped cells around 1 aperture. **Fibres** distributed in ± regularly repeating series of bundles both ad- and abaxially, usually attached to the surface. **Mesophyll** including sparsely scatted central solitary fibres and small vbs. **Veins** (Fig. 3A–E) attached to both surfaces by girders of 8–14 fibre layers; each vb with 1 large mxy vessel; phl strand of larger vbs enclosed in a cylinder of I.S. fibres that sometimes partitions it adaxially, the main strand then appearing accompanied by 2 small phl strands situated near the bundle periphery. Segment margin including a large marginal vein. **Ad-** and **abaxial ribs** of similar size; the abaxial rib including 15–20 large, independent vbs, the adaxial somewhat fewer. **Transverse veins** short, with a 2–3-layered fibre sheath.

PETIOLE

Central vbs with 2 wide mxy elements and single phl strand (Fig. 62K).

Stem

Central vbs with mostly 2 wide mxy elements and single phl strand, phl sheath well-developed, ground tissue somewhat lacunose (Fig. 62L).

Satranala (Fig. 62B, D, G)

A single sp., *S. decussilvae* only recently discovered in Madagascar. Single stemmed to 15 m high with costapalmate leaves. The palm is contrasted with other Hyphaeninae in its forest, rather than savannah habitat (Dransfield and Beentje 1995).

Leaf

Costapalmate, with adaxial hastula. Petiole ± unarmed.

LAMINA (Fig. 62D, G)

Dorsiventral (Fig. 62D). **Trichomes** absent. Wax layer present, occluding stomata. **Epidermis**: without stomata adaxially; cells shallow, narrowly rectangular but often with sl. oblique end walls; costal bands not differentiated; otherwise of **long cells** with narrow lumina in files above the outer periclinal walls of outermost adaxial hypodermal layer, and **short cells** with wider lumina, in single files above the longitudinal anticlinal walls of hypodermal cells. Abaxial epidermis with well-differentiated costal and intercostal bands, the costal cells like those of the upper epidermis. **Stomata** sunken, the g.c. depressed below the epidermal level; neighbouring cells small and apparently further subdivided to form an abrupt stomatal depression; t.s.c. short, 2–3 at each end. **Hypodermis** 2-layered below each surface; outer adaxial cells hexagonal; transversely extended to cubical, sl. thick-walled; cells surrounding s.st.ch. irregular and often extending across the chamber. **Mesophyll** compact; palisade scarcely developed.

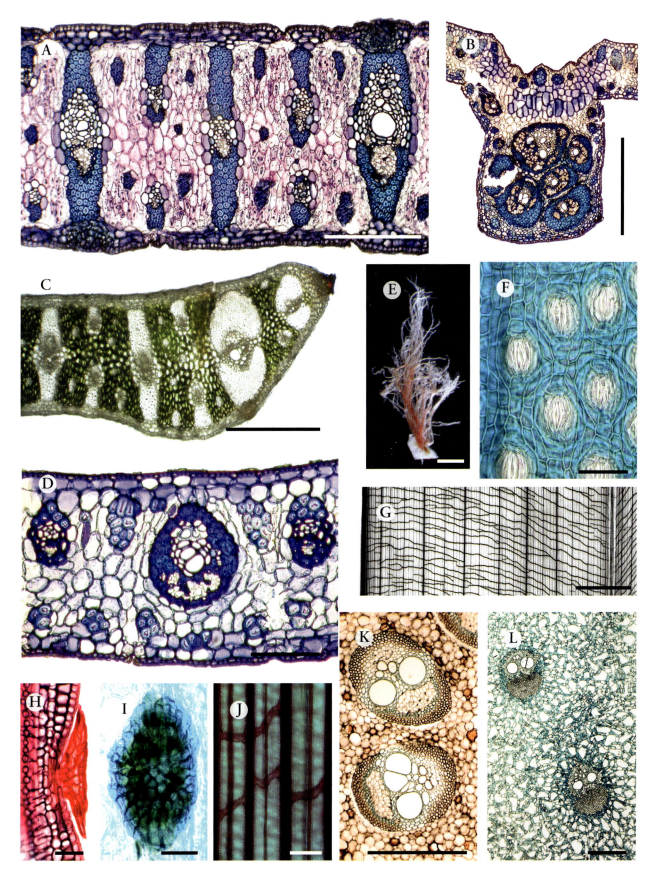

Fig. 62 Coryphoideae (Borasseae: Hyphaeninae). **A**. *Medemia argun*, lamina, TS (bar = 250 μm). **B**. *Satranala decussilvae*, midrib, TS (bar = 500 μm). **C**. *Medemia argun*, lamina margin, fresh unstained, TS (bar = 250 μm). **D**. *Satranala decussilvae*, lamina, TS (bar = 100 μm). **E**. *Bismarckia nobilis*, trichome removed from mature petiole, SV (bar = 1 mm). **F**. *Hyphaene dichotoma*, adaxial epidermis, SV (bar = 50 μm). **G**. *Satranala decussilvae*, venation in cleared lamina, SV (bar = 10 mm). **H**. *Hyphaene* sp., lamina trichome, LS (bar = 50 μm). **I**. *Hyphaene dichotoma*, lamina trichome, SV (bar = 500 μm). **J**. *Hyphaene thebaica*, venation in cleared lamina, SV (bar = 100 μm). **K**. *Bismarckia nobilis*, vbs in petiole, TS (bar = 500 μm). **L**. *Bismarckia nobilis*, vbs in stem centre, TS (bar = 1 mm).

Fibres (Fig. 62D) conspicuous in bundles below each surface, the fibres unlignified, thick-walled, but with wide, septate lumina. Adaxial strands anticlinally extended, alternating with veins; abaxial strands smaller, not anticlinally extended. **Veins** (Fig. 62D) all in contact with adaxial hypodermis ± alternating with larger bundles, only major veins with well-developed sclerotic I.S. **Ribs** (Fig. 62B) with several separate vbs and a single band of expansion cells with included fbs. **Transverse veins** (Fig. 62G) long, often extending from midrib to segment margin; sheath cells fibrous, well developed. **Stegmata** mostly next to vascular fibres, including those of the transverse veins; less commonly next to non-vascular fibres; silica bodies small, irregularly spherical. **Tannin cells** conspicuous and restricted mainly to adaxial hypodermis.

Hyphaene (Figs. 6C, D and 62F, H, I, J)

A genus of about 8 spp. distributed in regions of arid climate throughout much of Africa, with a disjunct sp. in W India. Most spp. have aerial axes that are equally dichotomously branched, which imparts a distinctive physiognomy to these palms that allows for immediate recognition so that they become juju trees in Africa. *Hyphaene* spp. are moderate to large, solitary or clustering palms; seeds without ruminate endosperm.

Leaf

Strongly costapalmate, with adaxial hastula. Petiole armed with a regularly repeating series of stout spines. Leaf sheaths often persistent.

LAMINA
Isolateral (Fig. 6C, D). **Trichomes** (Fig. 62H, I) massive, peltate, sunken in depressions; the bases partly included within the lamina tissue, but not interrupting the hypodermis; protruding above the lamina surface, with the stalk and shield cells suberolignified. **Epidermis** (Fig. 62F) with surfaces alike and with well differentiated costal regions of rectangular or irregularly polygonal cells with straight anticlinal walls. **Stomata** sunken; the l.s.c. often enveloping the g.c. **Hypodermis** 3-layered below each surface, cells hexagonal, those of the intercostal regions often extended transversely; s.st.ch. delimited by 3–5 curved or linear, thin-walled cells commonly defining 2–4 apertures of variable size. **Mesophyll** between the large vbs appearing **transversely partitioned** by regularly repeating series of uniseriate cells with conspicuously thicker walls than intervening mesophyll cells (Fig. 6D); the cells commonly accumulating tannin. **Fibres** distributed in usually narrow bundles, with the ad- and abaxial series in contact with the surface. Small fbs also occurring in central mesophyll, irregularly alternating with the vbs. **Veins** almost always connected to both surfaces by narrow fibre girders (Fig. 6C). Segment margin including a large, marginal vein. **Ad- and abaxial ribs** of similar size, including 5–7 large vbs centrally and few peripherally. **Transverse veins** (Fig. 62J) short, sheathed by sclerotic parenchyma.

Medemia (Fig. 62A, C)

Medemia argun is a palm of remote desert oases in Egypt and Sudan, and was for many years thought extinct until its relatively recent rediscovery in 1995 (Gibbons and Spanner 1996). This monotypic genus is quite similar to *Hyphaene*, but differs from it in lacking an adaxial hastula and in having a seed with ruminate endosperm.

Leaf

Strongly costapalmate, without an adaxial hastula. Petiole armed with a series of stout spines.

LAMINA (Fig. 62A, C)
Isolateral (Fig. 62A). Lamina surfaces similar, with well-differentiated costal regions. **Trichomes** each with a massive base, deeply included within the lamina tissue to the level of the fibre girder and interrupting the hypodermis but not sunken in a depression, with little tissue persisting beyond the lamina surface. **Epidermis** of rectangular to irregularly polygonal cells with straight anticlinal walls. **Stomata** sunken. **Hypodermis** 3(–4)-layered adjacent to each surface, cells hexagonal, the intercostal cells strongly transversely extended. **Mesophyll** with many cells accumulating tannins, and without transverse partitions of thick-walled cells. **Fibres** thick-walled and non-septate; grouped in bundles of up to 15 cells fbs free of surface and ± regularly distributed in 3 series (adaxial, central, abaxial) usually alternating with the vbs (Fig. 62A). **Veins** with well-developed fibre girders of 2–12 layers, connecting each bundle to both surfaces. Smaller vbs in 2 opposed series, both ad- and abaxial, with each bundle attached to the adjacent surface. **Segment margin** (Fig. 62C) including 1 massive vb. **Transverse veins** extending between adjacent large vbs only, with a 1–2-layered sheath of sclerotic parenchyma; stegmata absent.

SUBTRIBE LATANIINAE (Figs. 63; 64; and 65)

Key to genera of Lataniinae based on leaf anatomy

1. Stomata deeply sunken; lateral subsidiary cells forming a flask-shaped receptacle that envelopes the g.c.; outer cuticular ledges not exposed. [Weakly isolateral]. . . . . . . *Latania*

1A. Stomata sunken or not; lateral subsidiary cells just partially enveloping g.c., outer cuticular ledges clearly exposed. [Dorsiventral or isolateral] 2

2. Adaxial hypodermis 3(4)-layered; not including stegmata. Abaxial hypodermis 2-layered; both layers with sclerotic, conspicuously-pitted, lignified cells. Outer tier of unpartitioned s.st.ch. formed by 2 sclerotic hypodermal cells. Trichomes exclusively abaxial, peltate. [Dorsiventral] . . . .*Lodoicea*

2A. Adaxial hypodermis 1–2-layered; including large, thick-walled stegmata. Abaxial hypodermis 2-layered, cells thick-walled, but either unlignified, or only outermost layer lignified. Outermost tier of s.t.ch. with 4, thin-walled, U-shaped cells organized in 2 opposing pairs, forming a central partition. Trichomes either absent or present on both surfaces, not peltate. [Dorsiventral or isolateral] . . . 3

3. Stomatal aperture overarched externally by a pair of cuticular canopies (the canopy entire with a scalloped margin in

B. heineanus; strongly fimbriate in all other spp.). Longest transverse veins extending much less than half the distance from the midrib to the segment margin; unbranched (or rarely branched in *B. heineanus*). [Dorsiventral (*B. heineanus*) or isolateral] . . . . *Borassus*

3A. Stomatal aperture without overarching cuticular canopy. Longest transverse veins extending greater than half the distance from midrib to segment margin; with occasional, but prominent branches. [Dorsiventral] . . . . . . *Borassodendron*

Latania (Figs. 63D, G, H, J and 65B)

An endemic genus of the Mascarene Islands, containing 3 spp. that are all of high conservation priority.

Leaf

Lamina moderately costapalmate, with a symmetrical adaxial hastula. Petiole unarmed.

LAMINA (Fig. 63D, G, H, J)
Weakly isolateral (Fig. 63J), the 2 surfaces similar, but adaxially with more numerous and wider costal regions; abaxial palisade weakly developed. **Trichomes** absent. **Epidermis** with rectangular to irregularly polygonal cells, the anticlinal walls straight. **Stomata** (Fig. 63D) deeply sunken, g.c. situated well below the level of the epidermis; t.s.c. with an extra division narrowly crescentic, not overarching the g.c. L.s.c. narrowly crescentic, the pair flask-shaped in TS, with the aperture lined with wax deposits; entirely enveloping the g.c. **Hypodermis** 3-layered below each surface of thick-walled, transversely-extended cells (Fig. 63J); s.st.ch. (Fig. 63G, H) defined by 2 tiers of hypodermal cells with thinner walls than those of adjacent hypodermal cells. External s.st.ch. tier of 4 U-shaped cells either delimiting 2 apertures, or equally common, of 3–5 irregularly-shaped cells usually delimiting 3 apertures. Internal s.st. ch. tier of 2 C-shaped, transversely-extended or 4 L-shaped cells defining a single aperture. **Fibres** (Fig. 63J) adaxially as bundles irregularly alternating with the vbs and often not connected to the surface; narrowly elliptical in TS outline, but variable in size. Abaxial opposed to the adaxial series, usually free of the surface, ± circular in TS, of up to 20 cells. **Veins** all attached to adaxial surface by girders 5–10 cells deep and 3–6 wide. Many larger vbs also attached to adaxial surface by girders of relatively narrower fibres resembling I.S. fibres adjacent to the phl, and similar to those of the adaxial girder in *L. loddigesii*. Larger vbs with 1 large mxy vessel; phl sometimes irregularly partitioned by groups of sclerotic cells. **Transverse veins** short, with a 2–4-layered fibre sheath.

PETIOLE
Central vbs massive, with well-developed discontinuous fibre sheath, mxy vessels 1 or more, phl a single strand (Fig. 65B).

Lodoicea (Figs. 12I; 63E; 64F; and 65A, F–H)

A monotypic genus, famous for producing the most massive seed of any plant. The double coconut or coco-de-mer (*L. maldivica*) is a rare endemic of the Seychelles. Mature plants are massive palms, attaining or exceeding the height of the surrounding forest canopy. Equally as bizarre as the enormous fruits and seeds of

the genus is its mode of germination and establishment growth (Tomlinson 1990). Juvenile plants may remain without an aerial axis for decades, and when grown in shade, have the amazing ability to display their foliage at the level of the forest canopy by producing champion-length petioles claimed to be up to 15 m long (Edwards et al. 2002).

Leaf

Lamina strongly costapalmate; adaxial hastula absent. Petioles unarmed.

LAMINA (Figs. 63E and 64F)
Dorsiventral (Fig. 64F). **Trichomes** peltate, elliptical, the bases sunken and accompanied by 1–2 extra layers of hypodermal cells; base short, c. 3 cells deep; shield cells not persisting at maturity. Costal regions differentiated on abaxial surface only. **Epidermis**: cells of both surfaces with sinuous anticlinal walls. **Stomata** (Fig. 63E) not sunken, minute. **Hypodermis** (Fig. 64F) adaxially of 3–4-layers of colourless, unlignified, but thick-walled parenchyma cells, quadrangular and isodiametric to transversely extended. Abaxial hypodermis 2-layered, cells sclerotic, irregularly rectangular and transversely extended; s.st.ch. delimited by 2 tiers, both of 2 C-shaped sclerotic cells; those of the outer tier longitudinally oriented, the inner transversely oriented. **Fibres** (Fig. 64F) adaxially non-lignified, wide-lumened, septate, in large bundles up to 5–8 cells wide in contact with surface and irregularly alternating with vbs; abaxially in bundles of up to 15 lignified cells, ± circular in TS outline, mostly free of surface. **Veins** always attached to adaxial surface by girders of fibres 5–8 cells wide, these fibres resembling those of the adaxial non-vascular fbs; always free of abaxial surface. O.S. of thick-walled parenchyma. Large vbs with phl strand divided in half by fibres continuous with those of the I.S. Ribs of both surfaces ± equivalent in size and shape, elliptical in TS, each with c. 10 large, independent vbs; few small bundles at rib periphery. **Transverse veins** extending between ribs, or from midrib to segment margin, with a thick fibre sheath of up to 8 layers; accompanying stegmata rare, with spherical silica bodies.

PETIOLE
Adaxial vbs progressively narrower and more congested to adaxial surface, vbs inverted each with 1 wide mxy vessel and an undivided phl strand (Figs. 12I and 65A).

Stem (Fig. 65F–H)

Cortex (Fig. 65F) wide with numerous fbs and narrow vbs, outer **central cylinder** with narrow vbs (Fig. 65G), central vbs with massive phl sheath, ground parenchyma cells not yet expanded (Fig. 65H).

Note: silica bodies of stegmata of transverse veins said by Tomlinson (1961) to be hat-shaped.

Borassodendron (Figs. 11B, D; 63A; 64H; and 65I)

Consisting of 2 spp. of large rainforest palms distributed from S Thailand to Borneo. The genus is distinguished from *Borassus* most readily by its sharp-edged, but spineless petiole margins and in the comparatively fewer number of flowers present in the pits of the staminate rachillae.

Fig. 63 Coryphoideae (Borasseae: Lataniinae). **A**. *Borassodendron machadonis*, adaxial epidermis, SV (bar = 50 μm). **B**. *Borassus flabellifer*, abaxial epidermis, SV (bar = 100 μm). **C**. *Borassus flabellifer*, abaxial epidermis, stomata, SV (bar = 50 μm). **D**. *Latania verschaffeltii*, adaxial epidermis, stomata, TS (bar = 25 μm). **E**. *Lodoicea maldivica*, abaxial epidermis, stomata, TS (bar = 25 μm). **F**. *Borassus flabellifer*, lamina trichome, SV (bar = 50 μm). **G**. *Latania verschaffeltii*, abaxial hypodermis, outer tier of cells delimiting s.st.ch., SV (bar = 25 μm). **H**. *Latania verschaffeltii*, abaxial hypodermis, internal tier of cells delimiting s.st.ch., SV (bar = 25 μm). **I**. *Borassus flabellifer*, adaxial epidermis, stomata, TS (bar = 25 μm). **J**. *Latania verschaffeltii*, lamina, TS (bar = 200 μm). **K**. *Borassus heineanus*, lamina, TS (bar = 250 μm).

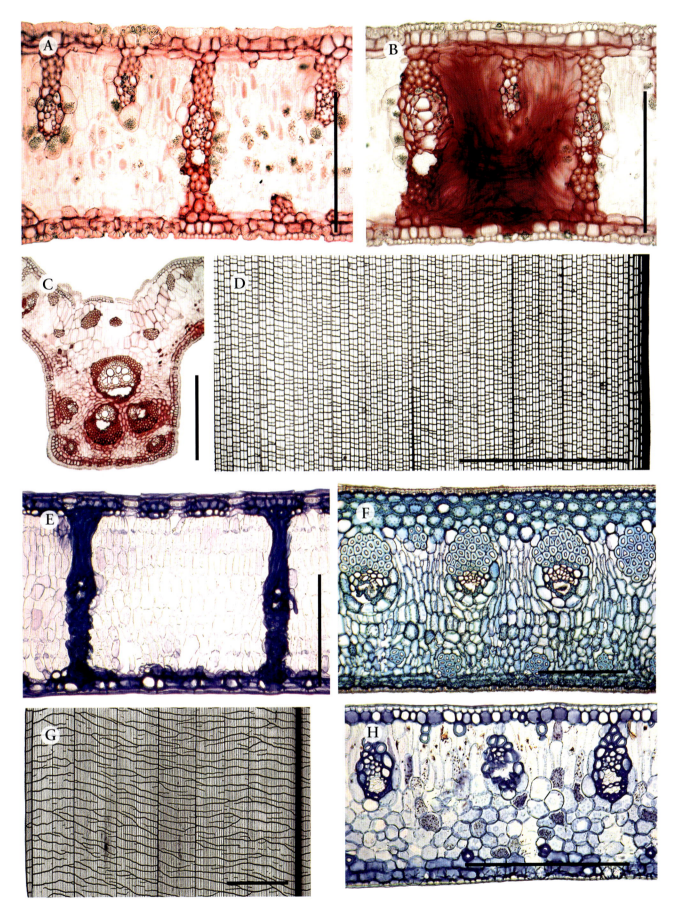

Fig. 64 Coryphoideae (Borasseae: Lataniinae). **A**. *Borassus flabellifer*, lamina, TS (bar = 250 μm). **B**. *Borassus flabellifer*, lamina, sclerenchyma partition in face view, TS (bar = 250 μm). **C**. *Borassus flabellifer*, lamina midrib, TS (bar = 500 μm). **D**. *Borassus flabellifer*, lamina venation, clearing, SV (bar = 10 mm). **E**. *Borassus flabellifer*, lamina LS, partition in TS, (bar = 250 μm). **F**. *Lodoicea maldivica*, lamina, TS (bar = 250 μm). **G**. *Borassodendron machadonis*, lamina venation, clearing, SV (bar = 10 mm). **H**. *Borassodendron machadonis*, lamina, TS (bar = 250 μm).

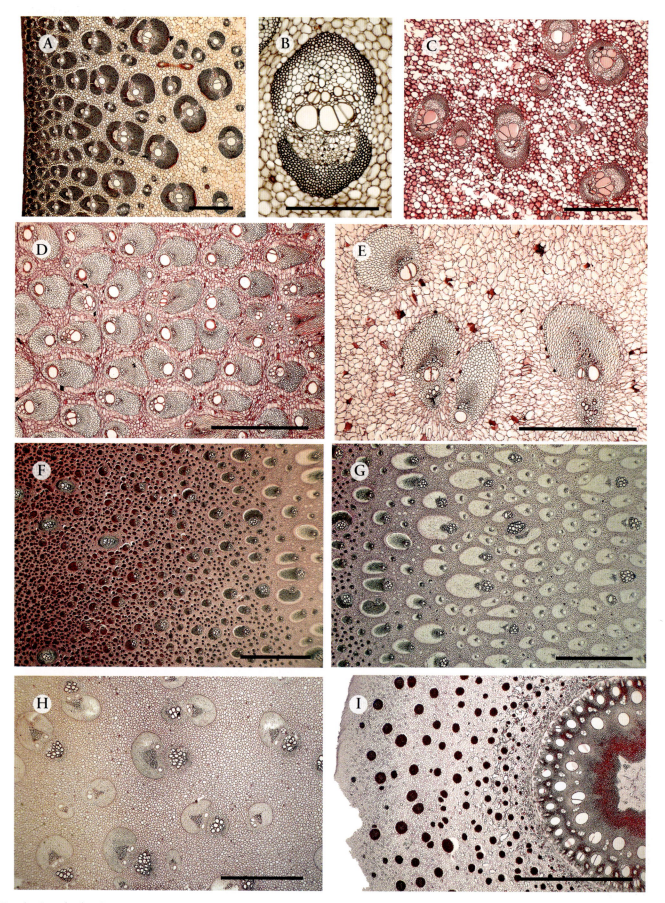

Fig. 65 Coryphoideae (Borasseae; Lataniinae). **A**. *Lodoicea maldivica*, petiole periphery, TS (bar = 1 mm). **B**. *Latania verschaffeltii*, petiole vb, TS (bar = 500 μm). **C**. *Borassus flabellifer*, petiole centre, TS (bar = 1 mm). **D**. *Borassus aethiopum*, stem periphery, TS (bar = 2 mm). **E**. *Borassus aethiopum*, stem centre, TS (bar = 2 mm). **F**. *Lodoicea maldivica*, stem periphery, TS (bar = 2 mm). **G**. *Lodoicea maldivica*, stem periphery to subperiphery, TS (bar = 2 mm). **H**. *Lodoicea maldivica*, stem centre, TS (bar = 2 mm). **I**. *Borassodendron machadonis*, root, TS (bar = 2 mm).

Leaf

Costapalmate. Petiole as above.

LAMINA (Figs. 11B, D; 63A; and 64G, H)

Dorsiventral (Fig. 64H). **Trichomes** present in *B. borneense* only; minute, not sunken, costal, with up to 4 cells. **Epidermis** (Fig. 63A) with costal regions differentiated only in *B. borneense*. Epidermal cells with thin, sinuous anticlinal walls. **Stomata** not sunken, g.c. with outer pair of cuticular ledges much more prominent than the inner pair (*B. machadonis*). **Hypodermis** (Fig. 64H) adaxially 1-layered, mostly of longitudinally extended, rectangular cells with thickened sinuous anticlinal walls, but also including a discontinuous series of non-vascular fibres. Abaxial hypodermis of lignified cells, mostly 2-layered, the outer layer of irregularly rectangular or polygonal cells, the inner layer of transversely extended elliptical cells; s.st.ch. surrounded by 4 U-shaped cells delimiting 2 apertures. **Fibres** adaxially non-lignified with wide lumina, septate; solitary or in bundles of up to 6 cells, with 1–6 such strands present next to adaxial hypodermis and between each pair of vbs. Abaxial fibres mostly solitary (*B. machadonis*) (Fig. 64H) or in bundles of up to 8 cells (*B. borneense*); lignified, the strands attached to surface. **Veins** occasionally free from surface in *B. machadonis*, otherwise bridged to adaxial surface by moderately-developed fibre girders to 2 layers deep. Larger vbs without specialized girder fibres; with 1 large mxy vessel, the phl strand with an irregular network of sclerotic cells. **Adaxial rib** narrowly elliptical, including 4–7 large vbs. **Abaxial rib** much stouter, widely elliptical, including at least 7 large, independent central vbs, and with many smaller peripheral vbs. Peripheral vbs with a sheath of phl fibres as much as 30 layers thick collectively ± confluent to form a sclerotic cylinder interrupted at the level of the expansion tissue. **Transverse veins** (Figs. 11D and 64G) stout, with a thick fibre sheath of up to 10 layers on abaxial side of minor longitudinal veins (Fig. 11B).

Root (Fig. 65I)

Surface layers lost. **Cortex** with numerous wide 'Raphia-type' fbs. **Stele** with peripheral conjunctive tissue including an array of wide mxy vessels and phl strands.

Borassus (Figs. 11J; 24J; 63B, C, F, I, K; 64A–E; and 65C–E)

One of the most widespread of all palm genera, with 6 spp. distributed from Africa to mainland SE Asia, and Indonesia east of Wallace's line, but possibly introduced in Australia. All spp. are large, solitary palms, usually of savannahs or open forest types in seasonally arid climates. *Borassus heineanus*, sister to the remaining spp., is a rainforest palm of N New Guinea, and shares several presumably plesiomorphic character states with *Borassodendron*.

Leaf

Costapalmate. Petiole margins armed with an irregular, sawblade-like series of small spines (Bayton 2007).

LAMINA (Figs. 11I, J; 63B, C, F, I; and 64A–E)

Isolateral (Fig. 64A, B). [Excluding *B. heineanus* (Fig. 63K)]. **Trichomes** (Fig. 63F) present costally on both surfaces, of a unique type, with 4–16 central, sunken cells grouped in 2 longitudinal files encircled by 3–4 layers of curved cells. **Epidermis** (Fig. 63B), with costal regions 3–5 cells wide delimiting intercostal regions with 4–7 irregular files of stomata. Epidermal cells rectangular, anticlinal walls straight. **Stomata** (Fig. 63C, I) sunken, the walls of the l.s.c. adjacent to the stoma producing a fimbriate canopy of epicuticular wax overarching the aperture; t.s.c. with an extra division. **Hypodermis** 2-layered below each surface (Fig. 64A); the outer layer of irregularly quadrangular or polygonal cells usually strongly transversely extended in intercostal regions, and costally with files of longitudinally rectangular cells; s.st.ch delimited by 4 U-shaped cells surrounding 2 apertures. **Fibres** restricted to abaxial half of the lamina and in contact with the surface, in bundles of 4–15 cells mostly opposite the smaller vbs. **Veins** (Fig. 64A, B) of 3 main types, but all continuous to 1 or both surfaces: 1) few but largest vbs attached by short fibre girders to the innermost hypodermal layer below each surface; 2) smaller vbs attached by tall fibre girders to each hypodermis or the girder completed by colourless parenchyma; 3) small vbs attached only to the adaxial hypodermis but each opposed to an independent abaxial fb (see Tomlinson 1961). **Transverse veins** (Fig. 64D, E): the isolateral sp. of *Borassus* distinctive among all palms in producing transverse sclerotic partitions (Fig. 11I, J), in which fibres continuous with those of the fibre girders and non-vascular fbs form a complete partition 2–4 layers thick ad- and abaxial to the transverse veins and always in contact with both surfaces. The fibres of these partitions have an orientation in the ad/abaxial plane of the lamina, i.e. ± at right angles to the 1–3-layered fibre sheath of the transverse vein itself (cf. Fig. 64B, E for 2 views of this partition). **Ribs** (Fig. 64C) with 1 large and several small independent vbs; expansion cells a single series, including fbs. **Stegmata** in longitudinal files associated with all veins, also present in the hypodermis next to the transverse sclerotic partitions but with larger silica bodies.

PETIOLE (Fig. 65C)

Vbs with single phl strand.

Stem (Fig. 65D–E)

Most vbs with single large mxy element; ground tissue with limited late expansion not forming aerenchymatous tissue (Fig. 65E).

Note: the lamina anatomy of *B. heineanus* (Fig. 63K) is similar to that of *Borassodendron* in all fundamental respects but is much less robust in terms of fibre development.

SUBFAMILY CEROXYLOIDEAE

INTRODUCTION

The subfamily Ceroxyloideae shows an unusually large amount of structural diversity for a clade that contains but eight genera distributed among three tribes. Like its sister group Arecoideae, Ceroxyloideae are characterized by their reduplicate pinnate leaves and in having relatively inconspicuous bracts subtending the primary branches of the inflorescence. With flowers in monads, these often spirally arranged on the rachillae, Ceroxyloideae differ from Arecoideae, which have flowers in triads or derivatives of this floral arrangement. Molecular phylogenetic data shape the current circumscription of Ceroxyloideae, which indeed is one of the most surprising advances to come out of the consensus of such studies. The Ceroxyloideae clade is resolved in many studies with moderate to high statistical support. Trénel et al. (2007), sampling all genera and most spp. within the subfamily, provides the phylogenetic framework for our analysis and discussion. The three tribes of the subfamily—Cyclospatheae (Fig. 66B, C), Ceroxyleae (Fig. 66D, F) and Phytelepheae (Fig. 66A)—are each structurally well defined but strongly disparate from one another, particularly in floral morphology. To a great extent, this pattern of variation is reflected in their vegetative anatomy. From both an anatomical and morphological perspective, Ceroxyloideae are without any unique synapomorphies. However, character analysis optimizes the presence of adaxial subepidermal fibres (here included within the hypodermis) as a synapomorphy for the subfamily, with a subsequent loss of these fibres within Ceroxyleae (Fig. 66E).

Tribe Cyclospatheae (Fig. 67), containing only *Pseudophoenix*, is sister to the other two tribes. The lamina anatomy of *Pseudophoenix* is among the most distinctive of all palms, as it possesses a unique suite of autapomorphies. The adaxial hypodermis is exclusively composed of a multiseriate array of septate fibres, usually two to five layers deep. *Pseudophoenix* and *Ceroxylon* share an unusual abaxial lamina surface topography of stomatal furrows alternating with fibrous ridges, as indicated by Read (1968). Character state reconstruction shows the furrowed surface is independently derived in each of the two genera (Fig. 67). Other lamina characters closely associated with the stomatal furrows differ between the two genera, and *Pseudophoenix* is further distinctive in having extended, cylindrical rods of wax lining the surface of the furrows (Fig. 68F). *Pseudophoenix* is also exemplified by its (adaxial) midrib structure, which protrudes

abaxially and contains a massive, complete sclerotic cylinder that includes only a single, large vascular bundle within; it is without any ground parenchyma (Fig. 68B).

The clade of Ceroxyleae and Phytelepheae (Fig. 67) is without clear anatomical synapomorphies, though the lamina anatomy of *Ravenea* and Phytelepheae share several character states (probably plesiomorphies) that are restricted to these taxa within Ceroxyloideae. The hypodermal layers in both are discontinuously fibrous, and, in all but the smaller *Ravenea* species, each pinna has a well-developed submarginal vein to either side of the midrib, the vein distinctive in having histology like that of a depauperate midrib. Taxa with submarginal veins also have bands of expansion tissue associated with non-vascularized folds in the pinnae. Outside of Ceroxyloideae, this combination of states has a limited distribution, present just in some Euterpeae and Areceae.

A synapomorphy for Ceroxyleae (Fig. 67) is the bilobed terminal subsidiary cell (the lobes often papillose) adjacent to the stomata (Fig. 69F, H, J). The tribe is also distinctive within the subfamily in possessing petiole vascular bundles with one phloem strand (Fig. 70B, C). Character-state reconstruction of petiole vascular bundle phloem strand number is, however, equivocal for both the basal node of the subfamily and that of the Ceroxyleae/Phytelepheae clade. The clade containing *Oraniopsis*, *Juania*, and *Ceroxylon* is the most distinctive of the subfamily in leaf anatomy, characterized by at least two synapomorphies (Fig. 67). Alone within the subfamily the longitudinal veins of the pinnae are bridged to the adaxial surface layers by fibre girders (Fig. 69C, D); and veins are always free in the mesophyll in other taxa (Fig. 68A). Also optimized as a synapomorphy in these genera, is the loss of adaxial subepidermal fibres, so that the adaxial hypodermis is uniformly parenchymatous. However, the deeply-embedded *C. echinulatum* (Trénel et al. 2007) does possess hypodermal fibres adaxially, a probable reversal or independent gain. Trichome structure likely also distinguishes this clade. *Oraniopsis*, *Juania*, and *Ceroxylon* have numerous, massive, peltate trichomes with multiseriate, elliptical bases of thick-walled cells; these always positioned below the longitudinal veins of the pinnae on the abaxial lamina surface (Fig. 69B, C). Trichomes are present on the pinnae of some species of *Ravenea* examined (e.g. *R. rivularis*), but are scarce, with small, uniseriate bases, and scattered on both surfaces without a clear positional relationship to the veins. Observations of other *Ravenea* species may clarify trichome evolution within the tribe. *Juania* and *Ceroxylon*

Fig. 66 Ceroxyloideae. **A**. *Phytelephas aequatorialis*, mature tree with inflorescence. **B**. *Pseudophoenix vinifera*, mature tree. **C**. *Pseudophoenix vinifera*, mature tree with inflorescence. **D**. *Ravenea rivularis*, mature tree. **E**. Summary phylogeny of Arecaceae showing position of Ceroxyloideae; anatomical synapomorphy for the subfamily indicated; formal optimization of character states not shown. **F**. *Ceroxylon quindiuense*, mature trees, 1.7-m-tall man on left for scale (photo by L. Raz).

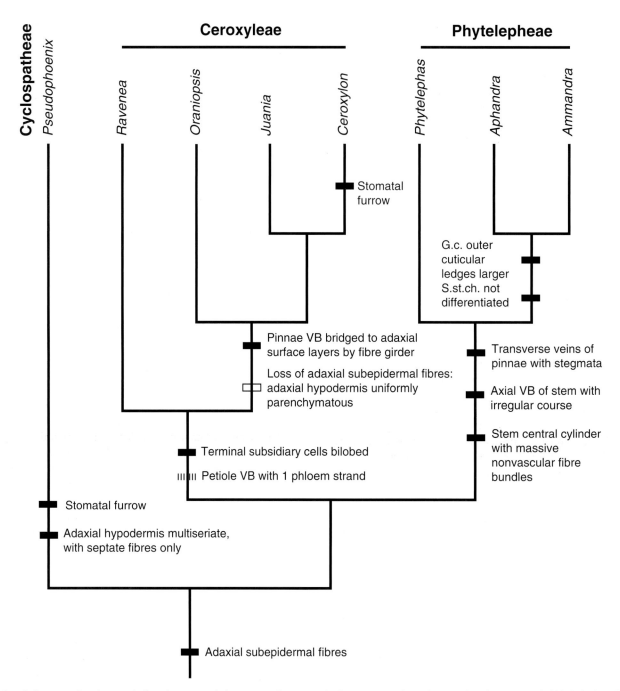

Fig. 67 Phylogeny and evolution of selected anatomical characters within Ceroxyloideae. Tree topology after Trénel et al. (2007). Solid black hashmarks indicate gain of character state; white hashmark indicates reversal; dashed hashmark indicates state with ambiguous optimization. Ancestral states at basal node: adaxial hypodermis uniseriate; abaxial surface of lamina planar, without ridges and furrows; terminal subsidiary cells with entire margin adjacent to stoma (in SV); pinnae vbs free of surface layers; stem central cylinder without fbs; axial vbs of stem with regular course; transverse veins of pinnae without stegmata; abaxial hypodermis with differentiated s.st.ch.; g.c. with outer cuticular ledge pair equivalent in size and shape (in TS). Petiole vbs of *Pseudophoenix* and Phytelepheae with 2 wide mxy vessels and 2 phl strands.

differ from *Oraniopsis* in sharing distinctly thicker, more strongly protruding trichome bases, 6–10 cell layers deep, as opposed to 1–3 layers in *Oraniopsis* (cf. Fig. 69B, C).

Phytelepheae (Fig. 67) are strongly divergent from other palms in their floral and inflorescence structure and have fewer distinctive anatomical features. Within the pinnate crown clade of Ceroxyloideae and Arecoideae, they are one of two clades to have stegmata associated with the transverse veins of the pinnae

(a core group of Attaleinae is the other); this state is optimized as a clear synapomorphy of the tribe. As mentioned, the lamina anatomy of Phytelepheae is quite like that of *Ravenea*. In addition to those shared features, characters that are distinctive in a combination that seems useful in diagnosing this tribe are glabrous laminae, relatively very small stomata and pinnae vascular bundles with a well-developed inner sheath of fibres that form a complete cylinder. Lamina anatomy is rather uniform within Phytelepheae; a hierarchy of character states is summarized in the tribal key below and indicated in Fig. 67. Stems of Phytelepheae show several distinctive character states. Axial vascular bundles have a strongly irregular course and are associated with massive non-vascular fibre bundles (Fig. 70J). *Phytelephas macrocarpa* and *Ammandra decasperma* are the only palms known without vessels in their stems (but present in roots and leaves). *Phytelephas aequatorialis*, in contrast, does have vessels in its stem, suggesting a complex pattern of tracheary element evolution in Phytelepheae. The presence of tracheids as the only water-conducting cell type in stems of Phytelepheae is almost certainly derived.

Key to tribes of Ceroxyloideae based on lamina anatomy

1. Lamina with a continuous fibrous adaxial hypodermis, 2–5(–25) cell layers deep. Midrib with thick cylinder of fibres including only a large, solitary vb and without ground parenchyma. Stomatal furrows bordered by extended wax rods; trichomes absent. [Petiole vbs with 2 phl strands]. . . . . **Cyclospatheae** (*Pseudophoenix*)
1A. Lamina hypodermis either uniformly parenchymatous or including at most a discontinuous series of fbs. Midrib with thick fibre cylinder including several vbs and ground parenchyma. Stomatal furrows absent or, if present, beset with trichomes and without wax rods. [Petiole vbs with 1 or 2 phl strands]. . . . . .2.
2. Stegmata absent from transverse veins of lamina. Terminal subsidiary cells bilobed. [Petiole vbs with 2–many wide mxy vessels and 1 phl strand. Central cylinder of stem without non-vascular fbs]. . . . . **Ceroxyleae**
2A. Stegmata present next to sclerenchyma of transverse veins of lamina. Terminal subsidiary cells without lobes. [Petiole vbs with 2 wide mxy vessels and 2 phl strands. Central cylinder of stem with numerous, often massive, non-vascular fbs] . . . . **Phytelepheae**

TRIBE CYCLOSPATHEAE (Figs. 23M and 68)

Containing only 1 genus.

Pseudophoenix (Figs. 23M, N, and 68)

An isolated genus of 4 spp. in the northern Caribbean from Mexico to Hispaniola and S Florida. In varying degrees all spp. must be regarded as threatened (especially *P. sargentii* in Florida). It is distinguished by the presence of long (pseudo)pedicellate flowers, the proximal ones perfect, the distal staminate, and the 1–3-lobed fruit. Stems are always solitary and ringed by annular leaf scars on a glaucous waxy trunk, which is often somewhat bottle-shaped (Fig. 66B). There is a distinct but short crownshaft, the presumed precursor to the condition in many Arecoid palms. Although the leaf is not shed before the inflorescence expands, the sheath has a distinct cleft opposite the petiole

which facilitates leaf abscission, but this originates mechanically and without a precise zone of tissue weakness.

Leaf

Pinnate.

LAMINA

Reduplicate, dorsiventral, abaxial surface finely corrugated (Fig. 68A). **Trichomes** absent. Continuous wax layer present. **Epidermis** wholly cutinized (Fig. 68C), outer wall thickened, anticlinal walls straight. Adaxial epidermis uniform, cells rectangular, longitudinally extended. Abaxial epidermis markedly undulate in TS, with costal ridges alternating with depressed, stomatal furrows scarcely wider than costal ridges. Furrows bordered by extended wax rods (Figs. 23M, N, and 68F). Costal ridges with long, narrowly rectangular cells. Stomatal furrows with cells shorter and less regular than costal ridges. Cells in same file as stomata short and wide (Fig. 68D). **Stomata** mostly abaxial, in 1(2) irregular file(s) within the furrows. T.s.c. narrowly crescentic or rectangular, transversely extended, overarching g.c. sl. (Fig. 68F). G.c. sl. sunken; outer cuticular ledges poorly differentiated from strongly thickened outer periclinal cell walls of l.s.c.; inner cuticular ledges small and delicate. L.s.c. deeply intruding into s.st.ch. **Hypodermis** largely fibrous (Fig. 68A, C). Adaxial hypodermis 2–5 cells deep, uniformly of thin-walled, unlignified, septate fibres with wide lumina. Abaxial hypodermis with bundles of similar but sl. lignified fibres forming the prominent abaxial ridges; furrows including a uniseriate layer of short thick-walled cells irregularly arranged around s.st.ch. **Mesophyll** without fbs, with a well-developed adaxial palisade of anticlinally extended cells, 2–3 layers deep (Fig. 68A). Abaxial mesophyll of looser, ± isodiametric cells. **Veins** in mesophyll below palisade; almost always independent of surface layers, at most the largest veins in contact with adaxial fibrous hypodermis, but never bridged by fibres. O.S. inconspicuous, always incomplete below veins and incomplete above larger veins. I.S. usually completely fibrous below all veins, but also fibrous above larger veins. Phl of larger veins sometimes divided into 2 separate strands. **Raphide sacs** conspicuous in mid-mesophyll (Fig. 68A). **Stegmata** with spherical silica bodies, restricted to the longitudinal veins and hypodermal fibres of both leaf surfaces. **Transverse veins** few, narrow, sheathed by thin-walled, angular sclerotic cells; mostly adaxial to longitudinal veins. **Midrib** (Fig. 68B) prominent abaxially with a massive sclerotic cylinder enclosing a massive, solitary vb; ground parenchyma absent from within the cylinder. **Expansion cells** in abaxial bands on each side of midrib (Fig. 68B).

PETIOLE

Trichomes absent. **Stomata** sunken. **Epidermis** wholly cutinized, outer wall thickened. **Hypodermis** incompletely fibrous but including fbs and lignified colourless cells; interrupted below stomata by loose, thin-walled cells. Peripheral vbs somewhat congested and with well-developed fibrous sheaths, but rarely confluent. Central V-arrangement well developed, enclosing a region of small vbs. Adaxial vbs inverted. Ground parenchyma uniform, with infrequent small vbs. Central **vbs** usually with 2 wide mxy vessels and 2 distinct phl strands (Fig. 68E). Bundle sheath fibres only well developed next to phl; regularly septate.

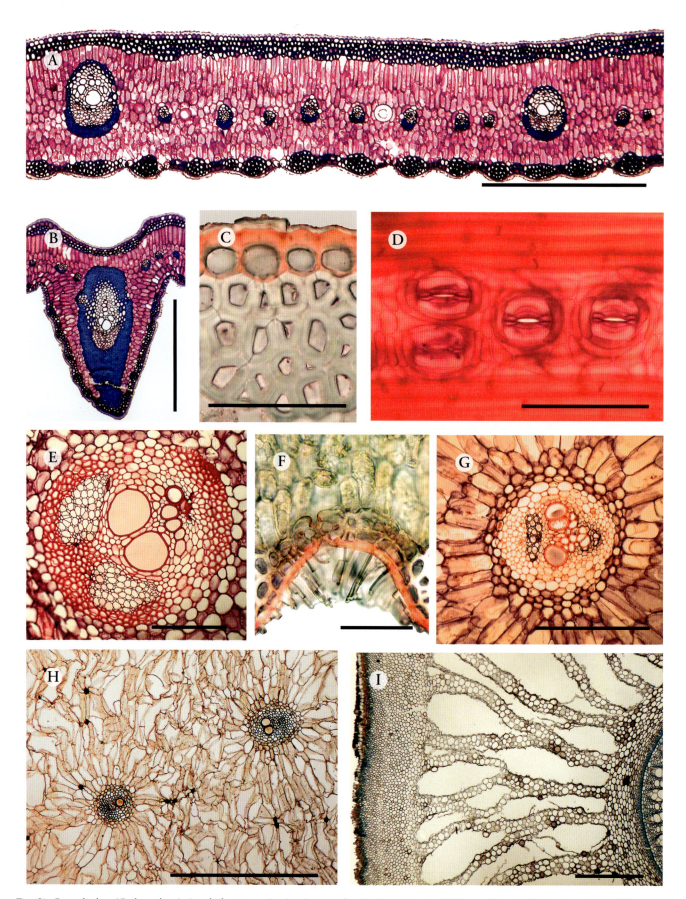

Fig. 68 Ceroxyloideae (Cyclospatheae). *Pseudophoenix sargentii* (**A–F**); *P. vinifera* (**G, H**). **A**. Lamina, TS (bar = 500 μm). **B**. Lamina midrib, TS (bar = 500 μm). **C**. Adaxial epidermis and hypodermal fibres, TS (bar = 50 μm). **D**. Lamina abaxial epidermis with stomata, SV (bar = 100 μm). **E**. Petiole vb, TS (bar = 200 μm). **F**. Lamina abaxial epidermis, stomatal furrow with wax rods, TS (bar = 50 μm). **G**. Young stem central vb, TS (bar = 500 μm). **H**. Old stem central vbs, TS (bar = 2 mm). **I**. Root, epidermis to outer stele, TS (bar = 500 μm).

Raphide sacs common, with densely staining (mucilaginous) contents, typically forming linear series within the ground parenchyma. **Mxy vessels** with long elements and with long oblique scalariform perforation plates on oblique end walls. Sieve tubes with short, oblique compound sieve plates.

LEAF SHEATH

Abscission zone formed coincident with the region of leaf attachment.

Note: in material of *P. vinifera* taken nearer the leaf base vbs are more congested, with a well developed fibrous sheath on the xyl as well as the phl side, peripheral vbs are often confluent whereas small vbs are virtually absent. This illustrates the problem of establishing a 'standard level' for comparison of different taxa.

Stem

PSEUDOPHOENIX SARGENTII

Only relatively immature material examined. **Cortex** wide, fibrous but outer layers forming a sclerotic periderm. **Central cylinder** with somewhat diffuse peripheral vbs, the central vbs then uncongested. Central **vbs** usually with 2 wide mxy vessels and a single phl strand, but with much variation. Wide mxy elements of **xyl** short, with oblique scalariform perforation plates and few thickening bars; sometimes reticulate. **Phl** with short sieve tube elements, the sieve plates v. oblique and compound. **Ground parenchyma** in young stages showing evidence of late expansion. Raphide sacs numerous, with densely-staining contents.

PSEUDOPHOENIX VINIFERA (Fig. 68G, H)

Older material with more mature structure. Vb construction more conventional and with well-developed bundle sheath fibres. A **distinctive feature** is the presence of abundant narrow pxy elements and abundant development of late but narrow mxy elements which make contact with the wide mxy elements (Fig. 68G). Ground parenchyma spongy, expanded cells producing a somewhat lacunose appearance (Fig. 68H).

Root (Fig. 68I)

Epidermis of large, thin-walled cells; not persistent. **Outer cortex** of many layers of sclerotic cells. **Inner cortex** without fibres, differentiated into 3 zones; the middle zone extensively aerenchymatous (Fig. 68I). **Endodermis** with massive U-shaped pitted (inner and radial) wall thickenings, the lumen almost occluded. **Pericycle** 1–2-layered, conspicuously pitted. **Stele** with congested polyarch vascular tissues, the medulla thin-walled. **Mxy vessels** wide, with simple transverse perforation plates, xyl and phl poles congested; medullary vessels infrequent. **Phl** poles with wide angular sieve tubes, the sieve plates long, oblique with numerous sieve areas little differentiated from the lateral sieve areas. **Tannin** cells frequent. **Raphide** canals absent. **Stegmata** absent.

Notes: the Hispaniolan endemic *P. ekmanii* differs sharply from the 3 other spp. in the genus in the extreme development of hypodermal fibres: 15–25 layers thick adaxially, occupying about half the depth of the lamina. Also extreme is its massive sheath of phl fibres of large vbs, to 30 layers thick. All fibres are lignified in this sp.

TRIBE CEROXYLEAE (Fig. 69)

The tribe contains 4 genera distributed exclusively in the southern hemisphere. While the current distribution of Ceroxyleae had long been thought to exemplify Gondwanic vicariance, Trénel et al. (2007) support a hypothesis of mid-Tertiary long-distance dispersal. Ceroxyleae are solitary palms, often very tall, without a crownshaft and dioecious. Inflorescences often bear several large, densely tomentose peduncular bracts. Flowers are pedicellate, unisexual, but otherwise scarcely dimorphic; immature flowers are open, never forming a closed bud. Familiar palms of this tribe include the Andean wax palms (*Ceroxylon* spp.) and the Majesty palm, *Ravenea rivularis*, which is widely cultivated in tropical climates (Fig. 66D). *Ceroxylon* was considered by Alexander von Humboldt to be the tallest of all trees, a height exaggerated when the surrounding forest is cleared (Fig. 66F); the tallest specimen is over 50 m tall.

Key to genera of Ceroxyleae, based on lamina anatomy

1. All longitudinal veins of lamina free in the mesophyll, not bridged to adaxial surface layers by fibres. [Petiole vbs with 1–2-many wide mxy vessels] . . . . *Ravenea*.
1A. All but smallest longitudinal veins of lamina bridged to adaxial surface layers by fibres [Petiole vbs with 2 to more often many wide mxy vessels] 2
2. Adaxial epidermal cells rectangular in SV, longitudinally elongated. Adaxial hypodermal cells appearing inflated in TS; each at least 2× as deep and wide as the adjacent epidermal cells . . . . *Oraniopsis*
2A. Adaxial epidermal cells predominantly hexagonal in SV, transversely elongated. Adaxial hypodermal cells not appearing inflated in TS; similar in size and shape to the adjacent epidermal cells, uncommonly smaller 3
3. Stomatal furrows absent, abaxial surface planar, not ridged. Trichome stalk cells and fibres of bridge bundles unlignified . . . . *Juania*
3A. Stomatal furrows present between abaxial ridges. Trichome stalk cells and fibres of bridge bundles strongly lignified . . . . *Ceroxylon*

Ravenea (Figs. 69A, E, F, M and 70A, B)

A genus of at least 18 spp. restricted to Madagascar and the adjacent Comoro Islands. *Ravenea* is the most ecologically diverse genus of the tribe, growing in dry, lowland spiny forest to wet, montane forest. Spp. vary in stature from treelets to quite large trees.

Leaf

Pinnate, with regularly-arranged reduplicate leaflets, each with a prominent adaxial rib; sometimes also with abaxially protruding submarginal veins. The leaf base becoming fibrous ventrally, splitting, but not forming a distinct crown shaft.

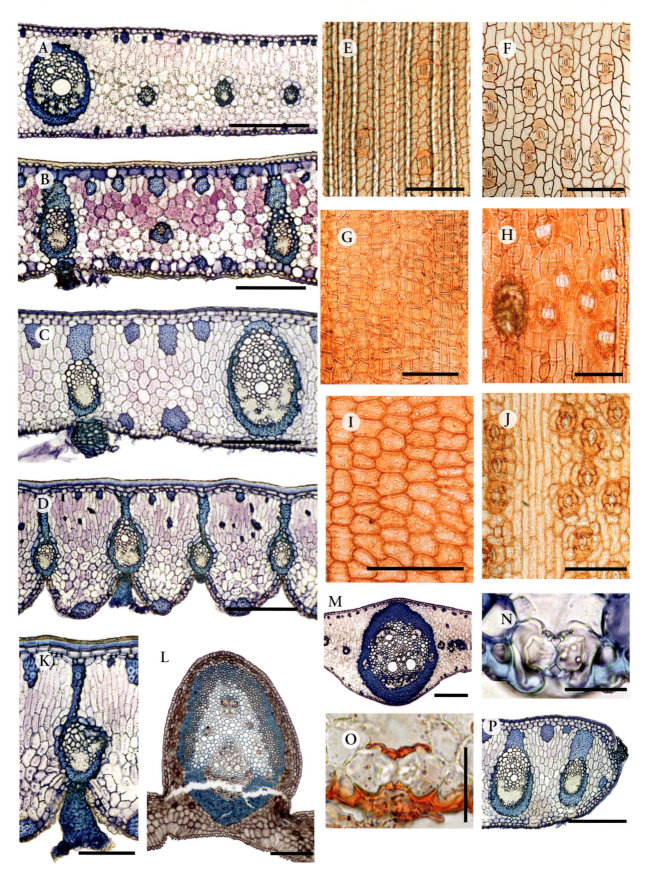

Fig. 69 Ceroxyloideae (Ceroxyleae). **A**. *Ravenea rivularis,* lamina, TS (bar = 200 μm). **B**. *Oraniopsis appendiculata*, lamina, TS (bar = 200 μm). **C**. *Juania australis*, lamina, TS (bar = 200 μm). **D**. *Ceroxylon quindiuense*, lamina, TS (bar = 200 μm). **E**. *Ravenea glauca*, adaxial epidermis, SV (bar = 100 μm). **F**. *Ravenea glauca*, abaxial epidermis, SV (bar = 100 μm). **G**. *Oraniopsis appendiculata*, adaxial epidermis, SV (bar = 100 μm). **H**. *Oraniopsis appendiculata*, abaxial epidermis, SV (bar = 100 μm). **I**. *Juania australis*, adaxial epidermis, particles are artefacts of clearing, SV (bar = 100 μm). **J**. *Juania australis*, abaxial epidermis, SV (bar = 100 μm). **K**. *Ceroxylon quindiuense*, trichome in abaxial lamina furrow, TS (bar = 100 μm). **L**. *Juania australis*, midrib, TS (bar = 200 μm). **M**. *Ravenea rivularis*, submarginal vein, TS (bar = 200 μm). **N**. *Oraniopsis appendiculata*, stomatal complex, TS (bar = 25 μm). **O**. *Juania australis*, stomatal complex, TS (bar = 25 μm). **P**. *Juania australis*, pinna margin, TS (bar = 200 μm).

LAMINA

Dorsiventral (Fig. 69A). **Epidermis** with outer walls sl. thickened and cutinized; cells sometimes with undulate or sinuous anticlinal walls (e.g. *R. hildebrandtii*). **Trichomes** absent from most spp. examined, but rare and diffuse on both surfaces of *R. rivularis*; small, with a mostly uniseriate base that distally bears a limited expanse of thin-walled cells. Adaxial **epidermis** (Fig. 69E) with few stomata. Cells variable, but with oblique end walls, and usually wider than long, ± hexagonal. Abaxial epidermis (Fig. 69F) with costal bands of narrow, longitudinally-extended cells and wider intercostal bands of shorter, wider and less regular cells, including 1–3 series of stomata. **Stomata** not sunken, l.s.c. elongated in some spp., t.s.c. always bilobed. **Hypodermis** 1-layered below each surface, but not well differentiated in *R. rivularis*, cells ± cubical or sl. longitudinally extended; 2–4 cells outline the s.st.ch. **Mesophyll** without distinct palisade layers. **Fibres** unlignified, in hypodermal region, solitary or in 2–4-celled fbs, adjacent to each hypodermis, fewest abaxially; sometimes within mesophyll and unconnected to hypodermis; distribution limited in *R. rivularis*. **Veins** narrow, independent of surface layers, the largest separated from hypodermis by at least 2 mesophyll layers. O.S. indistinct, usually interrupted above and below, I.S. a single fibre layer. Submarginal veins (Fig. 69M) including several vbs within a sclerotic sheath. **Midrib** with complete sclerotic cylinder including 5–many vbs and ground parenchyma; small vbs often embedded within the cylinder. **Transverse veins** few, inconspicuous; ± at same adaxial level as the longitudinal veins. **Stegmata** with druse-like spherical silica bodies, v. large in widest veins.

PETIOLE (Fig. 70A, B)

With broad V-arrangement of central vbs (Fig. 70A). Outer layers with a sclerotic zone of complex vbs, i.e. 1 or more large plus smaller vbs enclosed within a common sclerotic sheath, but least developed in *R. hildebrandtii*. **Central vbs** (Fig. 70B) with 2 or more wide mxy vessels, usually forming a xyl complex; phl a single broad strand. Mxy vessels with long oblique scalariform perforation plates and numerous thickening bars. Sieve tubes with compound sieve plates on transverse to sl. oblique end wall. Ground tissue uniform, without fbs.

Stem

Vbs at periphery with a single wide mxy vessel, but 2 or more in central vbs. Phl a single strand. Mxy vessels with oblique, scalariform perforation plates with numerous narrow thickening bars. Ground parenchyma becoming lacunose in stem centre and radiating around vbs.

Root

Epidermis large-celled, somewhat papillate, the outer wall thickened in some spp. Hypodermis a 1–2 celled layer of large sl. thick-walled cells. **Cortex** including 3 distinct regions: 1) outermost narrow **sclerotic layer**, inner and outer cells maturing before a middle layer of apparently long-meristematic cells; 2) wide middle layer including numerous **raphide canals**, which may become thick-walled with age, together with thick-walled angular cells which form a reticulum; 3) inner cortex of

undifferentiated cells, including radially extended air-canals. **Stele** sometimes irregular in outline, but never conspicuously fluted. **Endodermis** with U-shaped wall thickenings. **Pericycle** 1(–2)-layered becoming sl. thick-walled. Central medulla sometimes including isolated mxy vessels. Wide vessels with scalariform perforation plates on oblique end walls, thickening bars numerous. Phl sieve tubes with long oblique compound sieve plates; the sieve areas narrow.

Note: the critically-endangered *R. xerophila*, unique within the genus in inhabiting the dry spiny forests of southern Madagascar, differs from the above description in the following ways: mesophyll with palisade well-developed both ad- and abaxially, stomata sunken, and hypodermis uniformly parenchymatous.

Oraniopsis (Figs. 69B, G, H, N and 70C)

A single sp., *O. appendiculata*, restricted to coastal ranges of northeastern Queensland, where it grows in premontane to low montane rainforest. *Oraniopsis* attains a moderate size, with a large crown of semi-erect, pinnate leaves. It is reported to be very slow to mature, remaining without a visible aerial axis for decades.

Leaf

Pinnate, reduplicate leaflets regularly arranged with a prominent adaxial midrib.

LAMINA

Dorsiventral (Fig. 69B). **Epidermis**: outer wall thickened to half total cell depth, cutinized, abaxial cells sl. smaller and papillose. Adaxial epidermis (Fig. 69G) without trichomes or stomata. Cells uniform, longitudinally elongated and ± rectangular, short, the end wall often somewhat oblique; walls sometimes sl. undulate. Abaxial epidermis (Fig. 69H) with numerous **trichomes** situated below wider costal region, each trichome with a basal group of thick-walled cells extending into a distal shield of thin-walled cells with a somewhat stellate pattern, collectively forming a continuous, scurfy indumentum. Costal cells narrow, rectangular; intercostal cells irregular but generally longitudinally extended. **Stomata** (Fig. 69N) sl. sunken, outer wall of short t.s.c. sl. thickened, and somewhat overarching g.c. **Hypodermis** 1-layered below each surface. Adaxial cells conspicuous, square to sl. transversely extended and then somewhat hexagonal; without fibres (Fig. 69B). Abaxial cells narrow, rectangular to square but not transversely extended; s.st.ch. surrounded by 4 L-shaped cells (these sometimes further subdivided) with simple rectangular pore. **Mesophyll** with poorly defined palisade, cells compact, rather small. **Fibres** adaxially in bundles of up to 12 cells, the cell lumen wide; these fbs never within adaxial hypodermis, but commonly in contact with it. Abaxial fbs less frequent, in smaller strands and commonly replacing hypodermal cells so as to abut the epidermis. **Veins** in abaxial mesophyll; larger veins continuous to each hypodermis via a bundle sheath extension of fibres like those of the mesophyll, smaller veins either independent of surface layers or in contact with the abaxial hypodermis. O.S. indistinct, usually incomplete both above and below; I.S. ± continuous, mainly fibrous. **Transverse veins** few; sheathing

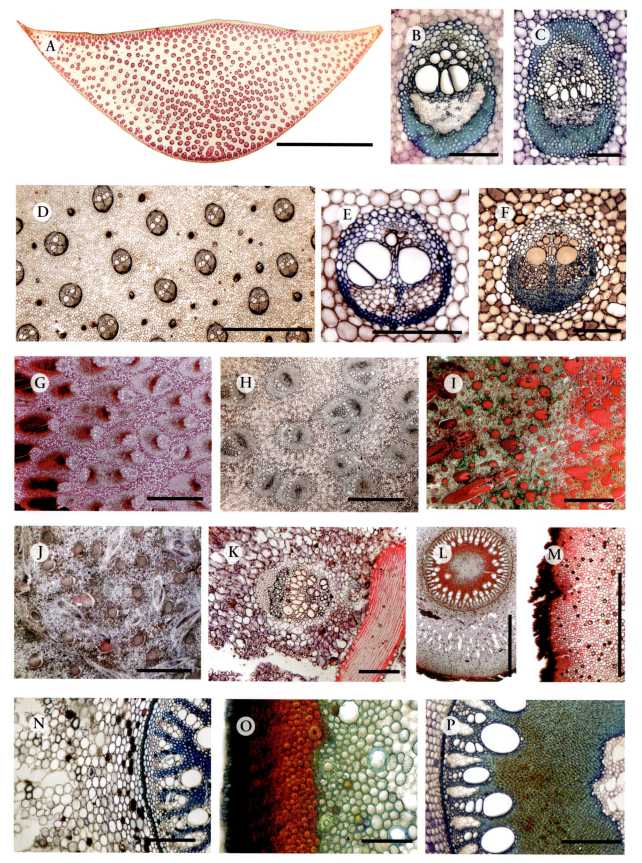

Fig. 70 Ceroxyloideae (Ceroxyleae and Phytelepheae). **A**. *Ravenea rivularis*, petiole, TS (bar = 10 mm). **B**. *Ravenea rivularis*, petiole vb, TS (bar = 200 μm). **C**. *Oraniopsis appendiculata*, petiole vb, TS (bar = 200 μm). **D**. *Phytelephas aequatorialis*, petiole centre, TS (bar = 2 mm). **E**. *Phytelephas* sp., petiole vb, TS (bar = 200 μm). **F**. *Ammandra decasperma*, petiole vb, TS (bar = 200 μm). **G**. *Juania australis*, stem periphery, TS (bar = 2 mm). **H**. *Juania australis*, stem centre, TS (bar = 2 mm). **I**. *Phytelephas aequatorialis*, stem periphery, TS (bar = 2 mm). **J**. *Phytelephas* sp., stem centre, TS (bar = 2 mm). **K**. *Phytelephas* sp., stem centre, detail of vb and fb, TS (bar = 200 μm). **L**. *Juania australis*, root, TS (bar = 2 mm). **M**. *Juania australis*, root outer cortex, TS (bar = 1 mm). **N**. *Juania australis*, root inner cortex and stele, TS (bar = 200 μm). **O**. *Phytelephas macrocarpa*, root outer cortex, TS (bar = 100 μm). **P**. *Phytelephas macrocarpa*, root stele, TS (bar = 200 μm).

cells few, sclerotic; ± at same level as the adaxial longitudinal veins. **Stegmata** with spherical silica bodies; present next to longitudinal vbs and fbs, absent from transverse veins. **Midrib** with well-developed sclerotic sheath including 1–few large vbs, the phl extensively sclerotic, plus smaller, often inverted vbs.

PETIOLE

Distinct V-configuration of central vbs present. Central vbs (Fig. 70C) with a well-developed phl sheath; pxy abundantly developed, mxy including a series of many wide vessels continuous with innermost pxy elements. Phl a single strand.

Note: in addition to the key characters, *Oraniopsis* is distinct from *Juania* and *Ceroxylon* in its relatively short trichome bases, 1–3 layers deep, and wide-lumened fibres.

Juania (Figs. 8H; 9D; 23C; 69C, I, J, L, O, P; and 70G, H, L, M, N)

A single sp., *J. australis* restricted to the Juan Fernandez Islands, where it is the only native palm. A massive single-stemmed, dioecious palm, with basal stem diameter of over 1 m.

Leaf

Pinnate. Leaflets reduplicate with a prominent adaxial midrib.

LAMINA

Dorsiventral (Fig. 69C). **Trichomes** restricted to abaxial surface, regularly positioned below longitudinal veins, each with a massive elliptical base of thick-walled, pitted, but unlignified cells extending into a shield-like weft of thin-walled cells that are lost from older leaflets (Fig. 8H). **Adaxial epidermis** (Fig. 69I) uniform, cells irregularly hexagonal and obliquely-transversely extended, the angle of obliquity changing over short distances; outer wall somewhat thickened and wholly cutinized. **Abaxial epidermis** (Fig. 69J) with smaller cells, somewhat papillate around stomata. Costal bands of narrow ± rectangular elongated cells alternating with wider intercostal bands of short irregular cells; stomata diffuse in 1–3 series per band. **Stomata** (Fig. 69O) with g.c. sl. sunken, overarched somewhat by lobes of the t.s.c. to form an outer shallow sometimes wax-filled cavity (Fig. 9D). **Hypodermis** (Fig. 69C) shallow, 1(2)-layered adaxially; only the inner layer interrupted by fbs; outer layer of sl. thick-walled, unlignified cells cubical to transversely extended. Abaxial hypodermis 1-layered, thin-walled and inconspicuous, cell size similar to epidermis. **Mesophyll** with an indistinct palisade transitional abaxially to compact cells. **Fibres** thick-walled, unlignified, non-septate; in large bundles typically in contact with either the ad- or abaxial surface layers, rarely free. Adaxial fbs irregular, large (20+ fibres), never included with the hypodermis; abaxial fbs fewer, next to abaxial hypodermis or epidermis, the larger alternating with the veins. **Veins** in contact with abaxial hypodermis, but only larger veins with a fibrous extension of the I.S. to the adaxial hypodermis, O.S. indistinct; I.S. complete around all veins but only fibrous below. **Raphide sacs** conspicuous in mid-mesophyll; wide, often in longitudinal series. **Transverse veins** short, infrequent, sheathed by sclerotic parenchyma; ± at same level as the adaxial longitudinal veins. **Stegmata** with spherical silica bodies (Fig. 23C); absent from transverse veins. **Midrib**

(Fig. 69L) with 1 large and 3 small vbs within a sclerotic sheath. **Margin** (Fig. 69P) not tapering, obtuse in outline in TS; without a prominent marginal rib.

PETIOLE

Narrow V-configuration of central vbs distinct, enclosing a region of mostly inverted small vbs. Abaxial vbs with a pronounced fibrous sheath, outermost cell layer including small fbs but never in contact with hypodermis. Central ground tissue including numerous fbs and occasional small vbs. **Central vbs** with a well-developed phl fibre sheath; pxy abundantly developed, mxy including a series of wide vessels continuous with innermost pxy elements. **Xyl** including vessels with long oblique perforation plates and numerous scalariform thickening bars. **Phl** a single strand with incipient median sinus; the sieve plates transverse to sl. oblique, the few sieve areas not much wider than the lateral sieve areas.

Stem (Fig. 70G, H)

Epidermis long persistent, surface smooth; epidermal cells anticlinally extended, wholly cutinized, but retaining cell contents. **Stomata** numerous, deeply sunken, but often aborted. **Cortex** narrow, indistinctly differentiated from central cylinder, but including fbs and narrow vbs. **Central cylinder** with outer congested bundles developing with age a massive fibrous phl sheath (Fig. 70G). **Central vbs** (Fig. 70H) each with a well-developed phl fibre sheath; abundant pxy continuous with a series (4–7) of wider mxy vessels via late pxy tracheids. Phl a single wide strand without included sclerenchyma. Outgoing leaf traces associated with numerous satellite bundles (i.e. inflorescence traces). **Ground parenchyma** becoming somewhat lacunose in older parts. **Raphide sacs** abundant. **Mxy vessels** with oblique scalariform perforation plates and numerous thickening bars. **Sieve tubes** with compound sieve plates on transverse to sl. oblique end walls, the sieve areas somewhat stellately arranged.

Root (Fig. 70L–N)

Epidermis of large, somewhat papillose cells, the outer wall thickened, but eroding from older roots. **Outer cortex** (Fig. 70M) of many layers of sclerenchyma, this mostly as fibres; not sharply demarcated from the inner cortex. **Inner cortex** 3-zoned (Fig. 70L), with solitary or paired, wide-lumened fibres mostly throughout. Outer zone with thick-walled ground tissue including a high concentration of isolated fibres; middle zone with aerenchyma v. weakly developed; most internal layers of inner zone without fibres and with many cells with tanniferous contents. **Endodermis** with U-shaped, pitted wall thickenings, the cell lumen almost occluded (Fig. 70N). Pericycle 3–4 (–5)-layered. **Stele** medullated with polyarch arrangement of alternate xyl and phl strands embedded in thick-walled conjunctive tissue (Fig. 70L), the cells septate with thick lamellate walls. Occasional isolated inner mxy vessels. **Mxy vessels** wide, with simple, transverse perforation plates. **Sieve tubes** with compound sieve plates, the sieve areas wide and with conspicuous pores.

Note: unlignified fibre girders in the lamina are autapomorphic.

Ceroxylon (Figs. 66F and 69D, K)

A genus of about 11 dioecious spp. occurring at high elevations in Andean South America (Venezuela to Bolivia). All are tall, single-stemmed palms and commonly are the largest plants at high elevations; Humboldt saw them as the tallest of all trees (Fig.66F). They are much threatened, either by deforestation or disease but have been important sources of palm wax (hence 'wax palms').

Leaf

Pinnate, reduplicate leaflets with prominent adaxial midribs. The abaxial surface is deeply furrowed or corrugated and obscured by a scurfy indumentum.

LAMINA (Fig. 69D, K)

Dorsiventral, with a prominent and regular series of longitudinal ridges and furrows (Fig. 69D). Furrows are strictly positioned opposite the longitudinal veins of the lamina (Fig. 69D). **Epidermis** outer wall v. thick and strongly cutinized, with an additional wax layer (Fig. 69K). Adaxial epidermis without hairs or stomata; cells transversely extended and somewhat hexagonal in surface view. Abaxial epidermis deeply furrowed, the costal (rib) cells shallow, narrow and rectangular, thinly cutinized; intercostal cells (within furrows) wider, shorter, thinner-walled and irregular. **Trichomes** abaxial only; large, peltate; each with massive, multicellular, lignified base positioned below the veins within a furrow (Fig. 69K). Trichome bases supporting an often extensive distal expanse of thin-walled, unlignified cells; these sometimes wholly occluding the furrow (e.g. *C. echinulatum*). **Hypodermis**: 1 layer of parenchyma with sl. transverse extension adaxially, and both parenchyma and fibres abaxially. **Stomata** restricted to both sides of abaxial intercostal furrows, g.c. thin-walled with 2 cutinized ledges. T.s.c. short; l.s.c. thin-walled and deeper than adjacent epidermal cells. **Mesophyll** usually with a well differentiated, 2–3-layered palisade. **Fibres** included in girders always strongly lignified, those elsewhere not or poorly lignified. Adaxial fibres usually not included within the hypodermis; in few-celled fbs or forming a ± continuous several-layered expanse. Few-celled fbs (or solitary fibres) also free in the adaxial mesophyll. Large, non-vascular fbs prominent abaxially at the base of the ridges in all spp. (Fig. 69D). **Veins** corresponding to furrows and mostly attached to both surfaces, with a girder-like extension of the I.S. to the adaxial hypodermis, the O.S. consequently discontinuous but well-developed. **Transverse veins** narrow, infrequent and short (most between 2 veins) in adaxial mesophyll, often running above small longitudinal veins. **Stegmata** associated only with vascular fibres; with spherical silica bodies. **Midrib** with a sclerotic cylinder enclosing a large vb and 1 or more smaller ones; **expansion cells** in 2 bands abaxially.

PETIOLE

Abaxial epidermis furrowed, as in the lamina, with outermost vbs in contact with the base of the furrows; ridges between furrows including irregular hypodermal fbs. **Trichomes** like those of the lamina, common in the base of the furrows opposite a vb. Adaxial epidermis without furrows; trichomes absent; hypodermis and sub-hypodermis including fbs. **Central vbs** with a conspicuous narrow V-shaped complex enclosing a region of few small vbs. Vbs otherwise crowded, mostly with 2 wide mxy vessels and a single broad phl strand. Bundle sheath next to xyl well developed but thin-walled, the phl sheath thicker-walled. **Ground tissue** uniform with few raphide sacs and diffuse tannin cells. Additional small vbs and fbs few and restricted to the V-shaped region. **Mxy vessels** with long, oblique scalariform perforation plates with many thickening bars. **Sieve tubes** with sl. oblique compound sieve plates.

Stem

Cortex abundantly fibrous. **Central cylinder** with congested peripheral vbs, progressively developing massive fibre sheaths. Vbs with 2–many wide mxy vessels, usually forming a complex associated with late-developing narrow tracheary elements. Outgoing leaf traces usually associated with numerous satellite vbs as presumed inflorescence traces. **Mxy vessels** with oblique scalariform perforation plates with few, well-separated thickening bars, sometimes associated with elements with reticulate perforation plates of considerable complexity. Material of a mature stem of *C. ferrugineum* (D. Fairchild s.n.) shows lacunose ground tissue appearing lace-like in TS, the ground parenchyma cells radiating away from the vbs and frequent, diffusely distributed fbs of varying diameter. Other taxa are not recorded with this feature.

Notes: vbs show particularly abundant pxy associated with outgoing leaf traces, consisting of narrow tracheids, with additional narrow late mxy elements in contact with both pxy and the central vascular complex. This somewhat elusive character seems diagnostic for the whole subfamily (q.v.). Well-developed inflorescence traces presumably reflect mature stems at reproductive age.

Stomatal furrows and lignified trichome bases are unique to the genus within Ceroxyleae. Lamina material identified as either *C. andicola* or *C. alpinum* in previous descriptions (Tomlinson 1961) is in all respects like that of *Allagoptera caudescens*, and is excluded from the present description on this basis.

TRIBE PHYTELEPHEAE (Figs. 70D–E, I–K, O, P; 71)

The phytelephantoid or vegetable ivory palms, based on the genus *Phytelephas* Ruiz & Pavon, have long been recognized as a distinctive group of palms (currently 3 genera) whose placement in a natural classification has been problematical (Barfod 1988, 1991). In Uhl and Dransfield (1987), they were accorded subfamilial status as Phytelephantoideae, indicative of an isolated position within palms. Molecular phylogenetic data now place them within Ceroxyloideae, an unexpected result, which adds considerably to the structural diversity of this subfamily.

The tribe occupies a restricted area in tropical American from Panama to Peru. Palms are usually single-stemmed and often with short trunks (even acaulescent), dioecious, and with large, pinnately-compound, reduplicate leaves. Male and female inflorescences are markedly dimorphic, each within a single complete peduncular bract. Male flowers have very numerous (to over 1000) stamens and a reduced perianth. Female flowers have

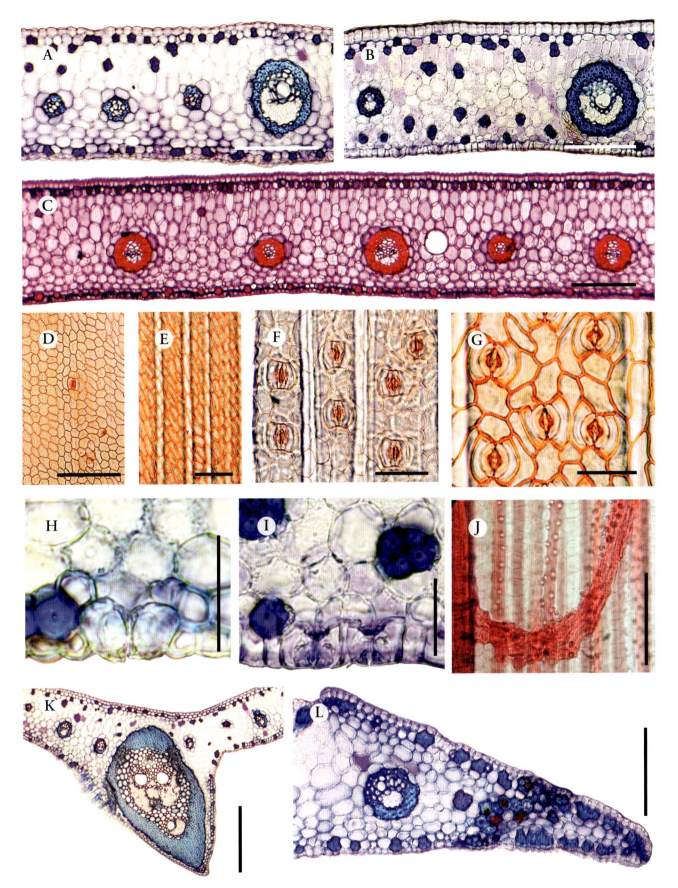

Fig. 71 Ceroxyloideae (Phytelepheae). **A**. *Phytelephas macrocarpa*, lamina, TS (bar = 100 μm). **B**. *Aphandra natalia*, lamina, TS (bar = 100 μm). **C**. *Ammandra decasperma*, lamina, TS (bar = 200 μm). **D**. *Phytelephas macrocarpa*, adaxial epidermis, SV (bar = 100 μm). **E**. *Ammandra decasperma*, adaxial hypodermis, SV (bar = 50 μm). **F**. *Phytelephas macrocarpa*, abaxial epidermis, SV (bar = 50 μm). **G**. *Ammandra decasperma*, abaxial epidermis, SV (bar = 25 μm). **H**. *Phytelephas macrocarpa*, stomatal complex, TS (bar = 25 μm). **I**. *Aphandra natalia*, stomatal complex, TS (bar = 25 μm). **J**. *Aphandra natalia*, transverse vein with stegmata, SV of cleared pinna (bar = 100 μm). **K**. *Phytelephas macrocarpa*, submarginal (major) vein, TS (bar = 200 μm). **L**. *Phytelephas macrocarpa*, pinna margin, TS (bar = 200 μm).

linear, fleshy perianth members and numerous, filamentous staminodes; gynoecia are multicarpellate (5–10), with long styles and stigmas. Fruit a syncarp with numerous pointed warts.

Key to genera of Phytelepheae based on lamina anatomy

1. G.c. in TS with outer cuticular ledge equivalent in size to the inner. Abaxial hypodermis with s.st.ch. differentiated, in surface view each with 4 L-shaped cells flanking an elliptical or lobed aperture . . . . *Phytelephas*
1A. G.c. in TS with outer cuticular ledge pair at least 2× the size of the inner pair. Abaxial hypodermis without regularly differentiated st.ch. 2
2. Fibres free in mesophyll either absent or rare as solitary fibres. Phl of larger longitudinal vbs irregularly partitioned into many strands by sclerotic cells . . . . *Ammandra*
2A. Fibres free in mesophyll common, mostly in bundles of up to 6 cells; in 2 series, 1 adaxial, the other abaxial. Phl of larger longitudinal vbs a single strand with few sclerotic cells . . . . *Aphandra*

Phytelephas (Figs. 12J; 70D, E, I–K, O, P; and 71A, D, F, H, K, L)

A tropical American genus of 6 spp., concentrated in the western Amazon Basin and northwestern South America (Panama to NW Venezuela, Colombia, Ecuador). They are rain forest palms with solitary or clustered stems that are of short to moderate height, erect or procumbent (Bernal 1998). The trunk is obscured by persistent leaf base fibres, but these are not elaborated usefully, unlike piassava.

Leaf

Regularly pinnate and pinnae in 1 plane, or pinnae grouped and with multiplanar divergence (*P. aequatorialis*). Leaflets reduplicate with a conspicuous adaxial midrib and a pair of moderately conspicuous, abaxially protruding submarginal veins.

LAMINA (Fig. 71A, D, F, H, K, L)
Dorsiventral (Fig. 71A). **Trichomes** absent. **Epidermis** with thin, cutinized outer wall, cells somewhat papillose. Adaxial epidermis (Fig. 71D) with occasional stomata; cells irregularly rectangular or polygonal, sl. elongated longitudinally or transversely; in distinct longitudinal files. Abaxial epidermis (Fig. 71F) with costal bands of narrow cells below hypodermal fibres; intercostal bands with shorter, wider and less regular cells, the stomata in 1(–2) series. **Stomata** (Fig. 71H) not sunken. T.s.c. irregularly rectangular or elliptical, extended transversely or longitudinally; l.s.c. deep but narrow; g.c. small, ± circular, with small, similarly sized inner and outer ledges. Neighbour cells generally closely paralleling the l.s.c., longitudinally extending to sl. beyond the base of the t.s.c. **Hypodermis** 1–(2)-layered adaxially, 1-layered abaxially, with bands of cubical or longitudinally extended thin-walled cells irregularly alternating with solitary or few-celled bundles of poorly lignified, thick-walled fibres. Abaxial hypodermis

with 4 L-shaped cells around the s.st.ch. **Mesophyll** with adaxial palisade poorly to moderately defined, 2–3-layered; many cells with tanniferous contents. **Fibres** both hypodermal, as described, and free in the mesophyll in bundles of up to 5 cells; the latter moderately lignified. Free non-vascular fbs principally within the palisade mesophyll and often in contact with the adaxial hypodermis. **Veins** dense, in mid-mesophyll or with a sl. abaxial position; not in contact with surface layers. O.S. cells sl. inflated, containing chloroplasts; only complete around small veins. I.S. a complete sclerotic cylinder of fibres 1–2 layers thick in most veins; sheath thicker in few largest veins. Larger veins with a single wide mxy vessel; phl strand commonly including sclerotic cells. Submarginal veins (Fig. 71K) with histological configuration like that of depauperate midrib; containing 3 major vbs. **Midrib** protruding above both surfaces, but especially adaxially; with a complete sclerotic cylinder enclosing 8–12 moderate to small, often inverted vbs embedded in sclerotic parenchyma; 8–10 vbs of variable orientation also embedded within the sclerotic cylinder itself. **Transverse veins** sparse, each with a well-developed sclerotic sheath; often positioned abaxial to longitudinal veins. **Stegmata** including spherical silica bodies; common along large longitudinal veins, less abundant along transverse veins and hypodermal fibres. **Expansion cells** as abaxial strands on each side of the midrib; also in longitudinal bands associated with non-vascularized ad- or abaxial folds. **Raphide sacs** conspicuous in mid-mesophyll. **Margin** without a marginal vein, tapering to only scattered hypodermal fibres (Fig. 71L).

PETIOLE (Figs. 12J and 70D, E)
Usually short; the V-arrangement of vbs pronounced or not. **Central vbs** (Fig. 70E) with a complete sclerotic sheath; mxy including 2 wide vessels, phl in 2 distinct strands. **Ground parenchyma** (Figs. 12J and 70D) uniform, including scattered, small vbs and wide fbs. **Peripheral vbs** crowded with well-developed fibrous sheaths, abaxial vbs large, compound with smaller vbs included within sheath of an enlarged bundle. **Mxy vessel** elements v. long, with long oblique scalariform perforation plates and many thickening bars. **Phl sieve tubes** long; sieve elements with transverse to sl. oblique compound sieve plates.

LEAF SHEATH
Leaf base including numerous fbs, like those of the stem, the leaf base eventually shredding into persistent fibres.

Stem (Fig. 70I–K)

Characterized by an irregular course of axial vbs and massive, cylindrical fbs (Fig. 70I, J). **Cortex** (Fig. 70I) wide, with numerous fbs of varying diameter; indistinctly differentiated from central cylinder; the transition region in the stem base most easily recognized by the insertion of adventitious roots. **Fibres** in numerous, massive, cylindrical fbs, each with many rather thin-walled helically organized fibres and surrounded by a complete sheath of large **stegmata** (Fig. 70K). **Vbs** diffuse and irregularly distributed, running obliquely or even horizontally; each wide, with a well developed sheath of thin-walled fibres next to the single wide phl strand (Fig. 70K). **Mxy** usually of numerous wide, long **tracheids** forming a transition from narrower pxy elements. In *P. aequatorialis* however, mxy includes 2 distinct, wide vessels, the elements with oblique to v. oblique end walls; perforation plates scalariform, with c. 10 thickening bars. **Phl** with sieve tube elements much shorter than the mxy tracheids, with transverse to sl. oblique, usually compound but occasionally simple sieve plates. **Ground tissue** with abundant **starch**, often including wide air-lacunae established by collapse of parenchyma cells. **Raphide sacs** wide, numerous and conspicuous, with densely staining mucilage.

Root (Fig. 70O, P)

Epidermis thin-walled, but persistent. **Exodermis** fibrous, not distinct from wide fibrous outer cortex (Fig. 70O). **Inner cortex** with 3 zones, the outer zone of dense, thick-walled, parenchymatous ground tissue that includes isolated sclereids, tannin cells, and raphide canals. **Endodermis** of cells with U-shaped wall thickenings. **Pericycle** 1–2 layered. Outer **medulla** (Fig. 70P) with 20–25 layers of thick-walled fibres, central medulla including lacunose parenchyma with tannin cells and starch.

Note: the adaxial lamina hypodermis consistently has 2 layers of parenchymatous cells in *P. aequatorialis*. Free fbs in the mesophyll are principally adaxial in distribution among *Phytelephas* spp. In contrast, free fbs are about evenly distributed both ad- and abaxially in *Aphandra*. *Ammandra* is almost without free mesophyll fibres.

Ammandra (Figs. 23A; 70F; and 71C, E, G)

A single sp., *A. decasperma*, in wet tropical forests of Colombia. Palms single-stemmed, the visible stem short or absent. It is distinguished from other genera in the tribe by a long petiole that is terete distally.

Leaf

Pinnate, reduplicate leaflets are regularly arranged and each with a prominent adaxial midrib and a pair of abaxially prominent submarginal veins.

LAMINA

Reduplicate, dorsiventral (Fig. 71C). **Epidermis:** cells including fine granular (crystalline) material, outer wall somewhat thickened; cuticle extended deeply into anticlinal walls. **Trichomes** absent. Adaxial epidermis (Fig. 70E) without stomata. Cells irregularly rectangular to spindle-shaped, markedly obliquely elongated, the longitudinal files thus obscured. Abaxial epidermis (Fig. 71G) with irregularly rectangular costal cells in longitudinal files; intercostal cells v. irregular. **Stomata** not sunken with g.c. appearing circular in surface view; cell walls bordering the aperture strongly thickened. T.s.c. often longitudinally elongated. **Hypodermis** uniseriate below each surface and including mainly solitary thick-walled, unlignified fibres alternating with files of ± cubical thin-walled cells. Abaxial hypodermis with smaller cells and fewer fibres separated from each other by up to 5 files of thin-walled cells. S.st.ch. not regularly demarcated. **Mesophyll** without a distinct adaxial palisade, cells compact, isodiametric except below stomata. **Fibres** in adaxial mesophyll rarely independent of those in hypodermis. **Veins** in mid- to abaxial mesophyll, never in contact with

surface layers. O.S. of small cubical cells, rather indistinct and rarely forming a complete sheath. I.S. (Fig. 71C) a complete and uniform cylinder of thick-walled, lignified fibres (2)3–4 layers thick; I.S. better developed in largest veins. **Phl** of larger veins divided into several strands by irregular sclerotic cells. Submarginal veins with histological configuration like that of depauperate midrib; containing 3 major vbs; ground tissue with scanty, sclerotic parenchyma and fibres. **Midrib** protruding from both surfaces, but v. pronounced adaxially, relatively narrow in TS; with a thick, complete sclerotic cylinder to 12 cells deep enclosing an ad/abaxially linear series of about 10 vbs. Ground tissue is mostly sclerotic parenchyma, v. scanty; also including few fibres. **Transverse veins** rather sparse; in mid to adaxial mesophyll, with a sheath of short sclerotic cells. **Stegmata** mainly restricted to longitudinal veins as large cells interrupting the O.S. and scattered along transverse veins; silica bodies large, spherical (Fig. 23A). **Expansion tissue** as a wide band on each side of the midrib, abaxially including fbs. **Raphide sacs** conspicuous in mid-mesophyll.

PETIOLE (Fig. 70F)

Central vbs with 2 wide mxy vessels, 2 phl strands, and a well developed phl fibre sheath. Peripheral vbs with a more complete fibre sheath, crowded but not fused to form complex bundles. Ground tissue including massive fbs, their position artefactually often represented by a hole because they are lost. **Mxy vessels** with long oblique scalariform perforation plates. **Sieve tubes** with ± transverse, compound sieve plates and few sieve areas.

Stem

Cortex and central cylinder indistinctly separated. **Ground tissue** starch-filled, including numerous wide to massive lignified fbs. **Central vbs** without regular order including numerous narrow tracheids and a single wide phl strand with a surrounding fibre sheath. **Leaf traces** frequently with numerous attendant 'satellite' vbs as presumed inflorescence traces, the satellite bundles also massive and with numerous narrow tracheary elements. **Mxy** tracheary elements imperforate as tracheids. **Phl** strand rather narrow; sieve tubes with transverse compound sieve plates, the sieve areas few and large.

Note: see note under *Phytelephas*.

Aphandra (Fig. 71B, I, J)

Including a single sp., *A. natalia*, from premontane and lowland rainforest of the upper Amazon Basin in Ecuador, Peru, and Brazil. Palms with a solitary, short stem; leaf sheaths disintegrating into abundant fibres.

Leaf

Regularly pinnate. Reduplicate pinnae with a prominent, adaxially protruding midrib and a pair of less prominent, abaxially protruding submarginal veins.

LAMINA

Dorsiventral (Fig. 71B). **Trichomes** absent. **Epidermis** with thickened outer periclinal cell wall to half the total depth of the cell. Adaxial epidermis without differentiated costal regions, cells irregularly polygonal and commonly obliquely elongated. Abaxial epidermis with smaller regularly cuboidal cells. **Stomata** (Fig.71I) not sunken, rare adaxially, common abaxially; g.c. with outer cuticular ledges larger than inner, g.c. pair circular in surface view. **Hypodermis** 1-layered on both surfaces, cells rectangular or hexagonal; adaxial cells larger and more clearly defined than abaxial. S.st.ch. not regularly circumscribed. **Fibres** hypodermal 1–3 (rarely 5) in bundles, densest on adaxial surface, frequently free in mesophyll. **Mesophyll** with 1–2 poorly defined palisade layers. **Veins** always free in mesophyll, O.S. indistinct and thin-walled, I.S. a complete cylinder of fibres (1–)2–3 layers thick; sheath thicker in few largest veins. Larger vbs including a single wide mxy vessel and a single phl strand. Submarginal vein with histology like v. depauperate midrib, or appearing intermediate between midrib and longitudinal vb structure. Sheathed by a complete cylinder of fibres, to 12 layers thick; 2–3 poorly differentiated vbs included within sheath, along with v. scanty amount of ground parenchyma. **Midrib** protruding principally adaxially, semicircular in outline, with complete cylinder of fibres to 12 layers thick enclosing about 5 irregularly arranged vbs. Few small vbs embedded within fibre cylinder. Ground parenchyma scanty, sclerotic. **Transverse veins** with sclerenchymatous sheath; at same level as or adaxial to the longitudinal veins. **Expansion tissue** a single wide band to each side of the midrib, abaxial, including fbs. **Stegmata** with spherical silica bodies; in discontinuous, single files along all non-vascular fibres and vb; in irregular groups along transverse veins (Fig. 71J). **Raphide sacs** absent from lamina.

Note: stem anatomy would repay investigation. See notes under *Phytelephas*.

SUBFAMILY ARECOIDEAE

INTRODUCTION

Arecoideae is the largest subfamily of Arecaceae, containing about 1300 spp. within 107 genera (Dransfield et al. 2008b). The group includes the great majority of arborescent palms with reduplicately pinnate leaves (Fig. 72A, C, G), as well as many palms of low stature (also reduplicately pinnate) that are important components of rainforest understorey vegetation. The distribution of the subfamily is primarily concentrated in the wet tropics of the New World and SE Asia to the South Pacific; few radiations occur outside these areas, with *Dypsis* and related genera in Madagascar being the largest of these (c. 150 spp.; Fig. 72B). Subtribe Attaleinae of Cocoseae, containing the familiar coconut palm (Fig. 72E), is notable for being the only lineage to have achieved any appreciable net diversification in regions with cool or semiarid climates.

The subfamily is most consistently characterized morphologically by a basic inflorescence unit called a triad, which is a sympodial aggregation of a central female flower and two, lateral male flowers. The few groups within Arecoideae that do not bear triads have more complex monochasia or monads as inflorescence units, which likely represent derivatives of this basic type (as in Chamaedoreeae). Less prevalent within Arecoideae, but nevertheless characteristic of the group, are leaf sheaths that are tubular and elongate at maturity, and densely packed below the leafy crown so as to form an obvious column, termed a crownshaft (Figs. 13I, J and 72F). Although crownshafts are also present in *Pseudophoenix* (Ceroxyloideae), those of Arecoideae often differ in that well-defined, mechanically-weak separation layers develop within each sheath circumferentially at the point of leaf attachment and longitudinally along the ventral side (i.e. opposite the petiole), effecting a precise method of abscission (Fig. 13A–D, I, J; *Veitchia*-type sheath; Tomlinson, 1962b). Inflorescences produced within such crowns often mature after the abscission of the subtending leaf, resulting in an infrafoliar presentation (Fig. 72F).

Although Arecoideae as a whole are structurally diverse, just a few groups within the clade are highly variable or strikingly disparate (e.g. Chamaedoreeae, Cocoseae, Geonomateae, Iriarteeae), and in many lineages the systematically critical variation largely consists of subtle, technical features of the reproductive organs. Such challenges to systematic practice, along with appreciable size of the subfamily, understandably lead to a research emphasis on defining species and genera, as epitomized by H.E. Moore's many contributions to the taxonomy of this subfamily.

Building on this foundation, molecular phylogenetic data have been critical to developing an understanding of the broader relationships within Arecoideae. Recent estimates of global palm phylogeny support the recognition of 14 tribal clades within the subfamily (Asmussen et al. 2006; Baker et al. 2009). Within the largest tribe, Areceae, the results of several molecular phylogenetic studies (Lewis and Doyle 2001; Hahn 2002; Loo et al. 2006; Norup et al. 2006), in combination with morphological information, make possible the formulation of a preliminary classification in which 11 subtribes and 10 unplaced genera are recognized (Dransfield et al. 2008b). Recent studies of other major clades of Arecoideae robustly resolve generic relationships, enabling analyses of their morphological evolution and historical biogeography (Cuenca et al. 2008, 2009; Meerow et al. 2009). However, any comprehensive view of the comparative biology of Arecoideae is hindered by the lack of a detailed and robust phylogenetic hypothesis of the relationships among the tribal clades. Structural evolution within the subfamily is complex, and many important character states have homoplasious distributions (Norup et al. 2006). Therefore, we think it premature to use our anatomical data for character reconstructions at the subfamilial level, and they are not emphasized.

Anatomical synapomorphies for Arecoideae remain obscure (Fig. 72D). Although existing phylogenetic uncertainty precludes the unambiguous optimization of many characters that have complex distributions, there are also no anatomical character states that could be considered obvious candidate synapomorphies for Arecoideae. However, two character states merit further investigation in this regard because their distributions are largely restricted to Arecoideae, each having a wide but, at present, insufficiently documented distribution within the group. The sieve tube elements of the petioles and stems of many Arecoideae have transversely-oriented, compound sieve plates with a stellated arrangement of wedge-shaped sieve areas (Fig. 22F). In this way they differ from those of nearly all Coryphoideae and many Ceroxyloideae, which generally have oblique sieve plates with an irregularly reticulate or scalariform arrangement of sieve areas (Parthasarathy 1968; though the sieve plates of the stem metaphloem of Phytelepheae are commonly of this 'arecoid' type). Also potentially distinguishing Arecoideae is a distinctive vascular bundle type found in the median abaxial position of the petioles in many taxa (Fig. 87H; petiolar 'keel' bundles).

A summary of major leaf anatomical characters and states that contribute to an understanding of the systematics of Arecoideae at the intertribal level is detailed in Table 3.

Fig. 72 Arecoideae. **A**. *Roystonea regia*, avenue of palms at Aburi Botanic Garden, Ghana. **B**. *Dypsis madagascariensis*, crown of mature tree. **C**. *Elaeis guineensis*, mature tree. **D**. Summary phylogeny of Arecaceae with anatomical synapomorphies for Arecoideae undetermined because of poor resolution within the subfamily. **E**. *Cocos nucifera*, mature tree. **F**. *Roystonea oleracea*, crown of mature tree. **G**. *Acrocomia crispa*, mature trees.

Labels in D: Calamoideae, Nypoideae, Coryphoideae, Ceroxyloideae, Arecoideae, Anatomical Synapomorphies Uncertain

Table 3 Systematic distribution of major leaf anatomical character state variation within Arecoideae

	Trichome	Epidermal cell shape	Stomata	Substomatal chamber architecture	Hypodermis: number of layers at each surface; shape in SV; distinctive features	Non-vascular fibres: presence and distribution	Veins: position in mesophyll and other noteworthy features	Pinna margin: presence of stout vein	Stegmata: silica body shape	Petiole: phloem strand number
Iriarteeae	Sunken; of few, basal sclerotic cells that are often superimposed, and distal expanse of ephemeral, thin-walled cells	Spindle-shaped; ± papillose abaxially	Sunken in most taxa	2–4 thin-walled cells surrounding 1 aperture	1(2):1; hexagonal; lignified adaxially in *Iriartea*; mucilaginous adaxially in *Dictyocaryum*	In bundles free of surfaces (absent from *Socratea*)	Abaxial	Absent	Hat-shaped	1 strand
Chamaedoreeae	Absent	Spindle-shaped	Superficial	Indistinct	1(2):1; not or poorly differentiated from mesophyll in all genera except *Hyophorbe*, where hexagonal	In scattered bundles that are usually free of surfaces	Central	Absent	Hat-shaped	1 strand
Podococceae	Sunken; of few superimposed, discoid, sclerotic cells	Spindle-shaped to, infrequently, rectangular	Superficial	Usually 4 L-shaped cells	1:1; hexagonal	In bundles free of surfaces; bundles often Λ-shaped in TS	Central	Absent	Spherical	2 strands
Oranieae	Sunken; of 2–3 sclerotic cells	Spindle-shaped; ± papillose abaxially; sometimes with thickened, lenticular-shaped deposits of cutin on anticlinal walls	Sl. sunken; the g.c. overarched by papillae of neighbouring cells	Without uniform arrangement around aperture	1(2):1; hexagonal; cells often appearing inflated in TS	In bundles, with series in contact with surfaces ad- and abaxially; also bundles free of surfaces	Abaxial	Absent	± hemispherical	2 strands
Sclerospermeae	Superficial, of 2–6 weakly sclerotic cells basal to a shield-like expanse of thin-walled cells	Spindle-shaped	Sunken	Without uniform arrangement around aperture	1:1; hexagonal	Ad- and abaxial series of small bundles free of surfaces	Central	Absent	Hat-shaped	?

(*continued*)

Table 3 Continued

	Trichome	Epidermal cell shape	Stomata	Substomatal chamber architecture	Hypodermis: number of layers at each surface; shape in SV; distinctive features	Non-vascular fibres: presence and distribution	Veins: position in mesophyll and other noteworthy features	Pinna margin: presence of stout vein	Stegmata: silica body shape	Petiole: phloem strand number
Roystoneeae	Deeply sunken in pits; broadly peltate	Spindle-shaped	Superficial	4 L-shaped cells	1:1; hexagonal	Ad- and abaxial series of bundles in contact with surfaces; also large bundles free of surfaces	Abaxial	Absent	Spherical	1 strand, with extensive, irregular network of sclerotic cells
Reinhardtieae	Sunken in pits; 1–2 sclerotic cells basal to few, thin-walled, ephemeral cells	Spindle-shaped	Superficial	4 L-shaped cells	1:1; hexagonal	Ad- and abaxial series of bundles in contact with surfaces; also bundles free of surfaces	Central	Absent	About hemispherical to roughly spherical	2 strands
Cocoseae	Bactridinae: superficial; simple or branched. Attaleinae & Elaeidinae: sunken in pits; few sclerotic cells terminating distal filament of thin walled-cells; large and much branched in *Parajubaea*	Spindle-shaped, or rectangular in several genera of Attaleinae; anticlinal walls sinuous in some Attaleinae	Superficial; in some Attaleinae sl. to deeply sunken or sometimes included within furrows	Attaleiniae: 4 L-shaped cells delimiting 1 aperture to 4 U-shaped cells delimiting 2 apertures Bactridinae: usually 4–8 irregularly shaped cells delimiting 1 aperture Elaeidinae: 4 L-shaped cells	Attaleinae: 1:1–3(4):1; hexagonal to rectangular Bactridinae: 1:1; hexagonal Elaeidinae: 1:1–2:2; hexagonal	Ad- and abaxial series of bundles in contact with surfaces; bundles free of surfaces often present	Abaxial. In several genera of Attaleinae: fibre girders present; transverse veins sheathed by fibres & accompanied by stegmata	Present in several genera of Attaleinae; otherwise absent	Spherical: Attaleinae, Elaeidinae Hat-shaped: Bactridinae	2 strands
Manicarieae	Absent	Broadly spindle-shaped	Superficial	4 transversely elongate cells with C-shaped notches that delimit a single, dumbbell-shaped aperture	1:1; narrowly hexagonal; cells large and thick-walled	Adaxial series of fibres solitary or in small bundles and largely free of surfaces; abaxial series of small bundles in contact with surface layers	Central; transverse veins accompanied by stegmata	Absent	Spherical	2 strands

(*continued*)

Euterpeae	Sunken; bases multicellular, narrowly elliptical in SV; with distal expanse of contorted cells in *Oenocarpus*	Rectangular to spindle-shaped (*Neonicholsonia*); anticlinal walls frequently sinuous	Superficial	Indistinct	1:1; mostly rectangular; often hexagonal in *Euterpe*; indistinct in *Hyospathe*	Ad- and abaxial series of solitary fibres or bundles in contact with surfaces, commonly subepidermal; also bundles in mesophyll free of surfaces	Central to abaxial	Present	Spherical	2 strands
Geonomateae	Superficial, uniseriate; of 6–8 cells, with a single basal sclerotic cell	Spindle-shaped	Superficial	Indistinct or, in *Welfia*, usually 4 L-shaped cells	1:1; hexagonal	Scattered in mesophyll and free of surfaces; mostly solitary, but grouped in bundles in *Asterogyne* and *Welfia*	Abaxial	Absent	Spherical	2 strands
Leopoldinieae	Absent, or consisting of persistent bases of many sclerotic cells; elliptical in SV	Rectangular; including files of fibres opposite the outer periclinal wall of the hypodermis	Superficial; t.s.c. often sclerotic	4 L-shaped cells	1:1; hexagonal	Widely scattered solitary fibres or bundles free of surfaces	Abaxial: I.S. well developed; transverse veins sheathed by fibres and accompanied by stegmata	Absent	Spherical	2 strands
Pelagodoxeae	Superficial, simple; with basal most few cells sclerotic	Rectangular to spindle-shaped with sinuous anticlinal walls	Superficial; t.s.c. and lateral neighbouring cells sclerotic	4 L-shaped cells	3:1 and hexagonal in *Pelagodoxa*; 1:1 and rectangular in *Sommieria*	Adaxial series of nearly continuous to discontinuous subepidermal fibres; bundles in mesophyll free of surfaces absent	Abaxial	Absent	Spherical	2 strands
Areceae	Superficial to deeply sunken in pits; of few cells to many cells and peltate	Spindle-shaped; commonly rectangular in Linospadicinae, Oncospermatineae, some unplaced genera; with sinuous anticlinal walls in Linospadicinae and some unplaced genera	Superficial	4 L-shaped cells	1(2):1, hexagonal or, in Linospadicinae and some unplaced genera, rectangular	Mostly in bundles free of the surfaces; series of bundles uncommonly also present in contact with surfaces; subepidermal in a few unplaced genera	Usually abaxial	Present in several lineages, notably Archontophoenicinae and Dypsidinae	Spherical	2 strands

TRIBE IRIARTEEAE (Fig. 73)

A group of New World palms restricted to Central America and northern South America, mainly in lowland or montane forests. Except for *Iriartella*, a slender palmlet of the forest understorey with single or multiple stems, the group is largely characterized by a robust, single-stemmed habit, massive crowns but including few individual leaves; all with a pronounced crownshaft. They are the 'stilt-palms' par excellence because of the massive development of a basal cone of ultimately wide aerial roots which support a stem that is otherwise narrowly obconical below. This habit in which the seedling axis in its phase of establishment develops extended internodes of progressively increasing diameter would appear to preclude a multiple-stemmed condition; the stems of robust spp. are therefore rarely branched. *Iriartella* is the chief exception and may be considered a possible paedomorphic version of its larger relatives which have slender basal (seedling) axes. The palms are unarmed except for second-order spine roots forming linear series on the stilt roots. Leaflets typically have praemorse apices, most pronounced in the broad deltoid leaflets of *Iriartella*; the major ribs radiating from a common base.

Note: distinguishing features that might be useful in segregating the large genera have not been demonstrated, apart from differences between *Iriartella* and other genera determined by the considerable differences in palm size. There may be as much variation within as between genera.

Anatomical features of the tribe
Iriarteeae (Figs. 7M; 8E, J; 16F, G; and 73)

1) **Trichomes** each with a basal, usually sunken, group of 1–2(–5) large, somewhat thick-walled densely pitted cells (Figs. 8E, J and 73D), the distal thin-walled cells rarely persistent and seemingly forming a uniseriate filament but sometimes terminating in a thick-walled pointed cell.

2) Adaxial **epidermis** of usually obliquely-extended rhombohedral to ± hexagonal or spindle-shaped cells, the walls never sinuous (Figs. 7M and 73A); abaxial epidermal cells smaller, less regular and with frequent costal bands of rectangular to rhombohedral cells (Fig. 73B).

3) **Stomata** (Fig. 73C) restricted to abaxial intercostal regions, t.s.c. not usually well differentiated, but in some spp. of *Iriartea* and *Wettinia* stomata overarched by papillae from t.s.c. and neighbouring cells.

4) **Hypodermis** uniseriate or apparently biseriate adaxially (*Dictyocaryum*) because of overlapping anticlinal walls; conspicuously large-celled (Fig. 73F) the walls often somewhat thickened and with conspicuous pits on the vertical walls. Adaxial cells hexagonal and markedly transversely extended, thus contrasting with the obliquity of the epidermal cells (Figs. 7M and 73A). Abaxial hypodermis with smaller cells, the cells surrounding the s.st.ch. thin-walled and not very regularly arranged (Fig. 73B).

5) **Mesophyll** either with well-differentiated palisade layers (Fig. 73F) or without in thinner leaves (e.g. *Iriartella*).

6) **Fibres** few, solitary or in small strands scattered in the mesophyll but rarely in contact with the hypodermal layers.

7) **Veins** (Fig. 73F) equidistant from each surface or situated abaxially and usually without contact with either surface, except for few large veins with small, lignified intervening mesophyll cells. O.S. inconspicuous and rarely complete around the vein. I.S. only complete around largest veins.

8) **Transverse commissures** at same level as longitudinal veins, few, narrow, and sheathed by thin-walled cells.

9) **Ribs** often prominent abaxially, with a well-developed sclerotic cylinder enclosing a larger abaxial vb and smaller mostly adaxial vbs (Fig. 73H).

10) **Petiole** typically with wide central vbs, each with 2 wide mxy vessels and a single wide phl strand (Fig. 73I).

11) **Stem:** (i) in narrow stems of *Iriartella* of conventional 'Rhapis-type' anatomy, the central ground tissues not expanding with age (Fig. 73J); (ii) in wide stems of larger genera with extensive late development of massive fibre sheaths to vbs and ground tissue, initially uniformly small-celled in stem centre (Fig. 16F), becoming extremely lacunose with age (Fig. 16G); central vbs usually with 2 wide mxy vessels and a single phl strand (Fig. 73K). Ground tissue sometimes with numerous narrow fibre strands.

12) **Root** inner cortex (Fig. 73G) including fbs accompanied by stegmata, each containing a central wide cell (a possible raphide canal). These fibres are clearly contrasted with 'Kentia-type' fibres of many other Arecoideae.

13) **Aerial stilt roots** with wide fluted stele and extensive development of medullary vessels (Figs. 20L, M and 73L, M).

14) **Silica bodies** hat-shaped (Fig. 73E).

We here describe *Wettinia* in detail as an example, with some commentary on the other genera.

Wettinia (Figs. 8J and 73B–E, K, M)

A group of about 20 spp. of New World palms mostly in the lowland humid tropics from Central to Andean America, the centre of diversity in Colombia. A number of segregate genera (e.g. *Acrostigma, Catoblastus, Catostigma*), in the absence of consistent distinctive characters, are no longer recognized. Plants are usually large, single- or multiple-stemmed, with a crown containing few leaves, the leaflets varying from lanceolate to deltoid with praemorse apices and numerous ribs. Stems commonly have basal stilt roots.

LAMINA

Dorsiventral. **Trichomes** each with a few large sclerotic pitted basal cells (Figs. 8 J and 73D), typically extended into a distal, pointed uniseriate filament. **Epidermis:** adaxial cells obliquely extended; abaxial cells (Fig. 73B, C) less regular, usually rectangular but with oblique end walls, costal bands infrequent. **Stomata** in irregular files corresponding to the anticlinal wall of the hypodermis within (Fig. 73B); g.c. sometimes partly overarched by lobes from surrounding cells. **Hypodermis** 1-layered below each surface cells large and somewhat thick-walled; adaxially hexagonal, transversely extended (Fig. 73A). **Fibres** few to

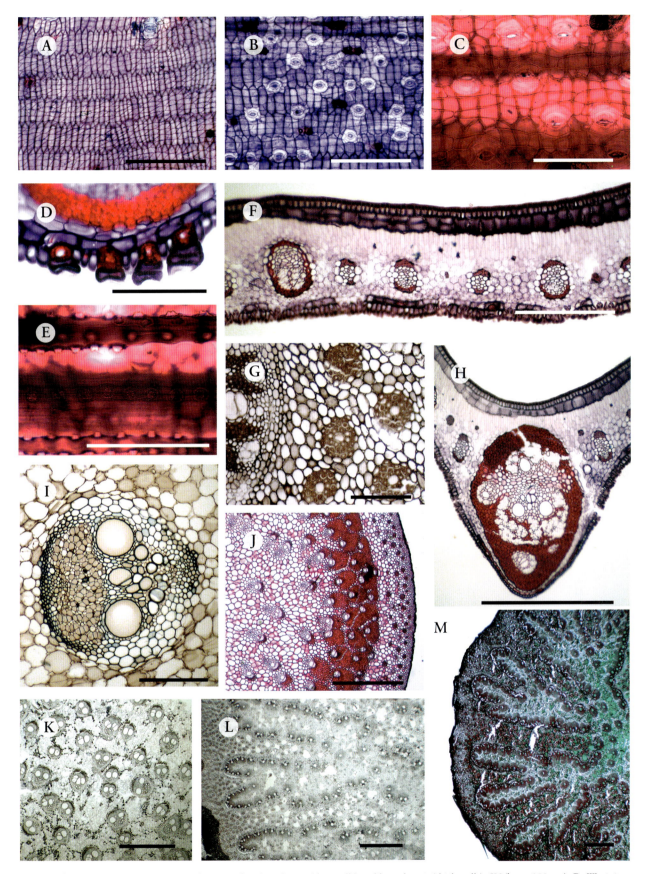

Fig. 73 Arecoideae (Iriarteeae). **A**. *Socratea* sp., lamina, adaxial epidermis (thin walls) and hypodermis (thick walls), SV (bar = 200 μm). **B**. *Wettinia* sp., lamina, abaxial epidermis, SV (bar = 200 μm). **C**. *Wettinia aequalis*, lamina, abaxial epidermis (thin walls) and hypodermis (thick walls), SV (bar = 100 μm). **D**. *Wettinia aequalis*, lamina, abaxial epidermis with trichomes below large vein, TS (bar = 100 μm). **E**. *Wettinia aequalis*, lamina, stegmata associated with fbs, SV (bar = 100 μm). **F**. *Iriartea deltoidea*, lamina, TS (bar = 250 μm). **G**. *Socratea exorrhiza*, root, outer stele and inner cortex with fbs, TS (bar = 200 μm). **H**. *Iriartea deltoidea*, lamina, midrib, TS (bar = 500 μm). **I**. *Socratea exorrhiza*, petiole vb, TS (bar = 200 μm). **J**. *Iriartella* sp., stem periphery, TS (bar = 1 mm). **K**. *Wettinia* sp., stem centre, TS (bar = 2 mm). **L**. *Socratea exorrhiza*, root stele, TS (bar = 2 mm). **M**. *Wettinia* sp., root periphery to centre, TS (bar = 2 mm).

frequent, solitary or in strands with up to 15 cells, in mesophyll but rarely in contact with hypodermis. **Veins** in abaxial mesophyll, not in contact with hypodermis.

Key to species of *Wettinia* examined

1. Abaxial epidermis partly to wholly papillose, stomata occluded by papillae from t.s.c. 2
1A. Abaxial epidermis without papillae, stomata not occluded 3
2. Papillae well-developed on all epidermal cells, the stomata largely occluded; mesophyll fibres numerous . . . . *Wettinia fascicularis*
2A. Papillae restricted to epidermal cells surrounding stomata; stomata partly occluded; mesophyll fibres few . . . . *Wettinia anomala*
3. Mesophyll fibres abundant, strands often wide, with up to c .15 cells, outer wall of g.c. appreciably thickened . . . . *Wettinia radiata*
3A. Mesophyll fibres moderate, strands with c. 10 cells, outer wall of g.c. unthickened 4
4. Lamina thick, mesophyll c. 10 cells deep; veins closely spaced. *Wettinia quinaria*
4A. Lamina thin, mesophyll <6 cells deep; veins not closely spaced. . . . . *Wettinia augusta*

Iriartea (Figs. 8E and 73F, H)

LAMINA

Dorsiventral (Fig. 73F). Trichomes each with 2 or more large densely pitted sunken basal cells (Fig. 8E) extending into a uniseriate filament. **Epidermis** with outer wall thickened, adaxial cells obliquely extended, spindle-shaped, abaxial cells irregular and often rectangular. **Stomata** in some spp. sunken, partly to wholly occluded by lobed extensions of the t.s.c. and neighbour cells; g.c. with outer wall thickened. **Hypodermis** 1-layered below each surface; adaxial cells narrowly hexagonal in surface view, somewhat thick-walled, the anticlinal walls conspicuously pitted, transverse walls wedge-shaped so as to appear 2-layered in TS (Fig. 73F). Abaxial hypodermis shallower, with 2–4 thin-walled cells surrounding each s.st.ch. **Mesophyll** with or without a 2-layered palisade. **Fibres** few, as small strands at the level of the veins together with occasional solitary fibres or 2–3-celled strands scattered in the mesophyll (Fig. 73F). Veins in mid or abaxial mesophyll, independent of surface layers except for few large veins continuous with the hypodermis. **Midrib** (Fig. 73H) projecting abaxially, with a sclerotic tissue enclosing 1 large abaxial vb and smaller abaxial vbs, expansion cells not conspicuous.

Note: the variation among spp. recorded here for *Wettinia* might be expected in *Iriartea* but has not been reported.

Socratea and *Dictyocaryum*

Information is summarized in the tribal description. This has provided no anatomical lamina characters that can be seen as diagnostic, but rather a representation of the variation seen in other large-stemmed spp. Stilt root of *Socratea* has a lobed, dissected stele (Fig. 20L, M).

Iriartella (Fig. 73J)

The 2 spp. of this genus are small understorey palms with narrow stems, the leaves with broadly deltoid leaflets the outer margins coarsely toothed or lobed. They provide an example where anatomical comparison with other genera must take into account size differences.

LAMINA

Dorsiventral. **Trichomes** with a basal pair (rarely 3) of large, pitted sclerotic cells or forming a column of such cells extended distally as an elongated thick-walled and pointed cell or into a filament of thin-walled cells and terminating in such a cell. **Epidermis** thin-walled, outer wall sl. thickened, except the cells below the veins developed as elongated sclereid-like cells. **Hypodermis** well-developed, 1-layered below each surface, adaxial cells largest. **Mesophyll** v. shallow 4 compact cell layers deep; cells somewhat transversely extended, palisade absent. **Fibres** common in irregular, often flat, strands of 2–7 cells scattered in the mesophyll but not often in contact with the hypodermis and so restricted to 3 mesophyll layers. **Veins** situated abaxially and all in contact with abaxial epidermis via a sclerotic layer of small cells. O.S. well developed but always interrupted below; vascular tissue v. reduced.

Stem (Fig. 73J)

Scarcely 1–2 cm diameter. **Epidermis** with numerous trichomes. **Cortex** including numerous fbs and few vbs. **Central cylinder** with few outer congested vbs, the fibrous sheath well developed and forming a continuous sclerotic layer (Fig. 73J); transition to diffusely-distributed vbs of central stem abrupt. **Ground tissue** cells somewhat horizontally extended with age, but not becoming lacunose. Central vbs each with 1 wide mxy vessel, the elements with long oblique end walls with scalariform perforation plates, the thickening bars few and narrow.

Root

Cortex narrow, with few or no air lacunae; including strands of fibres, each surrounding a thin-walled cell. **Stele** fluted, the endodermis interrupted and with continuity between cortex and medulla.

TRIBE CHAMAEDOREEAE (Figs. 4; 13A, D, I; 15; 22B; and 74)

A tribe of 5 monoecious or dioecious genera, of which 4 are widely distributed in tropical America, but the fifth (*Hyphorbe*) is restricted to the remote Mascarene Islands in the Indian Ocean, a remarkably disjunct condition. Leaves are pinnately reduplicate. Plants are unarmed and, except in *Chamaedorea*, single-stemmed. Only *Hyophorbe* and *Gaussia* form sizeable trees. *Chamaedorea*, the largest genus of over 100 spp. includes mostly cane-like and multiple-stemmed palms. One sp., *C. elatior*, is scandent. The leaf sheath is rather unspecialized and only in *Hyophorbe* (Fig. 13A,

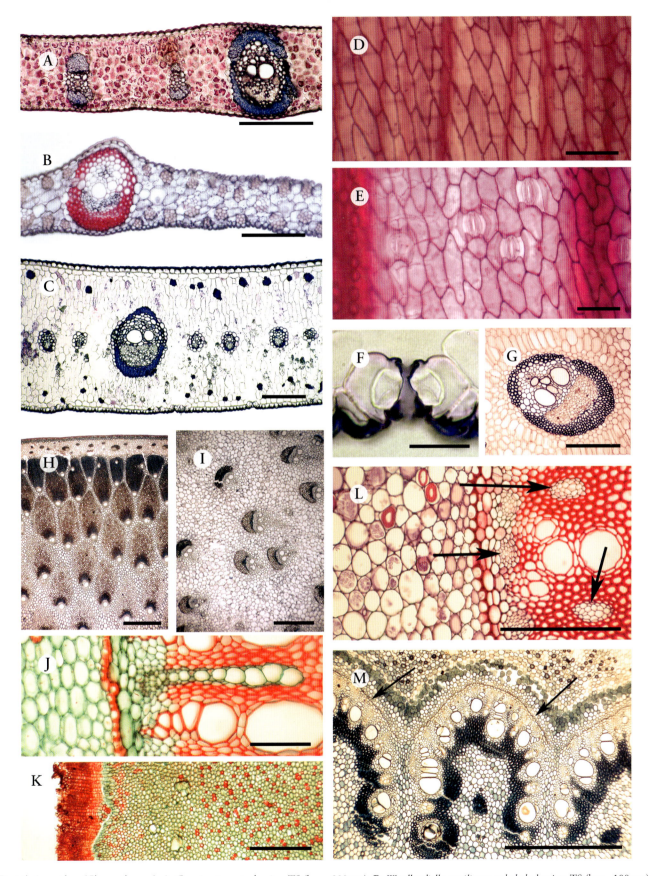

Fig. 74 Arecoideae (Chamaedoreeae). **A**. *Gaussia attenuata*, lamina, TS (bar = 200 μm). **B**. *Wendlandiella gracilis* var. *polyclada*, lamina, TS (bar = 100 μm). **C**. *Hyophorbe verschaffeltii*, lamina, TS (bar = 200 μm). **D**. *Synechanthus fibrosus*, lamina adaxial epidermis, SV (bar = 50 μm). **E**. *Chamaedorea arenbergiana*, lamina abaxial epidermis, SV (bar = 50 μm). **F**. *Hyophorbe verschaffeltii*, lamina abaxial epidermis with stomata, TS (bar = 25 μm). **G**. *Hyophorbe verschaffeltii*, petiole, central vb, TS (bar = 250 μm). **H**. *Chamaedorea brachypoda*, stem periphery, TS (bar = 500 μm). **I**. *Chamaedorea brachypoda*, stem centre, TS (bar = 500 μm). **J**. *Gaussia maya*, root, outer stele and inner cortex, TS (bar = 100 μm). **K**. *Gaussia maya*, root, outer cortex and root periphery, TS (bar = 500 μm). **L**. *Chamaedorea alternans*, root, outer stele; arrows = peripheral and subperipheral phloem strands, TS (bar = 200 μm). **M**. *Hyophorbe* sp., root, outer part of fluted stele; arrows = endodermis, TS (bar = 1 mm).

D, I) is a clear crownshaft developed by precise abscission of the oldest leaves so that inflorescences are decidedly infrafoliar. A distinctive reproductive feature is the linear cincinnate aggregate of flowers, the acervulus, associated with the monoecious spp. found in all genera except *Chamaedorea*, which is also dioecious.

Anatomical features of the tribe Chamaedoreeae

LAMINA

Dorsiventral throughout (Fig. 74A–C). **Trichomes** absent. **Epidermis** (Fig. 74D, E) with a tendency for oblique cell divisions producing elongated rhombohedral cells, the longitudinal cell files then obscured (Fig. 74D). **Hypodermis** absent from most genera. **Stomata** (Fig. 74E) tend to lack differentiated terminal subsidiary cells; stomata in *Synechanthus* uniquely specialized with a single cuticular ledge rather than 2 as is the usual condition (Fig. 74F). **Mesophyll** rarely developing palisade layers. **Fibres** little developed, or absent and mostly restricted to the mesophyll (except *Synechanthus*). **Veins** except for the few largest, mostly lacking contact with the surface layers. **Transverse veins** not well developed. Major ribs with a continuous peripheral sclerotic cylinder; and associated expansion cells in 2 abaxial bands.

PETIOLE

Vbs (Fig. 74G) usually with 2 wide mxy vessels and a single phl strand.

Stem (Figs. 15A–F and 74H, I)

Stem in *Chamaedorea* spp. showing clearly the *Rhapis*-model of vascular construction, with narrow cortex, congested peripheral vbs with massive phl fibre sheaths (Figs. 15A and 74H), more diffuse central vbs and without ground tissue expansion in smaller-stemmed spp. (Figs. 15B and 74I) and an essentially 2-vessel mxy configuration with conspicuous inflorescence traces (Fig. 15C–F). *Synechanthus* and *Wendlandiella* correspond to this construction closely. *Guassia* and *Hyophorbe* have not been studied but could be expected to show changes related to their wider stems and likely sustained primary growth.

Root (Figs. 19G; 22B; and 74J–M)

Cortex with uniformly solitary thick-walled **fibres** throughout (Fig. 74K). **Endodermis** with U-shaped wall thickenings. **Stele** fluted in larger roots (Fig. 74M); sometimes including long radiating phl strands (Fig. 74J) and consistently in all genera including **subperipheral phl strands** within the outer medulla (Figs. 22B and 74L: arrows). Surface roots may show elaborate sclerotic layers, an inner cortex without air spaces, and a wide stele with medullary vessels (Fig. 19G). **Stegmata** with **hat-shaped** silica bodies.

Key to genera of Chamaedoreeae based on lamina anatomy

1. Hypodermis absent (Fig. 74A, B), palisade never developed; mesophyll fibres usually present, sometimes in contact with epidermis 2
1A. Hypodermis present (Fig. 74C); palisade sometimes distinct; mesophyll fibres, if present not in contact with epidermis . . . . *Hyophorbe*
2. Fibres commonly adjacent to epidermis (Fig. 74B) 3
2A. Fibres, if present, not adjacent to epidermis 4
3. G.c. with a single cutinized ledge . . . . *Synechanthus*
3A. G.c. with 2 equal cutinized ledges– . . . . *Wendlandiella*
4. Mesophyll fibres absent. 5
4A. Mesophyll fibres present . . . . some *Chamaedorea* spp.
5. Lamina >10 cells thick . . . . *Gaussia*
5A. Lamina <10 cells thick . . . . .many *Chamaedorea* spp.

Chamaedorea (Figs. 4; 15; and 74E, H–L)

A large genus (more than 100 spp.; Hodel 1992), made familiar by the use of several representatives as indoor potted palms (e.g. *C. elegans*, *C. seifritzii*), a reflection of the characteristic shaded understorey habit of most spp. Identification of spp. can be difficult and since our account is based on so few samples we simply make reference to our limited study of variation without concern for systematically useful characters. Leaves of smaller spp. have provided models for the study of the details of leaflet separation, which involves apoptosis (Nowak et al. 2007, 2008, 2009) and can produce a distinct cellular configuration at the leaflet margin (Fig. 4).

All are small palms, mostly single- or multiple-stemmed, the axis slender and as little as 5 mm in diameter, but a number of spp. are acaulescent (e.g. *C. radicans*). Some are rhizomatous or stoloniferous, with axes dimorphic (*C. rhizomatosa*), or simply becoming horizontal as they recline with age (*C. stolonifera*). *Chamaedorea cataractarum* has an unusual creeping, dichotomously branched stem (Fisher 1974) related to its rheophytic habitat. *Chamaedorea elatior* is a rather unspecialized climber, scandent by virtue of its long internodes and weakly developed distal leaflets, scarcely acanthophylls. Inflorescences are initially interfoliar, but persist and become infrafoliar with age.

Leaf

Blade commonly simple, but with a deeply bifid apex, or pinnately compound with reduplicate leaflets of varying width and arrangement (Hodel 1992).

LAMINA

Dorsiventral. **Trichomes** absent. **Epidermis**; cells thin-walled, sometimes sl. papillose, walls not sinuous; adaxial epidermis with rhombohedral to rectangular, often elongated, cells; costal bands only differentiated above larger veins. Abaxial epidermis with more frequent costal bands. **Stomata** infrequent adaxially, rather diffuse in abaxial intercostal regions (Fig. 74E), t.s.c. not differentiated from other epidermal cells, but overarching lateral subsidiary cells sl.; l.s.c. often with chloroplasts. **Hypodermis** absent. **Mesophyll** uniform and without palisade layers. **Fibres** varying from absent, or present as narrow strands adjacent to each epidermis (e.g. *C. elatior*) or in mesophyll at the level of the small veins. **Veins** ± equidistant from and independent of each surface; larger veins sometimes in contact with lower epidermis; O.S. often indistinct, I.S. complete around larger veins but otherwise only well developed above and below small veins. **Phl** undivided. **Transverse veins** infrequent, at the level of the longitudinal veins, sheathed only by lignified parenchyma. **Midrib** with a continuous sclerotic cylinder enclosing a large abaxial and smaller adaxial vbs.

PETIOLE

Central vbs commonly with a distinct V-arrangement, the vbs usually with 2 wide mxy vessels and 1 phl strand.

Stem (Figs. 15 and 74H, I)

Stems usually small and cane-like, with standard *Rhapis*-model anatomy easily observed (Fig. 15A–G). Outer vbs developing massive fibre sheaths in wide stems. Central **ground tissue** not undergoing marked late expansion (Fig. 15B).

Root (Fig. 74L)

Surface layers becoming sclerotic after loss of epidermis. **Cortex** with frequent raphide canals in outer layers, becoming thick-walled with age; ground tissue within developing numerous solitary fibre-like cells with thick walls, air spaces often present in inner cortex. **Endodermis** with U-shaped wall thickenings. Isolated **phl strands** present in outer conjunctive tissue of stele (Fig. 74L).

Vascular tissues

Mxy with scalariform perforation plates in all organs.

Gaussia (Fig. 74A, J, K)

Four spp. (including the 2 formerly included in *Opsiandra* in Central America and the Caribbean), all characteristic of calcareous soils or limestone (Quero and Read 1986). All are single-stemmed palms of moderate stature, the trunk either tapered distally (*G. attenuata*) or swollen basally (*G. princeps*).

Leaf

Pinnate, reduplicate, the leaflets with a single fold and adaxially prominent midrib. The sheath splits ventrally.

LAMINA

Dorsiventral (Fig. 74A). **Trichomes** absent, **Epidermis** with irregular rectangular or rhombohedral cells; costal bands on both surfaces, but indistinct. **Stomata** occasional adaxially, diffuse and not in regular files abaxially, t.s.c. not well differentiated. **Hypodermis** absent. **Mesophyll** without palisade layers. **Fibres** absent. **Veins** equidistant from each surface, mostly not in contact with surface layers; **phl** undivided but including sclerenchyma. **Transverse** veins infrequent, sheathed by thick-walled cells.

Note: Zapatilla (1985) suggests that *G. maya* and *G. gomezpompae* may be distinguished by quantitative lamina characters, e.g. the relative frequency of adaxial stomata. Otherwise the genus appears uniform anatomically.

Hyophorbe (Figs. 13A, D, I and 74C, F, G, M)

A genus of 5 spp. restricted to the Mascarene Islands (Moore 1978) but now virtually extinct in the wild. It persists because of the horticultural value of the spp. as attractive stout-trunked

and solitary-stemmed feather palms, sometimes forming basally swollen trunks as bottle palms (e.g. *H. lagenicaulis*).

Leaf

Pinnately compound, with precise abscission so that a distinct crownshaft is present (Fig. 13A, D, I).

LAMINA (Fig. 74C, F, G)

Dorsiventral (Fig. 74C) or incipiently isolateral. **Trichomes** absent, but with long scales on the abaxial side of the ribs. **Epidermis** with a well-developed cuticle, the outer wall scarcely to somewhat distinctly thickened. Adaxial epidermis with or without stomata; cells irregularly rhombohedral, elongated. Abaxial epidermis with narrow bands of elongated costal cells. **Stomata** often numerous on adaxial surface (*H. lagenicaulis*), diffusely distributed on abaxial surface in wider intercostal bands; t.s.c. usually shorter than other epidermal cells, overarching guard cells sl.; g.c. with 2 equal cutinized ledges (Fig. 74F). **Hypodermis** inconspicuous 1–2-layered adaxially, 1-layered abaxially, the cells ± cubical. **Mesophyll** usually with a 2–3-layered adaxial palisade (Fig. 74C), but abaxial cells also somewhat palisade like in *H. lagenicaulis*. **Fibres** either absent, as in *H. lagenicaulis*, or, if present in mesophyll, not well developed. **Veins** in abaxial mesophyll, almost always independent of surface layers or, at most, the largest veins extending to each surface. O.S. incomplete above and below most veins, the cells elongated. I.S. sclerotic with fibres above and below. **Phl** undivided. **Transverse veins** infrequent at same level as longitudinal veins and often sheathed only by thin-walled parenchyma rather than sclerenchyma. **Ribs** prominent, including a peripheral sclerotic cylinder enclosing small peripheral vbs with 1 or more phl strands, and 1 or more large central vbs. **Expansion cells** in conspicuous abaxial bands on each side of the rib. **Raphide sacs** usually conspicuous, sometimes exclusively next to surface layers. **Stegmata** with hat-shaped silica bodies.

The following key to spp. is based on the account in Uhl (1978b), modified by our own observations. It is suggested, however, that floral anatomy is more distinctive in spp. identification (Uhl 1978a).

Key to species of *Hyophorbe* based on lamina anatomy

1. Lamina incipiently isolateral with stomata equally numerous on both surfaces; t.s.c. of stomata c. 3× longer than wide; mesophyll consistently without fibre strands . . . . *H. lagenicaulis*
1A. Lamina dorsiventral, stomata few or absent adaxially, t.s.c. of stomata short 2
2. Fibre strands (= non-vascular fibres) absent, or if present then only towards adaxial surface, raphide sacs hypodermal . . . . *H. indica*
2A. Fibre strands present equally or unequally towards each surface. 3
3. Fibre strands narrow, more or less uniformly scattered . . . . *H. vaughanii*

3A. Fibre strands wide, mostly or exclusively in adaxial mesophyll . . . . *H. amaricaulis, H. verschaffeltii*

Root

Stele wide and fluted in wider roots of larger individuals, the medulla and cortex seemingly continuous through an interrupted endodermis (Fig. 74M).

Synechanthus (Figs. 22B and 74D)

Two spp. but with some morphological diversity, ranging from southern Mexico to northern South America. Palms have slender single or multiple stems, the leaf morphology varying in the shape and width of the leaflets. As in *Gaussia* the leaf base is dorsally cleft, with the ventral side becoming fibrous. Stomatal g.c are distinctive among all palms because they have only 1 (outer) cuticular ledge (Tomlinson 1961). Root has features of tribe (Fig. 22B).

Leaf

Pinnately compound, reduplicate, the leaflet insertion very irregular, each leaflet with 1 or more prominent adaxial ribs.

LAMINA

Dorsiventral. **Trichomes** absent. **Epidermis** with outer wall only sl. thickened, the cuticle thin. Costal bands inconspicuous. Adaxial intercostals cells (Fig. 74D) extended but with oblique end walls and so rhombohedral. **Stomata** diffuse, not in distinct files, t.s.c. not differentiated. G.c. appearing collectively circular in surface view, each with a **single** prominent outer cutinized ledge. **Hypodermis** absent. **Mesophyll** without distinct palisade layers. **Fibres** frequent, solitary or in small strands, mostly immediately below each epidermis. **Veins** mostly independent of surface layers, but largest in contact with 1 or both surfaces. O.S. usually incomplete above and below. **Transverse veins** diffuse, narrow, sheathed by thin-walled cells. **Midrib** with a well-developed peripheral sclerotic sheath.

Wendlandiella (Fig. 74B)

A diminutive understorey palm (to 2 m) with the leaf blade entire and bifid or with a few basal pinnae; 3 spp. in Amazonian Peru.

LAMINA

Dorsiventral (Fig. 74B). **Trichomes** absent. **Adaxial epidermis** with occasional stomata near larger veins, but without differentiated costal bands. Cells elongated but with markedly oblique end walls, the longitudinal walls curved and the cell files thus obscured. **Abaxial epidermis** similar but with few and v. diffusely distributed stomata not in distinct files, cells elongated with curved oblique end walls, sl. papillose, the outer wall only sl. thickened; cuticle thin. G.c. with 2 prominent ledges, l.s.c. wide, thin-walled, the regular configuration lost because of oblique extension of cells. **Hypodermis** not differentiated as colourless layers, the subsurface mesophyll cells somewhat smaller than those elsewhere. **Mesophyll** uniformly chlorophyllous, the central cells enlarged; s.st.ch. not conspicuous. Non-vascular **fibres** forming numerous wide strands, mostly next to each surface (Fig. 74B), the fibres unlignified. **Veins** independent of surface layers except for major veins that make contact with each surface; the adjacent cells files of the midrib ± rectangular. O.S. not or scarcely differentiated; I.S. of unlignified fibres resembling those of the non-vascular strands except around largest veins. **Raphide sacs** frequent; wide and conspicuous. **Stegmata** abundant next to vascular and non-vascular fibres, thick-walled, each enclosing a conspicuously hat-shaped silica body.

   Note: in its leaf anatomy this genus is closest to *Synechanthus*.

TRIBE PODOCOCCEAE (Figs. 23B and 75H–Q)

Includes the single genus *Podococcus* with 2 spp. in a restricted coastal area of Nigeria and adjacent Gabon. It is a diminutive understorey palm of the rainforest with spicate inflorescences and sometimes trunkless. The fruits are distinctive with 1–3 lobes, each lobe containing a crescent-shaped seed. The palm is anatomically unusual in the absence of a differentiated hypodermis, a feature possibly related to its habitat. The trichomes are also unusual in their conspicuously pitted basal cells.

LEAF

Pinnately compound with few broad leaflets, their outline rhombic and the apex toothed or praemorse. Each has several divergent ribs.

LAMINA (Fig. 75H-Q)

Dorsiventral (Fig. 75H, I). **Trichomes** (Fig. 75O, P) uniformly distributed on both surfaces but sometimes more numerous abaxially. Each with a sl. sunken column (Fig. 75O) of 4–5 plate-like cells, the proximal thick-walled and conspicuously pitted, the pits elongated and forming a radiating series in SV (Fig. 75P); distal cells few and thin-walled but not indicative of any shield-like structure in the mature leaf. **Epidermis** (Fig. 75J, L, M) relatively large-celled on both surfaces, the outer wall sl. thickened and with a thin cuticle. Adaxial epidermis (Fig. 75J) without distinct costal regions, cells irregular but longitudinally extended, rectangular but with oblique end walls, the cell files remaining distinct. Abaxial epidermis (Fig. 75L) similar, but with indistinct costal regions of somewhat narrower cells. **Stomata** (Fig. 75L, M) v. diffuse and widely separated, not in distinct files; t.s.c. not usually differentiated as short cells; the neighbour cells short or long but not a consistent feature; s.st.ch. usually surrounded by 4 L-shaped cells. **Guard cells** each with 2 conspicuous cutinized ledges, back wall thin, front wall distinctly thickened. **Hypodermis** not differentiated as a specialized colourless layer, but cells compact (Fig. 75K). **Mesophyll** shallow and consisting of 4–5 layers of large cells, the hypodermal cells somewhat transversely extended; palisade layers absent. **Fibres** unlignified; mostly in the sub-hypodermal layer and not in contact with the epidermis; including solitary fibres or in few-celled (2–7) strands with an irregular outline in TS. **Veins** (Fig. 75H, I) uniformly narrow and independent of surface layers; O.S. incomplete above and below but laterally represented by a few files, or even a single

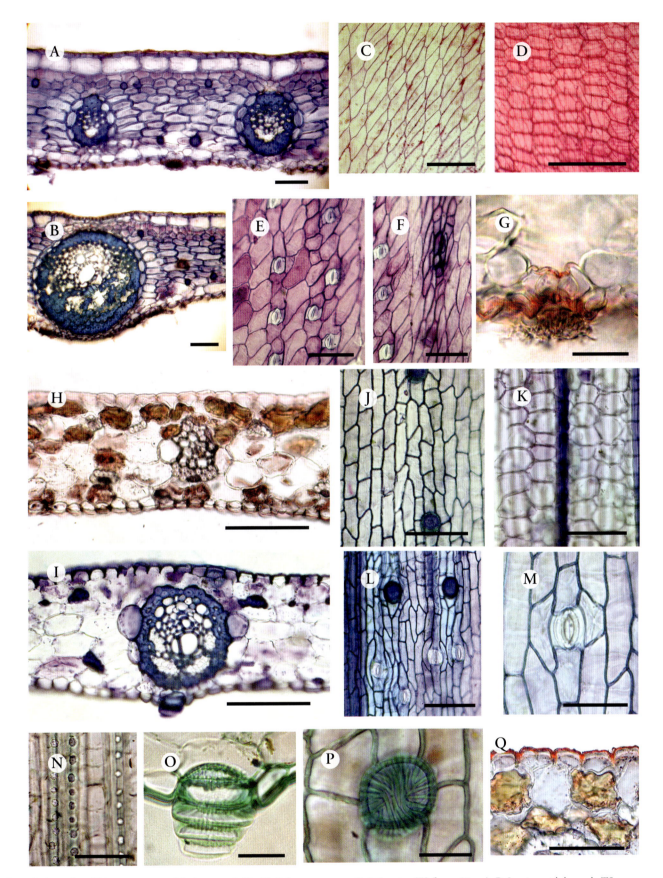

Fig. 75 Arecoideae (Sclerospermeae and Podococceae). (**A–G**). *Sclerosperma mannii*. **A**. Lamina, TS (bar = 50 μm). **B**. Lamina with large vb, TS (bar = 50 μm). **C**. Lamina, adaxial epidermis, SV (bar = 50 μm). **D**. Lamina, adaxial thin-walled epidermis, thick-walled hypodermis, SV (bar = 100 μm). **E**. Lamina, abaxial epidermis, SV (bar = 50 μm). **F**. Lamina, abaxial epidermis with trichome, SV (bar = 50 μm). **G**. Lamina, abaxial stomata occluded with wax, TS (bar = 50 μm). (**H–Q**). *Podococcus barteri*. **H**. Lamina, TS (bar = 100 μm). **I**. Lamina with large vb, TS (bar = 100 μm). **J**. Lamina, adaxial epidermis, SV (bar = 100 μm). **K**. Lamina, adaxial hypodermis, SV (bar = 100 μm). **L**. Lamina, abaxial epidermis, SV (bar = 100 μm). **M**. Lamina, adaxial stomata, SV (bar = 50 μm). **N**. Lamina, stegmata associated with fbs, SV (bar = 50 μm). **O**. Lamina, TS, with trichome, LS (bar = 25 μm). **P**. Lamina, trichome, SV (bar = 25 μm). **Q**. Adaxial epidermis and sclereids, TS (bar = 100 μm).

file of large cells. I.S. of unlignified fibres, resembling those of the mesophyll, situated above and below the vein, but only the larger veins with a continuous lignified sheath of pitted sclereids. Vascular tissues very reduced. **Transverse veins** diffuse and inconspicuous at the same level as the longitudinal veins, each with a narrow sheath of elongated sclerotic cells. Ribs projecting adaxially, with a continuous sclerotic cylinder enclosing a large abaxial vb and smaller sometimes inverted vbs. **Stegmata** (Figs. 23B and 75N) common in short longitudinal files next to vascular and non-vascular fibres, each with an irregular spherical silica body enclosed by a conspicuously lignified and pitted thick basal wall.

PETIOLE

Not examined in detail.

Stem and root

Not examined in detail.

Note: see key in Note under Sclerospermeae.

TRIBE ORANIEAE (Fig. 76A–D, F, G, J)

Includes the single genus *Orania* (25 spp.) with a geographic disjunction between Madagascar and SE Asia. Palms are all single-stemmed and often tall components of the canopy. A number of spp. have distichous phyllotaxis. A crownshaft is not developed and inflorescence and flowers are described as unspecialized within the Areceae (Dransfield et al. 2008b). The seed and 'cabbage' are said to be poisonous, an unusual condition in palms, but which remains uninvestigated.

Leaf

Leaflets usually with a praemorse apex.

LAMINA

Dorsiventral (Fig. 76A). **Trichomes** (Fig. 76D) abundant on abaxial surface, largely in intercostal regions, each with a sl. sunken peltate group of 2(–4) columnar epidermal pitted cells capped by a tier of thinner-walled cells extending into a shield of often filamentous cells producing a thin-walled indumentum. **Epidermis** (Fig. 76B, C) somewhat thick-walled, especially the outer wall, the outer anticlinal walls penetrated by cuticular material appearing as conspicuous lens-shaped (discoid) structures in SV (Fig. 76B). Adaxial epidermal cells large, obliquely extended and rhombohedral, sometimes with files of wider cells above anticlinal hypodermal cell walls. Abaxial epidermis (Fig. 76C) with smaller cells but with similar discoid structures, especially in costal bands below larger veins; intercostals cells irregular and somewhat papillose. **Stomata** restricted to intercostal regions, sl. sunken and overarched by papillate protuberances from t.s.c. and neighbour cells forming an almost closed canopy over the g.c. (Fig. 76C); g.c. small, thin-walled, each with 2 cutinized ledges. **Hypodermis** 1(–2)-layered beneath each surface, the abaxial hypodermis shallow, thin-walled and rather

inconspicuous (Fig. 76A). Adaxial hypodermal cells sl. thick-walled, transversely obliquely extended and hexagonal in surface view. **Fibres** thick-walled varying from frequent to infrequent or rarely absent, often within the same leaflet; sometimes common in wide strands in the mesophyll (Fig. 76A), often at the level of the vbs (*O. macropetala*) or occasionally as single fibres in some parts of the lamina; rarely hypodermal (*O. trispatha*). **Mesophyll** with or without palisade layers but at least with adaxial 1–2 layers somewhat anticlinally extended and more compact than the rest of the mesophyll. **Veins** in abaxial mesophyll, usually independent of surface layers, the largest veins producing distinct abaxial ribs; almost all veins separated from hypodermis by at least 1 shallow mesophyll layer. O.S. usually incomplete below, inconspicuous and often with chloroplasts; I.S. of smaller veins only fibrous below and completed by sclerotic parenchyma. **Transverse veins** diffuse, equidistant from each surface but often extending above smaller longitudinal veins, sheathed by 1–2 layers of sclerotic parenchyma. **Midrib** prominent adaxially and including a sclerotic cylinder enclosing a large abaxial vb and several smaller adaxial vbs.

PETIOLE (Fig. 76F)

Hypodermis differentiated as a sclerotic layer. Peripheral vbs on abaxial side congested, each with a massive fibrous sheath, flattened towards the surface and enclosing 1 or more vbs. **Central vbs** not showing a distinct 'V' in section, each with 1(–2) wide mxy vessels and 2 separate phl strands. **Ground parenchyma** (Fig. 76F) including short air-canals as discontinuous cavities produced by collapse and separation of groups of cells; small fibre strands and small vbs common.

Stem (Fig. 76G)

Cortex with peripheral layer of sclerenchyma produced from an etagen-cambium. Leaf traces in cortex associated with groups of vbs, often inverted and with a massive phl sheath, probably representing inflorescence traces. Central cylinder with outermost vbs each developing a massive fibrous sheath, the vascular tissue reduced. Stem centre (Fig. 76G) with aerenchymatous ground tissue separating widely-spaced vbs with a limited phl sheath and usually of the 1-vessel type.

Root (Fig. 76J)

Surface layers losing epidermis and developing a sclerotic ring of thick-walled, pitted sclereids, transitional internally to a narrow, thin-walled layer that may proliferate in older or damaged roots. Outer cortex with a sclerotic layer of sclerenchymatous thick-walled cells transitional to middle cortex (Fig. 76J) with solitary 'Kentia-type' fibres or small fibre aggregates (but with blunt ends). Air canals becoming radially extended by breakdown of ground parenchyma cells. Stele not fluted. Endodermis with somewhat thickened inner walls, U-shaped in TS. Vascular tissue polyarch and embedded in thick-walled conjunctive tissue but with thin-walled medulla. Wider inner vessels embedded in conjunctive tissue but separated from peripheral mxy; medullary vessels absent.

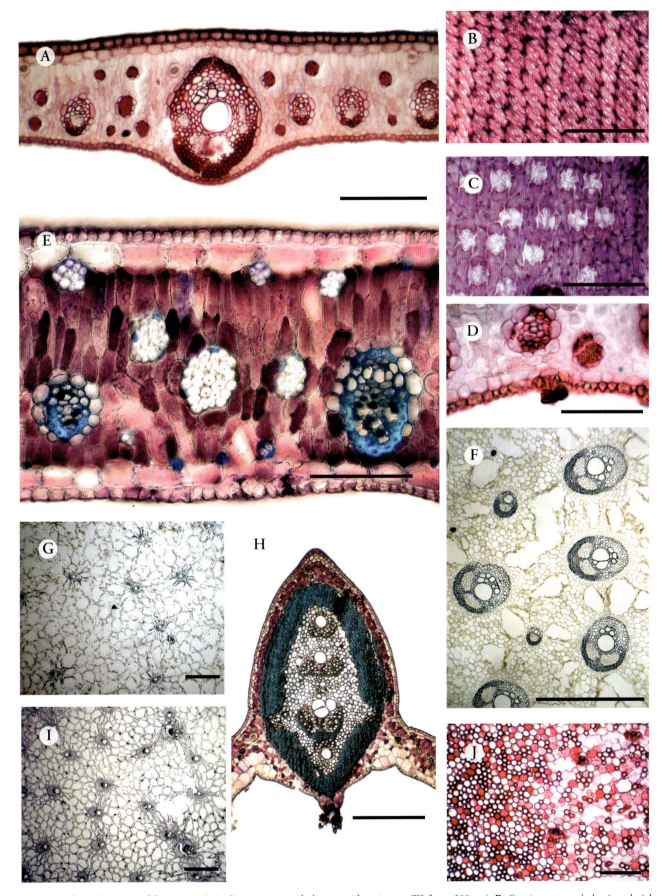

Fig. 76 Arecoideae (Oranieae and Roystoneeae). **A**. *Orania macropetala*, lamina with main vein, TS (bar = 200 μm). **B**. *Orania macropetala*, lamina adaxial epidermis with anticlinal wall thickenings, SV (bar = 100 μm). **C**. *Orania macropetala*, lamina abaxial epidermis, with anticlinal wall thickenings and lobed subsidiary cells, SV (bar = 100 μm). **D**. *Orania macropetala*, lamina abaxial epidermis with trichome, TS (bar = 100 μm). **E**. *Roystonea oleracea*, lamina, TS (bar = 100 μm). **F**. *Orania paraguanensis*, petiole vbs, TS (bar = 1 mm). **G**. *Orania paraguanensis*, stem centre, TS (bar = 2 mm). **H**. *Roystonea oleracea*, lamina midrib, TS (bar = 500 μm). **I**. *Roystonea regia*, stem centre, TS (bar = 2 mm). **J**. *Orania macropetala*, root cortex with scattered fibres, TS (bar = 200 μm).

Vascular tissue

Mxy vessels with scalariform perforation plates in all organs, those in the root with few thickening bars, in the stem with more numerous bars, and in the petiole with many bars. Sieve tubes with transverse to oblique end walls and compound sieve plates, those in the root on the longest and most oblique end walls, in the stem often of the 'arecoid-type.'

Cell inclusions

Stegmata with spherical silica bodies. Raphide sacs or raphide canals inconspicuous.

Note: the isolated status of this genus is supported by its many distinctive anatomical features, notable the peculiar lenticulate epidermis and stomata, aerenchyma type development in the petiole and massive 'satellite' vbs of the outgoing leaf traces of the stem. The relative uniformity and unspecialized nature of conducting elements could also support its early divergent status within the Arecoideae.

TRIBE SCLEROSPERMEAE (Figs. 23G, H and 75A–G)

Includes the single genus *Sclerosperma* with 1–3 spp. in W Africa (S Nigeria, Cameroon, and adjacent Gabon). It is an isolated genus but basal in the Arecoideae. The palm is adapted to the forest understorey, with a short erect stem, the leaves relatively large (to 3 m) and either simple but deeply emarginate or subdivided into broad multiplicate segments. Distinctive features are the trichomes, the thin-walled epidermis, the mesophyll without a palisade and hat-shaped silica bodies.

Leaf

Reduplicate, with a fibrous leaf sheath.

LAMINA

Dorsiventral (Fig. 75A). **Trichomes** (Fig. 75F) frequent on abaxial costal regions, each with a sunken base of 1–2 irregular tiers of cutinized cells extended into filaments of thinwalled cells, producing a scurfy covering to the leaf surface. **Epidermis** shallow, with a thin cutinized outer wall; adaxial epidermis (Fig. 75C, D) uniform, cells mostly rhombohedral and obliquely transversely extended. Abaxial epidermis (Fig. 75E, F) somewhat papillose, with obliquely extended narrow costal bands of rectangular cells, the intercostal bands of irregularly elongated cells with oblique end walls alternating with the stomatal files. **Stomata** in linear series within files of short narrow cells, with sl. sunken g.c., the outer chamber wax-occluded (Fig. 75G), they and the l.s.c. rather small, the t.s.c. also narrow. **Hypodermis** 1-layered below each surface (Fig. 75A, B), adaxial cells somewhat thick-walled, uniform, hexagonal, and transversely extended (Fig. 75D); abaxial cells less regular, smaller and thinner-walled, with small cells not uniformly arranged around s.st.ch. **Mesophyll** without distinct palisade layers but the adaxial 2 uppermost layers differentiated as a small-celled layer; remaining mesophyll of larger cells characteristically transversely extended and somewhat radially arranged around small veins (Fig. 75A, B). **Fibres** few, solitary or in pairs towards each surface but infrequently in contact with hypodermis. **Veins** equidistant and largely independent of surface layers except for 1–2 layers of small thin-walled sub-hypodermal cells. O.S. of conspicuous cubical cells, incomplete above and below. I.S. well-developed, fibrous above and below but completed laterally by sclerotic parenchyma. **Phl** of larger veins (Fig. 75B) divided into several strands by sclerotic partitions. **Transverse veins** at same level as longitudinal veins, few, narrow, and sheathed by a single layer of sclerotic parenchyma. **Ribs** with a sclerotic cylinder surrounding the central vascular tissue. **Expansion cells** in abaxial bands adjacent to adaxial ribs or occasional next to abaxial ribs.

Stem

Narrow, internodes short, congested; cortex narrow, with numerous massive fibrous strands and few small vbs; central cylinder not well differentiated from cortex and without marked concentration of peripheral vbs. Mxy with many rather narrow vessels or with 1 wide mxy vessel at stem centre. Phl undivided. Ground tissue not expanded; central fibre strands absent.

Root

Cortex lacunose; mucilage canals frequent in outer cortex; inner cortical cells often sclerotic but fibres absent. Stele with a fluted medulla including occasional isolated vessels.

Cell inclusions

Stegmata (Fig. 23G, H) with hat-shaped silica bodies. **Tracheary elements**: mxy elements in stem short, the perforation plates either simple or scalariform with few bars on irregular oblique to v. oblique end walls.

Notes: the leaf is unusual in the absence of a palisade mesophyll but with a lateral extension of the middle mesophyll cell layers. The abaxial trichomes require study in young leaf material.

The 2 genera of small palms in West Africa, *Sclerosperma* and *Podococcus*, occupy similar habits and confluent areas but can readily be distinguished as follows

1. Leaflets if present broad, linear, with a scurfy abaxial indumentums; trichomes with narrow thin-walled basal cells; epidermal cells conspicuously obliquely extended; hypodermis well-developed; mesophyll cells transversely extended; mesophyll fibres mostly solitary or in pairs . . . . *Sclerosperma*
2. Leaflets rhombohedral. without a permanent scurfy indumentums; trichomes with a single basal series of broad, pitted cells; epidermal cells rectangular, not markedly obliquely extended; hypodermis absent; mesophyll cells large but not transversely extended; mesophyll fibres in irregular strands including up to 8 fibres . . . . *Podococcus*

TRIBE ROYSTONEEAE (Figs. 8M; 17E–L; 72A, F; and 76E, H, I)

Including only the distinctive and horticuturally valuable genus *Roystonea* (Royal palms) with 10 spp. throughout the greater Antilles to northern South and Central America and considered native to S Florida (Zona 1990). The trees are imposing single-stemmed palms with a striking crownshaft (Fig. 72F) from which the leaves abscise precisely to leave a smooth columnar trunk with annular leaf scars. Infrafoliar inflorescences are conspicuous. Although fruits are used as foodstock for farm animals, the greatest value of royal palms lies in their use as ornamentals, especially as colonnades in urban landscaping and iconic of the tropical botanic garden (Fig. 72A).

Spp. delimitation is often imprecise, possibly because of hybridization.

Leaf (Figs. 8M and 76E, H)
Regularly and reduplicately pinnate, the leaflets extended in one or more planes.

LAMINA
Dorsiventral (Fig. 76E). **Trichomes** present, mostly abaxial, each with a sunken base expanded into a distal shield of thin-walled cells (Fig. 8M). **Epidermis:** outer wall sl. thickened, the cuticle thin. Adaxial cells rhombohedral to spindle-shaped, abaxial epidermis with narrow costal bands of similar cells and wider intercostal bands with less regular and somewhat more rectangular cells. **Stomata:** restricted to intercostal regions, t.s.c. not well-differentiated, but overarching g.c. somewhat. **Hypodermis** (Fig. 76E): conspicuously 1-layered below each surface, abaxial cells hexagonal, transversely extended, abaxial cells smaller. **Mesophyll:** with a well-developed 2-layered palisade. **Fibres:** abundant, usually in massive strands in the mesophyll and commonly next to the hypodermis, the strands larger and more numerous adaxially. **Veins** in abaxial mesophyll, independent of surface layers, O.S. well developed but incomplete around larger veins; I.S. fibrous above and below but with sclerotic parenchyma laterally; phl of largest veins sclerotic. **Transverse veins** at same level as smaller longitudinal veins, infrequent, narrow, sheathed by a single layer of sclerotic parenchyma. **Midrib:** (Fig. 76H) with a continuous sclerotic sheath enclosing several vbs. Expansion cells in abaxial strands 1 each side of midrib, inconspicuous elsewhere.

Note: the abundance of non-vascular fibres in the lamina of this genus is rather unusual compared with later divergent members of the subfamily.

PETIOLE
Central vbs with a distinct V-shaped arc, each usually with 1 wide mxy element, the phl indistinctly divided into 2 separate strands by sclerenchyma cells.

Stem (Figs. 17E–L and 76I)
Provides an extreme example of sustained primary growth, the peripheral outer layers of the central cylinder densely sclerotic in contrast to the highly aerenchymatous centre. **Surface layers** progressively replaced by sclerotic cells without forming a phellogen.

Cortex narrow and not delimited sharply from the central cylinder, the ground tissue cells becoming markedly extended tangentially as a result of overall stem expansion (Fig. 17F). **Central cylinder** with long-delayed maturation of bundle sheath fibres of outer axial vbs, the fibres early differentiated (Fig. 17E), becoming expanded but only gradually becoming thick-walled starting with the fibres next to the phl (Fig. 17F). This sequence also repeated in vbs in the subperipheral region (Fig. 17G) and the widely separated central vbs (Fig. 17H). **Ground tissue** cells of stem centre considerably extended horizontally except for tannin cells (Figs. 17H and 76I). **Vbs** mostly with a single wide mxy vessel and phl strand (Fig. 17J–L). This developmental sequence thus results in a diversity of vb and ground tissue cell types (Fig. 17I–L).

Root
Phellogen of etagen-type reported in hypodermal layers of above-ground roots of *R. oleracea* (Tomlinson 1961). **Fibres** sometimes present as solitary cells or fbs in inner cortex; innermost cortex including an incomplete cylinder of sclerotic cells; stele without vascular anomalies.

TRIBE REINHARDTIEAE (Figs. 5G; 12G; and 77A–F)

Includes only the genus *Reinhardtia* with 6 spp., mainly in Central America, all of them small to diminutive and characterized by undivided leaves. Splits along the folds in some spp. may develop incompletely as narrow perforations, hence 'window palms' (Fig. 5G). The smallest spp., with stems little more than pencil-thick, are often cultivated.

Leaf
Pinnately ribbed and apically notched. Leaf sheath marcescent, not forming a crownshaft.

LAMINA (Fig. 77A–C, F)
Dorsiventral (Fig. 77A). **Trichomes** (Fig. 77C) common on both surfaces, each with a large, deeply sunken top-shaped basal cell, the distal apparent shield of lobed thin-walled cells little persistent. **Epidermis** thin-walled, the outer wall sl. thickened and cutinized. Adaxial epidermis (Fig. 77B) without stomata; cells obliquely extended, the walls sometimes sl. sinuous. Abaxial epidermis (Fig. 77C) without costal bands, cells smaller, narrowly rectangular but with oblique end walls; not sinuous. **Stomata** v. diffusely distributed (Fig. 77C); g.c. long; l.s.c. conspicuous in surface view; t.s.c. short and wider than other epidermal cells. **Hypodermis** 1-layered and large-celled below each surface (Fig. 77A), the cells transversely extended and hexagonal (Fig. 77B), with 4-L-shaped cells around each s.st.ch. **Mesophyll** small-celled, palisade absent, the uppermost layer compact but scarcely anticlinally extended. **Fibres** unlignified, in mesophyll in 2 bands towards each surface and in contact with hypodermis, mostly solitary but also in larger strands of up to 8 cells at the level of the smaller veins (Fig. 77A). **Veins** diffuse, independent of surface layers, mostly narrow and with little vascular tissue; O.S. well differentiated but

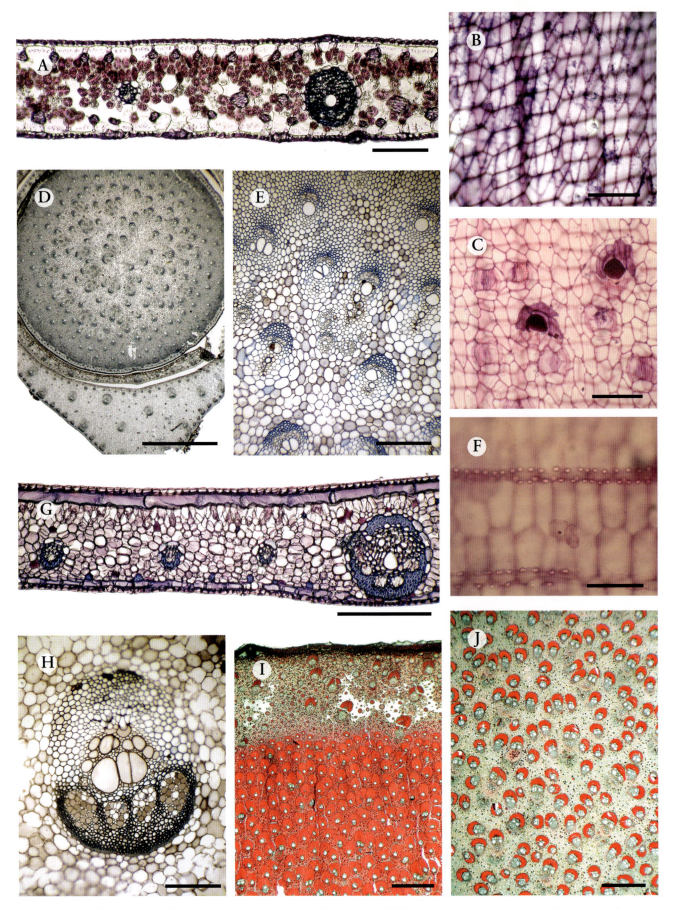

Fig. 77 Arecoideae (Reinhardtieae and Manicarieae). **A–F.** *Reinhardtia gracilis*. **A.** Lamina, TS (bar = 2 mm). **B.** Lamina, adaxial epidermis thin-walled, hypodermis thick-walled, SV (bar = 50 μm). **C.** Lamina, abaxial epidermis with stomata and trichomes, SV (bar = 50 μm). **D.** Stem with two surrounding leaf sheaths, TS (bar = 2 mm). **E.** Stem, centre to subperipheral regions, TS (bar = 200 μm). **F.** Lamina, adaxial stegmata over fb, SV (bar = 50 μm). **G–J.** *Manicaria saccifera*. **G.** Lamina, TS (bar = 200 μm). **H.** Petiole vb, TS (bar = 200 μm). **I.** Stem periphery, TS (bar = 2 mm). **J.** Stem centre, TS (bar = 2 mm).

chlorophyllous, usually incomplete below. I.S. reduced, mainly of thin-walled sclerenchyma but with few abaxial fibres. **Ribs** represented by a large vb with subdivided phl; sclerotic tissue of well-developed fibres above and below and including small adaxial vbs. **Transverse veins** few, narrow with a thin-walled sclerotic sheath. **Raphide sacs** elongated, in mid-mesophyll.

PETIOLE (Fig. 12G)
Very slender. Surface layer including a colourless hypodermis associated with frequent fbs. Central vbs forming an indistinct V-shape. Peripheral vbs becoming enclosed in a discontinuous sclerotic tissue of fused bundle sheaths, especially distal and in the 'keel' vb (Fig. 12G). Each central vb with a single narrow mxy vessel and 2 phl strands.

Stem (Fig. 77D, E)

Cortex v. narrow and with fbs, the hypodermal layers developing a phellogen. **Central cylinder**: outer vbs with massive fibrous sheaths commonly fused to form a continuous sclerotic cylinder (Fig. 77D). Vbs with a single narrow mxy vessel, the single phl strand often subdivided by a median sclerotic sinus. Ground tissue uniform, but peripheral cells sometimes inflated tangentially in outer layers. 'Satellite' vbs of leaf traces conspicuous (Fig. 77E).

Root

Epidermis of thin-walled large cells. **Cortex**: outer sclerotic layers including mucilage canals. Inner cortex including only solitary thick-walled fibres with wide lumina and radiating air-canals. **Endodermis** with U-shaped thickenings, the outer wall thin. Stele including a poorly differentiated medulla, with some central vessels.

Cell inclusions

Silica bodies (Fig. 77F) small usually ellipsoid, the base somewhat flattened, but not 'hat-shaped' in the usual sense, although this may be disputed (cf. Fig. 25A).

Vascular elements

Sieve tubes of stem with a mixture of both simple and compound sieve plates on transverse to sl. oblique end walls, exemplifying a transitional condition. Mxy **vessels** of leaf and stem with scalariform perforation plates on long oblique end walls.

TRIBE COCOSEAE

SUBTRIBE ATTALEINAE (Figs. 11K; 78; and 79)

Except for *Jubaeopsis* in South Africa and 2 genera in Madagascar this is a wholly American group; *Cocos* in somewhat problematic because of its pan-tropical distribution in cultivation. The assumption that it must have a New World origin is frequently challenged (Harries 1978), although Meerow et al. (2009) provide substantial support for this view and further indicate that *Cocos* diverged from

a common ancestor with *Syagrus* about 35 million years ago. Palms are mostly single-stemmed, tall to massive (*Jubaea*). Many spp. of the largest genus, *Syagrus* (31 spp.) are described as acaulescent, in relation to their occupancy of dry open areas. Here we briefly illustrate only the important genus *Cocos*, other genera are segregated on the basis of leaf anatomy in a synoptic key, but our sampling, especially of *Syagrus* is small. There is great need for more detailed study of this important group, using different approaches.

This subtribe is one of the more distinctive and diverse groups within the subfamily Arecoideae and of considerable economic importance. Although the tribe Cocoseae is well recognized by its fruit with a consistently developed stony endocarp bearing 3 surface pores or 'eyes', of which only 1 is functional in germination, its 3 constituent subgroups are much contrasted in structural features, (e.g. the spiny Bactridinae, the unarmed Attaleinae, and the Elaeidinae with condensed unisexual inflorescences), the tribe is clearly monophyletic. However, the Attaleinae retain some structural features that recall those of earlier diverging palm groups (non-vascular fibres with an association with surface layers, robust transverse veins, an occasional epidermis with sinuous walls and rectangular rather than rhombohedral cells).

Common diagnostic features for subtribe Attaleinae

1) **Lamina** thick, the mesophyll often many layers thick.
2) **Trichomes** with 1 or few basal cells, sunken within the epidermis to greater or lesser degree (Fig. 8N, P).
3) **Epidermis** usually with a thickened outer wall, especially adaxially.
4) **Hypodermis** well-developed, often 2–3-layered.
5) **Mesophyll** usually with a well developed palisade up to 4 cell layers deep.
6) Non-vascular **fibres** usually in contact with or within the hypodermis.
7) Longitudinal **veins** commonly in contact with one or both surfaces, the smaller veins abaxial and in contact with the abaxial hypodermis, the larger veins sometimes in contact with one or both surfaces via well-developed fibre girders.
8) **Transverse veins** wide, in mid mesophyll, extending above smaller longitudinal veins, with a well-developed fibre sheath (Fig. 11K).
9) **Stegmata** with spherical silica bodies.
10) **Petiole** vbs with 1 wide mxy vessel and 2 phl strands.
11) **Stem** (often massive) not showing extensive cell expansion of the ground tissue.
12) **Root** with an extensive sclerenchymatous outer cortex but without cortical fibres.

Diagnostic key to genera of Attaleinae based on leaf anatomy

1. Lamina isolateral, with equal numbers of stomata on each surface; larger vbs attached to each hypodermis by broad, tall fibre girders. . . . . *Butia* (Fig. 78E)

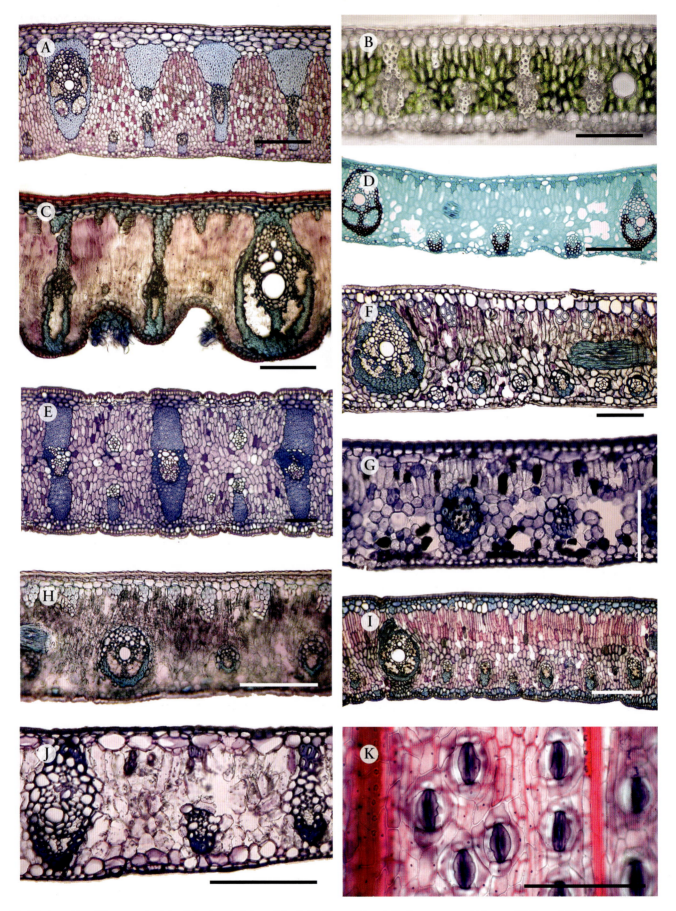

Fig. 78 Arecoideae (Cocoseae: Attaleinae), lamina. **A**. *Jubaea chilensis*, TS (bar = 200 μm). **B**. *Lytocaryum weddelianum*, TS (bar = 100 μm). **C**. *Parajubaea cocoides*, TS (bar = 200 μm). **D**. *Allagoptera caudescens*, TS (bar = 200 μm). **E**. *Butia capitata*, TS (bar = 100 μm). **F**. *Voanioala gerardii*, TS (bar = 200 μm). **G**. *Beccariophoenix madagascariensis*, TS (bar = 100 μm). **H**. *Syagrus orinocensis*, (bar = 100 μm). **I**. *Attalea speciosa*, TS (bar = 200 μm). **J**. *Jubaeopsis caffra*, TS. (bar = 200 μm). **K**. *Syagrus coronata*, abaxial epidermis, SV (bar = 100 μm).

1A. Lamina dorsiventral, stomata absent from or few on adaxial surface, large vbs variously attached to the hypodermis; smaller vbs to the abaxial surface layers 2

2. Abaxial surface deeply ribbed, abaxial vbs within ribs, alternate with furrows . . . . *Parajubaea* (Fig. 78C)

2A. Abaxial surface not ribbed. 3

3. Adaxial hypodermis 3–4 cell layers deep; larger vbs attached to both leaf surfaces by fibre girders, adaxial girders often massive; non-vascular fbs absent . . . . *Jubaea* (Fig. 78A)

3A. Adaxial hypodermis 1–2 cell layers deep, larger vbs without massive fibre girders, non-vascular fbs present. 4

4. Adaxial hypodermis mostly 2 cell layers deep, including fbs . . . . *Attalea* (Fig. 78I)

4A. Adaxial hypodermis ± 1 cell layer deep, fb adjacent to, not within hypodermis 5

5. Adaxial hypodermis strictly 1 layer deep, abaxial fibres immediately below and with wide septate lumina, abaxial fibres few or absent . . . . *Voanioala* (Fig. 78F)

5A. Adaxial fibres with narrow non-septate lumina, abaxial fibres varying from frequent to few 6

6. Hypodermis strictly uniseriate; fibres few 7

6A. Adaxial hypodermis 1–2 cell layers deep, outer layer continuous and conspicuous, somewhat thick-walled, inner layer discontinuous thin-walled, including fibre strands 8

7. Mesophyll shallow, c. 6 cells wide, smaller veins abaxially in contact with the hypodermis. . . . . . *Lytocaryum* (Fig. 78B)

7A. Mesophyll deeper, fibres scattered in mesophyll, adaxial hypodermal cells ± rectangular . . . . . *Beccariophoenix* (Fig. 78G)

8. Veins mostly attached to adaxial hypodermis by narrow bands of fibres with wide septate lumina, non-vascular fibres similar . . . . . . (Robertson 1978) *Jubaeopsis* (Fig. 78J)

8A. Veins infrequently attached to hypodermis, if so the fibres without wide lumina. 9

9. Mesophyll cells anticlinally extended, palisade 3–4-layered . . . . *Cocos* (Fig.79A, B)

9A. Mesophyll cells not anticlinally extended, palisade 1–2-layered . . . . *Allagoptera* (Fig. 78D), *Syagrus* (Fig. 78K)

Note: this key will separate the more distinctive genera, as in the early part of the key, but inaccuracies are likely found in the later entries, notably and obviously in larger genera where little material has been examined and variation among spp. is likely to overlap that between genera. On the other hand, the key should provide a guide for use in later detailed study.

The extent of variation within a genus is best exemplified by *Syagrus*, the largest genus in the Attaleinae with over 30 spp. The following features illustrate differences among the material we have examined. Glassman (1972) studied the lamina anatomy of *Syagrus* extensively, emphasizing many quantitative features, but without taking into account the variation within leaflets of a single leaf or different leaves of 1 sp., least of all different individuals of 1 sp. Nevertheless this work is still usefully consulted by anyone working on this genus.

Anatomical features that vary in Syagrus

1) Abundance and distribution of trichomes.

2) Adaxial epidermal cells varying from rectangular, with distinctly sinuous anticlinal walls to rhombohedral, obliquely extended cells with straight or at most undulate walls.

3) The thickness of the outer epidermal wall, varying from very thick with the cell lumen scarcely one tenth its depth, to thinner walls only 2–3 times thicker than the remaining walls, the cell lumen large.

4) The abundance and location of adaxial non-vascular fbs, varying from an almost continuous layer within the hypodermis, to few fibres.

5) The extent and location of the abaxial non-vascular fibres.

6) The extent to which the smaller abaxial veins are in contact with the abaxial hypodermis.

7) The degree to which the inner sheath of larger veins develops fibrous extensions to the upper surface layers.

Despite this uncertainty, associations of features may be indicative either of relationships or isolated taxa. These include:

1) The surprising structural divergence between the sister genera *Butia* and *Jubaea*.

2) Relative similarity of most of the South American genera, but not *Lytocaryum* (perhaps on account of its relatively small size).

3) The many leaf anatomical character states that are shared between *Cocos* and *Syagrus*.

4) The way in which *Jubaeopsis* (South Africa) resembles *Voaniola* from Madagascar.

5) The basic features in which the Madagascan genera conform to the essential features of the whole subtribe, i.e. are among themselves less distinct from each other than from some of the geographic congeners in South America, as in (1) above.

This leads to the conclusion that lamina anatomy could be useful in explaining the evolutionary history of the subtribe, and indeed our observations are highly congruent with the phylogenetic hypothesis of Meerow et al. (2009).

Cocos (the coconut palm) (Fig. 5H; 8N; 12H; 72E: and 79)

The single sp. *Cocos nucifera* L., the most iconic of all tropical trees (Fig. 72E), a major contributor to the economy of tropical countries as a source of vegetable oil, food, building materials, and as an ornamental, but with many other local uses. It is distinguished within the Attaleinae by its large female flower and fruit. Many cultivars exist, but its origin and natural range are unknown (Harries 1978) even though it is clearly adapted to coastal and frequently saline conditions because of its large fibrous drupe, appropriate for long-distance marine dispersal. Despite its importance many aspects of coconut biology are unknown.

Leaf

Large. Regularly pinnate with opposite, single-fold reduplicate leaflets; seedling leaves entire and often somewhat fenestrate.

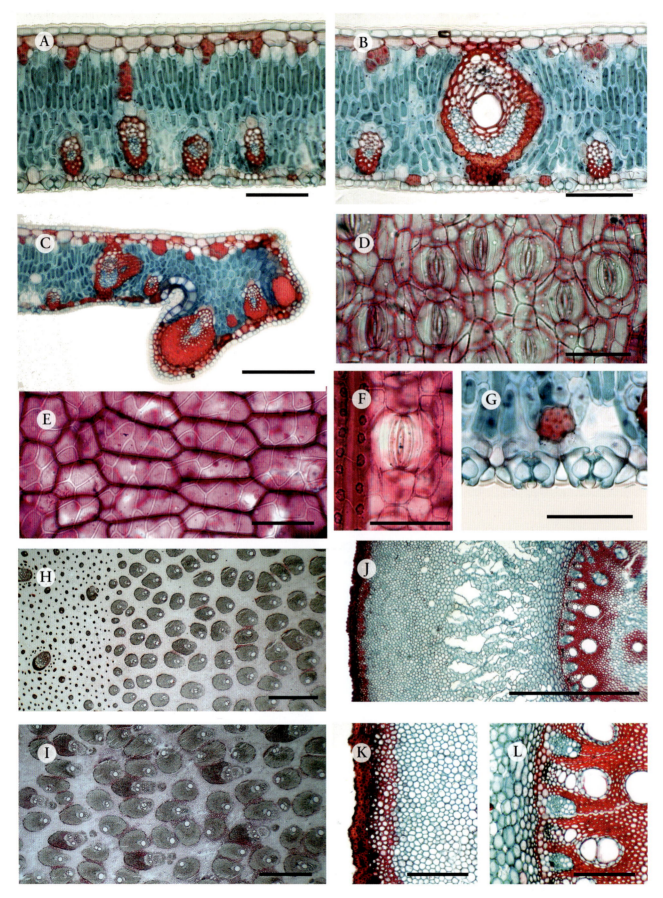

Fig. 79 Arecoideae (Cocoseae: Attaleinae). *Cocos nucifera*. **A**. Lamina, TS (bar = 100 μm). **B**. Lamina with major vein, TS (bar = 100 μm). **C**. Lamina margin, TS (bar = 250 μm). **D**. Lamina, abaxial epidermis, SV (bar = 50 μm). **E**. Lamina, adaxial epidermis (light walls) and hypodermis (dark walls), SV (bar = 50 μm). **F**. Abaxial epidermis with stomata and stegmata above fb, SV (bar = 50 μm). **G**. Stomata in abaxial epidermis, TS (bar = 50 μm). **H**. Stem, peripheral region, (bar = 2 mm). **I**. Stem, central region, TS (bar = 2 mm). **J**. Root, periphery to stele, TS (bar = 1 mm). **K**. Root, periphery and outer cortex, TS (bar = 200 μm). **L**. Root, inner cortex and stele, TS (bar = 200 μm).

Leaves abscising to leave a clean scar, the leaf-base becoming fibrous ventrally and splitting along the margin of the dorsal part of the axis to produce a material resembling sacking, the free portion thus becoming a pseudopetiole (Tomlinson 1964)

LAMINA (Figs. 8N and 79A–G)

Dorsiventral (Fig. 79A, B). **Trichomes** restricted to abaxial costal regions, each with a sunken base of 3–4 cutinized inflated cells extending into thin-walled ephemeral cells (Fig. 8N). **Epidermis** with outer wall thickened, wholly cutinized; adaxial cells (Fig. 79E) irregular but cubical to longitudinally extended, often with oblique end walls. Abaxial cells shallower with narrow bands of longitudinally extended cells and shorter irregular cells with narrow associated stomata in 1–3 series (Fig. 79D). **Stomata** (Fig. 79F, G) not sunken; g.c. thin-walled each with 2 prominent cutinized ledges plus possibly smaller inner ledges. **Hypodermis** mostly 2-layered large-celled adaxially, the inner layer somewhat discontinuous, the files often replaced by fbs; 1-layered, small-celled abaxially. Adaxial outer layer of transversely-extended, hexagonal cells (Fig. 79E). **Mesophyll** with a well-developed 3-layered adaxial palisade, other mesophyll cells often somewhat palisade-like. **Fibres** frequent, non-septate and in irregular strands mostly within the adaxial hypodermis, infrequent abaxially and often replacing the hypodermal layer. **Veins** mostly in abaxial mesophyll and often in contact with the hypodermis (Fig. 79A), but the largest (Fig. 79B) attached to each hypodermis by either unlignified extensions of the I.S. or completed by lignified sclerenchyma. O.S. indistinct, incomplete below small veins, incomplete above and below the large veins, the cells cubical or anticlinally extended. I.S. incompletely fibrous around large veins but of sclerotic cells above small veins. Phl of large veins divided by sclerotic partitions into 2 larger abaxial strands and often 2 smaller lateral strands. **Transverse veins** equidistant from each surface and usually extending above small longitudinal veins; frequent, wide and sheathed by numerous fibres. **Midrib** with a well-developed fibrous cylinder; expansion cells in conspicuous abaxial bands 1 each side of midrib; **marginal vein** somewhat enlarged (Fig. 79C).

PETIOLE

Central vascular vbs without a distinctive V-arrangement, each usually with 1 wide mxy vessel and 2 phl strands. Ground parenchyma uniform, with fibrous strands common.

LEAF SHEATH (Figs. 12H and 79H)

Large vbs on ventral side forming the major strands of the sacking (Fig. 12H).

Stem (Figs. 5H and 79H, I)

Solitary, tall, and frequently curved, with pronounced dorsal leaf scars, internodes short. Texture relatively uniform (Fig. 5H), the ground parenchyma not secondarily expanded but including scattered narrow fibrous strands. **Cortex** with few vbs but numerous fibre strands, with an abrupt transition to the central cylinder (Fig. 79H). **Stele**: central vbs not widely spaced (Fig. 79I), each with a well-developed fibrous phl sheath; commonly with 2 wide mxy vessels centrally, but 1 peripherally; phl undivided.

Root (Fig. 79J–L)

Surface layers (Fig. 79J, K) eroding with age, leaving a well-developed sclerotic layer; outer cortical cells narrow. **Cortex** lacunose (Fig. 79J), without cortical fibres, mucilage canals frequent but narrow; innermost cortical cells becoming sclerotic. **Stele** (Fig. 79J) medullated with some medullary vessels; endodermis with U-shaped wall thickenings (Fig. 79L).

Cell inclusions

Stegmata conspicuous in lamina (Fig. 79F) and stem, the silica body irregularly spherical; absent from roots.

Vascular elements

Vessels in all parts; in roots with simple ± transverse perforation plates; in stem with simple or scalariform plates on somewhat oblique end walls; in leaf with vessel elements up to 8 mm long and scalariform oblique perforation plates with many thickening bars. **Sieve tubes** in all parts with compound sieve plates, but most oblique in the root and least oblique in the petiole.

SUBTRIBE BACTRIDINAE (Figs. 23D, E and 80)

This group of exclusively American palms is probably the most heavily-armed of all members and so is sharply contrasted with the sister-group Attaleinae, which is unarmed. **Spines** are outgrowths of the surface layers, usually directed apically, but with a basal cushion of parenchyma cells that erects them, usually leaving a significant shallow imprint in any surface to which they are initially impressed. They are typically black and so conspicuous at maturity. Although they are not the homologue of any other organ, the larger spines are frequently vascularized. They can occur on the surface of any part of the palm except the roots, i.e. lamina, rachis, petiole, outer surface of the leaf sheath, stem internode, inflorescence branches, fruits, but not flowers. Only in *Desmoncus* can spines serve a function other than for defence in the exceptional **acanthophylls** (Fig. 24N, O), which are the reflexed distal leaflets of the mature foliage leaves that are reduced to spines but function as grapnels and so facilitate the climbing habit. Our account outlines anatomical features for the group as a whole, but with more detailed descriptions of *Bactris*, the largest genus (73 spp.), which serves to contrast the anatomy of other genera, particularly *Desmoncus* which is its sister group of climbing palms.

Acrocomia, Aiphanes, and *Astrocaryum* are of similar habit and differ from each other only in small quantitative details, but more comparative study is necessary.

Anatomical features of the subtribe Bactridinae

1) Trichomes with a narrow base of linear cutinized or sclerotic cells extended into a distal filament.
2) Adaxial epidermal cells obliquely extended, usually thin-walled.

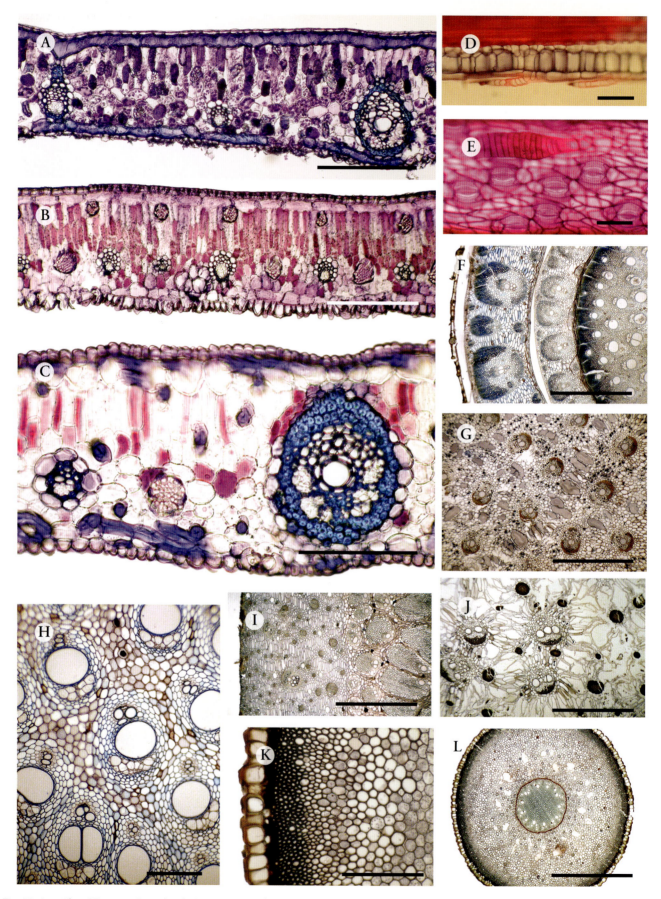

Fig. 80 Arecoideae (Cocoseae: Bactridinae). **A**. *Astrocaryum alatum*, lamina, TS (bar = 200 μm). **B**. *Acrocomia aculeata*, lamina, TS (bar = 200 μm). **C**. *Bactris major*, lamina, TS (bar = 100 μm). **D**. *Bactris campestris*, lamina epidermal trichomes, TS (bar = 50 μm). **E**. *Bactris gasipaes*, lamina epidermis with trichome, SV (bar = 50 μm). **F**. *Desmoncus prestoei*, stem and leaf sheaths, TS (bar = 1 mm). **G**. *Bactris* sp., petiole centre, TS (bar = 1 mm). **H**. *Desmoncus polyacanthos*, stem centre, TS (bar = 200 μm). **I**. *Bactris gasipaes*, stem periphery with radial cell expansion, TS (bar = 1 mm). **J**. *Bactris acanthocarpoides*, stem centre with fbs and vbs, TS (bar = 1 mm). **K**. *Bactris simplicifrons*, root periphery, TS (bar = 200 μm). **L**. *Bactris simplicifrons*, root, TS (bar = 1 mm).

3) Hypodermis 1-layered below each surface, the adaxial cells conspicuous, hexagonal and transversely extended; s.st.ch. commonly surrounded by several small cells.
4) Mesophyll fibres mostly in many-celled strands of narrow fibres with a v. narrow lumen, the strands never within the hypodermal layers.
5) Veins mostly free of the hypodermal layers.
6) Petiole with pronounced V-configuration, central vbs each with 1 wide mxy element and 2 phl strands.
7) Stem commonly with wide fbs among the central vbs (Fig. 80J).
8) Stegmata with hat-shaped silica bodies (Fig. 23D, E).
9) Root possibly without cortical fibres (Fig. 80L).

Note: *Bactris*, as would be expected from its large numbers of spp., shows most variation in lamina characters, including the frequent presence of fibre sclereids in the mesophyll, as indicated later. *Desmoncus*, despite its climbing habit is not particularly distinctive in its leaflet anatomy.

Acrocomia (Fig. 80B)

LAMINA

Dorsiventral (Fig. 80B). **Trichomes** with a distal multiseriate filament of thin-walled but rigid cells. **Epidermis** with transversely-obliquely extended cells, abaxial epidermis with somewhat papillose cells in the intercostal regions. **Stomata** with short t.s.c. overarching the g.c. somewhat. **Hypodermis** 1-layered beneath each surface, the adaxial cells hexagonal and transversely extended (Fig. 80B). **Mesophyll** with a 2–3-layered palisade. **Fibres** (Fig. 80B) in 2 distinct series, adaxially as large multicellular strands next to the adaxial hypodermis and a series of somewhat larger strands at the level of and alternating with the smaller veins. **Veins** in abaxial mesophyll independent of surface layers except larger veins attached to each surface by small somewhat thick-walled mesophyll cells.

Astrocaryum (Fig. 80A)

LAMINA

Resembles *Acrocomia*; as illustrated (Fig. 80A) it may differ in the thicker-walled hypodermis and greater frequency of non-vascular fbs.

Aiphanes

LEAF

Leaflets have praemorse tips, a distinctive generic character.

LAMINA

Differs from the above 2 genera in the relative absence of the adaxial series of fbs.

Bactris (including *Guilielma*, *Yuyba*)

(Fig. 80C–E, G, I–L)

A large genus of at least 70 spp. (Henderson 2000). It ranges from the West Indies to Central and South America. *Bactris*

(*Guilielma*) *gasipaes* (peach palm) is widely cultivated for its fleshy fruit, eaten after being boiled (Ferreira 1999). Habit and size is variable, but most spp. have multiple stems, which are often cane-like. Spines are abundantly developed and may be transitional to filamentous microscopic hairs.

LEAF

Pinnately compound, but sometimes simple with a deeply bifid apex (e.g. *B. militaris*) (Salzman and Judd 1995).

LAMINA (Figs. 8A and 80C–E)

Dorsiventral (Fig. 80C). **Trichomes** (Fig. 8A; 80D, E) common, either with a short uniseriate base of thick-walled cells extending into a thin-walled filament or with a larger multiseriate base with a distal filament. **Epidermis** usually thin-walled and with a thin cuticle. Adaxial epidermis with rhombohedral cells and files of wider cells above anticlinal hypodermal walls, alternating with 2–3 files of narrower cells. Abaxial epidermis with costal bands of rectangular or rhombohedral cells below the few large veins, intercostals cells irregular. **Stomata** infrequent adaxially, diffuse and not in distinct files abaxially; t.s.c. short, wide and sometimes enlarged; g.c. large, with 2 usually equivalent cutinized ledges. **Hypodermis** 1-layered below each surface. Cells uniformly hexagonal and transversely extended, but irregular abaxially; s.st.ch surrounded by several thin-walled cells. **Mesophyll** usually without a distinct palisade layer, the number of layers varying from as few as 4 to 9 or more, this largely determining lamina thickness. **Fibres** always narrow, unlignified, the lumen almost occluded, rarely single, but usually forming strands of up to 50 cells (Fig. 80C), the larger circular in TS. Fibre strands restricted to the mesophyll and vary in abundance and position (e.g. whether or not in contact with the hypodermis) in diagnostic ways (see later). Independent **fibre sclereids** (Fig. 80C) in some spp. abundantly developed and ramifying throughout the mesophyll. **Veins** diffuse, situated in middle or abaxial mesophyll, largely independent of surface layers, i.e. without bundle sheath extensions. O.S. well-developed, usually chlorenchymatous, complete only around smaller veins. I.S. lignified, with fibres completed laterally by sclerotic pitted cells in larger veins, in smaller veins usually a single fibre layer, phl of larger veins often with a reticulum of sclerotic cells. **Transverse veins** infrequent, with a narrow sheath of sclerotic cells; equidistant from each surface. **Midrib** with a well-developed sclerotic cylinder enclosing a central large abaxial vb and smaller peripheral vbs inverted adaxially. **Expansion cells** in 2 bands each side of the midrib, infrequent elsewhere. **Raphide sacs** large and frequent.

Specific diagnoses

As a large genus *Bactris* is little known anatomically, but specific differences do exist. The earlier supposition that fibre sclereids were diagnostic for the genus as a whole (Tomlinson 1961) is incorrect, as is made most obvious by the inclusion of *Guilielma* within *Bactris* (Henderson 2000). There is appreciable anatomical diversity illustrated by the following synoptic key based on our limited sampling

1. Epidermis and hypodermis large-celled; mesophyll > 8 cells deep, adaxial palisade well-developed . . . . *B. campestris*
(as *B. savannarum*)
1A. Epidermis and hypodermis small-celled; mesophyll < 4–8 cells deep, palisade absent or inconspicuous 2
2. Fibre sclereids abundant in mesophyll. . . . *B. acanthocarpa,*
B. cuspidata (as *B. mitis*), *B. guineensis* (as *B. minor*), *B. major*
2A. Fibre sclereids absent. 3
3. Fbs uniformly large, not in contact with hypodermis
. . . . *B.* (= *Guilielma*) *gasipaes*
3A. Fbs of varying diameter, sometimes in contact with hypodermis. 4
4. Fbs narrow (2–12 cells), almost all in contact with each hypodermis. . . . . *B.* (*Yuyba*) *simplicifrons* (as *B. trinitensis*)
4A. Fbs ranging widely in diameter, often large (40+ cells), infrequently in contact with each hypodermis 5
5. Fbs in central mesophyll large, plus an adaxial series (2–4 cells wide) next to adaxial hypodermis . . . . *B.* sp. (HEM 8055)
5A. Fbs ± uniformly distributed from narrow (few cells) to wide (30–40 cells) . . . . *B. longiseta*

Other features which show specific variation include flat versus papillose epidermal cells and the frequency and size of raphide sacs.

PETIOLE (Fig. 80G)

V-pattern well developed. Central vbs with 1 wide mxy vessel and 2 phl strands. Ground tissue sometimes including irregular air lacunae produced by cell necrosis (Fig. 80G).

Stem (Figs. 23I and 80I, J)

Cortex with frequent fbs, cortical ground tissue in older stems becoming tangentially extended (Fig. 80I). **Central cylinder** with the collective massive sheaths of the peripheral vbs becoming, with age, extremely sclerotic through lignification. Central vbs diffuse, each with 2 or more wide mxy vessels and 1 phl strand, **ground tissue**, including numerous v. wide fibrous strands (Fig. 80J) and mucilage canals (Fig. 23I), becoming lacunose through expansion and collapse of parenchyma cells.

Root (Fig. 80K, L)

Epidermis (Fig. 89K) large-celled, the outer wall sl. thickened, hypodermal layer small-celled tanniniferous, **outer cortex** sclerotic and including wider cells as presumed raphide canals, the sclerotic layer gradually transitional to the uniform **inner cortex**. Fibres absent. Ground tissue in the material illustrated with abundant starch. Innermost cortex with irregular air lacunae. **Stele** without a parenchymatous medulla (Fig. 80L). **Endodermis** with thick U-shaped wall thickenings almost occluding the cell lumen.

Desmoncus (Figs. 8B; 23K, L; 24N, O; and 80F, H)

A genus of New World palms with a climbing habit that parallels the rattans of the Old World. The spp. number is not established precisely but is upwards of 30. Aerial parts are spiny and generally unsympathetic of botanical collectors.

Leaf

Pinnate, the leaflets somewhat irregularly arranged, often narrowed at the insertion; rachis terminating in a cirrus of modified paired leaflets (acanthophylls) that become reflexed as grapnels (Fig. 24N, O). This parallels the similar structure in the West African subtribe Ancistrophyllinae. Leaf sheath persistently tubular and extending into a long tubular ligule that shreds with age.

LAMINA

Dorsiventral. **Trichomes** infrequent in costal regions of abaxial surface, occasional adaxially, varying from those with a uniseriate cutinized and thick-walled base (Fig. 8B), to a similar but multiseriate and sl. sunken base, both types extended into distal thin-walled filaments. **Epidermis** thin-walled, cutinized layer thin, but deeply penetrating the anticlinal walls, undulate or at most sl. sinuous. **Adaxial epidermis** uniform, mostly without distinct costal regions, cells elongated, rectangular or usually rhombohedral, with oblique end walls; cells above anticlinal hypodermal walls sometimes forming files of sl. wider and shorter cells. **Abaxial epidermis** with narrow bands of elongated rectangular to rhombohedral cells alternating with less regular stomatal bands. **Stomata** mostly absent adaxially; diffusely distributed and not in distinct files; neighbouring cells evident; t.s.c. short, wide. **Hypodermis** thin-walled, 1-layered below each surface, the abaxial cells smallest. Adaxial hypodermal cells hexagonal and transversely extended. Abaxial hypodermis shallow, the s.st.ch. surrounded by 4–6 small cells. **Mesophyll** uniform, palisade scarcely developed. **Fibres** narrow, unlignified and in small strands, circular in TS, forming 2 distinct series towards each surface, but rarely in contact with hypodermis. **Veins** diffuse, always independent of surface layers; small veins situated abaxially; larger veins few. O.S. complete around smaller veins but interrupted above and below larger veins. I.S. represented in smaller veins by limited sclerotic parenchyma; forming a discontinuous sclerotic layer around larger veins. Vascular tissue reduced, except in larger veins, the phl then sometimes divided by a median sclerotic partition. **Transverse veins** infrequent, narrow, equidistant from each surface and often extending above smaller longitudinal veins, sheathed by narrow thin-walled sclerotic cells. **Stegmata** with hat-shaped silica bodies, numerous and conspicuous in files next to non-vascular fibres, but v. small, the basal wall thickened and lignified in contrast to adjacent fibres. **Raphide sacs** locally common, often next to hypodermis (Fig. 23L); conspicuously larger than mesophyll cells (Fig. 23K). Expansion cells in shallow bands, represented by inflated hypodermal cells of both surfaces. **Midrib**, including a complete sclerotic cylinder with 1 larger vb and 1 smaller inverted bundle.

LEAF SHEATH

With a single series of large vbs, the bundle sheath well-developed on the abaxial side (Fig. 80F).

Stem (Fig. 80F, H)

Although the stems of *Desmoncus* show many parallel features to rattans (narrow cortex, single wide mxy vessels in the vbs) they are not used commercially in the same way because they

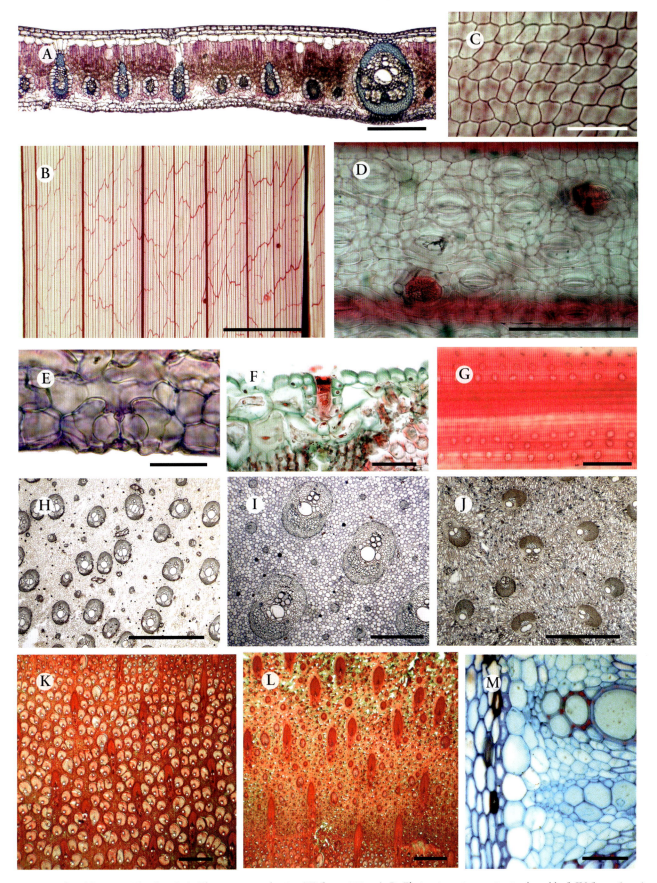

Fig. 81 Arecoideae (Cocoseae: Elaeidinae). **A**. *Elaeis guineensis*, lamina, TS (bar = 200 μm). **B**. *Elaeis guineensis*, venation in cleared leaf, SV (bar = 5 mm). **C**. *Elaeis guineensis*, lamina adaxial epidermis, SV (bar = 50 μm). **D**. *Elaeis guineensis*, lamina abaxial epidermis, SV (bar = 100 μm). **E**. *Elaeis guineensis*, lamina abaxial epidermis with stomata, TS (bar = 25 μm). **F**. *Elaeis oleifera*, lamina adaxial epidermis with trichome, TS (bar = 50 μm). **G**. *Elaeis guineensis*, lamina fbs with silica bodies, SV (bar = 50 μm). **H**. *Elaeis guineensis*, petiole centre, TS (bar = 2 mm). **I**. *Elaeis guineensis*, rachis central vbs, TS (bar = 500 μm). **J**. *Elaeis guineensis*, stem centre, TS (bar = 2 mm). **K**. *Elaeis guineensis*, stem centre–cortex boundary, TS (bar = 2 mm). **L**. *Elaeis guineensis*, stem cortex, TS (bar = 2 mm). **M**. *Elaeis guineensis*, root, periphery of stele, TS (bar = 50 μm).

show more limited modification. Consequently the texture of the stems is not constant throughout.

As shown by Tomlinson and Zimmermann (2003) these differences are:

1) Outer vbs with massive fibrous sheaths (Fig. 80F).
2) Central vbs more diffusely distributed than peripheral vbs.
3) Central vbs with a single phl strand (Fig. 80H).
4) Central ground tissue not becoming uniformly sclerotic.
5) Transverse commissures connecting axial vbs absent.
6) Most significantly, the '*Rhapis*-model' of vascular continuity is still recognizable, unlike the discontinuities reported for *Calamus* (Tomlinson et al. 2001; Tomlinson and Spangler 2002).

SUBTRIBE ELAEIDINAE (Fig. 81)

Includes 2 genera, *Barcella* (Brazil) and *Elaeis*, with 2 spp., *E. oleifera* in the Americas and *E. guineensis* (the African oil palm) in equatorial Africa. The 2 spp. of *Elaeis* are contrasted in quantitative features, e.g. few or no non-vascular fibres in *E. oleifera*, but in the absence of more extensive study this distinction may not be constant. They share distinctive features of unisexual flowers borne in deep pits in the rachillae and similar fruits with the persistent woody 3-lobed stigmas. Our understanding of them is contrasted as *Barcella* is little studied and unfamiliar to most palm biologists, whereas *Elaeis* is investigated extensively at research institutions devoted exclusively to its study. *Elaeis* has spiny parts, e.g. basal leaf spines, rachilla tips and even the woody recurved persistent stigmas.

Economic uses

Elaeis guineensis (West African oil palm) is extensively cultivated outside its native Africa as a plantation crop. It produces the largest amount of vegetable oil of any tropical plant. Oil is extracted from both the endosperm (kernel oil) and mesocarp (pericarp oil), which are usually kept separate. The palm has other local uses apart from a source of cooking oil. Unfortunately most of the cultivated area is at the expense of natural primary vegetation and has resulted in the destruction of vast areas of rain forest. Its natural occurrence seems to have been determined by human activity for a very long time. In African rain forests its presence may be interpreted as a sign of previous cultivation, whereas its true ecological status is uncertain, but possible it is a riparian sp.

Elaeis (Figs. 19A–E; 24K; and 81)

LEAF

Pinnately compound; leaflets reduplicate and lanceolate, held at differing angles to give a plumose aspect, each with an adaxial midrib. Basal leaflets reduced to bulbous-based midrib spines, which are continuous below with the marginal spines resulting from persistent large vbs of the leaf sheath to produce a short but effectively armed pseudopetiole (Fig. 24K).

LAMINA (Fig. 81A–G)

Dorsiventral (Fig. 81A). **Trichomes** (Fig. 81D, F) frequent on both surfaces, but especially abaxial, each with a sunken flask-shaped base of 1 or more pitted cells extending into a short distal filament of ephemeral thin-walled cells. **Epidermis** thin-walled, the outer wall only sl. thickened; cuticle thin. Adaxial epidermis (Fig. 81C) with isodiametric or polygonal cells with undulate walls, the transverse end walls commonly oblique, the angle of obliquity often changing over short domain distances. Abaxial epidermis (Fig. 81D) of smaller, shallower cells with few costal regions of rectangular or polygonal cells, the intercostal cells irregular. **Stomata** (Fig. 81E) restricted to abaxial intercostal regions, t.s.c. short wide, g.c. each with 2 prominent cutinized ledges, the wall adjacent to the pore sl. thickened. **Hypodermis** 1–2-layered beneath each surface; adaxial outer layer of more or less hexagonal transversely extended cells, inner layer of cubical cells with thinner wall; abaxial cells cubical in costal regions, intercostal cells irregular and longitudinally extended, with 4 L-shaped cells around each s.st.ch. **Mesophyll** with a well-developed mostly 2-layered palisade. **Fibres** variable, but commonly in large strands mostly at the level of the veins, or as solitary fibres elsewhere, but fibres almost absent from some samples. **Veins** (Fig. 81A) in abaxial mesophyll and independent of surface layers except for few larger veins attached to each surface by 1–3 layers of small colourless cells. O. S. of cubical cells complete around smaller veins but incomplete above and below largest veins. I. S. fibrous above and below larger veins; smaller veins with fewer fibres or even sheathed only by sclerotic parenchyma. **Phl** of largest veins irregularly sclerotic or divided into several separate strands. **Transverse commissures** (Fig. 81B) equidistant from each surface, sheathed by short fibres, with an irregular transverse course. **Midrib** with a sclerotic cylinder, including small embedded vascular strands, enclosing a large abaxial and smaller adaxial vbs. **Expansion cells** in 2 abaxial bands, 1 each side of the midrib.

LEAF SHEATH

Long persistent but ventrally becoming loosely fibrous with age, the dorsal side persistent but eventually abscising.

PETIOLE (Fig. 81H, I)

Peripheral vbs not confluent but each with a massive fibre sheath; central V-configuration absent. Central vbs including 1 wide mxy vessel and 2 phl strands (Fig. 81I). Ground parenchyma uniform, without air lacunae but including frequent fibre strands and small vbs; transverse commissures common. Rachis anatomy is similar (Fig. 81I).

Stem (Fig. 81J–L)

Internodes congested, with persistent woody remains of leaf bases. Surface layers sclerotic, eroding irregularly from old stems. **Cortex** with numerous narrow vascular and fbs (Fig. 81L), the transition to the central cylinder not v. abrupt (Fig. 81K), although the boundary is especially clear in region of adventitious root initiation (Fig. 19E). Central **vbs** with 1–2 wide mxy vessels; phl undivided. Ground parenchyma cells not expanding markedly with age and so never markedly lacunose (Fig. 81J).

Root (Figs. 19A–E and 81M)

Surface layers eroding from old roots but initially with a large-celled epidermis, the inner wall thickened, inner cells conspicuously tanniniferous (Fig. 19A, B). **Outer cortex** narrow-celled, thick-walled, with conspicuous mucilage canals including raphide crystals (Fig. 19B). **Middle cortex** with well-developed radial air-canals (Fig. 19C). **Fibres** absent. **Stele** medullated, the medulla with occasional included mxy vessels, phl strands radially elongated (Fig. 19D). **Endodermis** becoming uniformly thickened (Fig. 81M), pericycle (1)2-layered. **Pneumathodes:** described by Yampolsky (1924).

Cell inclusions

Wide **raphide sacs**, common in mesophyll of lamina. **Stegmata** (Fig. 81G) abundant in discontinuous files adjacent to fibres of leaf and stem, with spherical silica bodies.

Barcella

A single sp. of limited distribution in Amazonian Brazil, a small unarmed palm with bisexual inflorescences compared to *Elaeis* with unisexual inflorescences.

LAMINA

Very similar in the 2 genera but *Barcella* may be distinguished anatomically from *Elaeis*, as follows:

1. Lamina hypodermis commonly 2-layered; hairs with the basal cell or cells inflated; smaller veins independent of surface layers . . . . *Elaeis*
1A. Lamina hypodermis 1-layered, adaxial cells large and conspicuous; hairs with a deeply sunken narrow uniseriate base; smaller veins commonly continuous with surface layers via fibrous extensions of the I.S. completed by narrow thin-walled cells . . . . *Barcella*

TRIBE MANICARIEAE (Figs. 12P and 77G–J)

The single genus *Manicaria*, regarded either as 1 variable sp. (*M. saccifera*) or several closely related spp., ranging in coastal freshwater swamps where it may form dense stands; from Central America to the Amazon basin. The palm is distinctive in habit, with a short single trunk supporting massive leaves, which can be simple to irregularly pinnate; where separate leaflets are developed these are reduplicate. Female flowers have a trilocular ovary and fruits, with a distinctive warted surface, and up to 3 seeds (cf. *Pelagodoxa* and Phytelepheae).

Leaf (Fig. 77G, H)

Simple or irregularly dissected, the margin somewhat praemorse.

LAMINA

Dorsiventral (Fig. 77G). **Trichomes** absent except for large abaxial scales on ribs. **Epidermis** with thin walls and a thin cuticle. Adaxial epidermis uniform, cells rhombohedral, obliquely extended. Cell files above transverse anticlinal walls somewhat wider than files elsewhere. Abaxial epidermis with occasional costal bands of almost rectangular cells below main veins, elsewhere as in adaxial epidermis but cells smaller, more irregular especially in association with stomatal files. **Stomata** restricted to abaxial epidermis but diffuse, t.s.c. not well differentiated from other epidermal cells. **Hypodermis** 1-layered below each surface, adaxial cells large, transversely extended and almost fibre-like, with thick walls, the end walls interlocked; abaxial hypodermis similar but with smaller, less regular and thinner-walled cells; s.st.ch. surrounded by U- or L-shaped cells. **Mesophyll** without distinct palisade, the adaxial layers somewhat compact and not transversely extended as with other mesophyll cells. **Fibres** diffuse, short, sl. lignified; solitary or 2–3 together in mesophyll towards each surface and often in contact with hypodermis. **Veins** equidistant from and mostly independent of surface layers except for larger veins attached to abaxial hypodermis; O.S. of cubical or somewhat anticlinally extended cells; I.S. fibrous especially around larger veins. Phl of larger veins divided into either 2 or 4 strands. **Transverse veins** wide but infrequent, sheathed by sclerotic parenchyma. **Ribs** including a sclerotic cylinder surrounding a central vasculated region. **Stegmata** with spherical silica bodies but silica reported elsewhere (Tomlinson 1961) as small bodies or silica sand in epidermis and hypodermis.

PETIOLE (Figs. 12P and 77H)

Central V-arrangement conspicuous. Vbs; peripheral with sclerotic sheaths forming a sclerotic layer; central vbs (Fig. 77H) large, with usually 2 wide mxy vessels, phl of either 2 or 4 strands, the sieve tubes with compound sieve plates on sl. oblique end walls, the sieve areas often widely separated. Mxy vessels with long scalariform perforation plates and many thickened bars.

Stem (Fig. 77I, J)

From a mature specimen. **Cortex** (Fig. 77I) with sclerotic outer layers and numerous fbs and vbs. **Central cylinder**: outer layers (Fig. 77I) densely sclerotic with congested vbs, stem centre with more diffuse vbs (Fig. 77J), the fibrous bundle sheath also well developed around the phl, the fibres becoming v. thick-walled. Central vbs usually with 2 wide mxy vessels, the phl undivided except in leaf traces. **Ground tissue** sclerotic in outer congested layer, in stem centre uniformly spongy, with lobed cells delimiting wide intercellular spaces; tannin cells common. Mxy vessels with scalariform perforation plates on oblique end walls but with few thickening bars.

Root

Epidermis persistent, the cells wholly lignified, with markedly unequally thickened walls; the inner tangential walls remaining thin. **Exodermis** 1-layered, of unlignified cells with walls thickened like the epidermis. **Outer cortex** 3–4-layered, the cells weakly lignified, but hardly different in shape and size from those of the exodermis. **Inner cortex** 3-zoned. The outer zone of compact cells with sl. and equally thickened, lignified walls; these occasionally with tanniniferous contents. Middle zone aerenchymatous and including large, circular bundles of 'Raphia-type' fibres (i.e. lignified and accompanied by stegmata). Inner zone with compact occasionally thick-walled cells, but never lignified. **Endodermis** of lignified U-shaped cells with unequally thickened walls, the outer tangential walls remaining thin. Pericycle 1–3-layered, the cells thin-walled and unlignified. Conjunctive tissue of stele lignified. **Medulla** consisting of only thin-walled parenchyma; vascular tissue absent.

Note: in the lamina the most distinctive feature is the transversely-extended, almost fibre-like cells of the hypodermis. The root is distinctive in possessing cortical fibre strands of the 'Raphia-type', otherwise mostly known in early diverging clades (Calamoideae and Coryphoideae).

TRIBE EUTERPEAE (Fig. 82)

A tribe of 5 genera from the Caribbean to Central and South America. Plants are unarmed and range in size from treelets to stately trees, especially when multi-stemmed. The group is not recognizeable by any obvious morphological feature, but does include the relatively advanced character of a peudomonomerous ovary. *Oenocarpus* is recognizably distinct from the other genera in its leaf anatomy and is described in some detail, but the other genera share a suite of characters, fairly distinctive for the tribe, but not readily discriminating among them (Henderson and de Nevers 1988). Their combined features are outlined below.

Euterpe, Hyospathe, Neonicholsonia, and Prestoea

LAMINA (Fig. 82A–D)
Dorsiventral (Fig. 82A, B). **Trichomes** common in some spp., with a multicellular sclerotic base. **Epidermis**: adaxial epidermal cells commonly with undulate to distinctly **sinuous** anticlinal walls (Fig. 82D), typically with oblique end walls, rhombohedral. Abaxial epidermis (Fig. 82C) without sinuous walls. **Hypodermis** usually present as a layer of small cubical colourless cells alternating with files of fibres or narrow fbs, together with similar fibres or fibre strands in the mesophyll. Palisade layer of **mesophyll** present but not in *Neonicholsonia*. **Stomata** without well differentiated t.s.c. (Fig. 82C). **Stegmata** in lamina next to non-vascular fibres with silica bodies much smaller than those next to the vascular fibres. **Veins** in abaxial mesophyll independent of surface layers except the few largest touching the abaxial epidermis. Transverse veins narrow, infrequent and sheathed by sclerotic parenchyma.

PETIOLE (Fig. 82H)
Vbs with 1 wide mxy vessel and 2 phl strands.

Stem (Figs. 15H; 16I; and 82K, L)
Outer vbs of central cylinder developing massive fibre sheaths with age, forming an extremely **sclerotic** layer in old stems (Figs. 15H and 82K). Stem centre with age developing a **ground tissue** of enlarged cells with wide intercellular spaces supporting vbs via radiating plates of parenchyma (Figs. 16I and 82L).

Root (Fig. 82I, J)
Outer layers (Fig. 82I) with persistent epidermis of wide cells, the outer wall somewhat thickened, underlying layers variously thick-walled and transitional to inner cortex. without fbs of 'Kentia-type', but including scattered solitary fibres; **endodermis** (Fig. 82J) with U-shaped cells in TS; pericycle 1-layered.

Key to genera of *Euterpeae* based on lamina anatomy

1. Epidermis with rectangular cells in SV, the walls straight. Outer epidermal wall as thick as one-half total cell depth and wholly cutinized. Trichomes abundant, sunken, with sinuous shield cells. Fbs wide, rarely hypodermal . . . . *Oenocarpus*
1A. Epidermis largely with rhombohedral cells, the walls commonly sinuous. Outer epidermal wall thin and thinly cutinized. Trichomes infrequent, not sunken, without sinuous shield cells. Fibres mostly hypodermal and either solitary or as narrow fbs. 2
2. Epidermal cells with mostly straight cell walls. Fbs narrow in hypodermis, up to 12 cells wide in mesophyll . . . . *Prestoea acuminata*
2A. Epidermal cells with mostly sinuous cell walls. Fbs uniformly narrow 3
3. Mesophyll with a distinct palisade of (1–) 2 or more cells deep . . . . *Euterpe*, some *Prestoea* spp.
3A. Mesophyll lacking a distinct palisade 4
4. Fibres of hypodermal layer not associated with colourless cells . . . . *Neonicholsonia*
4A. Hypodermis including fibres and colourless cells . . . . *Hyospathe, Prestoea* spp.

Note: this key is contrasted with that in Henderson and Galeano (1996) but is intended to demonstrate the variation seen by us in the material we examined.

Oenocarpus (including *Jessenia*) (Fig. 82E, F, M)

Includes 9 spp. of solitary or multiple-stemmed large palms with a wide range from Central to South America as far as Bolivia. Leaves either spirally-arranged, or distichous in *O. distichus*. *Oenocarpus bacaba* is a potential commercial source of palm oil (Balick and Gershoff 1981; Balick 1986).

Leaf
Pinnate with fibrous leaf sheaths, the vbs of the sheath often persisting as spiny structures.

LAMINA (Fig. 82E, F)
Dorsiventral (Fig. 82E). **Trichomes** uniformly distributed and forming a complete indumentum abaxially, but more diffuse adaxially; each hair with a deeply sunken circular to elliptic base of pitted sclerotic cells, the central cells thin-walled and producing a distal flat plate of elongated thick-walled but unlignified and elongated cells with a sickle-shaped outline, these cells interlocked and resembling spaghetti in SV. **Epidermis** (Fig. 82F) with the outer wall thickened to over half the cell depth and wholly cutinized; without distinct costal bands. **Adaxial epidermis** with rectangular cells and straight walls. **Abaxial epidermis**

Fig. 82 Arecoideae (Euterpeae). **A**. *Euterpe* sp., lamina main vein, TS (bar = 100 μm). **B**. *Euterpe* sp., lamina minor veins, TS (bar = 100 μm). **C**. *Euterpe broadwayi*, abaxial epidermis thin-walled and hypodermis thick-walled, SV (bar = 100 μm). **D**. *Euterpe broadwayi*, adaxial epidermal and hypodermal cells, SV (bar = 100 μm). **E**. *Oenocarpus* sp., lamina TS with major and minor veins,TS (bar = 200 μm). **F**. *Oenocarpus* sp., lamina adaxial surface with fbs dispersed in mesophyll, TS (bar = 100 μm). **G**. *Euterpe* sp., petiole central vbs, TS (bar = 1 mm). **H**. *Euterpe* sp., petiole central vb (bar = 200 μm). **I**. *Prestoea pubigera*, root, surface layers, TS (bar = 100 μm). **J**. *Prestoea pubigera*, root endodermal region, TS (bar = 100 μm). **K**. *Prestoea acuminata*, outer part of central cylinder, vbs with massive fibre sheaths, TS (bar = 2 mm). **L**. *Prestoea acuminata*, stem, middle part of central cylinder, TS (bar = 2 mm). **M**. *Oenocarpus bataua* var. *oligocarpus*, stem, middle part of central cylinder, TS (bar = 2 mm).

with diffusely distributed stomata, not in regular files. **Stomata** with short t.s.c., each overarching the somewhat sunken g.c. **Hypodermis** of shallow cells but not differentiated as a colourless layer except where enlarged as expansion cells below either surface. **Mesophyll** with or without well-differentiated palisade, cells compact, with limited s.st.ch. **Fibres** (Fig. 82E) abundant in bundles of varying size, occasionally solitary; mainly towards adaxial surface, but rarely in contact with surface layers. **Veins** in abaxial mesophyll when palisade is differentiated, but not in contact with surface layers. O.S. incomplete, chlorophyllous. I.S. mostly fibrous abaxially. **Transverse veins** few, narrow, and sheathed with elongated pitted sclerotic cells. **Midrib** with complete sclerotic cylinder enclosing several vbs; marginal rib conspicuous, similar but smaller. **Stegmata** with irregularly spherical silica bodies.

Note: *Oenocarpus* is easily distinguished from the other genera in its lamina anatomy. It is described in Uhl and Dransfield (1987) as differing from *Jessenia* in the absence of the sickleshaped cells of the trichomes, but they are clearly developed in the material studied here. Such trichomes are described for *Jessenia* in Tomlinson (1961).

Stem (Fig. 82M)

Central **vbs** fairly densely arranged with mostly 1 wide mxy element maturing with well-developed fibrous phl sheaths. Ground tissue becoming aerenchymatous by expansion of cells and appearing in TS as if radiating from the vbs (Fig. 82M).

Root

Not examined

TRIBE GEONOMATEAE (Figs. 14 and 83)

A tribe long-recognized as a natural group because the triads of flowers, typical of the Arecoideae, are each sunken in a deep pit in the surface of the rachilla. Floral parts are thus variously extended to reach the surface at anthesis. The group is wholly American and includes an appreciable range from taller midstorey, single-stemmed palms (*Pholidostachys, Welfia*), to moderate sized palms (*Calyptronoma*) but most typically small solitary or multiple-stemmed, often cane-like palms, (*Calyptrogyne*, many but not all *Geonoma* spp.) to short, narrow-stemmed or even acaulescent understorey palms (*Asterogyne*, some *Geonoma* spp.). Smaller spp. with simple bifid leaves can make attractive ornamentals, e.g. *Asterogyne*. The largest genus *Geonoma* (c. 60 spp.) shows the biggest range in size and leaf morphology. A crown-shaft is not developed.

Note: this tribe is anatomically recognizable in ways which reflect the generalist habitat of understorey palms in forest environments, i.e. humid, shady situations, with *Welfia* the most exceptional. All its members are easily recognized by their filamentous trichomes. The range of variation seen in other characters further reflects this uniformity because it does not clearly delimit the individual genera. Within the group there is in the lamina a typological reduction series with loss of hypodermal

layers and mesophyll fibres, which may be size or habitat-related. Detailed study especially of *Geonoma*, the most diverse genus, would prove of interest to ecologists.

Diagonostic anatomical features for the tribe Geonomatae

LAMINA

Dorsiventral (Fig. 83A–D). **Trichomes** (Fig. 83I) represented by a uniseriate filament of short cells, the base often sunken and cutinized. **Epidermis** with adaxial cells obliquely extended, rhombohedral to spindle shaped, abaxial cells less regular. **Hypodermis** often absent but at most 1-layered below each surface, the adaxial cells either cubical or transversely-extended and hexagonal in surface view. **Mesophyll** in thinner leaves shallow and without a palisade. **Fibres** with wide septate lumina, restricted to mesophyll and rarely in contact with surface layers, varying from frequent to scarce. **Veins** diffuse (Fig. 83B) to congested (Fig. 83D), situated in the abaxial mesophyll, smaller veins independent of surface layers, the largest veins at most continuous with the surface layers via short colourless mesophyll cells. **Midrib** projecting abaxially, including a sclerotic cylinder enclosing a large abaxial vb and smaller adaxial vbs; phl of abaxial bundle subdivided into many strands by sclerotic partitions (Fig. 83E, F). **Stegmata** commonly abundant next to non-vascular fibres, but small and with small silica bodies.

PETIOLE (Fig. 83H, K)

Vbs with either 1 (Fig. 83K) or 2 (Fig. 83H) mxy vessels and 2 phl strands.

Stem

Smaller-stemmed spp. are good examples of the '*Rhapis*-model' (Fig. 14); without pronounced expansion of ground tissue cells.

Root

Lacking cortical fbs (Fig. 83L).

The following account contrasts smaller and larger spp., with a summary of intermediate taxa.

Asterogyne (Fig. 83A, E)

Six spp. of diminutive palms of the forest understorey, with mostly undivided leaves.

Leaf

Small and usually undivided with a deeply emarginated apex.

LAMINA

Dorsiventral (Fig. 83A). **Trichomes** abundant mainly on abaxial surface as a uniseriate filament arising from a single epidermal cell. **Epidermis** thin-walled, cuticle thin; adaxial epidermis uniform, with hexagonal but obliquely extended and ± isodiametric cells, abaxial epidermis similar, with costal bands only below the largest veins. **Stomata** diffusely distributed abaxially, t.s.c. short, not markedly differentiated, g.c. each with 2 thin cutinized ledges. **Hypodermis** (Fig. 83A) of cubical colourless

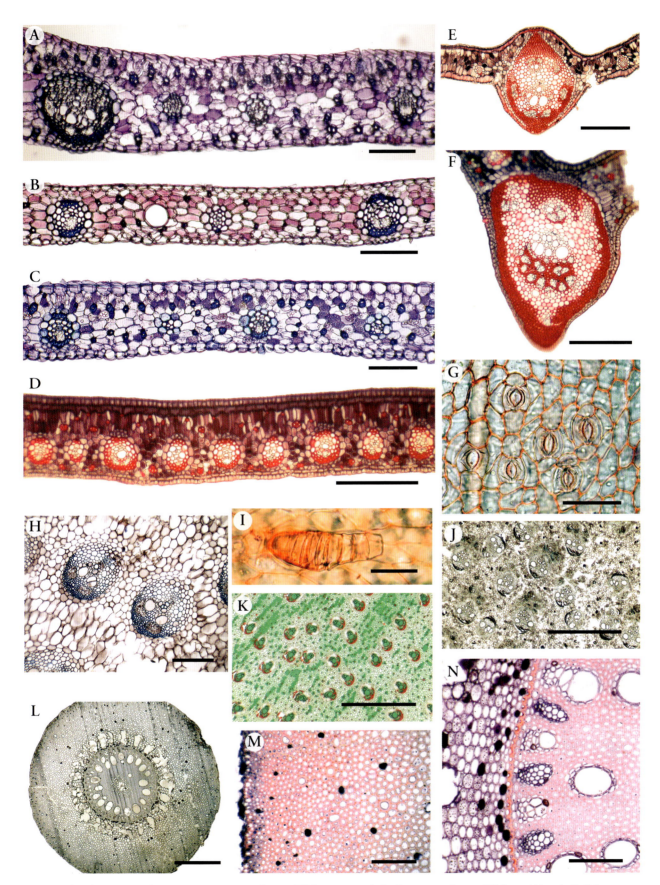

Fig. 83 Arecoideae (Geonomateae). **A**. *Asterogyne martiana*, lamina, TS (bar = 100 μm). **B**. *Geonoma* sp., lamina, TS (bar = 100 μm). **C**. *Calyptrogyne ghiesbreghtiana*, lamina, TS (bar = 100 μm). **D**. *Welfia regia*, lamina, TS (bar = 200 μm). **E**. *Asterogyne martiana*, lamina midrib, TS (bar = 200 μm). **F**. *Welfia regia*, lamina midrib, TS (bar = 200 μm). **G**. *Welfia regia*, lamina, abaxial epidermis, SV (bar = 50 μm). **H**. *Asterogyne* sp., petiole, central vbs., TS (bar = 200 μm). **I**. *Welfia regia*, lamina, abaxial epidermis with trichome, SV (bar = 25 μm). **J**. *Welfia regia*, stem centre, TS (bar = 2 mm). **K**. *Pholidostachys pulchra*, petiole centre, TS (bar = 2 mm). **L**. *Geonoma* sp., root, TS (bar = 1 mm). **M**. *Geonoma interrupta*, root periphery, TS (bar = 100 μm). **N**. *Geonoma interrupta*, root, outer stele, TS (bar = 100 μm).

cells with thin walls but only abaxially clearly differentiated from the inner mesophyll cells. **Mesophyll** shallow, as few as 5 cells deep, palisade not differentiated except for compact adaxial cells. **Fibres** frequent, with a wide septate lumen, mostly solitary or few-celled, scattered but most abundant adaxially near the surface and within the hypodermal layer; especially abaxially. **Veins** abaxial in the mesophyll, independent of surface layers except the largest but then separated from epidermis by at least 2 mesophyll layers; O.S. complete only around smallest veins, I.S. poorly developed, vascular fibres only common abaxially, the fibres with wide lumina like the non-vascular fibres. Phl of largest veins with several strands separated by sclerotic tissue. **Transverse veins** v. few, narrow, sheathed by a single layer of thick-walled cells. **Midrib** (Fig. 83E) with a continuous sclerotic cylinder enclosing a large abaxial vb, with several separate phl strands, and smaller abaxial vbs. **Stegmata** abundant in continuous files next to non-vascular fibres, especially conspicuous immediately below the epidermis on both surfaces, silica bodies small, irregularly spherical.

PETIOLE

Shallow V of central vbs evident. Each vb with 2 wide mxy vessels (or a complex representing 2 vessels) the phl either as 2 strands, each associated with a mxy vessel, or the phl further subdivided by sclerotic tissue.

Calyptrogyne (Fig. 83C)

Resembles *Asterogyne* but with the following distinctive features. **Epidermis** with outer adaxial cell wall sl. thickened. **Hypodermis** absent or indistinctly developed. **Fibres** (Fig. 83C) mostly solitary or in pairs and towards each surface, the cell lumen wide; sometimes in contact with the hypodermal mesophyll layer, but never with the epidermis. **Mesophyll** with at most an incipient palisade, the adaxial cell layer as a layer of compact somewhat anticlinally-extended cells. **Midrib** projecting abaxially, with a sclerotic cylinder enclosing a central bundle complex and adaxial vbs. **Stegmata** few, small, mostly associated with the vascular fibres, the silica bodies v. small.

Calyptronoma

A genus of 3 spp. of moderate-sized palms of wet places within the Greater Antilles, distinguished by its distally much-branched inflorescence. Not always distinguished from *Calyptrogyne*.

Leaf

Pinnately erect with pendulous leaflets (cf. *Welfia*).

LAMINA

Dorsiventral. **Trichomes** v. small, often represented only by the basal cell. **Epidermis** thin to somewhat thick-walled. Abaxial epidermis with markedly obliquely extended and almost spindle-shaped cells. Abaxial epidermis with costal bands below larger veins, intercostal bands with short, irregular, and somewhat papillose cells. **Stomata** congested, t.s.c. papillose and overarching g.c. somewhat, l.s.c. cells wide; g.c. thin-walled and with 2 prominent cutinized ledges. **Hypodermis** 1-layered below

each surface, adaxial cells ± cubical. **Mesophyll** c. 10 cells deep compact with a distinct 2-layered palisade of shortly anticlinally-extended cells. **Fibres** few, with wide lumina, solitary or in pairs in the mesophyll towards each surface. **Veins** in the mid mesophyll, not in contact with the surface layers; O.S. conspicuous but usually discontinuous; I.S. with well developed fibres above and below most veins.

Note: this description is based on *C. rivularis*, other un-named material differs in the thinner-walled epidermis, with somewhat sinuous walls, absence of epidermal papillae, and more numerous fibres which are restricted to surface layers.

Geonoma

A genus of palms with a wide distribution in tropical America including about 60 spp., most typically small plants of the forest understorey but very diverse in size and appearance.

Leaf

Pinnately compound and variously divided but sometimes entire with a deeply emarginate apex (cf. *Asterogyne*).

LAMINA (Fig. 83B)

Examination of limited material shows a range of structural features that fall well within those of the whole tribe but without generic diagnostic characters. This range involves variation in the thickness of the lamina, the extent to which the hypodermis is differentiated as a discrete colourless layer, the presence or absence of a palisade layer in the mesophyll, and the abundance and size of the non-vascular fibres. A description of a representative sp. is provided.

Dorsiventral (Fig. 83B). **Trichomes** mostly below veins, each with a basal filament of thin-walled cells extending distally into thin-walled cells; if persistent branched and terminating in inflated and bladder-like cells. **Epidermis** thin-walled, the outer wall scarcely thickened and with a thin cuticle. Abaxial epidermis uniform, with few costal bands above larger veins, cells obliquely extended and hexagonal, the walls often somewhat undulate. Abaxial epidermis with narrow costal bands of narrow cells and irregularly polygonal cells in wide intercostal regions. **Stomata** v. diffuse, with short t.s.c. and wide l.s.c., the g.c. thin-walled with 2 equal cuticular ledges. **Hypodermis** not differentiated or represented adaxially by a layer of hexagonal subsurface cells, oblique in orientation to the epidermal cell. **Mesophyll** shallow, c. 6 cells deep, without a discrete palisade layer, the cells transversely extended. Fibres few, mostly solitary towards each surface but not in contact with the epidermis. **Veins** in the abaxial mesophyll, independent of the surface layers. **Midrib** with a configuration similar to that of *Asterogyne*.

PETIOLE

Vbs in a shallow V-arrangement (Fig. 12 D).

Stem (Fig. 14A, B)

Cane-like, with prominent leaf scars. Used to demonstrate the '*Rhapis*-principle' of palm stem vascular anatomy (Fig. 14). **Cortex** narrow with few fbs. **Central cylinder** with crowded peripheral axial bundles and leaf traces (Fig. 14A). Central ground tissues not undergoing sustained primary growth, vbs mostly with a single wide mxy vessel, phl 2-stranded (Fig. 14B).

Root (Fig. 83L–N)

Outer surface layers eroded with age (Fig. 83L). Wide region of outer cortex (Fig. 83M) becoming sclerotic and thick-walled with age, including scattered wide-lumened cells (probable raphide canals); fibres absent. Innermost cortex becoming narrowly lacunose by breakdown of cells. **Endodermis**: thick-walled, cells with U-shaped thickening (Fig. 83N). **Stele** with regular peripheral vascular tissue, peripheral conjunctive tissue and medulla becoming sclerotic, infrequently with medullary vessels.

Pholidostachys (Fig. 83K)

An understorey palm with rather narrow stems showing conspicuous leaf scars.

Leaf

The anatomy is somewhat distinctive for the tribe as follows.

LAMINA

Dorsiventral. **Trichomes** with uniseriate filamentous construction as for other genera. **Epidermis** thin-walled, with outer wall sl. thickened. Adaxial epidermis with large, obliquely and somewhat transversely extended rhombohedral cells, the walls distinctly sinuous, abaxial cells smaller, with less conspicuously sinuous walls and irregularly polygonal. **Stomata** v. diffusely distributed, t.s.c. short, l.s.c. wide. **Hypodermis** absent. **Mesophyll** shallow, 6–7 cells deep, without palisade layers. **Fibres** with wide septate lumina, solitary or in few-celled strands in the mesophyll but not in contact with the epidermis. **Veins** not in contact with epidermis. **Stegmata** small but numerous in association with non-vascular fibres, silica body irregular.

PETIOLE (Fig. 83K)

Central vbs each with 1 wide mxy vessel and 2 phl strands.

Welfia (Fig. 83D, F, G, I, J)

The largest palm in the tribe, reaching the upper canopy of the forest, the habit distinct in the suberect leaves with pendulous leaflets.

Leaf (Fig. 83D, F, G, I)

Leaflets with a conspicuous midrib, the lamina with several prominent ribs.

LAMINA

Dorsiventral (Fig. 83D). **Trichomes** (Fig. 83I) each with a deeply sunken uniseriate filament of cutinized cells extending into an ephemeral filament of thin-walled cells. **Epidermis** with outer wall sl. thickened, the cuticle thin. Adaxial epidermis uniform except for costal regions above few larger veins, cells rhombohedral, obliquely extended; abaxial epidermis with more frequent costal bands, intercostal regions with irregular cells (Fig. 83G). **Stomata** not in regular files, t.s.c. short, l.s.c. wide. Hypodermis (Fig. 83D) 1–2-layered adaxially, 1-layered abaxially; adaxial

cells large transversely extended and hexagonal, abaxial cells smaller more nearly cubical with usually 4 L-shaped cells surrounding the s.st.ch. **Mesophyll** with a well-developed 2-layered palisade. **Fibres** (*W. regia*, Fig. 83D) solitary or in 2–3-celled strands in the mesophyll, largely independent of the surface layers in 2 series, most abundant adaxially in the palisade, but also forming a series abaxially. **Veins** congested (Fig. 83D), in the abaxial mesophyll and largely independent of the surface layers. O.S. inconspicuous, I.S. fibrous above and below larger veins, but fibres few or absent above smaller veins; phl of larger veins irregularly sclerotic. **Midrib** (Fig. 83F) projecting abaxially, with a continuous sclerotic cylinder enclosing a large abaxial vb, the phl much subdivided, together with smaller adaxial vbs.

Note: the earlier description in Tomlinson (1961), based on material identified as *W. georgii* indicates the consistent presence of a 2-layered adaxial hypodermis and non-vascular fibres mostly as wide multicellular strands at the level of and alternating with the veins. The above description referring to *W. regia* is more consistent in its distribution of non-vascular fibres with the distribution seen in other members of the tribe.

Stem (Fig. 83J)

Vbs with 2 or more wide mxy vessels, the fibre sheath little developed. Ground tissue not becoming enlarged with age.

TRIBE LEOPOLDINIEAE (Fig. 84)

Consists of the taxonomically isolated genus *Leopoldinia*, with 3 or 4 spp. restricted to the middle reaches of the Amazon Basin. The genus is unusual in the Arecoideae in its triovulate ovary but most distinctively in the **fibrous epidermis** of the lamina. The 2 commonest species (*L. piassaba* and *L. pulchra*) differ strikingly in leaf sheath morphology and lamina anatomy. According to Putz (1979) the stems are at first horizontal before turning erect.

Leaf (Fig. 84A–H)

Pinnate, the leaf sheath with a pronounced ligule opposite the petiole, the method of sheath disintegration contrasted in different species.

LAMINA

Dorsiventral (Fig. 84A, B). **Trichomes** absent from *L. pulchra*, occasional below the veins as elliptical groups of sclerotic cells in *L. piassaba*, but long brown scales with a sclerotic base abaxially on the basal part of the midrib and transitional to other smaller scales occasional at the abaxial leaf margin and below the larger veins. In *L. piassaba* the leaf margin may retain obscure tooth-like projections of presumed scale bases. **Epidermis** (Fig. 84B, D) unique within palms with alternating files of 1) wide rectangular and thin-walled short cells alternating with 2) overlapping fibres (Fig. 84F), the fibres with a well-developed unlignified secondary wall almost occluding the cell lumen but separating from the cutinized primary wall. Short cells always situated above anticlinal walls of hypodermal cells (Fig. 84D, E). Abaxial epidermis (Fig. 84G, H) with stomata restricted to short-cell files;

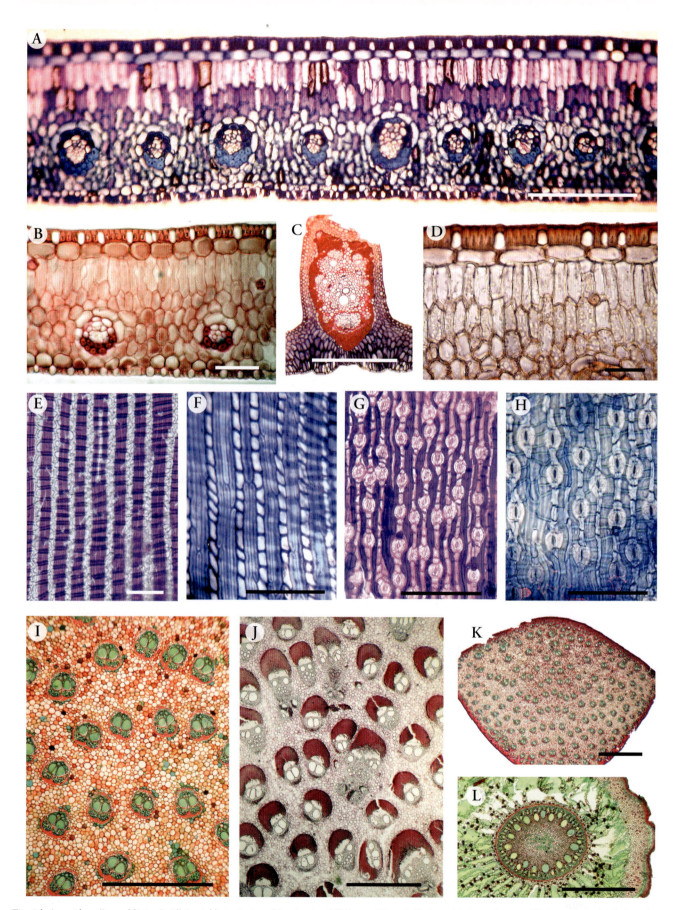

Fig. 84 Arecoideae (Leopoldinieae). All *Leopoldinia*. **A**. *L. pulchra*, lamina, TS (bar = 200 μm). **B**. *Leopoldinia* sp., lamina, TS (bar = 50 μm). **C**. *Leopoldinia* sp., midrib, TS (bar = 1 mm). **D**. *L. pulchra*, lamina, adaxial surface layers, TS (bar = 50 μm). **E**. *L. pulchra*, adaxial surface, hypodermal level, SV (bar = 100 μm). **F**. *L. pulchra*, adaxial surface, epidermal level, SV (bar = 100 μm). **G**. *L. pulchra*, abaxial surface, epidermal level, SV (bar = 200 μm).
H. *L. pulchra*, abaxial surface, hypodermal level, SV (bar = 100 μm). **I**. *Leopoldinia* sp., petiole centre, TS (bar = 1 mm). **J**. *Leopoldinia* sp., stem centre, TS (bar = 1 mm). **K**. *Leopoldinia* sp., petiole, TS (bar = 2 mm). **L**. *Leopoldinia* sp., root, TS (bar = 1 mm).

fibres somewhat sinuous in SV. **Veins** (Fig. 84A) narrow, never in contact with surface layers. **Transverse veins** conspicuous, often short, with a massive fibrous sheath, commonly extending above the longitudinal veins. **Midrib** (Fig. 84C) with a complete sclerotic cylinder, expansion cells in 2 abaxial bands.

For other characters the 2 studied species can be distinguished readily as follows:

Leopoldinia pulchra

LEAF SHEATH

Drying into a mesh of interwoven broad flattened fibres.

LAMINA (Fig. 84A)

Eleven to 16 cells deep. **Trichomes** absent. Colourless **hypodermis** present, 1-layered below each surface; palisade well differentiated, biseriate. Mesophyll **fibres** mainly solitary, infrequent unlignified. T.s.c. of **stomata** rarely thick-walled. Short epidermal cells without rod-like bodies. **Stegmata** common next to fibres of longitudinal and transverse veins; silica bodies spherical.

Leopoldinia piassaba

LEAF SHEATH

Drying into extremely long pendent (over 1 m long) black fbs ('piassava' of commerce), clothing and obscuring the stem.

LAMINA

Eight to 9 cells deep. **Trichomes** occasional as warty structures below larger veins, leaflet margin with short blunt spines. Colourless **hypodermis** absent, palisade undifferentiated. **Fibres** frequent, unlignified, solitary or in wider fbs. T.s.c. of **stomata** together with neighbouring cells of stomatal bands thick-walled and pitted. Short epidermal cells with non-crystalline rod-like bodies.

PETIOLE (Fig. 84I, K)

Central V present but not pronounced (Fig. 84K). **Vbs** with 2 wide mxy elements and 2 phl strands (Fig. 84I). **Ground tissue** without air spaces or fbs.

Stem (Fig. 84J)

Vbs with well-developed fibrous bundle sheaths, congested in stem centre; with mostly 2 wide mxy vessels, phl undivided. **Ground tissue** without late cell expansion.

Root (Fig. 84L)

Outer cortex sclerotic. **Inner cortex** without fibres, radiating air lacunae separated by radial cell plates, tannin cells conspicuous. **Endodermis** with U-shaped wall thickening. **Stele** without medullary vessels, medulla not well differentiated.

Ecological note: Gustavo Romero (pers. comm.) suggested that the contrasted anatomical differences between species may relate to ecological differences; *L. pulchra* is characteristic of the seasonally inundated varzea and so submerged part of the year, *L. piassaba* is rarely submerged. The difference thus may relate to water with few nutrients, low pH and low levels of dissolved oxygen (Putz 1979).

Economic uses: the leaf sheath fibres ('piassaba' or 'piassava') of *L. piassaba* have local use and enter into commercial trade as ropes and brooms. The leaves are used as thatch and the thin flesh of the fruits soaked in water produces a refreshing drink (Putz 1979). The anatomy of the leaf sheath in this genus has not been studied.

TRIBE PELAGODOXEAE (Fig. 85)

An isolated tribe with 2 genera of dissimilar habit and geography, but with a shared distinctive anatomy. Molecular evidence has placed *Pelagodoxa* (Marquesas) and *Sommieria* (Western New Guinea) together, as is supported by their similar warty fruits and, despite their differing stature, by their largely undivided leaves.

Anatomical features of the tribe Pelagodoxeae

Trichomes uniseriate, the basal cells thick-walled, pitted, producing a continuous indumentum of thin-walled distal cells. **Stomata** with the 4 neighbouring cells thick-walled, lignified and pitted, the stomata complex thus clearly outlined (Fig. 85B). Colourless **hypodermis** absent, replaced by a discontinuous layer of thick-walled unlignified fibres.

Anatomically the 2 genera can be distinguished as follows:

1. Adaxial epidermal cells without distinctly sinuous anticlinal walls; adaxial hypodermal fibres solitary or in few-celled strands and alternating with files of thin-walled cells; abaxial fibres absent; palisade tissue not differentiated . . . . *Pelagodoxa* (Fig. 85A–C)
1A. Adaxial epidermal cells with distinctly sinuous anticlinal walls; abaxial hypodermal fibre strands occasional; palisade mesophyll well-differentiated . . .*Sommieria* (Fig. 85D–H)

Pelagodoxa (Fig. 85A–C)

One sp., *P. henryana* endemic to the Marquesas and almost extinct in the wild, but to some extent preserved in cultivation. The palm is single-stemmed and distinctive because of its large corky warted fruits, more than 30 mm in diameter.

Leaf

Massive but with an essentially undivided lamina except for the bifid apex; becoming secondarily split into irregular leaflets.

LAMINA

Dorsiventral (Fig.85A). **Trichomes** (Fig. 85B) uniformly abundant on abaxial surface, each consisting of a uniseriate filament of cells arising from a basal series of 2(–4) thick-walled, pitted columnar cells, the distal cells forming an initially continuous indumentum. **Adaxial epidermis** v. uniform, without trichomes or stomata, costal regions not differentiated. Cells obliquely-transversely extended, the anticlinal walls scarcely sinuous. **Abaxial epidermis** (Fig. 85B) with wide stomatal bands alternating with narrow costal bands resembling the upper epidermal

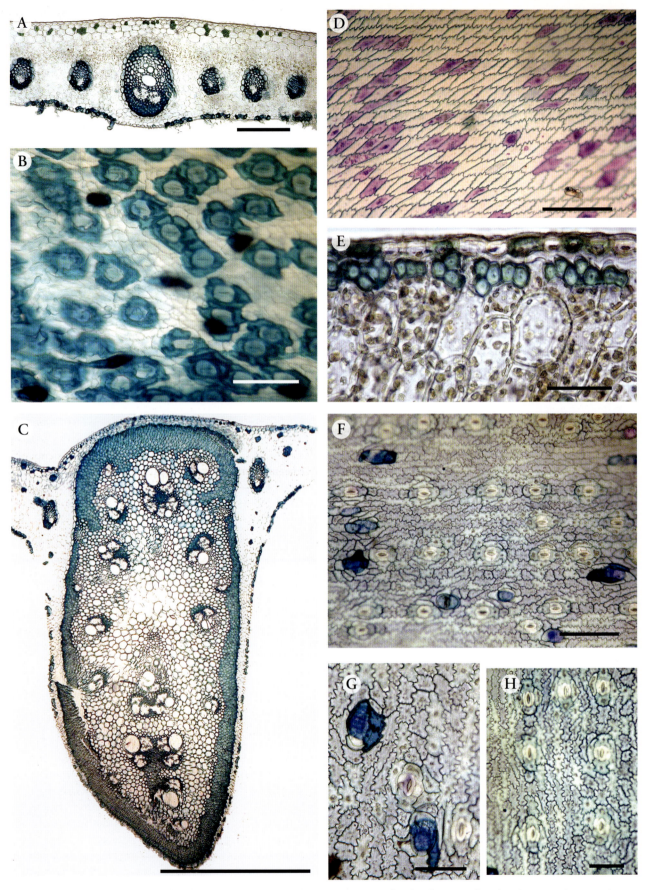

Fig. 85 Arecoideae (Pelagodoxeae). **A.** *Pelagodoxa henryana,* lamina, TS (bar = 200 μm). **B.** *Pelagodoxa henryana,* abaxial epidermis, trichomes and stomata with thick-walled subsidiary cells, SV (bar = 100 μm). **C.** *Pelagodoxa henryana,* midrib, TS (bar = 1 mm). **D.** *Sommieria leucophylla,* lamina adaxial epidermis, SV (bar = 100 μm). **E.** *Sommieria leucophylla,* lamina adaxial hypodermal fibres, TS (bar = 50 μm). **F.** *Sommieria leucophylla,* lamina abaxial epidermis, SV (bar = 100 μm). **G.** *Sommieria leucophylla,* lamina abaxial trichomes and stomata, SV (bar = 50 μm). **H.** *Sommieria leucophylla,* lamina, abaxial epidermis emphasizing sinuous anticlinal walls, SV (bar = 50 μm).

cells or somewhat longitudinally extended. **Stomata** not in regular files, each complex distinguished by the thick-walled, pitted, and lignified neighbouring cells in contrast to the thin-walled l.s.c. visible as an almost continuous sclerotic epidermis in TS. **Hypodermis** chlorenchymatous, adaxially the outermost layer with solitary or paired unlignified fibres alternating with files of thin-walled cubical or transversely extended cells. Abaxial hypodermis without fibres. **Mesophyll** uniform, without well-developed palisade layers, compact except near s.st.ch. **Veins** mostly in abaxial mesophyll and independent of surface layers except for largest veins in contact with lower hypodermis. O.S. incomplete below most veins, incomplete both above and below the few large veins; cells sl. thick-walled, chlorenchymatous. I.S. usually fibrous above and below, or absent above smallest veins; the sheath completed laterally by short pitted sclereids; phl of largest veins with a median sclerotic sinus or with irregular sclerotic partitions. **Transverse veins** at same level as longitudinal veins, few, narrow and sheathed by sclerotic parenchyma cells. **Midrib** (Fig. 85C) with a well-developed sclerotic sheath enclosing numerous independent or aggregated vbs.

Sommieria (Fig. 85D–H)

Three spp. of small single-stemmed palms from western New Guinea with undivided but apically bifid leaves, or at most a single basal pair of leaflets.

Leaf

Lower leaf surface typically with a silvery indumentum, but possibly not in all species.

LAMINA

Dorsiventral. **Trichomes** (Fig. 85G) numerous on abaxial surface, each with 1 or more thick-walled, pitted basal cells extended into a filament of thin-walled cells to form the collective indumentum. **Adaxial epidermis** (Fig. 85D) without distinct costal bands, cells in uniform files, the end walls sl. oblique; anticlinal walls thick, distinctly sinuous. **Abaxial epidermis** (Fig. 85F–H) with distinct costal bands below largest veins, the cells resembling those of the adaxial epidermis; intercostal bands with irregular obliquely extended cells; the walls thickened and markedly sinuous. **Stomata** in irregular files, the neighbouring and t.s.c. thick-walled, pitted, and lignified. **Hypodermis** chlorenchymatous; adaxially differentiated as a layer of narrow thick-walled unlignified fibres, the cell lumen almost occluded (Fig. 85E); abaxial hypodermis with fewer fibres, mostly in the form of narrow strands. **Mesophyll** differentiated adaxially as a 2(–3)-layered palisade, the abaxial mesophyll of compact isodiametric cells, the s.st.ch. narrow. Mesophyll fibres absent except for occasional narrow strands that may represent free vein endings. **Veins** compact, not continuous with surface layers, O.S. discontinuous below all veins; I.S. sclerotic, well-developed. Vascular tissues reduced; xyl elements narrow and inconspicuous, phl largely sclerotic. **Transverse veins** few, narrow, sclerotic, vascular tissues inconspicuous. Larger veins associated with narrow bands of **expansion cells.**

Note: *Pelagodoxa* shares with *Sommieria* the distinctive lignified stomatal complex, uniseriate hairs and fibrous adaxial hypodermis features which together with floral and fruit morphology and anatomy provide evidence for a close relationship.

TRIBE ARECEAE (Figs. 86, 87, and 88)

This group is the largest tribe in the palm family, with 59 genera and 660 spp. (Dransfield et al. 2008b), but its anatomical diversity is limited by the considerable uncertainty of the phylogenetic relationships among its constituent taxa, as outlined in our introduction to the Arecoideae. The tribe is referred to as the 'Indo-Pacific pseudomonomerous clade' indicating both its geographic distribution and the presence of a gynoecium derived by reduction (often evident developmentally) from an original tricarpellate condition. Nevertheless it is divided into 14 subtribes and extended by an artificial alliance of 'unplaced genera', i.e. unplaced as to their phylogenetic status that would otherwise indicate to which subtribe they might belong. The subtribes themselves are of uncertain phyletic position in relation to each other and are therefore here listed alphabetically, as in Dransfield et al. (2008b).

Anatomy itself seems unhelpful in providing evidence that might clarify relationships. Members of the Arecaeae show more limited anatomically diagnostic features than is found in other arecoid palms so that the demonstration of morphologically 'natural groups' is difficult compared with the existence of long-recognized tribes like Chamaedoreeae, Cocoseae, Geonomateae, and Iriarteeae, whatever their status in an evolutionary context. Habit diversity in the Arecaeae is also limited so that the typical *Gestalt* of many arecoids is constantly expressed, i.e. single-stemmed palms with pinnately reduplicate leaves, a distinct crownshaft, and infrafoliar inflorescences.

Few members can be described as massive trees, whereas many spp. in certain genera, e.g. *Iguanura* and *Pinanga*, qualify as 'palmlets' of the forest understorey.

Dypsis, the largest genus (~150 spp.), itself includes a range comparable to that of the whole group, with riverine, dichotomizing, climbing, and diminutive acaulescent or rhizomatous forms as extremes.

As the tribe is probably a relatively recent radiation, anatomical uniformity might be expected, but its overall expression in the lamina should be appreciated as being based on a quite different structural ground plan from that found in the earlier diverging subfamilies. This is easily perceived by comparing the illustrations in the final plates (Fig. 86–88) with examples from earlier pages, especially in the Coryphoideae. This difference contrasts epidermal and hypodermal cell shape, distribution of non-vascular fibres largely independent from rather than included within the surface layers, and the almost complete disassociation of the longitudinal veins from any contact with the surface layers, differences of presumed considerable mechanical significance. This overall shift in structural properties of the lamina is one of the evolutionary developments within the palms that our survey has documented with some precision (Horn et al. 2009).

Our treatment of this large group is regrettably brief because it has been rather skimpily sampled, and to present what we do know in detail would be highly repetitive. We frequently indicate that generic boundaries are not often discernable at the anatomical level. The limited documentation should rather be accepted as a guide to the more extensive study of these elegant palms that surely will produce worthwhile results.

SUBTRIBE ARCHONTOPHOENICINAE (Figs. 5D, F; 16A–E, H; and 86A, I, L–N)

Five genera of moderate to slender solitary-stemmed palms with a well-developed crownshaft, the leaves regularly pinnate with 1-fold leaflets (Fig. 5D). Flowers are uniformly multistaminate. *Archontophoenix* in eastern Australia includes several species, some common in cultivation; *Actinorhytis* is from New Guinea and the Solomon Islands and 3 genera (*Actinokentia, Chambeyronia,* and *Kentiopsis* including *Mackeea*) are endemic to New Caledonia.

Archontophoenix has provided the best example of sustained primary growth as exemplified in the classic study of Waterhouse and Quinn (1978) in which continuous measurements were made on a natural population over a period of 7 years. The stem undergoes continual changes in histology as it ages resulting in the expansion of cells in all parts at any given level and a consequent increase in stem diameter. Examples are provided in Fig. 16A–E.

Kentiopsis (Fig. 86A)
Provides a brief introductory example.

Leaf
Pinnately reduplicately compound, the leaflets lanceolate with a single fold.

LAMINA

Dorsiventral (Fig. 86A). **Trichomes** absent adaxially, common abaxially below veins, each sl. sunken, with 2–4 tiers of thin-walled, pitted, small basal cells. **Epidermis** thin-walled, outer wall sl. thickened. Adaxial cells obliquely-transversely extended, spindle-shaped, abaxial cells irregular. **Stomata** absent adaxially; densely distributed abaxially, in short files, the t.s.c. short; outer wall of g.c. much thickened except at the poles, the cell lumen dumbbell-shaped. **Hypodermis** conspicuously 2-layered adaxially, 1(–2)-layered and small-celled abaxially. **Mesophyll** with a well-developed palisade layer 3–4 cells deep. **Fibres** with v. narrow cell lumina, from single cells to small strands of up to 8 cells scattered in the mesophyll, mostly abaxial, but the adaxial fibres often in contact with the hypodermis. **Veins** (Fig. 86A) in abaxial mesophyll, not in contact with surface layers, O. S. usually incomplete; I. S. well-developed, fibrous above and below.

Key to the genera of the subtribe Archontophoenicinae

1. Hypodermis 1-layered below each surface, palisade not or sl. developed. Fibres mostly solitary or at most in few-celled strands (*Actinorhytis, Archontophoenix*) 2

1A. Hypodermis 2-layered adaxially, the cells sl. thickened, abaxial hypodermis 2(1)-layered. Palisade usually well developed. Fibres mostly in few to several- (up to 10) celled strands (*Actinokentia, Chambeyronia, Kentiopsis*) 3

2. Palisade at most 2-layered. Fibre strands solitary or in few (2–5)-celled strands, mostly towards or next to each hypodermis. Trichomes few, abaxially below veins, the base many-celled . . . . *Actinorhytis*

2A. Palisade varying from absent to distinctly 2–3-layered. Fibres few, mostly solitary or in 2-celled strands, adaxially rarely in contact with hypodermis. Trichomes numerous and uniformly distributed on abaxial surface, each with a small sunken base of 1–2 tiers of cells . . . . *Archontophoenix*

3. Outer wall of epidermis thick, g.c. much thickened, the lumen largely occluded . . . . *Actinokentia*

3A. Outer wall of adaxial epidermis thin, g.c. at most somewhat thickened 4

4. Fibre strands restricted to the abaxial mesophyll, few solitary, including massive wide strands pectinating with the veins . . . . *Chambeyronia*

4A. Fibres not restricted to the abaxial epidermis, often solitary, larger strands few-celled, not pectinating regularly with the veins . . . . *Kentiopsis*

Note: although there is a continuum of structural variation in this subtribe, including the extent of the development of non-vascular fibres and the degree of expression of the palisade mesophyll, further investigation might demonstrate more discrete features that could separate the genera. Trichome distribution and structure suggests itself. For the whole tribe the tendency to develop thick-walled g.c. is notable and the degree of development of the inner sclerotic sheath of the vbs seems somewhat diagnostic. The relative dense arrangement of the longitudinal veins is also distinctive.

Actinokentia

PETIOLE

Cylindrical in TS. Central deep V-arrangement of vbs distinct; 'keel' vb conspicuous. Central vbs each with numerous overlapping mxy vessels, 2 phl strands.

Root (Fig. 86M, N)
Surface layers eroded from old roots Outer **cortex** (Fig. 86M) extensively sclerotic, with numerous mucilage canals, transitional to inner cortex with abundant fbs of the 'Kentia-type', the cell lumen of the inner fibres tanniniferous; air lacunae developed in inner cortex by localized cell collapse. **Stele** of wider roots fluted (Fig. 86N), the inner cortex intruded into the histologically similar stele through an interrupted endodermis. **Endodermis** v. thick-walled, the lumen almost occluded.

Actinorhytis

PETIOLE

Transverse commissures abundantly developed. Central vbs each with 1 wide mxy vessel and 2 phl strands, the sclerotic partition often narrow.

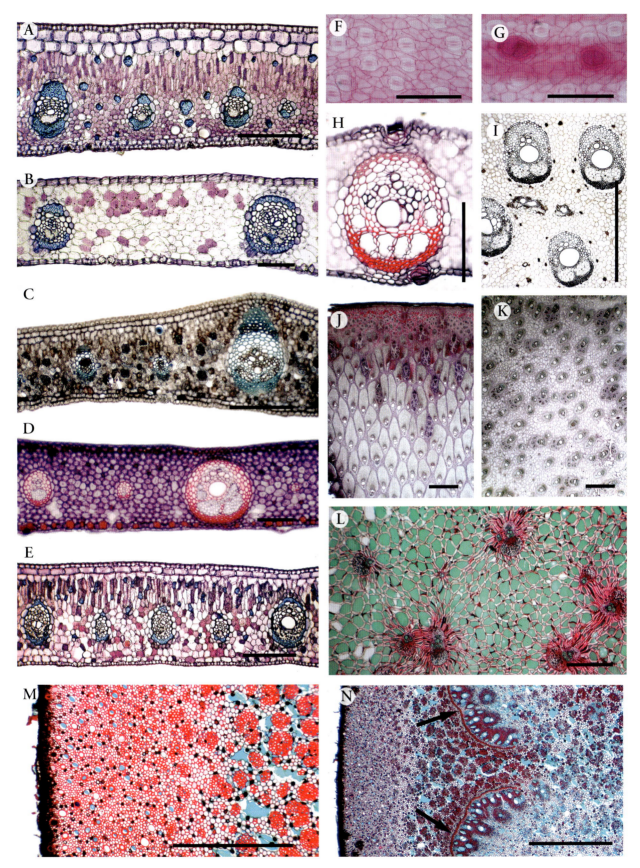

Fig. 86 Arecoideae (Areceae: Archontophoenicinae [Arch.], Arecinae [Arec.], Basseliniinae [Bass.], Carpoxylinae [Carp.]). **A**. *Kentiopsis oliviformis* [Arch.], lamina, TS (bar = 200 µm). **B**. *Areca triandra* [Arec.], lamina, TS (bar = 100 µm). **C**. *Lepidorrhachis mooreana* [Bass.], lamina, TS (bar = 200 µm). **D**. *Cyphosperma tanga* [Bass.], lamina, TS (bar = 100 µm). **E**. *Neoveitchia storckii* [Carp.], lamina, TS (bar = 200 µm). **F**. *Areca catechu* [Arec.], lamina, abaxial epidermis, SV (bar = 100 µm). **G**. *Areca catechu* [Arec.], lamina, abaxial epidermis, trichomes, SV (bar = 100 µm). **H**. *Areca catechu* [Arec.], lamina, vein and trichomes, TS (bar = 100 µm). **I**. *Actinorhytis calapparia* [Arch.], petiole centre, TS (bar = 1 mm). **J**. *Areca catechu* [Arec.], stem periphery, TS (bar = 2 mm). **K**. *Areca catechu* [Arec.], stem centre, TS (bar = 2 mm). **L**. *Archontophoenix alexandrae* [Arch.], stem centre, TS (bar = 2 mm). **M**. *Actinokentia divaricata* [Arch.], root periphery, TS (bar = 1 mm). **N**. *Actinokentia divaricata* [Arch.], root periphery to centre, TS (arrows = endodermis) (bar = 2 mm).

Root (from a narrow root sample)

Epidermal cells large, thin-walled. **Cortex**: outer layers somewhat sclerotic; inner cortex with numerous solitary fibres and scattered tannin cells, the fibres becoming aggregated internally. Innermost cortex including a series of radially-extended air lacunae.

Archontophoenix

PETIOLE

Shallow V-shaped arrangement of central vbs present; each vb usually with 1 wide mxy vessel and 2 phl strands.

Stem (Figs. 16A–E, H and 86L)

Central ground tissue (Fig. 86L) becoming extremely lacunose by cell expansion laterally and formation of intercellular lacunae, the vbs becoming suspended within radial plates of cells. Individual **vbs** ultimately develop wide fibre sheaths with the cells continuing to expand and thicken their walls (Fig. 16H).

Root

Cortex including abundant fibres, either solitary or in larger bundles, air-lacunae not well developed. **Stele** somewhat fluted, the endodermis interrupted.

Chambeyronia

As in the earlier key.

SUBTRIBE ARECINAE

Three genera, *Areca*, *Pinanga* (each with many spp.), and *Nenga* (5 spp.) widely distributed in Asia, from S China to the Solomon Islands but concentrated in Malesia. They are distinguished from most related arecoid palms in the single inflated bract (prophyll) enclosing the inflorescence, without additional large peduncular bracts. They vary in size from moderate to small or even diminutive palms, with both single- and multiple-stemmed species in all 3 genera. A crownshaft is present except for some of the smaller and especially the few acaulescent species. Included is the widely cultivated betelnut palm, *Areca catechu* whose endosperm (betel) is chewed together with lime and leaves of *Piper betle* as a mild stimulant or narcotic with putative medicinal properties. In the distribution of anatomical characters, groups of spp. seem more distinctive than differences among genera. The following account refers to *Areca catechu* with some of the differences between it and other taxa briefly summarized. This brief introduction clearly cannot do justice to the 2 large genera, which collectively include about 200 spp.

Areca (Fig. 86B, F–H, J, K)

LAMINA (Fig. 86B, F–H)

Dorsiventral (Fig. 86B). **Trichomes** frequent above and below veins on both surfaces (Fig. 86G, H) but most numerous abaxially, each with a sunken base of usually 2 tiers of sclerotic pitted cells, distal thin-walled cells not persistent. **Epidermis** with outer wall only sl. thickened, walls not sinuous; adaxial cells rhombohedral and obliquely extended; abaxial epidermis (Fig. 86F) with smaller, less regular cells and infrequent costal bands below larger veins. **Stomata** diffusely distributed, t.s.c. not well differentiated but overarching g.c. somewhat, each g.c. with 2 prominent cutinized ledges, but walls little thickened. **Hypodermis** 1-layered below each surface, thin-walled the adaxial cells cubical to sl. transversely extended, the abaxial cells rather indistinct and even with chloroplasts. **Mesophyll** without a distinct palisade layer. **Fibres** varying from few or even absent and at most represented by solitary fibres scattered in the mesophyll, to more numerous and then often included within the hypodermis. **Veins** in abaxial mesophyll and mostly independent of surface layers, O. S. rarely complete even around smallest veins; I. S. fibrous above and below; phl of larger veins divided into several separate strands by sclerotic partitions (Fig. 86H). **Transverse veins** at same level as longitudinal veins, few narrow, sheathed by sclerotic parenchyma.

PETIOLE

V-configuration present. 'Keel' bundle often conspicuous. Central vbs with 1 wide mxy vessel and 2 phl strands.

Stem (Fig. 86J, K)

Cortex narrow with abundant fibre strands; peripheral meristematic activity of an etagen-type producing a ligno-suberized cork. **Central cylinder** with outer layer of congested vbs with massive fibre sheaths (Fig. 86J), the fibres progressively thick-walled and lignified with age. Central vbs less congested (Fig. 86K), each usually with 1 mxy vessel; ground tissue becoming lacunose with age.

Root

Cortex lacunose with fibres and fibre strands of 'Kentia-type'. **Stele** becoming fluted and polyxylic in wider roots.

Vascular elements

Vessels: perforation plates of mxy elements in roots simple, transverse, in stem scalariform on oblique end walls, in leaf axis with scalariform or scalariform reticulate end walls on oblique to v. oblique end walls with many thickening bars. **Sieve tubes** with compound sieve plates, decreasing in obliquity from root to leaf.

Cell inclusions

Stegmata in short discontinuous files in leaf and stem, with spherical silica bodies.

Pinanga

LAMINA

Shows a different range of characters. **Trichomes** often with a radially filamentous development of distal thin-walled cells. **Epidermis** varying from thin- to quite thick-walled cells.

Hypodermis often absent or poorly differentiated. **Fibres** sometimes numerous and in strands of up to 8 cells, usually uniformly distributed in the mesophyll.

Root

Shows some tendency for development of fluted steles. 'Kentia-type' fibres in the cortex (and sometimes medulla) characteristic of most material examined, but seen as solitary thick-walled fibres in narrow roots.

Nenga

Material of *N. pumila*: lamina thin, the mesophyll v. uniform, 7 cells deep without differentiation of hypodermis or palisade. Fibres represented by occasional cells next to the hypodermis or narrow strands near the lamina margin.

Note: the range of variation in lamina anatomy in this subtribe possibly shows a correlation with stature and habitat, leading in the smaller spp. to a reduced version of that in larger spp. of *Areca*. This suggestion needs to be verified by much more extended study.

SUBTRIBE BASSELINIINAE (Fig. 86C, D)

A group including 6 genera of rather uniform appearance, with a centre of distribution in New Caledonia (4 genera, 2 endemic). The group is most precisely defined as having an incompletely encircling prophyll. *Physokentia* ranges from Fiji and Vanuatu to New Britain and the Solomon Islands. *Cyphophoenix* occurs in both the Loyalty Islands and New Caledonia. *Basselinia*, endemic to New Caledonia, is the largest genus (12 spp.). The outlying monotypic *Lepidorrhachis* is restricted to Lord Howe Island. All are moderate-sized palms, with single, rarely multiple stems. They have regular pinnate leaves except for a few diminutive spp. of *Basselinia* with entire leaves. They develop a crown shaft with infrafoliar inflorescences, except in *Cyphosperma*. Some spp. develop a basal cone of aerial stilt-roots.

The group shows a common suite of features, as summarized here, but there is appreciably variation which is clearly diagnostic. This is outlined later.

General features of lamina anatomy in subtribe Basseliniinae

LAMINA (Fig. 86C, D). Dorsiventral, often thick (Fig. 86C). **Trichomes** usually restricted to abaxial epidermis, most commonly below larger veins; each with a sunken (often deeply) base of sclerotic pitted cells, ellipsoidal in SV, but without a persistent distal series of cells in mature leaves. **Epidermis** usually shallow, but outer wall varying considerably in thickness, when thick then somewhat papillose. Adaxial epidermis uniform without costal bands, the cells obliquely extended, then varying from rhombohedral, hexagonal, to spindle-shaped; abaxial epidermis with narrow costal bands, intercostal regions with densely distributed stomata, the intercostal cells irregular. **Stomata** without variation in g.c. structure, each with 2 cutinized ledges, sl. sunken in leaves with a thick epidermal wall; t.s.c. often short, but not well differentiated. **Hypodermis** always present, either 1-layered below each surface and then not very conspicuous (Fig. 86D), but often 2-layered below each surface and conspicuous (Fig. 86C), sometimes including fbs. Cells of outer hypodermal layer most conspicuous, adaxial cells commonly hexagonal and transversely or obliquely extended, inner cells less well differentiated, thinner walled. **Mesophyll** usually without a differentiated palisade. **Fibres** varying considerably in abundance and distribution; either ± restricted to surface layers or in mesophyll and then uniformly distribute or most abundant towards each surface; either as solitary fibres or in associated few-celled strands. **Veins** usually in abaxial mesophyll, often quite densely distributed, independent of surface layers, except the largest veins attached to each surface by small thick-walled mesophyll cells. **Transverse veins** few, narrow, sheathed by sclerotic parenchyma cells. **Stegmata** with spherical silica bodies; raphide sacs common, varying much in size and abundance.

Note: the following brief statements outline differences among spp. based on our limited observations:

Basselinia

- *B.* (*Alloschmidia*) *glabrata*. Fibres few or absent.
- *B. deplanchei*, *B. tomentosa*, *B. velutina*. Fibres abundant, mainly in small strands throughout the mesophyll, or towards the surface and sometimes in contact with or within the innermost hypodermal layer.
- *B. iterata*. Outer wall of epidermis thick, abaxially somewhat papillose, the stomata sunken. Fibres abundant in small strands in the mesophyll and also adaxially within the hypodermis.
- *B. humboldtiana*. Fibres mostly restricted to and included within the hypodermis.

Burretiokentia

Hypodermis 2-layered below each surface. Fibres abundant within the mesophyll towards each surface and often in contact with the hypodermis.

Cyphophoenix

- *C. nucele*. Outer epidermal wall thick. Hypodermis 1-layered below each surface. Palisade well developed. Fibres almost exclusively hypodermal, adaxially a well-developed series of strands alternating with colourless thin-walled cells, abaxially fewer and more diffuse.
- *C.* (= *Campecarpus*) *fulcita*. Outer epidermal wall thick, adaxially the cell lumen v. shallow; hypodermis 2-layered below each surface; fibres abundant in mesophyll, often as large strands, commonly in contact with or included in the inner hypodermal layer.

- *C.* (= *Veillonia*) *alba.* Lamina thin. Mesophyll 6–7 cells deep. Epidermal cell walls thick. Hypodermis 2-layered below each surface, cells large and conspicuous. Fibres few, mainly solitary in mesophyll. Veins v. congested.

Cyphosperma (Fig. 86D)

- *C. humboldtensis.* Lamina thin, mesophyll 7–8 cells deep, Epidermal walls thin. Hypodermis shallow, including fibres alternating with files of colourless cells, other fibres scattered in mid mesophyll. Palisade absent.
- *C. tanga.* Lamina thicker, but hypodermal fibres as above (Fig. 86D).

Lepidorrhachis (Fig. 86C).

Lamina thick. Trichomes deeply sunken. Epidermis with outer wall somewhat thickened. Hypodermis 2-layered below each surface. Palisade absent. Fibres not in contact with hypodermis, fbs in 2 series in mesophyll towards each surface. Veins diffuse.

Physokentia

Epidermis thin-walled. Hypodermis 2-layered adaxially, 1-layered abaxially. Palisade absent. Fibres solitary or in small strands scattered in the mesophyll, not usually in contact with the hypodermis.

This synopsis should indicate the lack of agreement between lamina anatomy and the subdivision of these taxa into discrete genera. Although the New Caledonian taxa have already been examined in detail, as outlined in Uhl and Dransfield (1987), a more extensive analysis should be attempted as there is much diversity within the few species of this subtribe.

SUBTRIBE CARPOXYLINAE (Fig. 86E)

Three monotypic genera, all highly threatened, with a scattered distribution. *Carpoxylon*, 1 sp. from Vanuatu, was only rediscovered after a lapse of over 100 years; *Neoveitchia*, 2 spp., 1 in Fiji and 1 in Vanuatu and *Satakentia*, 1 sp. from the Ryukyu Islands, only described formally in 1969. All are moderate-sized single-stemmed palms. *Neoveitchia* is the most distinctive because it does not develop the crownshaft found in the other 2 genera. The fruit has a consistent structure with characteristic apical stigmatic remains. *Neoveitchia* has a unique longitudinal arrangement in the distal flower pairs. The genera share many features. Lamina quite thick, but not well distinguished from other Areceae; possibly the large elongated ellipsoidal trichome bases are unusual.

Neoveitchia (Fig, 86E)

LAMINA
Dorsiventral (Fig. 86E). **Trichomes** with an ellipsoidal elongated group of pitted cells, usually below veins. **Epidermis** sl. thick-walled; adaxial cells rhombohedral to spindle-shaped and obliquely extended, those of the abaxial epidermis irregular, with some differentiation of narrow costal bands. **Hypodermis** 1–2-layered adaxially, the outermost layer of hexagonal, transversely extended cells, the inner layer somewhat discontinuous and of cubical cells, the abaxial layer uniseriate and shallow. **Mesophyll** palisade well differentiated and 2–3 layered **Fibres** mostly solitary, scattered within the palisade and sometimes contacting the lower hypodermal layer, less abundant abaxially. **Veins** rather congested, in the abaxial mesophyll but free of the surface layer, the fibres of the I. S. not well developed. **Transverse veins** few, narrow, sheathed by sl. thick-walled sclerenchyma. **Stegmata** with lignified walls.

PETIOLE
Central vbs mostly with 1 wide mxy element and 2 phl strands.

Key to genera of the subtribe Carpoxylinae based on lamina anatomy

1. Outer epidermal wall thin. Adaxial hypodermis conspicuous 1–2-layered. Palisade well developed, 2–3-layered. Mesophyll fibres moderately developed. . . . . . . *Neoveitchia*
2. Outer epidermal wall thin. Adaxial hypodermis inconspicuous, 1-layered. Palisade absent. Mesophyll fibres abundant towards each surface and often in contact with the hypodermis . . . . *Carpoxylon*
3. Outer epidermal wall thickened, wholly cutinized. Hypodermis 1-layered. Palisade well developed. Mesophyll fibres few. Vascular fibres well developed, especially on adaxial side of veins . . . . *Satakentia*

Note: material of *Neoveitchia storkii* differing from the above descriptions probably represented juvenile material.

SUBTRIBE CLINOSPERMATINAE (Fig. 87A)

A group of New Caledonian palms at one time treated as 5 genera (see key) but currently (Dransfield et al. 2008b) represented as 2 genera, *Clinosperma* (4 spp.) and *Cyphokentia* (2 spp.). As a monophyletic group, they indicate adaptive radiation within the island from a single ancestor, an interpretation supported by their limited anatomical variability. They are moderate-sized, almost always single-stemmed palms of uniform appearance, although they show in varying degrees development of a crownshaft, so that inflorescences are variably inter- or infrafoliar. Fruits are rather uniform with lateral stigmatic remains and a homogenous endosperm.

Note: the group shares a common suite of leaf anatomical characters, as described here, with a key to the differences seen in our samples.

LAMINA
Dorsiventral, thick (Fig. 87A). **Trichomes** on both surfaces, in costal regions above or below larger veins but mostly abaxial, deeply sunken. **Epidermis** deep, with thick outer wall up to

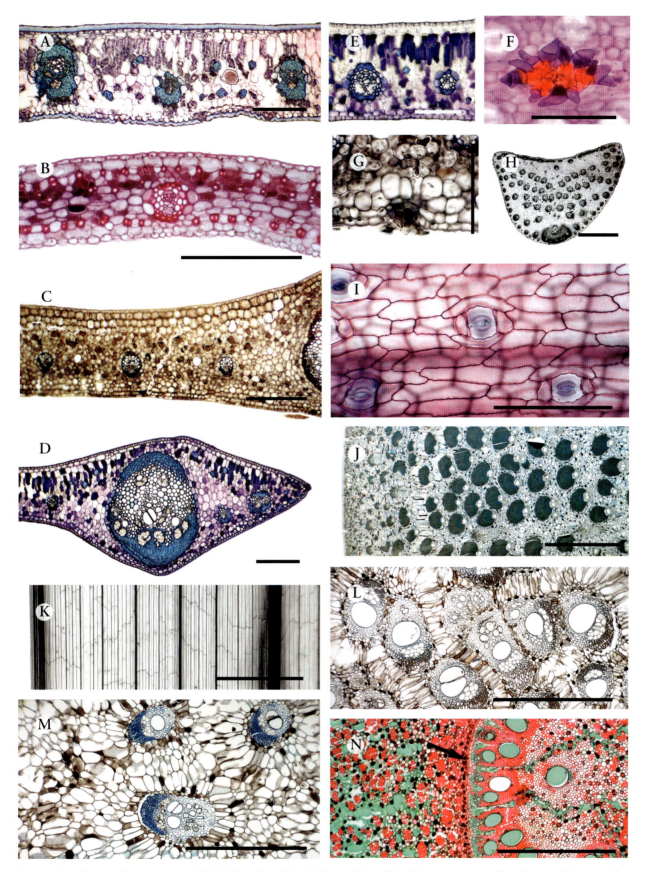

Fig. 87 Arecoideae (Areceae: Clinospermatinae [Clin.], Dypsidinae [Dyps.], Linospadicinae [Lino.], Oncospermatinae [Onco.]. **A**. *Cyphokentia cerifera* [Clin.], lamina, TS (bar = 200 µm). **B**. *Calyptrocalyx hollrungii* [Lino.], lamina, TS (bar = 200 µm). **C**. *Oncosperma fasciculatum* [Onco.], lamina, TS (bar = 200 µm). **D**. *Dypsis lutescens* [Dyps.], lamina margin, TS (bar = 200 µm). **E**. *Dypsis lutescens* [Dyps.], lamina, TS (bar = 100 µm). **F**. *Dypsis decaryi* [Dyps.], lamina abaxial trichome, SV (bar = 100 µm). **G**. *Oncosperma fasciculatum* [Onco.], lamina abaxial trichome (bar = 100 µm). **H**. *Dypsis lanceolata* [Dyps.], petiole, TS (bar = 1 mm). **I**. *Linospadix albertisianus* [Lino.], lamina abaxial epidermis, SV (bar = 100 µm). **J**. *Dypsis lutescens* [Dyps.], stem periphery, TS (bar = 2 mm). **K**. *Dypsis lutescens* [Dyps.], lamina, cleared, SV (bar = 5 mm). **L**. *Oncosperma tigillarium* [Onco.], stem centre, TS (bar = 1 mm). **M**. *Dypsis lutescens* [Dyps.], stem centre, TS (bar = 1 mm). **N**. *Dypsis madagascariensis* [Dyps.], root inner cortex and outer stele, TS (arrow = endodermis) (bar = 1 mm).

229

three-quarters total cell depth (Fig. 87A). Adaxial epidermis uniform with obliquely extended rhombohedral to hexagonal to spindle-shaped cells, abaxial epidermis with narrow costal regions and wide intercostal regions of irregular densely-arranged stomata. **Stomata** sunken or not, the stomatal cavity sometimes wax-filled. **Hypodermis** 1 or more-layered below each surface, rather indistinct, thin-walled, the cells hexagonal and transversely extended. **Mesophyll** with a 1–2-layered palisade. **Fibres** either absent or few, solitary or in 2–3-celled strands, in 2 series towards each surface. **Veins** dense and rather massive by virtue of well-developed sheath fibres above and below larger, but only below smaller veins (Fig. 87A), independent of surface layers, the smaller towards the abaxial surface, the largest separated from surface layers by small mesophyll cells. Larger veins with multi-stranded phl. **Transverse veins** few, narrow, sheathed by sclerotic parenchyma cells. **Midrib** with a massive sclerotic sheath enclosing several vbs. Marginal to submarginal vein well-developed.

Key to spp. of Clinospermatinae based on lamina anatomy

1. Hypodermis 2-layered . . . . .2.
. . . . Clinosperma (= Brongniartikentia)
1A. Hypodermis 1-layered 3
2. Fibres present, mostly 3–5-celled strands in middle mesophyll, raphide sacs numerous . . . . Cl. vaginata
2A. Fibres infrequent, mostly solitary or in small strands . . . . .Cl. lanuginosa
3. Outer epidermal wall thick, adaxially up to 4× depth of epidermal cell lumen, stomata sunken, the outer cavity wax-filled, fibres in mesophyll solitary or in up to 5-celled strands . . . . Cyphokentia (= Moratia) cerifera
3A. Outer epidermal cell wall moderate to thin, <2× depth of epidermal cell lumen; stomata not sunken, fibres various 4
4. Trichomes with a narrow, sunken, few-celled filamentous base . . . . Clinosperma (= Lavoixia) macrocarpa
4A. Trichomes with a broad, elliptical many-celled base 5
5. Trichomes and fibres few or absent . . . . Clinosperma bracteale
5A. Trichomes deeply sunken, frequent above and below larger veins . . . . Cyphokentia macrostachya

SUBTRIBE DYPSIDINAE (Figs. 72B and 87D–F, H, J, K, M, N)

A subtribe which shows extensive adaptive radiation in Madagascar, with a limited extension to the E coast of Africa on the island of Pemba (Tanzania). It includes over 200 spp., most of them in the genus *Dypsis*, into which a number of other genera have been condensed, so that only 3 other genera are now recognized. *Lemurophoenix* was only described in 1991; *Masoala* and *Marojejya*, although recognized much earlier, required considerable recent effort in relocating them so that a full description

could be provided (Dransfield and Beentje 1995). Flower triads include a large central female and 2 small lateral male flowers. Fruits typically have basal stylar remains, with an indication of an initial trilocular condition.

Our understanding of this assemblage, which dominates the Madagascan palm flora, has resulted from extensive field work that has shown the existence of many new taxa, but also their endangered status resulting from forest clearance. The need for revisionary work was clearly stated in Uhl and Dransfield (1987). The group includes a number of commonly cultivated palms, notably *Dypsis* (= *Chrysalidocarpus*) *lutescens*. Different taxa can be seen to represent a transition towards the development of a crownshaft (Fig. 72B) with infrafoliar inflorescences, depending on the precision of leaf abscission and the precise longitudinal splitting of a tubular leaf base through a mechanically weak zone opposite the petiole. Size and habit variation is considerable, from diminutive understorey palmlets to robust single-stemmed palms which can reach the forest canopy (Fig. 72B). Some are associated with water and, at least in early stages, might be described as rheophytes. Small palms may be acaulescent or rhizomatous and *Dypsis scandens* is a weakly climbing palm, a facsimile of the American *Chamaedorea elatior*. Species may have dichotomizing trunks, e.g. *Dypsis* (= *Vonitra*) *utilis*.

In view of the novelty of many spp. and this considerable diversity our account can give only a very superficial view of the anatomy of the subtribe and its possible application to systematic subdivision. *Dypsis* is treated largely on the basis of 1 sp., but the remaining 3 genera more completely because they are quite distinctive, as far as we know.

Dypsis (Figs. 7F, G; 11H; 72B; and 87D–F, H, J, K, M, N)

A large genus of over 150 spp., because it includes several early described genera that have been subsequently included within it. The genus is remarkable for its diversity in stature and habitat, occupying a wide range of ecological situations which represent much of the considerable floristic diversity of Madagascar. Our brief examination of a few spp. can give no indication of likely considerable variability. Consequently we provide a description of the commonly cultivated *Dypsis* (= *Chrysalidocarpus*) *lutescens* as an example with some commentary on the variability we have seen in other species (Rudall et al. 2003; Fisher 1973).

Leaf
Pinnately reduplicate compound.

LAMINA
Dorsiventral (Fig. 87D, E, F). **Trichomes** mostly abaxial, in costal regions, each with a sl. sunken base of tiers of irregular sclerotic cells, extended distally into a shield of thin-walled cells (Fig. 87F). **Epidermis** with outer wall somewhat thickened, walls not sinuous; adaxial epidermis uniform, cells rhombohedral, obliquely extended (Fig. 7F); abaxial epidermis with occasional costal bands below largest veins, intercostal wide, cells irregular. **Stomata** restricted to abaxial intercostal regions, but not in distinct files rather diffuse; t.s.c. short, overarching g.c. somewhat, each g.c. with 2 equal prominent ledges. **Hypodermis**

1-layered below each surface; adaxial cells uniformly hexagonal, transversely or somewhat obliquely extended (Fig. 7G); abaxial cells smaller and more cubical, with 4 L-shaped cells surrounding each s.st.ch. **Mesophyll** with a 1–2-layered palisade. **Fibres** varying from mostly large strands of many cells at the level of the veins, smaller strands or solitary fibres occasional elsewhere, to infrequent solitary or small strands of few cells scattered in the mesophyll, but even absent from some samples. **Veins** in abaxial mesophyll, independent of surface layers, with largest veins only separated from abaxial hypodermis by a single layer of small mesophyll cells. O.S. incomplete above and below most veins, the cells cubical; I.S. incompletely sclerotic around small veins, phl of large veins divided into separate strands by sclerotic partitions. **Transverse veins** (Fig. 11 H; 87K) infrequent, at same level as longitudinal veins, narrow, sheathed by sclerotic parenchyma. **Midrib** prominent adaxially with a sclerotic cylinder enclosing a large abaxial vb and smaller, often inverted adaxial vbs. **Marginal rib** (Fig. 87D) with a structure intermediate between larger veins and midrib.

PETIOLE (Fig. 87H)
From a small-leaved specimen. V-configuration conspicuous; keel vb well developed. Central vbs with 1 wide mxy vessel and 2 phl strands.

Stem (Fig. 87J, M)
Surface layers little modified with age. **Cortex** narrow with numerous fibrous strands. **Central cylinder** with congested vbs developing a wide fibre sheath around phl (Fig 87J). Central vbs diffuse, each mostly with 1 wide mxy vessel; phl divided into 2 strands by a median sclerotic partition. **Ground tissue** somewhat expanded in older stem, but not becoming lacunose (Fig. 87M).

Root (Fig. 87N)
Cortex developing a phellogen external to outer layer, middle cortex scarcely or irregularly lacunose. Inner cortex developing 'Kentia-type' fibres (Fig. 87N). **Stele** with a wide fibrous medulla, peripheral conjunctive tissue sclerotic, medullary vessels occasional.

Vascular tissues
Vessels: elements in roots with simple perforation plates mostly on ± transverse end walls; in stem with scalariform plates on oblique end walls, the elements in the leaf with v. oblique end walls and many thickening bars. **Sieve tubes** with compound sieve plates in all parts, those in the root on oblique end walls, in stem and leaf less oblique or transverse.

Note: the following features indicate some of the variation observed in limited material of other *Dypsis* species.

1) Lamina thickness in terms of number of mesophyll layers.
2) Thickness of outer epidermal wall and whether stomata are sunken or not.
3) Occasionally epidermal cell walls sinuous in surface view.
4) Stomata with somewhat thickened g.c. walls.
5) Hypodermis absent or indistinct or the abaxial layer chlorophyllous.
6) Palisade variously developed.

7) Fibre distribution in mesophyll varying from solitary to few- to many-celled and of different-sized strands.
8) Larger veins attached to adaxial hypodermis by a girder of large colourless anticlinally-extended cells.
9) Stem anatomy in relation to overall diameter, from as narrow as 8 mm to 50 cm.
10) Extent of cell expansion with age in ground tissue of stem.

Lemurophoenix

Endemic to Madagascar and an example of a palm with a crown structure that accumulates falling leaf material of presumed nutritional benefit ('litter-trapping').

LAMINA
Dorsiventral. **Trichomes** few, abaxial below larger veins, base massive with numerous small sclerotic cells, each with a narrow lumen. **Epidermis** small-celled and thin-walled, possibly wholly cutinized: adaxially v. uniform, cells rhombohedral, obliquely extended but isodiametric, cell files above anticlinal hypodermal walls wider than those elsewhere; abaxial epidermal cells similar in costal regions but irregular in intercostal regions. **Stomata** diffuse but within files of short cells, the t.s.c. correspondingly short, thin-walled, g.c. not sunken and with narrow cutinized ledges. **Hypodermis** 2-layered below each surface, the adaxial cells largest, outer adaxial layer thick-walled, narrowly hexagonal and markedly transversely extended. **Mesophyll** with 1–2-layered palisade. **Fibres** numerous, with narrow cell lumina, including large bundles below the palisade, smaller bundles or single bundles scattered in mesophyll, sometimes next to hypodermal layers. **Veins** independent of surface layers, below palisade.

Marojejya

LAMINA
Dorsiventral, thick, up to 18 cells deep. **Trichomes** infrequent or absent, abaxial only, restricted to costal regions below larger veins, sl. sunken, the base small, the distal thin-walled cells forming a short filamentous extension. **Epidermis** thin-walled, the outer wall scarcely thickened; adaxial epidermis uniform, the cells rhombohedral and obliquely extended, abaxial epidermis shallower and smaller-celled. **Stomata** not sunken, small, thinwalled with 2 cutinized ledges. **Hypodermis** indistinct but 1-layered below each surface, with cells little larger than those of the adjacent epidermis. **Mesophyll** palisade absent, but adaxial cells smaller and more compact than those elsewhere, abaxial cells larger, somewhat radiately arranged around veins. **Fibres** restricted to mesophyll in an upper and lower series, as solitary fibres or 2–3-celled strands, uniformly scattered and sometimes next to adaxial hypodermal layer. **Veins** independent of surface layers, mostly large and densely distributed; largest veins continuous with surface layers via small, thick-walled mesophyll cells. O.S. conspicuous with somewhat thickened walls; vascular tissues with 1 large abaxial vb and several smaller peripheral vbs, often with 2 mxy vessels and 2 phl strands. **Transverse veins** few, narrow sheathed by sclerotic, pitted cells.

Note: the 2 spp. in this genus may be distinguished by the distribution of non-vascular fibres as follows:

1. Non-vascular fibres restricted to surface layers, mostly as few-celled strands alternating with files of colourless hypodermal cells. (Trichomes present) . . . . *M. darianii*
1A. Non-vascular fibres not restricted to surface layers, mostly as 1–2-celled strands in the mesophyll towards each surface. (Trichomes absent?) . . . . *M. insignis*

Masoala

Leaf

Includes a suite of distinctive features.

LAMINA

Dorsiventral, thick, up to ~20 cells deep. **Trichomes** few, but massive, elliptically extended and restricted to costal regions below largest veins, each including a large group of sclerotic, pitted basal cells. **Epidermis** thin-walled and shallower than the hypodermis; adaxial epidermis v. uniform, cells rhombohedral to rectangular with oblique end walls. **Stomata** sl. sunken, g.c. with extensively thickened outer wall, almost occluding the cell lumen, the outer ledge extended to form the margin of a deep cupular chamber; outer wall of t.s.c. thickened. **Hypodermis** 2-layered below each surface. **Mesophyll** without a conspicuous palisade. **Fibres** with v. wide cell lumen, septate, as small to large strands in adaxial mesophyll, rarely elsewhere. **Veins** densely distributed in mid mesophyll, phl of largest veins multi-stranded.

Note: these last 3 genera thus can be distinguished by a combination of features, represented here in simple terms:

1. Hypodermis well-developed 2-layered the outer adaxial layer of transversely extended hexagonal cells thick-walled, 2-layered below each surface; g.c. of stomata with thin walls, non-vascular fibres non-septate, with narrow lumina . . . . *Lemurophoenix*
2. Hypodermis 1–2-layered but with thin walls, the cells small, thin-walled, not markedly transversely extended, g.c. of stomata thin-walled; non-vascular fibres wide, with wide septate lumina . . . . *Masoala*
3. Hypodermis 1-layered, indistinct, g.c. of stomata with thin-walls, non-vascular fibres with narrow cell lumina . . . .*Marojejya*

To what extent these anatomical differences are diagnostic within the subtribe can only be established by a thorough survey of *Dypsis* itself, an ideal problem in view of the ecological diversity of these palms.

SUBTRIBE LINOSPADICINAE (Fig. 87I)

A natural and monophyletic group of mostly small to diminutive palms of the forest understorey; only *Howea* (endemic to Lord Howe Island) is moderate sized; although familiar in the horticultural trade as a 'parlour palm' it is rarely seen grown to maturity. Otherwise the group occurs from the Moluccas (1 sp.

of *Cayptrocalyx*), throughout New Guinea and to eastern Australia. They vary from single- to multiple-stemmed, but lack a crownshaft so that the inflorescences are interfoliar; also spicate and sometimes several per leaf axil, with each flower triad sunken in a deep pit in the rachilla.

As with many subtribes in the Areceae lamina anatomy is very uniform and is not illustrated here.

Howea

LAMINA

Dorsiventral. **Trichomes** most frequent on abaxial surface, but associated with veins, each with a sunken base of irregular tiers of somewhat sclerotic, pitted cells extended distally into a distal shield of thin-walled cells. **Epidermis** with outer wall from thin to somewhat thickened, adaxial cells rhombohedral, transversely - obliquely extended, abaxial cells most regular in narrow costal bands. **Stomata** with usually short t.s.c. overarching the g.c. sl.; g.c. either with or without unequal wall thickening (see below). **Hypodermis** 1-layered below each surface, adaxial cells cubical or somewhat transversely extended; s.st.ch. surrounded v. regularly with 4 L-shaped cells. **Mesophyll** without well-developed palisade. **Fibres** few to many, scattered in the mesophyll and often in contact with the hypodermis. **Veins** in abaxial mesophyll and independent of surface layers; fibres of I.S. well-developed. Phl of larger veins divided into several separate strands by sclerotic partitions. **Transverse commissures** at same level as longitudinal veins, few narrow, sheathed by sclerotic parenchyma cells.

PETIOLE

V-shaped arrangement of central vbs distinct; central vbs each with a 1 or more wide mxy vessels, phl divided into 2 separate strands by a median sclerotic partition. Ground tissue becoming somewhat radially arranged around central vbs.

Stem

Conforms to general features of a 1-vessel palm. Phl sheath of central vbs well developed. Ground tissue cells somewhat expanded with age.

Root

'Kentia-type' fibres not recorded in cortex (but cf. Drabble 1904).

Note: the difference between the 2 spp. of *Howea* reported in Tomlinson (1961) is well supported by further examination.

1. Epidermis thin-walled, smooth; g.c. of stomata only sl. thick-walled within cuticular ledges, the cell lumen not constricted. Palisade absent. Fibres few, not conspicuously associated with adaxial hypodermis of large cells . . . . *H. belmoreana*
1A. Epidermis with outer wall somewhat thickened, somewhat papillose abaxially in intercostal regions; g.c. of stomata with outer wall thickened, the median part of the lumen somewhat constricted. Palisade somewhat differentiated. Fibres abundant especially adaxially in association with rather small hypodermal cells . . . . *H. forsteriana*

Laccospadix

LAMINA

Dorsiventral. **Trichomes** mostly abaxial. **Epidermis:** cells rhombohedral, but ± isodiametric, outer wall sl. thickened. Walls somewhat undulate. Abaxial cells irregular. **Stomata** with short t.s.c. **Hypodermis** 1-layered below each surface, the abaxial cells smallest. Adaxial cells somewhat hexagonal to sl. extended transversely. **Mesophyll** with indistinct palisade, the adaxial 2(–3) layers sl. extended anticlinally. **Fibres** abundant. Mostly in 2–3-celled strands in adaxial mesophyll, sometimes next to or even within the adaxial hypodermis; abaxial mesophyll with few fibres and rarely in contact with hypodermis. **Veins** in mid-mesophyll, remote from hypodermis; I.S. of well-developed somewhat wide-lumened fibres. **Transverse veins** few, at level of longitudinal veins, narrow sheathed by sclerotic parenchyma.

Note: this brief statement shows the similarity with *Howea*, but most significantly with *H. forsteriana* in terms of the fibre distribution.

Linospadix (Fig. 87I)

This is particularly distinctive in the shape of the **epidermal cells**, which are longitudinally extended but with oblique end walls, the anticlinal walls appreciably sinuous. Abaxial epidermal cells are narrow in the intercostal regions, but the t.s.c. still usually short (Fig. 87I). **Trichomes** quite deeply sunken. The lamina is thin, but with large **mesophyll** cells. The adaxial hypodermal cells are cubical, compact, and somewhat hexagonal in surface view. A palisade layer is absent. **Fibres** are scattered in the mesophyll usually in strands of 2–3 forming an indistinct layer towards each surface and in contact with the hypodermal layer.

Calyptrocalyx

From the description in Tomlinson (1961) this is distinguished by the absence of a palisade and the paucity of fibres.

Note: these limited statements suggest that more detailed examination could distinguish the genera as now recognized, with the caution that *Howea* itself shows appreciable differences in lamina anatomy among its 2 spp.

SUBTRIBE ONCOSPERMATINAE

Four genera of rather disparate distribution; 3 are monospecific, either in the Seychelles (*Deckenia*) or Mauritius (*Acanthophoenix*, *Tectiphiala*) and of moderate stature, mostly single-stemmed and quite rare, in contrast to the tall multiple-stemmed *Oncosperma* (5 spp.) ranging from Sri Lanka to Malesia and the Philippines with extensive populations (e.g. *O. tigillarium*) in the back mangrove. *Oncosperma* may include the tallest of multiple-stemmed palms. All are spiny, at least in juvenile stages, and often as adults, with a well-developed crownshaft. In *Acanthophoenix* the outermost leaf sheath is unusual in its dense covering of spines. Leaves are regularly pinnate and the leaflets have a single fold.

Anatomically this group forms a fairly natural unit, relatively uniform in the distribution of fibres, veins, and epidermal layers. The trichomes have a constant structure and arrangement, especially as they are usually deeply sunken. The veins are relatively dense, with well-developed fibres of the inner bundle sheath above and below larger veins; phl tissues in the larger veins are commonly multistranded. *Tectiphiala* is most distinctive in the peculiar coiled cells of the trichome shield, structure of the g.c. and massively thickened epidermis; *Deckenia* is distinct in the shape of the adaxial epidermal cells but resembles *Oncosperma* most closely in fibre and mesophyll structure and distribution.

Oncosperma (Fig. 87C, G, L)

Leaf

Pinnately reduplicate, the leaflets variably planar or multifarious, characteristically pendulous.

LAMINA

Dorsiventral (Fig. 87C). **Trichomes** (Fig. 87G) common, either on both surfaces or restricted to abaxial surface in some material of *O. tigillarium*, mostly in costal regions, each with a deeply sunken base of 2–4 persistent tiers of irregular sclerotic cells, the distal thin-walled cells ephemeral. **Epidermis** with outer wall sl. thickened and cutinized. Adaxial epidermis with rhombohedral, obliquely-extended cells. Abaxial epidermis with narrow costal bands of longitudinally extended, often rectangular cells, intercostals bands wider and composed of irregular cells. **Stomata** in intercostal bands; t.s.c. usually short, l.s.c. wide. Guard cells rather small, each with 2 equal cutinized ledges. **Hypodermis** conspicuous, 2–3-layered below each surface, adaxial cells cubical or somewhat transversely extended; abaxial cells smaller, longitudinally extended below veins, elsewhere ± cubical, irregular. **Mesophyll** with or without a distinct palisade, otherwise uniformly compact. **Fibres** solitary in the mesophyll mostly towards each surface. **Veins** equidistant from each surface or somewhat abaxial and mostly independent of surface layers; O.S. incomplete above and below all veins, cells cubical and somewhat chlorophyllous; I.S. only fibrous above and below most smaller veins; phl of larger veins separated into 2 or more strands by sclerotic parenchyma. **Transverse veins** few, at same level as longitudinal veins; sheathed by sclerotic parenchyma. **Midrib** with a sclerotic peripheral layer enclosing separate vbs. **Expansion cells** associated with midrib and other large veins (Fig. 87C).

Cell inclusions

Stegmata in discontinuous files; silica bodies spherical, each enclosed by massive basal wall of silica cell.

Petiole

V-arrangement of central vbs evident. Central vbs each with 1 or more mxy vessels seen often as a complex, phl in 2 strands. Small vbs and fbs common.

Stem (Fig. 87L)

Central ground tissue cells becoming horizontally extended (Fig. 87L), sometimes forming wide lacunae outlined by multilayered parenchyma bands; vbs mostly with 1 wide mxy vessel.

Root

Mucilage canals frequent in outer **cortex**, inner cortex lacunose. **Fibres** frequent, more or less solitary, thin-walled in outer cortex, inner cortical fibres thicker-walled and grouped.

Vascular elements

Vessels: in petiole mxy elements with scalariform perforation plates with numerous thickening bars on oblique end walls, in stem also with scalariform perforation plates, but fewer thickening bars, in root with simple transverse perforation plates. **Sieve tubes**: with sieve plates compound in all parts, the sieve areas often of the 'arecoid-type'; the lateral sieve areas sometimes wider than the sieve areas of the sieve plates.

Acanthophoenix

A single species, *A. rubra*, almost extinct in Mauritius and Reunion and uncommonly cultivated. The leaf sheath is fiercely armed with long spines; these appear as short bristles on the ribs of the lamina. A moderate-sized, single-stemmed palm.

Leaf

Pinnate, abscising as a single unit.

LAMINA (Fig. 11C)

Dorsiventral. Adaxial **epidermis** without stomata or hairs. Cells thin-walled with thinly cutinized outer layers; in distinct files, usually longitudinally extended but with oblique end walls, the costal bands not clearly differentiated. Cell walls sl. sinuous. Abaxial epidermis with narrow costal bands, the stomata in broad intercostal bands. **Trichomes** present, randomly distributed and conspicuously tanniniferous; each with a uniseriate base of thick-walled cells extending into 1–2 seriate groups of distal thin-walled cells. Intercostal cells shorter and less regular than costal cells, walls scarcely sinuous. **Stomata** with evident trapezoid cells and short t.s.c. **Hypodermis** 1-layered below each surface, with regular adaxial files of cubical to hexagonal sl. transversely-extended cells. Abaxial hypodermis thin-walled but not differentiated as a colourless layer with 4 L-shaped cells around the s.st.ch. **Mesophyll** extensively tanniniferous with 1–2 palisade layers, compact. Non-vascular fibres in mesophyll few, solitary in strands of 2–3, unlignified, mostly toward adaxial surface, sometimes in contact with adaxial hypodermis. **Veins** independent of surface layers, the largest sometimes extending to adaxial hypodermis. O.S. developed mostly laterally; I.S. sclerotic but distinctive with upper and lower fibres not or little lignified compared to median lignified sclerenchyma. **Transverse veins** few, narrow, but with a sclerotic sheath (Fig. 11C). **Raphide sacs** few but large and conspicuous in cleared material. **Stegmata** present next to vascular and non-vascular fibres; silica body small and irregularly spherical.

Notes: *Acanthophoenix* is distinguished by limited fibre development and the presence of unlignified vascular fibres. The short bristles on the lamina ribs are easily overlooked in microscope preparations. It shares a number of features with it putative relative *Deckenia*.

Tectiphiala

Endemic to Mauritius, relatively recently described and rare in cultivation. The lamina anatomy of this genus is distinctive in many features, but notably in the coiled thin-walled cells of the indumentum.

LAMINA

Dorsiventral. **Trichomes**; v. distinctive, restricted to abaxial intercostal regions and leaf margin. Each trichome with a sunken basal group of thick-walled sclerotic pitted cells expanded distally into elongated thin-walled cells, these cells **collectively coiled** in 1 plane as in a coiled rope, the indumentum consequently totally closed as a canopy over the intercostal abaxial bands. **Epidermis** with outer wall extensively thickened and lamellate to more than half its depth, the cutinized portion leaving uncutinized arc-shaped portions in TS. Adaxial cells without costal bands, the cells hexagonal but ± isodiametric or sl. obliquely extended; crystalline aggregates conspicuous at the boundary between epidermis and hypodermis. Abaxial cells smaller, less regular. **Stomata** diffuse within files of short cells, sunken, the outer cavity often wax occluded; t.s.c short and wide; g.c. thin-walled, the outer cuticular ledge pronounced. **Hypodermis** 2-layered adaxially, the inner layer thin-walled and discontinuous. **Mesophyll** with a 3-layered palisade of markedly anticlinally extended cells. **Fibres** v. infrequent as solitary or few-celled strands mostly adaxial and next to the inner hypodermis. **Veins** dense, in abaxial mesophyll, the smaller independent of surface layers, the larger connected to the adaxial surface by colourless palisade-like cells. O.S. incomplete above and below most veins; I.S. completely fibrous around large veins and well developed as fibres above and below small veins. Phl of large veins divided by sclerotic cells into separate strands. **Transverse veins** few, narrow, sheathed by sclerotic parenchyma.

Deckenia

A solitary stemmed palm with well-developed crown shaft, endemic to the Seychelles; frequently cultivated.

LAMINA

Dorsiventral. **Trichomes** restricted to abaxial surface, the base only sl. sunken. **Epidermis** with thickened outer wall half its depth; adaxially v. uniform the cells rectangular to rhombohedral, elongated but with oblique end walls; abaxial epidermis similar but less regular. **Stomata** diffuse, not sunken but l.s.c. deeper than epidermal cells, the g.c. thin-walled each with 2 cutinized ledges; t.s.c. sometimes short, but otherwise not well differentiated. **Hypodermis** 1-layered below each surface, adaxial cells large, often hexagonal and transversely extended, abaxial hypodermis less conspicuous. **Mesophyll** with a well-developed palisade, 1- or more-layered. **Fibres** scattered in the mesophyll, solitary or in 2- or more-celled strands, often next to hypodermis. **Veins** in abaxial mesophyll, dense, not in contact with surface layers, fibres of I.S. well developed above and below larger veins; phl of larger veins divided into several strands by sclerotic partitions. **Transverse veins** few, at the same level as the longitudinal veins, sheathed by sclerotic parenchyma.

SUBTRIBE PTYCHOSPERMATINAE

(Figs. 8D; 13B, C, J; 17 A–D; and 88A, F, M, N)

A group of 12 genera and 70 spp., many are commonly cultivated. The genera (as listed in Dransfield et al. 2008b) are: *Ptychosperma, Ponapea, Adonidia, Solfa, Balaka, Veitchia, Carpentaria, Wodyetia, Drymophloeus, Normanbya, Brassiophoenix,* and *Ptychococcus*. Most are small, often understorey and single-stemmed palms, but multiple-stemmed in some commonly cultivated species of *Ptychosperma* (e.g. *P. elegans* single-, *P. macarthurii* multiple-stemmed). The group has a restricted distribution in east Malesia as far as Fiji and northern Australia, with discontinuities in the Philippines and Ponapea. A crownshaft with infrafoliar inflorescences is always present and leaf morphology is frequently distinctive, with broad subdivided leaflets and especially a truncate apex, with ragged (praemorse) edges or the leaflets even trilobed (*Brassiophoenix*). Male flowers have numerous stamens and are consequently larger than female flowers. Fruits are rather uniform and always with a basal embryo, the endosperm either homogenous, or ruminate (e.g. *Normanbya*). In general these palms make ideal ornamentals if they can accommodate local climate and soils. *Wodyetia bifurcata* from Queensland is a striking example of a recently described species (Irvine 1983) rapidly adopted as a commercially successful landscape palm. *Adonidia merrillii* from the Philippines is an example of a palm successful in S Florida because its natural habitat is on limestone. The smaller stems can become quite woody and have value as walking-sticks (e.g. *Balaka* spp.).

We have examined some material of all genera but the subtribe is anatomically quite uniform so that variation within a single individual may be as much as between different taxa and is only outlined here. The general anatomical features include:

Leaf

Pinnate leaf with clean abscission (Fig. 13B) after splitting on ventral side of sheath (Fig. 13C, J).

LAMINA (Figs. 8D and 88A, F)
Always dorsiventral (Fig. 88A). **Trichomes** usually restricted to abaxial epidermis, each with a sunken base of 2 or more tiers of shallow cells with pitted walls; shield cells often filamentous (Fig. 8D). **Epidermal cells** thin-walled and with a shallow cuticle. Adaxial epidermis: cells with oblique end walls, rhombohedral, without well-developed costal bands; walls at most undulate or rarely sl. sinuous. Abaxial epidermis (Fig. 88F) without well-developed costal bands, the stomatal bands with irregularly oblique end walls, varying from rhombohedral to somewhat spindle-shaped; **stomata** not sunken, diffusely distributed the g.c. large (to 36 µm long); thin-walled and with 2 equal cutinized ledges; t.s.c. short and overarching g.c. somewhat. **Hypodermis** (Fig. 88A) thin-walled and 1-layered below each surface; adaxial cells ± hexagonal in surface view, transversely to obliquely extended, the angle of orientation contrasted with the obliquely-oriented epidermal cells; abaxial cells smaller ± cubical and commonly with 4 L-shaped cells surrounding the s.st.

ch. **Mesophyll** uniform, either without a palisade layer or with a 1–3-layered palisade. **Fibre strands** sometimes absent, or with inconspicuous strands of 1–5 cells scattered in the mesophyll, with or without contact with the epidermis. **Veins** (Fig. 88A) situated in the mid to lower mesophyll, not in contact with the hypodermis. O.S. inconspicuous, usually present only laterally; I.S. of fibres above and below. **Transverse veins** few, narrow, situated in the mid-mesophyll and sheathed by few sclerotic parenchyma cells. **Midrib** (and marginal rib where present) with a continuous sclerotic cylinder including 1 wide abaxial and smaller adaxial vbs.

Petiole

Surface with a sclerotic layer, abaxially with massive vbs and often a distinct 'keel' bundle. **Central** V-configuration of vbs not obvious, but shown by orientation of central vbs, each with 1 wide mxy vessel and 2 phl strands. **Fibrous strand**s present intermixed with narrow axial vbs, but probably only in proximal region.

Stem

Most central vbs with 1 wide mxy vessel, ('1-vessel type'). **Ground parenchyma** cells becoming variously enlarged and/or transversely extended, especially in wider diameter stems, often producing a lattice-like texture with vbs connected by plates of narrow cells (Fig. 88M). *Normanbya* may be unique in the very heterogenous texture, with large groups of many small cells infilling the space between the lattice of transversely extended cells (Fig. 17A–D).

Root

Stele of wider roots tending to become fluted (Fig. 88N), appearing lobed in TS with cortex and medulla continuous through an interrupted endodermis. **Cortex** often with complex outer layers and developing a periderm. Outer cortex frequently including wide raphide-sacs or raphide-canals. Outer sclerotic layer of cortex gradually transitional to fibrous strands of the 'Kentia-type.' **Endodermis** with U-shaped wall thickenings. **Vascular tissues** mostly confined to outer portion of the stele; medullary vessels consequently uncommon. **Medulla** including fibre-strands as in the cortex, the cells narrower and thinner-walled.

Vascular tissues

Mxy vessels in root with simple or short scalariform perforation plate; in stem and petiole with scalariform perforation plate on oblique end walls; those in the petiole with numerous narrow thickening bars. **Sieve tubes** in all organs with compound sieve plates on transverse to sl. oblique end walls, often with appreciable diversity in a single organ.

Cell inclusions

Stegmata with spherical silica bodies. Stegmata present as short files in root in some taxa, either next to the cortical fibres or more typically next to the inner surface cells of the thick-walled conjunctive tissue at its boundary with the thin-walled medulla.

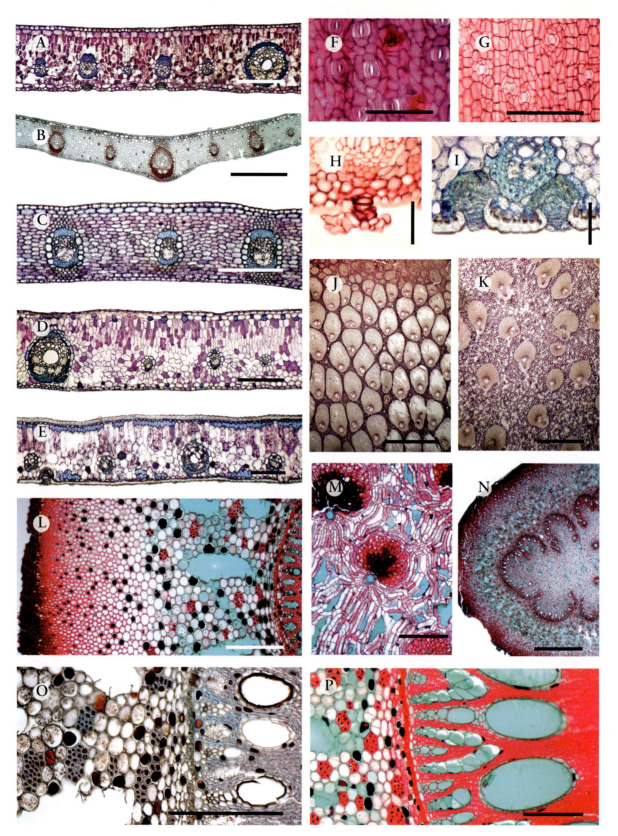

Fig. 88 Arecoideae (Areceae: Ptychospermatinae [Ptych.], Rhopalostylidinae [Rhop.], Verschaffeltiinae [Vers.], Unplaced genera [Unpl.]. **A**. *Ptychosperma macarthurii* [Ptych.], lamina, TS (bar = 200 μm). **B**. *Rhopalostylis sapida* [Rhop.], lamina, TS (bar = 500 μm). **C**. *Verschaffeltia splendida* [Vers.], lamina, TS (bar = 200 μm). **D**. *Bentinckia nicobarica* [Unpl.], lamina, TS (bar = 200 μm). **E**. *Cyrtostachys renda* [Unpl.], lamina, TS (bar = 100 μm). **F**. *Ptychococcus lepidotus* [Ptych.], lamina abaxial epidermis, SV (bar = 100 μm). **G**. *Rhopalostylis sapida* [Rhop.], lamina, both abaxial epidermis (elongated cells) and hypodermis (square cells) in focus, SV (bar = 200 μm). **H**. *Rhopalostylis sapida* [Rhop.], lamina, abaxial epidermis with trichome, TS (bar = 50 μm). **I**. *Cyrtostachys renda* [Unpl.], lamina, abaxial epidermis with trichomes, TS (bar = 50 μm). **J**. *Rhopalostylis sapida* [Rhop.], Stem periphery, TS (bar = 2 mm). **K**. *Rhopalostylis sapida* [Rhop.], stem centre, TS (bar = 2 mm). **L**. *Heterospathe humilis* [Unpl.], root periphery to stele, TS (bar = 200 μm). **M**. *Veitchia joannis* [Ptych.], stem centre, vb in old region, TS (bar = 1 mm). **N**. *Ptychococcus lepidotus* [Ptych.], root, fluted stele to periphery, TS (bar = 2 mm). **O**. *Heterospathe humilis* [Unpl.], Root, outer stele and inner cortex, TS (bar = 200 μm). **P**. *Hydriastele macrospadix* [Unpl.], root, outer stele and inner cortex, TS (bar = 500 μm).

SUBTRIBE RHOPALOSTYLIDINAE

A geographically isolated group (*Hedyscepe* endemic to Lord Howe Island, *Rhopalostylis* to New Zealand and outlying islands, including Norfolk Island). Trees are single-stemmed and distinguished from other Areceae by somewhat technical characters, e.g. the asymmetric male flowers. *Hedyscepe* is protogynous, an unusual condition in palms.

Rhopalostylis (Figs. 6E, F; 8G; 9A; 88B, G, J, K)

Two or 3 species known by the Maori name 'nikau'. The stiffly erect leaves and bulbous crownshaft produce a distinctive shuttlecock appearance.

Leaf (Figs. 6E, F; 8G; 9A; and 88B, G)

Pinnately reduplicate with subopposite leaflets, each with a prominent adaxial midrib and a small adaxial subsidiary rib halfway between the midrib and the unribbed margin.

LAMINA

Dorsiventral (Fig. 88B). **Trichomes** frequent, mostly restricted to abaxial costal regions, each with a sunken base of 2–4 tiers of sclerotic pitted cells (Figs. 6F and 8G) expanded distally into a shield of thin-walled cells, the marginal cells irregularly filamentous. **Epidermis** with outer wall sl. thickened, cutinized layer thin. Adaxial epidermis v. uniform, cells obliquely extended, rhombohedral to polygonal (Fig. 6E). Abaxial epidermis (Fig. 88G) with costal regions of somewhat longitudinally-extended cells below larger veins, the intercostal regions wide and of irregular cells. **Stomata** diffuse in intercostal regions (Fig. 6F), the t.s.c. cells usually short but not well differentiated (Fig. 9A). Guard cells thin-walled, each with 2 cutinized ledges. **Hypodermis** 2-layered adaxially, at least in the basal part, cells cubical or sl. transversely extended, 1(2)-layered abaxially, the cells less regular and usually with 4 L-shaped cells around each s.st.ch. **Mesophyll** without distinct palisade layers (Fig. 88B) but with the adaxial layers more compact than elsewhere, abaxial mesophyll cells loosely lobed next to s.st.ch. **Fibres** (Fig. 88B) few, scattered in the mesophyll, either solitary or in few-celled strands, mostly towards each surface but rarely in contact with hypodermis. **Veins** equidistant from both surfaces, always independent of surface layers, the largest separated from the abaxial hypodermis by at least 1 mesophyll cell layer and adaxially connected to the hypodermis by cells that resemble an extension of the O.S. O.S. forming a layer of cubical cells ± complete around most veins. I.S. well-developed with fibres above and below all veins but completed laterally by sclerotic parenchyma; phl of larger veins divided into separate strands by sclerenchyma. **Transverse veins** few, narrow and at same level as longitudinal veins and sheathed by sclerenchyma. **Midrib** with a sclerotic cylinder enclosing 1 large abaxial vb and smaller adaxial vbs. Subsidiary ribs intermediate in structure between midrib and larger veins of lamina, the complete hierarchy continued to the small veins. **Expansion cells** in abaxial bands each side of midrib and less commonly elsewhere.

PETIOLE

Central vbs including distinct V-shaped arcs adaxially. Each vb usually with a well-developed fibrous phl sheath, but parenchyma elsewhere, a single wide mxy vessel and the phl divided into 2 separate strands by a median sclerotic partition. Ground tissue including numerous scattered fibrous strands.

Stem (Fig. 88J, K)

Cortex (Fig. 88J) narrow, including crowded fibrous strands and narrow leaf traces. Outer part of central cylinder (Fig. 88J) with congested **axial vbs** with massive fibrous sheaths, but the fibres only becoming thick-walled with age. Gradual transition to the diffuse vbs of the stem centre but each still with a well-developed fibre sheath (Fig. 88K); most vbs with a single wide mxy element (1-vessel palm). **Ground parenchyma** of the stem centre little lignified and becoming somewhat lacunose with age.

Root

Outer **cortex** becoming sclerotic but inner layers not markedly differentiated. Inner cortex with regular and well-developed radially-extended air lacunae formed by cell breakdown, the intervening persistent parenchyma including numerous fibre strands of the 'Kentia-type'. **Stele** wide but without medullary vessels. **Endodermis** with U-shaped wall thickenings.

Cell inclusions

Stegmata in the lamina in continuous series, the basal wall becoming lignified; silica bodies ± spherical. **Raphide sacs** common in leaf and stem, usually conspicuous.

Hedyscepe

The lamina of this sp. corresponds closely to that described above, but the anatomy is somewhat more robust, with the hypodermis sometimes 3-layered and the mesophyll fibres more numerous.

SUBTRIBE VERSCHAFFELTIINAE

A subtribe of 4 monospecific genera of moderate-sized single-stemmed palms endemic to the Seychelles. They are distinctive in the development of stem and leaf spines on juveniles that are largely lacking from adults. Leaves are irregularly or regularly pinnately compound, that of *Phoenicophorium* initially almost entire. Although leaves abscise cleanly, only *Roscheria* develops a distinct crownshaft. *Verschaffeltia* is distinctive in its basal cone of aerial stilt-roots (Fig. 19F), although the above-ground axis is not tapered below as in, e.g. *Socratea*. Within the group there is only minor anatomical variation and only 1 genus is described in detail, but with a synoptic key to suggested differences from the other genera.

Verschaffeltia (Fig. 88C)
Leaf

Reduplicately pinnate but irregularly split into segments with several folds.

LAMINA

Dorsiventral (Fig. 88C). **Trichomes** present on both surfaces but mostly restricted to abaxial intercostals regions, each with a sunken base of 2–4 tiers of sclerotic, pitted cells. **Epidermis** with outer wall sl. thickened, cutinized, walls at most sl. sinuous. Adaxial epidermis with infrequent bands of elongated costal cells, the intercostal cells irregular but ± isodiametric, rhombohedral to polygonal. Abaxial epidermis with costal bands below larger veins, the cells like those of the adaxial epidermis. **Stomata** restricted to abaxial intercostals regions but not in v. distinct files. T.s.c. not well differentiated from other epidermal cells but overarching g.c. somewhat. **Hypodermis** 1(–2)-layered adaxially, adaxial cells somewhat transversely extended; mostly 1-layered abaxially, often with 4 L-shaped cells around s.st.ch. **Mesophyll** without an adaxial palisade; the cells somewhat transversely extended (Fig. 88C), abaxial cells adjacent to s.st.ch. rather loose, with small extensions. **Fibres** few, in small strands absent from hypodermal layers, most commonly adaxial, but solitary fibres also occasional within central mesophyll. **Veins** (Fig. 88C) mostly in abaxial mesophyll, smaller independent of surface layers, larger veins attached to hypodermis by small mesophyll cells; O.S. of cubical or somewhat axially extended cells; I.S. fibrous above and below veins, smaller completed laterally by sclerotic parenchyma. **Transverse veins** infrequent, at the same level as the longitudinal veins, narrow, sheathed by sclerotic parenchyma; only extending between minor veins.

Cell inclusions

Stegmata with spherical silica bodies, the basal wall thickened; stegmata tending to be in clusters rather than long linear series.

Notes: this subtribe shows considerable anatomical uniformity of hair type, epidermal cell shape, absence of a palisade, vein structure, transverse vein distribution, and clustered stegmata. The range of structural variation suggests that genera can be distinguished as follows:

1. Non-vascular fibres abundant in adaxial hypodermis, alternating with files of cubical and scarcely transversely extended cells 2
1A. Non-vascular fibres absent from hypodermal layers, adaxial hypodermal cells hexagonal and transversely extended 3
2. Non-vascular fibres abundant in hypodermis, the adaxial fibres alternating with files of cubical cells . . . . *Phoenicophorium*
2A. Non-vascular fibres less abundant in hypodermis, the adaxial fibres associated with files of somewhat transversely extended cells . . . .*Nephrosperma*
3. Lamina thin (<15 cells deep); g.c. of stomata with thickened outer walls . . . . *Roscheria*
3A. Lamina thick (20+ cells deep); g.c. of stomata with inner and outer walls equally thickened . . . .*Verschaffeltia*

Root

The stilt roots (Fig. 19F) have a massive fluted stele that is essentially polystelic, as illustrated in Tomlinson (1961). The cortex

includes fibres of the 'Kentia-type', with continuity into the medulla through the extensively interrupted endodermis.

UNPLACED ARECEAE (Fig. 88D, E, I, L, O, P)

A miscellany of 10 genera and about 150 spp., still unplaced phylogenetically within the tribe. It includes the monotypic *Dransfieldia, Dictyosperma,* and *Loxococcus* as well as sizeable genera with 30–40 spp. (*Heterospathe, Hydriastele,* and *Iguanura*). The other genera are *Bentinckia, Clinostigma, Cyrtostachys,* and *Rhopaloblaste.* They remain unresolved on the basis of several lines of evidence. We have examined limited material of all genera but insufficient to help in resolving their taxonomic position. Although they do show features indicative of a place within the Areceae and have some anatomically diversity, the general level of variation corresponds to that seen in all subtribes in the group. The approach for the future therefore requires a detailed study of more spp. in a much broader context. Here we present vignettes that should provide an introductory guide.

Geographically they range from as far west as *Dictyosperma* in the Mascarene Islands, *Loxococcus* in Sri Lanka, and *Bentinckia* in S India and the Nicobar Islands, eastwards through SE Asia to Fiji and Samoa. Some are commonly cultivated, as is *Cyrtostachys renda* (sealing-wax palm) and *Dictyosperma album* (Princess palm). All are small to moderate-sized, mostly single-stemmed palms with a well-developed crownshaft. There is some indication that lamina anatomy is correlated with stature and habitat (e.g. *Iguanura*).

Common anatomical features of unplaced Areceae

LAMINA

Dorsiventral (Fig. 88E). **Trichomes** common, mostly below the longitudinal veins; each with a sclerotic sl. to deeply (Fig. 88I) sunken base of a few tiers of pitted sclerotic cells extending distally into thin-walled ephemeral cells, sometimes filamentous. **Epidermis** with cells usually obliquely extended and varying from rhombohedral to spindle-shaped, the cell walls mostly straight, rarely cells rectangular or with sinuous walls. Cell walls varying from thin, with a thin cuticle, to cells with a thickened wholly cutinized outer wall. Hypodermis usually 1-layered below each surface, sometimes absent; variously associated with or replaced by fibres. Adaxial hypodermal cells often hexagonal and transversely or obliquely extended. Abaxial hypodermis usually with 4 L-shaped cells surrounding the s.st.ch. **Mesophyll** varying in depth from as few as 5–6 to as many as 15 cell layers, palisade absent to well-developed and 2–3-layered. **Fibres** varying from absent to few to many, sometimes restricted to the hypodermis or diffusely distributed in the mesophyll, either solitary or in strands, but never very large. **Veins** in the abaxial mesophyll except central in the thinnest leaves, not in contact with the surface layers, at most the larger veins continuous with the surface layers via small mesophyll cells or in a few taxa with continuity to the adaxial surface via an inconspicuous girder of anticlinally extended sl. thick-walled cells. **Transverse veins** at the level of the longitudinal veins, short, narrow and sheathed with sclerotic parenchyma. **Midrib** with a sclerotic cylinder enclosing vbs. Marginal rib sometimes including 1 or more large vbs.

PETIOLE
V-shaped arrangement of central bundles common; abaxial 'keel' bundle often pronounced; individual vbs with a single wide mxy vessel and 2 or sometimes several phl strands.

Stem
Not studied.

Root (Fig. 88L, O, P)
Surface layers becoming sclerotic with age (Fig. 88L). **Cortex**; outer layers with often conspicuous raphide canals in the outer cortex. 'Kentia-type' fibres commonly developed, the innermost fibre strands often with tanniniferous contents (Fig. 88O, P). **Endodermis** with U-shaped wall thickenings. **Medulla** with central parenchyma but vessels uncommon.

Cell inclusions
Raphide sacs sometimes large and prominent. **Stegmata** all with spherical silica bodies.

Some diagnostic anatomical features of individual genera of unplaced Areceae
Bentinckia (Fig. 88D)
This falls well within the range of variation of lamina anatomy of the Areceae, with rhombohedral adaxial epidermal cells, **hypodermis** with included fibres solitary or in few-celled groups, a 2–3-layered palisade and veins in the abaxial mesophyll and mostly independent of the abaxial surface, the larger **veins** with 4 or more separate phl strands (Fig. 88D).

Clinostigma
Epidermis with thick outer wall. **Hypodermis** 1-layered below each surface. Hypodermis with associated fibres, below adaxially and within abaxially. Palisade well-developed.

Cyrtostachys
LAMINA (Fig. 88E, I)
Dorsiventral (Fig. 88E). **Trichomes** frequent on both surfaces but most numerous abaxially, each with a deeply sunken flask-shaped base of 1–2 tiers of pitted sclerotic cells (Fig. 88I) extended into a filament of thin-walled cells. **Epidermis** with outer wall considerably thickened and wholly cutinized. Adaxial epidermis with ± rectangular cells and sinuous walls. Abaxial epidermis with similar costal cells, the intercostal cells shorter wider and often more rhombohedral. **Hypodermis** absent, but adaxially occupied by an almost continuous layer of fibres (Fig. 88E); abaxially fibres less common in small flattened strands. **Mesophyll** with a 1-layered palisade. **Veins** in abaxial mesophyll mostly independent of surface layers except for a few large veins in contact with abaxial epidermis.

Note: the presence of hypodermal fibres instead of colourless cells is unusual in lamina of the Areceae.

PETIOLE
Central vbs each with a single mxy vessel, phl subdivided by sclerotic partitions. Ground parenchyma including small air lacunae.

Root
Including cortical fibre strands of the 'Kentia-type'.

Dictyosperma
LAMINA
Dorsiventral. Adaxial **epidermal cells** rhombohedral but ± isodiametric. Hypodermis 1-layered, including wide fibre strands adaxially. Palisade present. Larger veins attached to adaxial surface by anticlinally extended cells.

Dransfieldia
LAMINA
Thin, mesophyll c. 5–6 cells deep, without a palisade. Hypodermis conspicuously 1-layered below each surface. Fibres infrequent, solitary, or in small strands of 2–3 cells scattered in mesophyll, mostly adaxially.

Heterospathe (Fig. 88L, O)
A large genus extending from the Philippines through New Guinea to the western Pacific.

H. (= Gronophyllum) pleurocarpum
Epidermis large-celled, thin-walled. Stomata v. diffuse. **Mesophyll** c. 7 cells deep, without a palisade. **Fibres** in irregular strands.

H. (= Gulubia) macrospadix
Lamina thick. **Epidermis** with outer wall somewhat thickened. Adaxial hypodermal cells hexagonal, obliquely-transversely extended. **Mesophyll** palisade present. **Fibres** in mesophyll as narrow strands, not in contact with surface layers. Larger veins attached to surface layers by sl. thick-walled anticlinally extended cells.

H. phillipsii
Hypodermis obscurely differentiated abaxially. **Fibres** in narrow strands next to ad- and abaxial hypodermis. Palisade absent.

H. solomonensis
Larger **veins** with multiple phl strands.

Hydriastele (Fig. 88P)
Another large genus with an even greater range than *Heterospathe* (Sulawesi and Australia). The following 2 examples indicate appreciable species differences:

H. rostrata

Lamina thin, 5–6 cells deep. **Hypodermis** absent. **Fibres** abundant, solitary, or in small strands mostly towards or in contact with each surface.

H. vitiensis

Lamina thick. Trichomes deeply sunken. Outer wall of epidermis thick. **Hypodermis** shallow, 1-layered below each surface. **Mesophyll** palisade well-developed. **Fibres** common, mostly in adaxial mesophyll, sometimes next to hypodermis. Larger veins attached to adaxial hypodermis by sl. thick-walled anticlinally extended cells.

Iguanura

Lamina thin, mesophyll c. 7 cells wide. **Trichomes** with extended apical filament of thin-walled cells. Epidermis thin-walled, cells rhombohedral to spindle-shaped. **Hypodermis** and palisade absent. **Mesophyll** with only 3–4 adaxial layers extensively chlorophyllous. **Fibres** absent except for occasional solitary cells. **Veins** occupying c. three-quarters of mesophyll depth but separated from surface by 1–2 layers of narrow cells.

Note: the reduced anatomical features may be associated with the small stature and understorey habitat of these palms.

Loxococcus

Lamina thick, to 10 cells deep. **Epidermis** thin-walled. **Hypodermis** 1-layered below each surface, including numerous fibres interspersed within files of small cubical cells plus scattered fibres within the mesophyll. **Mesophyll** palisade 2-layered. **Stegmata** often in clusters as well as linear series.

Rhopaloblaste

Epidermal cells variously rectangular to rhombohedral. **Hypodermis** 1-layered below each surface, cells ± cubical, with included fibres. **Mesophyll** with 2-layered palisade and further included fibres.

BIBLIOGRAPHY

With few exceptions we do not cite papers that were summarized in Tomlinson (1961) where all the literature before that date is covered. A further fairly complete bibliography before 1990 is found in Tomlinson (1990). The most recent comprehensive bibliography on palms is found in Dransfield et al. (2008b) which also covers anatomical literature to some extent. Our bibliographic treatment is then in accord with the principal objective of our book, which is to provide information that relates to structural changes that have occurred as palms have diversified in evolutionary time, rather than describe in encyclopaedic detail all palm anatomy within a rigid systematic context.

Abasolo, W., Yoshida, M., Yamamoto, H., and Okuyama, T. (1999). Internal stress generation in rattan canes. *International Association of Wood Anatomists Journal*, **20**, 45–58.

Anderson, A.B. and Balick, M.J. (1988). Taxonomy of the babassu complex (*Orbignya* spp.: Palmae). *Systematic Botany*, **13**, 32–50.

Arber, A. (1925). *Monocotyledons: A Morphological Study*. Cambridge University Press, Cambridge.

Asmussen, C.B., Dransfield, J., Deickmann, V., Barfod, A.S., Pintaud, J.-C., and Baker, W.J. (2006). A new subfamily classification of the palm family (Arecaceae): evidence from plastid DNA phylogeny. *Botanical Journal of the Linnean Society*, **151**, 15–38.

Bacon, C.D., Baker, W.J., and Simmons, M.P. (in subm.). Miocene dispersal drives island radiations in Trachycarpeae (Arecaceae/Palmae).

Baker, W.J., Dransfield, J., and Hedderson, T.A. (2000a). Phylogeny, character evolution, and a new classification of the Calamoid palms. *Systematic Botany*, **25**, 297–322.

Baker, W.J., Hedderson, T.A., and Dransfield, J. (2000b). Molecular phylogenetics of subfamily Calamoideae (Palmae) based on nrDNA ITS and cpDNA rps16 intron sequence data. *Molecular Phylogenetics and Evolution*, **14**, 195–217.

Baker, W.J., Hedderson, T.A., and Dransfield, J. (2000c). Molecular phylogenetics of *Calamus* (Palmae) and related rattan genera based on 5S nrDNA spacer sequence data. *Molecular Phylogenetics and Evolution*, **14**, 218–31.

Baker, W.J., Savolainen, V., Asmussen-Lange, C.B., Chase, M.W., Dransfield, J., Forest, F., Harley, M.M., Uhl, N.W., and Wilkinson, M. (2009). Complete generic-level phylogenetic analyses of palms (Arecaceae) with comparisons of supertree and supermatrix approaches. *Systematic Biology*, **58**, 240–56.

Balick, M.J. (1986). Systematics and economic botany of the *Oenocarpus-Jessenia* (Palmae) complex. *Advances in Economic Botany*, **3**, 1–140.

Balick, M.J. and Gershoff, S.N. (1981). Nutritional evaluation of the *Jessenia batua* palm: source of high quality protein and oil from tropical America. *Economic Botany*, **35**, 261–71.

Balick, M.J. and Johnson, D. (1994). The conservation status of *Schippia concolor* in Belize. *Principes*, **38**, 124–8.

Barfod, A.S. (1988). Leaf anatomy and its taxonomic significance in Phytelephantoid palms. *Nordic Journal of Botany*, **8**, 341–8.

Barfod, A.S. (1991). A monographic study of the subfamily Phytelephantoideae (Arecaceae). *Opera Botanica*, **105**, 1–73.

Barrow, S. (1998). A revision of *Phoenix*. *Kew Bulletin*, **53**, 513–75.

Barsikow, M. (1901). Über das sekundäre Dickenwachstum der Palmen in den Tropen. *Verhandlungen der Physikalisch-Medicinischen Gesellschaft in Wurzburg*, N.F. **35**, 213–45. See *Botanisches Zeitung* (1902), **60**, Abt. II, pp. 42–3.

Barthlott, W. and Thiesen, I. (1998). Epicuticular wax ultrastructures. In: Kubitzki, K. (ed.) *The Families and Genera of Vascular Plants III. Flowering Plants. Monocotyledons*, pp. 20–2. Springer-Verlag, New York.

Barthlott, W., Neinhuis, C., Cutler, D., Ditsch F., Meusel, I., Theisen, I., and Wilhelmi, H. (1998). Classification and terminology of plant epicuticular waxes. *Botanical Journal of the Linnean Society*, **126**, 236–60.

Bayton, R.P. (2005). *Borassus* L. and the borassoid palms: systematics and evolution. PhD. Thesis. The University of Reading.

Bayton, R.P. (2007). A revision of *Borassus* L. (Arecaceae). *Kew Bulletin*, **62**, 561–86.

Bernal, R. (1998). The growth form of *Phytelephas seemanii* – a potentially immortal solitary palm. *Principes*, **42**, 15–23.

Behnke, H.-D. (1981). Sieve element characters. *Nordic Journal of Botany*, **1**, 381–400.

Bhat, K.M., Liese, W., and Schmitt, U. (1990). Structural variability of vascular bundles and cell wall in rattan stems. *Wood Science and Technology*, **24**, 211–24.

Broschat, T.K. and Donselman, H.M. (1984). Root regeneration in transplanted palms. *Principes*, **28**, 90–1.

Burkill, I.H. (1935). *A dictionary of the economic products of the Malay Peninsula*, 2 vols. London.

Caissard, J.-C., Meekijjironenroj, A, Baudino, S., and Anstett, M.-C. (2004). Localization of production and emission of pollinator attractant on whole leaves of *Chamaerops humilis* (Arecaeae). *American Journal of Botany*, **91**, 1190–9.

Cardon, J.-O. (1978). Aerial roots in *Raphia*. *Principes* **22**, 136–41.

Cheadle, V.I. (1942). The occurrence and types of vessels in the various organs of the plant in the Monocotyledoneae. *American Journal of Botany*, **29**, 441–50.

Cheadle, V.I. (1943a). The origin and certain trends of specialization of the vessel in the Monocotyledoneae. *American Journal of Botany*, **30**, 11–17.

Cheadle, V.I. (1943b). Vessel specialization in the late metaxylem of the various organs in the Monocotyledoneae. *American Journal of Botany*, **30**, 484–90.

Cheadle, V.I. (1944). Specialization of vessels within the xylem of each organ in the Monocotyledoneae. *American Journal of Botany*, **31**, 81–92.

Cheadle, V.I. (1948). Observations on the phloem in the Monocotyledoneae. II. Additional data on the occurrence and phylogenetic specialization in structure of the sieve tubes in the metaphloem. *American Journal of Botany*, **35**, 129–31.

Cheadle, V.I. and Uhl, N.W. (1948). Types of vascular bundles in the Monocotyledoneae and their relation to the late metaxylem conducting elements. *American Journal of Botany*, **35**, 486–96.

Cheadle, V.I. and Whitford, N.B. (1941). Observations on the phloem in the Monocotyledoneae. I. The occurrence and phylogenetic specialization in structure of the sieve-tubes in the metaphloem. *American Journal of Botany*, **28**, 623–7.

Cormack, B.G. (1896). On polystelic roots of certain palms. *Transactions of the Linnean Society of London, Botany, ser 2*, **5**, 275–86.

Crisp, M.D., Isagi, Y., Kato, Y., Cook, L.G., and Bowman, D.M.J.S. (2009). Livistona palms in Australia: Ancient relics or opportunistic immigrants? *Molecular Phylogenetics and Evolution*, **54**, 512–23.

Cutler, D.F. (1969). *Juncales*. In: Metcalfe, C.R. (ed) *Anatomy of Monocotyledons IV*. Clarendon Press, Oxford.

Cuenca, A. Asmussen-Lange, C.B., and Borchsenius, F. (2008). A dated phylogeny of the palm tribe Chamaedoreeae supports Eocene dispersal between Africa, North and South America. *Molecular Phylogenetics and Evolution*, **46**, 760–75.

Cuenca, A., Dransfield, J., and Asmussen-Lange, C.B. (2009). Phylogeny and evolution of morphological characters in tribe Chamaedoreeae. *Taxon*, **58**, 1092–108.

Dahlgren, M.T. and Clifford, H.T. (1982). *The Monocotyledons, a comparative study*. Academic Press, New York, London.

Dassanayake, M.D. and Sivakadachchan, B. (1972). The vascular skeleton of the leaf base of *Caryota urens*. *Phytomorphology*, **22**, 296–304.

Dengler, N.G. and Dengler, R.E. (1984). Formation of plications in the pinnate leaves of *Chrysalidocarpus lutescens* and the palmate leaves of *Rhapis excelsa*. *Principes*, **28**, 31–48.

Dengler, N.G., Dengler, R.E., and Kaplan, D.R. (1982). The mechanism of plication inception in palm leaves: histogenetic observations on the pinnate leaf of *Chrysalidocarpus lutescens*. *Canadian Journal of Botany*, **60**, 2976–98.

Drabble, E. (1904). On the anatomy of the roots of palms. *Transactions of the Linnean Society of London, Botany Ser. 2*, **6**, 427–90.

Dowe, J.L. (2009). A taxonomic account of *Livistona* R.Br. (Arecaceae). *Gardens' Bulletin Singapore*, **60**, 185–344.

Dransfield, J. (1970). *Studies on the Malayan palms Eugeissona and Johannesteijsmannia*. PhD Thesis. Cambridge University.

Dransfield, J. (1978). Growth form of rain forest palms. In: Tomlinson, P.B. and Zimmermann, M.H. (eds) *Tropical trees as Living Systems*, pp. 247–68. Cambridge University Press, New York.

Dransfield, J. and Beentje, H. (1995). *The palms of Madagascar*. Royal Botanic Gardens, Kew, UK.

Dransfield, J., Uhl, N.W., Asmussen, C.B., Baker, W.J., Harley, M.M., and Lewis, C.C. (2005). A new phylogenetic classification of the palm family, Arecaceae. *Kew Bulletin*, **60**, 559–69.

Dransfield, J., Rakotoarinivo, M., Baker, W.J., Bayton, R.P., Fisher, J.B., Horn, J.W., Leroy, B., and Metz, X. (2008a). A new Coryphoid palm genus from Madagascar. *Botanical Journal of the Linnean Society*, **156**, 79–91.

Dransfield, J., Uhl, N.W., Asmussen, C.B., Baker, W.J., Harley, M.M., and Lewis, C.E. (2008b). *Genera Palmarum. The evolution and classification of palms*. Kew Publishing, Royal Botanic Gardens, Kew, UK.

Edwards, P.J., Kollmann, J., and Fleischmann, K. (2002). Life history evolution in *Lodoicea maldivica* (Arecaceae). *Nordic Journal of Botany*, **22**, 227–37.

Evans, R.J. (1995). Systematics of *Cryosophila* (Palmae). *Systematic Botany Monographs*, **46**, 1–70.

Ferreira, E. (1999). The phylogeny of pupunha (*Bactris gasipaes* Kunth, Palmae) and allied species. *Memoirs of the New York Botanical Garden*, **83**, 225–36.

Fisher, J.B. (1973). Unusual branch development in the palm, *Chrysalidocarpus*. *Botanical Journal of the Linnean* Society, **66**, 83–95.

Fisher, J.B. (1974). Axillary and dichotomous branching in the palm *Chamaedorea*. *American Journal of Botany*, **61**, 1045–6.

Fisher, J.B. (1981). A palm spine by any other name – is still a spine. *Fairchild Tropical Garden Bulletin*, **36**, 16–21.

Fisher, J.B. and Dransfield, J. (1977). Comparative morphology and development of inflorescence adnation in rattan palms. *Botanical Journal of the Linnean Society*, **75**, 119–40.

Fisher, J.B. and Dransfield, J. (1979). Development of axillary and leaf-opposed buds in rattan palms. *Annals of Botany, London*, **44**, 57–66.

Fisher, J.B. and Jayachandran, K. (1999). Root structure and arbuscular mycorrhizal colonization of the palm *Serenoa repens* under field conditions. *Plant and Soil*, **217**, 229–41.

Fisher, J.B. and Jayachandran, K. (2005). Presence of arbuscular mycorrhizal fungi in South Florida native plants. *Mycorrhiza*, **15**, 580–8.

Fisher, J.B. and Jayachandran, K. (2008). Beneficial role of arbuscular mycorrhizal fungi on Florida native palms. *Palms*, **52**, 115–25.

Fisher, J.B. and Maidman, K.J. (1999). Branching and architecture in palms: value for systematics. *Memoirs of the New York Botanical Garden*, **83**, 35–46.

Fisher, J.B and Mogea, J.P. (1980). Intrapetiolar inflorescence buds in *Salacca* (Palmae): development and significance. *Botanical Journal of the Linnean Society*, **81**, 47–59.

Fisher, J.B. and Moore, H.E. Jr. (1977). Multiple inflorescences in palms (Arecaceae): their development and significance. *Botanische Jahrbucher für Systematik, Pflanzengeschichte Pflanzengeographie*, **98**, 573–611.

Fisher, J.B. and Tomlinson, P.B. (1973). Branch and inflorescence production in saw palmetto (*Serenoa repens*). *Principes*, **17**, 10–19.

Fisher, J.B., Sanders, R.W., and Edmonson, N. (1987). The flowering and fruiting of *Corypha umbraculifera* in Miami, Florida. *Principes*, **31**, 68–77.

Fisher, J.B., Goh, C.J., and Rao, A.N. (1989). Non-axillary branching in the palms *Eugeissona* and *Oncosperma* (Arecaceae). *Botanical Journal of the Linnean Society*, **99**, 347–63.

Fisher, J.B., Burch, J.N., and Noblick, L.R. (1996). Stem structure of the Cuban belly palm (*Gastrococos crispa*). *Principes*, **40**, 125–28.

Fisher, J.B., Tan, H.T.W., and Toh, L.P.L. (2002). Xylem of rattans: vessel dimensions in climbing palms. *American Journal of Botany*, **89**, 196–202.

Fong, F.W. (1986). *Studies on the population structure, growth dynamics and resource importance of nipa palm (Nypa fruticans Wurmb.).* Ph.D. Thesis, University of Malaya, Kuala Lumpur.

Fong, F.W. (1987). An unconventional fuel alcohol crop. *Principes*, **31**, 64–67.

French, J.C. and Tomlinson, P.B. (1986). Compound vascular bundles in monocotyledonous stems: construction and significance. *Kew Bulletin*, **41**, 561–74.

Frey, A. (1929). Calciumoxalat-Monohydrat und Trihydrat. In: Linsbauer, K. (ed) *Handbuch der Pflanzenanatomie* Band 3, pp. 82–127. Gebrüder Borntraeger, Berlin.

Funk, V.A. and Specht, C.D. (2007). Meta-trees: grafting for a global perspective. *Proceedings of the Biological Society of Washington*, **120**, 232–40.

Gee, C.T. (2001). The mangrove palm *Nypa* in the geologic past of the New World. *Wetlands Ecology and Management*, **9**, 181–94.

Ghose, M. (1979). Ontogenetic study of stomata and trichomes in some palms. *Phytomorphology*, **29**, 26–33.

Ghose, M. and Davis, T.A. (1973). Stomata and trichomes in leaves of young and adult palms. *Phytomorphology*, **23**, 216–29.

Gibbons, M. and Spanner, T.W. (1996). *Medemia argun* lives! *Principes*, **40**, 65–74.

Glassman, S.F. (1972). Systematic studies in the leaf anatomy of the palm genus *Syagrus*. *American Journal of Botany*, **59**, 775–88.

Gomez-Navarro, C., Jaramillo, C., Herrera, F., Wing, S.L., and Callejas, R. (2009). Palms (Arecaceae) from a Paleocene rainforest of northern Colombia. *American Journal of Botany*, **96**, 1300–12.

Guttenberg, H. von. (1968). Die primäre Bau der Angiospermenwurzel. *Handbuch der Pflanzenanatomie*.Bd. VIII Teil 5. Gebrüder Borntraeger, Berlin.

Hahn, W.J. (2002). A phylogenetic analysis of the Arecoid line of palms based on plastid DNA sequence data. *Molecular Phylogenetics and Evolution*, **23**, 189–204.

Hallé, F. (1977). The longest leaf in palms? *Principes*, **21**, 18.

Hallé, F., Oldeman, R.A.A., and Tomlinson, P.B. (1978). *Tropical trees and forests: an architectural analysis.* Springer, Berlin, Heidelberg, New York.

Harries, H.C. (1978). The evolution, dissemination and classification of *Cocus nucifera* L. *Botanical Review*, **44**, 265–319.

Hanstein, J. (1870). Die Entwicklung des Keimes der Monokotylen und Dikotylen. In: Hanstein, J. (ed) *Botanische Abhandlungen aus dem Gebiet der Morphologie und Physiologie*. Band 1 Heft 1: 1–112. Adolf Marcus, Bonn.

Hastings, L.H. (2003). A revision of *Rhapis. Palms*, **47**, 62–78.

Heimsch, C. and Seago, J. L. (2008). Organization of the root apical meristem in angiosperms. *American Journal of Botany*, **95**, 1–21.

Henderson, A. (2000). *Bactris* (Palmae). *Flora Neotropica. Monograph*, **79**, 1–181.

Henderson, A. and Galeano, G. (1996). *Euterpe, Prestoea*, and *Neonicholsonia* (Palme:Euterpeinae). *Flora Neotropica. Monograph*, **72**, 1–90. New York Botanical Garden, New York.

Henderson, A. and de Nevers, G. (1988). *Prestoea* (Palmae) in Central America. *Annals of the Missouri Botanical Garden*, **75**, 203–17.

Hodel, D.R. (1992). *Chamaedorea palms. The species and their cultivation.* Allen Press. Lawrence, KS.

Holbrook, N.M., Putz, F.E., and Chai, P. (1985). Above-ground branching of the stilt-rooted palm *Eugeissona minor*. *Principes*, **29**, 142–6.

Holttum, R.E. (1955). Growth habits of monocotyledons – variations on a theme. *Phytomorphology*, **5**, 399–413.

Horn, J.W., Fisher, J.B., Tomlinson, P.B., Lewis, C.E., and Laubengayer, K.M. (2009). Evolution of lamina anatomy in the palm family (Arecaceae). *American Journal of Botany*, **96**, 1462–86.

Huard, J. (1967). Étude anatomique des épines de quelques palmiers. *Adansonia, Sér. 2.* 7, 221–35.

Huggett, B. and Tomlinson, P.B. (2010). Aspects of vessel dimensions in the aerial roots of epiphytic Araceae. *International Journal of Plant Science*, **171**, 362–9.

Irvine, A. (1983). *Wodyetia*, a new arecoid genus from Australia. *Principes*, **27**, 158–67.

Isnard, S. and Rowe, N.P. (2008a). The climbing habit in palms; biomechanics of the cirrus and flagellum. *American Journal of Botany*, **95**, 1538–47.

Isnard, S. and Rowe N.P. (2008b). The mechanical role of the leaf sheath in rattans. *New Phytologist*, **177**, 643–52.

Isnard, S., Speck, T., and Rowe, N.P. (2005). Biomechanics and development of the climbing palm habit in two species of the South American palm genus *Desmoncus* (Arecaceae). *American Journal of Botany*, **92**, 1444–56.

Kahn, F. and Mejia, K. (1987). Notes on the biology, ecology, and use of a small Amazonian palm: *Lepidocaryum tessmanii*. *Principes*, **31**, 14–19.

Kaplan, D.R., Dengler, N.G., and Dengler, R.E. (1982a). The mechanism of plication inception in palm leaves: problem and developmental morphology. *Canadian Journal of Botany*, **60**, 2939–75.

Kaplan, D.R., Dengler, N.G., and Dengler, R.E. (1982b). The mechanism of plication inception in palm leaves: histogenetic observations on the palmate leaf of *Rhapis excelsa*. *Canadian Journal of Botany*, **60**, 2999–3106.

Kauff, F., Rudall, P.J., and Conran, J.G. 2000. Systematic root anatomy of Asparagales and other monocotyledons. *Plant Systematics and Evolution*, **223**, 139–54.

Kaul, K.N. (1935). On a method of preparing large thin sections of plants by grinding. *Current Science*, **4**, 99–102.

Klotz, L.H. (1977). *A systematic survey of the morphology of tracheary elements in palms.* PhD. Thesis. Cornell University, Ithaca, NY.

Klotz, L.H. (1978a). Form of the perforation plates in the wide vessels of metaxylem in palms. *Journal of the Arnold Arboretum*, **59**, 105–28.

Klotz, L.H. (1978b). The number of wide vessels in petiolar vascular bundles of palms: an anatomical feature of systematic significance. *Principes*, **22**, 64–9.

Klotz, L.H. (1978c). Observations on diameters of vessels in stems of palms. *Principes*, **22**, 99–106.

Kuo-Huang, L., Huang, Y., Chen, S., and Huang, Y. (2004). Growth stresses and related anatomical characteristics in coconut palm trees. *International Association of Wood Anatomists Journal*, **25**, 297–310.

Lewis, C.E., and Doyle, J.J. (2001) A phylogenetic analysis of tribe Areceae (Arecaceae) using two low-copy nuclear genes. *Plant Systematics and Evolution*, **236**, 1–17.

Lewis, C.E., and Zona, S. (2008). *Leucothrinax morrisii*, a new name for a familiar Carribean palm. *Palms*, **52**, 84–8.

Liese, W. and Weiner, G. (1987). *Anatomical structures for the identification of rattan*. Paper presented at the International Rattan Seminar, Chiangmai, Thailand.

Loo, A.H.B., Dransfield, J., Chase, M.W., and Baker, W.J. (2006). Low-copy nuclear DNA, phylogeny and the evolution of dichogamy in the betel nut palms and their relatives (Arecinae; Arecaceae). *Molecular Phylogenetics and Evolution*, **39**, 598–618.

Maddison, W.P., and Maddison, D.R. (2009). *Mesquite: a modular system for evolutionary analysis*. Version 2.72. http://mesquiteproject.org

Martens, J. and Uhl, N.W. (1980). Methods for the study of leaf anatomy in palms. *Stain Technology*, **55**, 241–6.

Martius, C.F.P. von. (1823–1850). *Historia Naturalis Palmarum*. 3 vols. Munich.

McArthur, I.C.S. and Steeves, T. A. (1969). On the occurrence of root thorns on a Central American palm. *Canadian Journal of Botany*, **47**, 1377–82.

Meerow, A.W., Noblick, L., Borrone, J.W., Couvreur, T.L.P., Mauro-Herrera, M., Hahn, W.J., Kuhn, D.N., Nakamura, K., Oleas, N.H., and Schnell, R.J. (2009). Phylogenetic analysis of seven WRKY genes across the palm subtribe Attaleinae (Arecaceae) identifies *Syagrus* as sister group of the coconut. *PLoS ONE*, 4, e7353. doi: 10.1371/journal.pone .0007353.

Meneghini, G. (1836). *Ricerche sulla struttura del caule nelle piante Monocotiledoni*. Padua

Mettenius, G. H. (1864) Über die Hymenophyllaceae. *Abhandlungen der Mathematisch-Physisichen Klasse der Königlich Sachsischen Gesellschaft der Wissenschaften. Leipzig. Wissenschaft*, **7**, 403–504.

Mohl, H. von (1849). On the structure of the palm stem. Reports and papers of the Ray Society. C. and J. Adlard, London. [English translation of H. von Mohl (1831) *De Palmarum Structura*. Vol 1. In: von Martius, C.F.P. (ed) (1823–50) *Historia Naturalis Palmarum*. 3 vols. Munich.]

Moore, A.M. (1957). Permanent mounts from frozen sections. *Stain Technology*, **29**, 43.

Moore, H.E. (1973).The major groups of palms and their distribution. *Gentes Herbarum*, **11**, 27–141.

Moore, H.E. (1978). The genus *Hyophorbe*. *Gentes Herbarum*, **11**, 27–140.

Moore, H.E. and Uhl, N.W. (1978). The monocotyledons: Their evolution and comparative biology. VI. Palms and the origin and evolution of the monocotyledons. *Quarterly Review of Biology*, **48**, 414–36.

Muller, J. (1981). Fossil pollen records of extant angiosperms. *Botanical Review*, **47**, 1–142.

Norup, M.V., Dransfield, J., Chase, M.W., Barfod, A.S., Fernando, E.S and Baker, W.J. (2006). Homoplasious character combinations and generic delimitation: a case study from the Indo-Pacific arecoid palms (Arecaceae: Areceae). *American Journal of Botany*, **93**, 1065–80.

Nowak, J., Dengler, N.G., and Posluszny, U. (2007). The role of abscission during leaflet separation in *Chamaedorea elegans* (Arecaceae). *International Journal of Plant Sciences*, **168**, 533–45.

Nowak, J., Dengler, N.G., and Posluszny, U. (2008). Abscission-like leaflet separation in *Chamaedorea seifrizii* (Arecaceae). *International Journal of Plant Sciences*, **169**, 723–34.

Nowak, J., Nowak, A., and Posluszny, U. (2009). Developmental comparison of leaf shape variation in three *Chamaedorea* species. *Botany*, **87**, 210–21.

Parthasarathy, M.V. (1968). Observations on metaphloem in the vegetative parts of palms. *American Journal of Botany*, **55**, 1140–68.

Parthasarathy, M.V. (1974). Ultrastructure of phloem in palms. I. Immature sieve elements and parenchymatic elements. II. Structural changes, and fate of the organelles in differentiating sieve elements. III. Mature phloem. *Protoplasma*, **79**, 59–91; 93–125; 165–315.

Parthasarathy, M.V. (1980). Mature phloem of perennial monocotyledons. *Berichte der deutschen Botanischen Gesellschaft*, **93**, 57–70.

Parthasarathy, M.V. and Klotz, L.H. (1976). Palm 'Wood'. I. Anatomical aspects. II. Ultrastructural aspects of sieve elements, tracheary elements and fibres. *Wood Science and Technology*, **10**, 215–29, 247–71.

Parthasarathy, M.V. and Tomlinson, P.B. (1967). Anatomical features of metaphloem in stems of. *Sabal, Cocos* and two other palms. *American Journal of Botany*, **54**, 1143–51.

Prychid, C.J. and Rudall, P.J. (1999). Calcium oxalate crystals in monocotyledons: a review of their structure and systematics. *Annals of Botany*, **93**, 415–24.

Prychid, C.J., Rudall, P.J., and Gregory, M. (2003). Systematics and biology of silica bodies in Monocotyledons. *Botanical Review*, **69**, 377–440.

Putz, F.E. (1979). Biology and human use of *Leopoldinia piassaba*. *Principes*, **23**, 149–56.

Pykkő, M. (1985). Anatomy of the stem and petiole of *Raphia hookeri* (Palmae). *Annales botanici fennici*, **22**, 129–38.

Quero, H.J. (1991). *Sabal gretheriae*, a new species from the Yucatan Peninsula. *Principes*, **35**, 219–24.

Quero, H.J. and Read, R.W. (1986). A revision of the genus *Gaussia. Systematic Botany*, **11**, 145–54.

Read, R.W. (1966). New chromosome counts in the Palmae. *Principes*, **10**, 55–61.

Read, R.W. (1968). A study of *Pseudophoenix* (Palmae). *Gentes Herbarum*, **10**, 160–213.

Read, R.W. (1975). The genus *Thrinax* (Palmae:Coryphoideae). *Smithsonian Contributions to Botany*, **19**, 1–98.

Rich, P.M. (1987a). Mechanical structure of the stem of arborescent palms. *Botanical Gazette*, **148**, 42–50.

Rich, P.M. (1987b). Developmental anatomy of the stem of *Welfia georgii, Iriartea gigantea* and other arborescent palms, implications for mechanical support. *American Journal of Botany*, **74**, 792–802.

Robertson, B.L. (1978). Leaf anatomy of *Jubaeopsis caffra* Becc. *Journal of South African Botany*, **44**, 127–41.

Roncal, J., Zona, S., and Lewis, C.E. (2008). Molecular phylogenetic studies of Caribbean palms (Arecaceae) and their relationships to biogeography and conservation. *Botanical Review*, **74**, 78–102.

Rudall, P.J., Abranson, K., Dransfield, J., and Baker, W. (2003). Floral anatomy in *Dypsis* (Arecaceae - Areceae); a case of complex synorganization and stamen reduction. *Botanical Journal of the Linnean Society*, **143**, 115–33.

Rüggeberg, M., Speck, T., and Burgert, I. (2009). Structure-function relationships of different vascular bundle types in the stem of the Mexican fanpalm (*Washingtonia robusta*). *New Phytologist*, **182**, 443–50.

Salter, J.and Delmiglio, C. (2005). Exploratory study of petiolar scales in *Rhopalostylis* (Arecaceae). *New Zealand Journal of Botany*, **43**, 631–46.

Salzman, V. and Judd, W. (1995). A revision of the Greater Antillean species of *Bactris* (Bactridinaeae:Arecaceae). *Brittonia*, **47**, 345–71.

Schatz, G.E., Williamson, G.B., Cogswell, C.M., and Stam, A.C. (1985). Stilt roots and growth of arboreal palms. *Biotropica*, **17**, 206–9.

Schmitt, U., Weiner, G., and Liese, W. (1995). The fine structure of the stegmata in *Calamus axillaris* during maturation. *International Association of Wood Anatomists Journal*, **16**, 61–8.

Schoute, J.C. (1909). Über die Verästelung bei monokotylen Baumen. II. Die Verästelung von *Hyphaene*. *Recueil des travaux botaniques néerlandais*, **6**, 211–32.

Schoute, J.C. (1912). Über das Dickenwachstum der Palmen. *Annales du Jardin botanique de Buitenzorg, Sér 2*, **11**(26), 1–209.

Schoute, J.C. (1915). Sur la fissure médiane de la gaine foliare de quelques palmiers. *Annales du Jardin botanique Buitenzorg, Sér. 2*, **14** (29), 57–81.

Schuiling, E. (2009) *Growth and development of true sago palm [Metroxylon sagu Rottbøll]*. PhD.Thesis. Wageningen University.

Seubert, E. (1996a). Root anatomy of palms. II. Calamoideae. *Feddes Repertorium*, **107**, 143–59.

Seubert, E. (1996b). Root anatomy of palms. III. Ceroxyloideae, Nypoideae, Phytelephantoideae. *Feddes Repertorium*, **107**, 597–619.

Seubert, E. (1997). Root anatomy of palms. I. Coryphoideae. *Flora*, **192**, 81–103.

Seubert, E. (1998a). Root anatomy of palms. IV. Arecoideae, part 1, general remarks and description of roots. *Feddes Repertorium*, **109**, 89–127.

Seubert, E. (1998b). Root anatomy of palms. IV. Arecoideae, part 2 systematic implications. *Feddes Repertorium*, **109**, 231–47.

So.reder, H. and Meyer, F.J. (1928). *Systematische Anatomie der Monokotyledonen. Heft 3, Palmae*. 3–85. Gebrüder Borntraeger, Berlin.

Sperry, J.S. (1985). *Hydraulic architecture and xylem embolism in the palm Rhapis excelsa (Thunb.) Henry*. PhD Thesis, Harvard University.

Sperry, J.S. (1986). Relationship of xylem embolism to xylem pressure potential, stomatal closure, and shoot morphology in the palm *Rhapis excelsa*. *Plant Physiology*, **80**, 110–16.

Swamy, B.G.L. and Govindarajalu, E. (1961). Studies on the anatomical variability in the stem of *Phoenix sylvestris*. I.

Trends in the behaviour of certain cells and tissues. *Journal of the Indian Botanical Society*, **40**, 243–62.

Swofford, D.L. (2002). *PAUP*. *Phylogenetic analysis using parsimony (*and other methods) v. 4.0d106*. Sinauer Associates, Sunderland, MA.

Taylor, Y.R. (1999). *Systematic survey of Coryphoid palms using foliar epicuticular wax and anatomical characters*. Ph.D. Thesis. University of Florida.

Tomlinson, P.B. (1959a). Rapid dehydration for celloidin embedding by means of azeotropic distillation. *Stain Technology*, **34**, 15–18.

Tomlinson, P.B. (1959b). Structure and distribution of sclereids in the leaves of palms. *New Phytologist*, **58**, 253–66.

Tomlinson, P.B. (1961). *Anatomy of the Monocotyledons. Vol. II: Palmae* (ed. C. R. Metcalfe). Clarendon Press, Oxford.

Tomlinson, P.B. (1962a). Essays on the morphology of palms. A digression about spines. *Principes*, **6**, 46–52.

Tomlinson, P.B. (1962b). The leaf base in palms. Its morphology and mechanical biology. *Journal of the Arnold Arboretum*, **43**, 23–46.

Tomlinson, P.B. (1964). The vascular skeleton of coconut leaf base. *Phytomorphology*, **14**, 218–30.

Tomlinson, P.B. (1971a). Flowering in *Metroxylon* (the Sago Palm). *Principes*, **15**, 49–62.

Tomlinson, P.B. (1971b). The shoot apex and its dichotomous branching in the *Nypa* palm. *Annals of Botany (London)*, **35**, 865–79.

Tomlinson, P.B. (1974). Development of the stomatal complex as a taxonomic character in the monocotyledons. *Taxon*, **23**, 109–28.

Tomlinson, P.B. (1982). *Anatomy of the Monocotyledons VII. Helobiae (Alismatidae)*. (ed. C.R. Metcalfe). Clarendon Press. Oxford.

Tomlinson, P.B. (1986). *The botany of mangroves*. Cambridge University Press, Cambridge.

Tomlinson, P.B. (1990). *The structural biology of palms*. Clarendon Press, Oxford, U.K.

Tomlinson, P.B. (1995). Non-homology of vascular organization in monocotyledons and dicotyledons. pp. 589–602. In: Rudall, P.J., Cribb, P.J., Cutler D.F., and Humphries C.J. (eds) *Monocotyledons: systematics and evolution*. Kew Publishing, Royal Botanic Gardens, Kew, U.K.

Tomlinson, P.B. (2006). The uniqueness of palms. *Botanical Journal of the Linnean Society*, **151**, 5–14.

Tomlinson, P.B. and Esler, A.E. (1973). Establishment growth in woody monocotyledons native to New Zealand. *New Zealand Journal of Botany*, **11**, 627–44.

Tomlinson, P.B. and Fisher, J.B. (2000). Stem vasculature in climbing monocotyledons: a comparative approach. In: Wilson, K.L. and Morrison, D.A. (eds) *Monocots: Systematics and Evolution*, pp, 89–97. CSIRO, Melbourne, Australia.

Tomlinson, P.B. and Soderholm, P.K. (1975).The flowering and fruiting of *Corypha elata* in South Florida. *Principes*, **19**, 83–99.

Tomlinson, P.B., and Spangler, R. (2002). Developmental features of the discontinuous stem vascular system in the rattan palm *Calamus* (Arecaceae-Calamoideae-Calamineae). *American Journal of Botany*, **89**, 1128–41.

Tomlinson, P.B. and Vincent, J.R. (1984). Anatomy of the palm *Rhapis excelsa*, X. Differentiation of stem conducting tissue. *Journal of the Arnold Arboretum*, **65**, 191–214.

Tomlinson, P.B. and Zimmermann, M.H. (1966a). Anatomy of the palm Rhapis excelsa, II. Rhizome. *Journal of the Arnold Arboretum*, **47**, 248–61.

Tomlinson, P.B. and Zimmermann, M.H. (1966b). Anatomy of the palm *Rhapis excelsa*, III. Juvenile Phase. *Journal of the Arnold Arboretum*, **47**, 301–12.

Tomlinson, P.B. and Zimmerman, M.H. (2003). Stem vascular architecture in the American climbing palm *Desmoncus* (Arecaceae-Arecoideae-Bactridinae). *Botanical Journal of the Linnean Society*, **142**, 243–54.

Tomlinson, P.B., Fisher, J.B., Spangler, R.E., and Richer, R.A. (2001). Stem vascular architecture in the rattan palm *Calamus* (Arecaceae-Calamoideae-Calamineae). *American Journal of Botany*, **88**, 797–809.

Tomlinson, P.B., Horn, J.W. and Fisher, J.B. (2009). Palms and Ponce de Leon: The secret of eternal youth exemplified by sustained primary growth in Arecaceae. Page 93 In: Botany and Mycology 2009. Snowbird, Utah July 25–29.

Trénel, P., Gustafsson, M.H.G., Baker, W.J., Asmussen-Lange, C.B., Dransfield, J., and Borschenius, F. (2007). Mid-Tertiary dispersal, not Gondwanan vicariance explains distribution patterns in the wax palm subfamily (Ceroxyloideae: Arecaceae). *Molecular Phylogenetics and Evolution*, **45**, 272–88.

Tsai, J.H. and Fisher, J.B. (1993). Feeding sites of some leaf- and plant-hopper insects (Homoptera: Auchenorrhyncha) associated with coconut palms. *Principes*, **37**, 35–41.

Uhl, N.W. (1972). Leaf anatomy in the *Chelyocarpus* alliance. *Principes*, **16**, 101–10.

Uhl, N. W. (1978a). Floral anatomy of the five species of *Hyophorbe* (Palmae). *Gentes Herbarum*, **11**, 245–67.

Uhl, N.W. (1978b).Leaf anatomy in the species of *Hyophorbe* (Palmae). *Gentes Herbarum*, **11**, 268–83.

Uhl, N.W. (1988). Floral organogenesis in palms. In Leins, P., Tucker, S.C., and Endress, P.K. (eds) *Aspects of floral development*, pp. 25–44. J. Cramer, Berlin.

Uhl, N.W. and Moore, H.E. (1982). Major trends of evolution in palms. *Botanical Review*, **48**, 1–69.

Uhl, N.W. and Moore, H.E. (1973).The protection of pollen and ovules in palms. *Principes*, **17**, 111–49.

Uhl, N.W. and Dransfield, J. (1987). *Genera Palmarum*. L.H. Bailey Hortorium and International Palm Society, Lawrence, KS.

Waterhouse, J.T. and Quinn, C.J. (1978). Growth patterns in the stem of the palm *Archontophoenix cunninghamiana*. *Botanical Journal of the Linnean Society*, **77**, 73–93.

Weiner, G. (1992). *Zur Stammanatomie der Rattanpalmen*. Diss. University of Hamburg.

Weiner, G. and Liese, W. (1990). Rattan-stem anatomy and taxonomic implications. *International Association of Wood Anatomists Bulletin n.s.*, **11**, 61–70.

Weiner, G. and Liese,W. (1995). Wound response in the stem of the Royal palm. *International Association of Wood Anatomists Journal*, **16**, 433–42.

Weiner, G., Liese, W., and Schmitt, U. (1996). Cell wall thickening in fibres of the palms *Rhapis excelsa* (Thunb.) Henry and *Calamus axillaris* Becc. In: Donaldson, L.A., Singh, A.P., Butterfield, L.J., and Whitehouse, L.J. (eds) *Recent Advances in Wood anatomy*, pp. 191–7. New Zealand Research Institute, Rotorua, New Zealand.

Wellendorf, M. (1963). The microscopical structure of palm starches. *Botanisk Tidskrift*, **59**, 209–19.

Yampolsky, C. (1924). The pneumathodes on the roots of the oil palm *Elaeis guinieensis* Jaccq.). *American Journal of Botany*, **11**, 501–12.

Zapatilla, E.S. (1985). Anatomia del genero *Opsiandra* (*O. maya* O.F. Cook, *O. gomez-pompae* Quero). Tesis Biologio, Universidad Nacional Autonoma de Mexico, D.F. Mexico.

Zimmermann, M.H. (1973). The monocotyledons: their evolution and comparative biology. IV. Transport problems in arborescent monocotyledons. *Quarterly Review of Biology*, **48**, 314–21.

Zimmermann, M.H. (1983). *Xylem structure and the ascent of sap*. Springer-Verlag, Heidelberg.

Zimmermann, M.H. and Mattmüller, M. (1982). *The vascular system of the palm Rhapis excelsa. I. The mature stem, Film C 1404 and II. The growing tip, Film D 1418*. Institut für wissenschaftlichen Film. Göttingen.

Zimmermann, M.H. and Sperry, J. (1983). Anatomy of the palm *Rhapis excelsa*. IX. Xylem structure of the leaf insertion. *Journal of the Arnold Arboretum*, **64**, 599–609.

Zimmermann, M.H. and Tomlinson, P.B. (1965). Anatomy of the palm *Rhapis excelsa*. I. Mature vegetative axis. *Journal of the Arnold Arboretum*, **46**, 160–78.

Zimmermann, M.H. and Tomlinson, P.B. (1966). Analysis of complex vascular systems in plants: optical shuttle method. *Science*, **152**, 72–3.

Zimmermann, M.H. and Tomlinson, P.B. (1967). Anatomy of the palm *Rhapis excelsa*. IV. Vascular development in apex of vegetative aerial axis and rhizome. *Journal of the Arnold Arboretum*, **48**, 122–42.

Zimmermann, M.H., McCue, K.F., and Sperry, J.S. (1982). Anatomy of the palm *Rhapis excelsa*, VIII. Vessel network and vessel length distribution. *Journal of the Arnold Arboretum*, **63**, 83–95.

Zona, S. (1990). A monograph of *Sabal* (Arecaceae: Coryphoideae). *Aliso*, **12**, 583–666.

Zona, S. (2004). Raphides in palm embryos and their systematic significance. *Annals of Botany*, **93**, 415–24.

INDEX TO SCIENTIFIC NAMES

Plant genera and classification; figure numbers in **bold font**.

INDEX TO SUBJECT

Index to subject and common names; figure numbers in **bold face.**